Homotopy Theory

PURE AND APPLIED MATHEMATICS

A Series of Monographs and Textbooks

Edited by

PAUL A. SMITH and SAMUEL EILENBERG

Columbia University, New York

I: ARNOLD SOMMERFELD. Partial Differential Equations in Physics. 1949 (Lectures on Theoretical Physics, Volume VI)

II: REINHOLD BAER. Linear Algebra and Projective Geometry. 1952

III: HERBERT BUSEMANN AND PAUL J. KELLY. Projective Geometry and Projective Metrics. 1953

IV: STEFAN BERGMAN AND M. SCHIFFER. Kernel Functions and Elliptic Differential Equations in Mathematical Physics. 1953

V: RALPH PHILIP BOAS, JR. Entire Functions. 1954

VI: HERBERT BUSEMANN. The Geometry of Geodesics. 1955

VII: CLAUDE CHEVALLEY. Fundamental Concepts of Algebra. 1956

VIII: SZE-TSEN HU. Homotopy Theory. 1959

IX: A. OSTROWSKI. Solution of Equations and Systems of Equations. 1960

X: J. DIEUDONNÉ. Foundations of Modern Analysis. 1960

XI: S. I. GOLDBERG. Curvature and Homology. 1962

In preparation

XII: SIGURDUR HELGASON. Differential Geometry and Symmetric Spaces.

XIII: T. H. HILDEBRANDT. Introduction to the Theory of Integration.

Homotopy Theory

SZE-TSEN HU
Wayne State University, Detroit, Michigan

1959

ACADEMIC PRESS • New York and London

ACADEMIC PRESS INC.
111 Fifth Avenue
New York 3, N. Y.

United Kingdom Edition
Published by
ACADEMIC PRESS INC. (London) Ltd.
Berkeley Square House, London W. 1

First Printing, 1959
Second Printing, 1962

Library of Congress Catalog Card Number 59-11526

PRINTED IN THE UNITED STATES OF AMERICA

Preface

The recognition of the branch of mathematics now called homotopy theory took place in the few years after the introduction of homotopy groups by Witold Hurewicz in 1935. Since then, with numerous advances made by various workers, it has been playing an increasingly important role in the expanding field of algebraic topology. However, there exists no textbook on the subject at any level except the extremely condensed Cambridge tract of P. J. Hilton entitled "An Introduction to Homotopy Theory."

The present book is designed to guide a reader, who might be a beginning student or a newcomer to this branch of mathematics and who has a little knowledge of elementary algebraic topology, through the basic principles of homotopy theory. The author has aimed to provide the reader with sufficient detail for him to understand the fundamental ideas and master the elementary techniques so that he may be able to study the more advanced and more complicated results directly from the original papers.

The main problem in homotopy theory is the extension problem as formulated in Chapter I and illustrated in Chapter II. The fiber spaces, which are of fundamental importance, are defined and studied in Chapter III. Homotopy groups are constructed and axiomatized in Chapter IV while the elementary techniques of computation are given in Chapter V. Chapter VI gives an introduction to the obstruction theory of continuous maps, and Chapter VII contains an account of the cohomotopy groups. In the next three chapters, one will find an exposition of the spectacular results obtained mostly by the French school after Leray's discovery of the spectral sequence. The techniques developed in these chapters are applied to compute the first few homotopy groups of spheres in the final chapter.

As indicated in the second paragraph above, this book is by no means designed to be an exhaustive treatment of its subject; for example, the recent celebrated contribution of M. M. Postnikov is not included. Besides, homotopy theory is advancing so rapidly that any treatment of this subject becomes obsolete within a few years.

At the end of each chapter is a list of exercises. These cover material which might well have been incorporated in the text but was omitted as not essential to the main line of thought. The inexperienced reader should not be discouraged if he cannot work out these exercises. In fact, if he is interested in one of the exercises, he is expected to read the papers indicated there.

The bibliography at the end of this book has been reduced to the minimum essential to the text and the exercises. References to this bibliography are included for the convenience of the reader so that he can find more details concerning the material; the references are not intended to be a historical record of mathematical discovery. (These references are cited in the text by numbers enclosed in square brackets). Frequently, expository articles are preferred to the earlier original papers. Whenever no reference is given concerning some subject or in some exercise, it means only that the reader does not have to look for further details in order to understand the material or to work out the problem. Cross references are given in the form (II; 7.1), where II stands for Chapter II and 7.1 for the numbering of the statement in the chapter.

A list of special symbols and abbreviations used in this book is given immediately after the Table of Contents. Certain deviations from standard set-theoretic notations have been adopted in the text; namely, □ is used to denote the empty set and $A \setminus B$ the set-theoretic difference usually denoted by A–B. On the other hand, the symbol ▮ indicates the end of a proof and the abbreviation iff stands for the phrase "if and only if." Finally, for the algebraic terminology used in this book, the reader may refer to Claude Chevalley's "Fundamental Concepts of Algebra," published in this series.

The author acknowledges with great pleasure his gratitude to Professor Norman Steenrod who has read several versions of the manuscript and whose numerous suggestions and criticisms resulted in substantial improvements. The author also wishes to express his appreciation of the friendly care with which Dr. John S. Griffin, Jr., and Professor C. T. Yang have read the final manuscript, of the many improvements they suggested, and of their help in the proofreading.

It is a pleasure to acknowledge the invaluable assistance the author received in the form of partial financial support from the Office of Naval Research when he was at Tulane University and from the Air Force Office of Scientific Research while at Wayne State University.

SZE-TSEN HU

Wayne State University, Detroit, Michigan

Contents

List of Special Symbols and Abbreviations

The symbols and abbreviations listed below are followed by a brief statement of their meaning and by the number of the page on which they first appear. Symbols and abbreviations which are universally used in most branches of mathematics, such as $\in$, $\cup$, $\cap$, $\approx$, $\geqslant$, sup, inf, etc., are not listed.

▮	End of proof, *3*
$\setminus$	Set-theoretic difference, *5*
$\simeq$	Is homotopic to, *11*
$\square$	Empty set, *17*
$\otimes$	Tensor product, *261*
∂	Boundary of, *47*
δ	Coboundary of, *50*
I	Closed unit interval $[0,1]$, *2*
I^n	n-cube, *3*
R^n	Euclidean n-space, *4*
S^n	Unit n-sphere, *4*
Δ_n	Unit n-simplex, *7*
H_n	n-Dimensional homology group of, *3*
H^n	n-Dimensional cohomology group of, *5*
π_1	Fundamental group of, *40*
π_n	n-th homotopy group of, *109*
π^n	n-th cohomotopy group of, *205*
Z	Group of integers, *109*
Z_p	Group of integers mod p, *281*
$X = 0$	The set X consists of a single element, *13*
$f\|A$	Restriction of the map f on A, *1*
$[Y; A, B]$	Space of all paths $f: I \to Y$ such that $f(0) \in A$ and $f(1) \in B$, *78*
Y^X	Space of all maps $X \to Y$, *12*
$X \vee Y$	One-point union of X and Y, *145*
ACHEP	Absolute covering homotopy extension property, *62*
ACHP	Absolute covering homotopy property, *62*
AHEP	Absolute homotopy extension property, *13*
ANR	Absolute neighborhood retract, *26*
AR	Absolute retract, *26*
BP	Bundle property, *65*
CHEP	Covering homotopy extension property, *62*
CHP	Covering homotopy property, *24*

Coker	Cokernel of, *298*
deg	Degree of, *37*
dim	Dimension of, *49*
HEP	Homotopy extension property, *13*
iff	If and only if, *2*
Im	Image of, *215*
Int	Interior of, *8*
Ker	Kernel of, *215*
LPLP	Local path lifting property, *98*
NHEP	Neighborhood homotopy extension property, *30*
PCHEP	Polyhedral covering homotopy extension property, *62*
PCHP	Polyhedral covering homotopy property, *62*
PLP	Path lifting property, *82*
rel	Relative to, *17*
SSP	Slicing structure property, *97*
Tor	Torsion product of, *270*

CHAPTER I

MAIN PROBLEM AND PRELIMINARY NOTIONS

1. Introduction

There is a general type of topological problem which will be called the *extension problem*. One of the principal objectives of the book is to show that this problem is fundamental in topology. It will be shown that many theorems of topology and most of its applications in other fields of mathematics are solutions of special cases of the extension problem.

The objective of the first chapter is to formulate the extension problem precisely, and to study the problem in its most general terms. It will be shown that various other problems are fully equivalent to extension problems.

We shall begin with a restricted form of the extension problem, and one of its special cases, namely, the retraction problem. The solution is shown to depend only on the homotopy class of the map involved, and then only on the homotopy types of the spaces. These considerations lead naturally to a more general type of problem. The latter is then shown to be reducible to the simplest type of problem, namely, the retraction problem. Finally, dual problems of deformation and lifting (finding a cross-section) are discussed in an analogous fashion together with their interrelations with the extension problem. Thus it will appear that underlying these various questions is one central question, namely, the extension problem.

2. The extension problem

By a *map*, or *mapping*, $f:X \to Y$ of a space X into a space Y, we mean a single-valued *continuous* function from X to Y. The space X is called the *domain* of f or the *anti-image* of f; and the space Y is called the *range* of f.

We shall not recall the definition and the elementary properties of continuous functions, since these can be found in any textbook on general topology, for example, [K; pp. 84–88]. On the other hand, we assume that the reader is familiar with the popular notions and notations concerning maps such as given in [E–S].

Let $f:X \to Y$ be a map and A a subspace of X. Then f defines a unique map $g:A \to Y$ such that $g(x) = f(x)$ for each $x \in A$. This map g is called the *restriction* of f to A or the *partial map* of f on A and is denoted by

$$g = f \mid A;$$

f will be called an *extension* of g over X.

If $h:A \subset X$ denotes the *inclusion map* defined by $h(a) = a \in X$ for each $a \in A$, then the relation $g = f \mid A$ is equivalent to the commutativity relation $fh = g$ in the diagram:

$$\begin{array}{ccc} A & \xrightarrow{g} & Y \\ & {\scriptstyle h}\searrow \quad \nearrow{\scriptstyle f} & \\ & X & \end{array}$$

The (restricted) *extension problem* is concerned with whether or not a given map $g:A \to Y$ defined on a given subspace A of a space X has an extension over X. When X, A, Y and g are given in some reasonably effective manner, the problem is to find an effective procedure for deciding whether g has an extension over X, and finding one when it exists. As will be seen, solutions have been given in numerous special cases; these are quite varied in nature, and the situations in which they apply are very restricted. As yet there is no reasonably complete theory.

Actually it is convenient to concentrate on a broader problem which will be stated in § 9, but several important ideas arise in connection with this problem. Let us begin by considering a few simple examples.

1. Let X be a given space and A be a subspace of X which consists of two points x_0 and x_1. Let Y be a 0-sphere, say the boundary sphere of the closed unit interval I. Consider the map $g:A \to Y$ defined by $g(x_0) = 0$ and $g(x_1) = 1$. Then g has an extension over X iff x_0, x_1 lie in different quasi-components of X, [E–S; p. 254]. Hereafter, the symbol "iff" will stand for "if and only if".

2. Let $X = I$ and A be the boundary sphere of I. Let Y be any given space, and $g:A \to Y$ a given map. Then g has an extension over X iff $g(0)$, $g(1)$ lie in a compact, connected and locally connected subspace of Y satisfying the second countability axiom.

3. Let A be the union of two disjoint closed subspaces B, C of a normal space X, let $Y = I$, and let $g:A \to Y$ denote the map defined by

$$g(B) = 0, \quad g(C) = 1.$$

Then, by Urysohn's lemma [L_2; p. 27], g has an extension over X.

4. Let A be a closed subspace of a normal space X, let $Y = I$, and let $g:A \to Y$ denote any map. Then, by Tietze's extension theorem [L_2; p. 28], g has an extension over X. See Ex. D at the end of the chapter.

In the last example given above, the space $Y = I$ has the property that the extension problem is always trivial regardless of the domain (X, A) provided that X is normal and A is closed. The class of spaces having this property are the solid spaces. Precisely, a space Y is said to be *solid* if every map $g:A \to Y$ of any closed subspace A of an arbitrary normal space X has an extension over X.

Proposition 2.1. *Any topological product of solid spaces is solid.*

Proof. Let $\{ Y_\mu \mid \mu \in M \}$ be a collection of solid spaces and $Y = \mathbf{P}_\mu Y_\mu$ denote the topological product of this collection, [L_2; p. 10]. We are going to prove that Y is solid.

Let A be a closed subspace of a normal space X and $g:A \to Y$ any given map. Denote by $p_\mu: Y \to Y_\mu$, $\mu \in M$, the natural projection of Y onto Y_μ and set

$$g_\mu = p_\mu g:A \to Y_\mu.$$

Since Y_μ is solid, g_μ has an extension $f_\mu:X \to Y_\mu$. Define a map $f:X \to Y$ by taking

$$p_\mu f(x) = f_\mu(x), \quad (x \in X).$$

It is obvious that f is an extension of g over X. ∎

Since the closed unit interval I is solid as noted above, it follows from (2.1) that any compact parallelotope, [L_2; p. 19], is solid; in particular, the n-cube I^n and the Hilbert cube I^ω are solid. Their homeomorphs are likewise solid, hence the n-cell and the n-simplex are solid.

3. The method of algebraic topology

In the preceding section, we formulated the extension problem and gave examples in which the extension existed. It is natural to look for examples where the extension does not exist. The primary method of proving non-existense is to apply homology theory and derive an algebraic problem from the geometric one and, finally, show that the algebraic problem has no solution.

For this purpose, let us consider the triangle

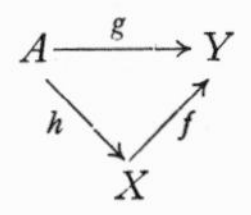

of maps as described in the preceding section. In any homology theory satisfying the Eilenberg-Steenrod axioms [E–S; pp. 10–12], the maps f, g, h induce for each m the homomorphisms f_*, g_*, h_* indicated in the following diagram:

$$\begin{array}{ccc} H_m(A) & \xrightarrow{g_*} & H_m(Y) \\ & {\scriptstyle h_*}\searrow \quad \nearrow{\scriptstyle f_*} & \\ & H_m(X) & \end{array}$$

According to Axiom 2, [E–S; p. 11], the relation $fh = g$ implies the commutativity relation

$$f_* h_* = g_*$$

in the triangle of homomorphisms given above. Hence, the existence of an extension $f:X \to Y$ of the map $g:A \to Y$ gives a solution of a *derived algebraic problem*, namely, to find a homomorphism

$$\phi:H_m(X) \to H_m(Y)$$

such that the commutativity relation $\phi h_* = g_*$ holds. Thus, the existence of the homomorphism ϕ is a necessary (though not generally sufficient) condition for the existence of an extension of the given map g over X. On many occasions, this necessary condition provides us a method to show that a particular given map $g:A \to Y$ fails to have an extension over X.

For example, let us take X to be the unit n-cell E^n of the n-dimensional euclidean space R^n, $A = Y$ to be the boundary $(n-1)$-sphere S^{n-1} of E^n, and $g:A \to Y$ to be the identity map on S^{n-1}. We shall prove the following

Proposition 3.1. *For each $n \geqslant 1$, the identity map $g:S^{n-1} \to S^{n-1}$ has no extension over E^n.*

Proof. Assume that there is some extension $f:E^n \to S^{n-1}$ of g. We shall deduce a contradiction as follows.

Assume $n > 1$ and consider the homology theory with the group Z of integers as the coefficient group. Take $m = n - 1$, then we have

$$H_m(E^n) = 0, \quad H_m(S^{n-1}) \approx Z.$$

Since g is the identity map on S^{n-1}, it follows from Axiom 1 that g_* is the identity automorphism of $H_m(S^{n-1})$. Since $H_m(S^{n-1}) \neq 0$, this implies that $g_* \neq 0$. On the other hand, since $H_m(E^n) = 0$, the inclusion map $h:S^{n-1} \subset E^n$ induces $h_* = 0$. Thus, we obtain $f_*h_* = 0$ and $g_* \neq 0$. This contradicts the relation $f_*h_* = g_*$. In fact, the derived algebraic problem has no solution.

It remains to dispose of the case $n = 1$. In this case, S^{n-1} consists of two points and hence is disconnected. On the other hand, E^n is connected and so is its continuous image $f(E^n)$. Since f is an extension of the identity map g, we have $f(E^n) = S^{n-1}$. This is a contradiction. ∎

Note. The case $n = 1$ of (3.1) can also be proved from the derived algebraic problem provided that one uses the reduced homology groups, [E–S; pp. 18–19].

As an important application of (3.1), we shall give the following

Theorem 3.2. (The Brouwer Fixed-Point Theorem). *Every map $f:E^n \to E^n$ has a fixed point, that is to say, there exists a point x of E^n such that $f(x) = x$.*

Proof. Assume that $f:E^n \to E^n$ is free of fixed points. Then, we may define a map $r:E^n \to S^{n-1}$ as follows. Let $x \in E^n$. Since f has no fixed points, we have $f(x) \neq x$. Draw the line from $f(x)$ to x and produce until it intersects S^{n-1} at a point $r(x)$. One verifies that the assignment $x \to r(x)$ defines a *continuous* function $r:E^n \to S^{n-1}$. If $x \in S^{n-1}$, it is obvious from the construction that $r(x) = x$. Hence r is an extension of the identity map on S^{n-1}. This contradicts (3.1). ∎

This application of (3.1) shows that the negative nature of a "non-existence" theorem may not diminish its interest. A reformulation sometimes gives it a positive aspect.

In the derived algebraic problem formulated above, one may of course use cohomology theory instead of homology theory.

4. The retraction problem

If $Y = A$ and $g = i$ is the identity map on A, then we obtain an important special case of the extension problem which will be referred to as the *retraction problem*. If i has an extension $r: X \to A$, then A is called a *retract* of X, r is called a *retraction* of X onto A, and we will write

$$r: X \supset A.$$

According to (3.1), the boundary $(n-1)$-sphere S^{n-1} of E^n is not a retract of E^n. On the other hand, if X denotes the space obtained by deleting from E^n an interior point which may be assumed to be the origin 0 without loss of generality, then S^{n-1} is a retract of X. In fact, a retraction $r: X \supset S^{n-1}$ is given by

$$r(x) = \left(\frac{x_1}{|x|}, \ldots, \frac{x_n}{|x|} \right)$$

for every point $x = (x_1, \ldots, x_n)$ of $X = E^n \setminus 0$, where $|x|$ denotes the distance between 0 and x. The same formula also gives a retraction r of $R^n \setminus 0$ onto S^{n-1}.

For another example of retracts, let us consider the topological product $X = A \times B$. Pick a point b_0 from B. Then A can be considered as a subspace of X by means of the homeomorphism $h: A \to X$ defined by $h(a) = (a, b_0)$ for each $a \in A$. This having been observed, it becomes clear that the natural projection $X = A \times B \to A$ gives a retraction of X onto A. In particular, a meridian of the torus $T_2 = S^1 \times S^1$ is a retract of T_2.

Observe that if A is a retract of X then the extension problem becomes trivial regardless of the range Y. Indeed, we have the following

Proposition 4.1. *A is a retract of X iff, for any space Y, every map $g: A \to Y$ has an extension over X.*

Proof. If A is a retract of X with a retraction $r: X \supset A$, then $gr: X \to Y$ is an extension of g. Conversely, assume that the condition holds and take $Y = A$. Then the identity map i on A should have an extension $r: X \to A$. ∎

The retraction problem gives rise to a *derived algebraic problem* as follows. In any homology theory or cohomology theory satisfying the Eilenberg-Steenrod axioms, the inclusion map $i: A \subset X$ induces for each m the homomorphisms

$$i_*: H_m(A) \to H_m(X), \quad i^*: H^m(X) \to H^m(A).$$

The derived algebraic problem is to determine whether or not there exist homomorphisms

$$\phi : H_m(X) \to H_m(A), \quad \psi : H^m(A) \to H^m(X)$$

such that ϕi_* and $i^*\psi$ are the identities on $H_m(A)$ and $H^m(A)$ respectively.

The existence of the homomorphisms ϕ and ψ is a necessary condition for A to be a retract of X. In fact, if $r : X \supset A$ is a retraction, then ri is the identity map on A and hence $\phi = r_*$ and $\psi = r^*$ are solutions of the derived algebraic problem. Furthermore, since $r_* i_*$ and $i^* r^*$ are the identities, it follows that i_*, r^* are monomorphosms, that r_*, i^* are epimorphisms, and that $H_m(X)$, $H^m(X)$ decompose into the following direct sums:

$$H_m(X) = \text{Image } i_* + \text{Kernel } r_*,$$
$$H^m(X) = \text{Kernel } i^* + \text{Image } r^*.$$

If the coefficient group of the cohomology theory is a ring, then the cohomology groups $H^m(X)$, $m = 0, 1, \cdots$, constitute a ring $H^*(X)$ with the cup product as multiplication. The inclusion $i : A \subset X$ and the retraction $r : X \supset A$ induce the ring homomorphisms

$$i^* : H^*(X) \to H^*(A), \quad r^* : H^*(A) \to H^*(X).$$

Since ri the identity map on A, it follows that $i^* r^*$ is the identity automorphism of the ring $H^*(A)$. Hence r^* is a monomorphism, i^* is an epimorphism, and $H^*(X)$ decomposes into the direct sum

$$H^*(X) = \text{Kernel } i^* + \text{Image } r^*$$

where Kernel i^* is an ideal and Image r^* is a subring isomorphic to $H^*(A)$ under r^*.

These necessary conditions can be used to prove that a particular given subspace A of a certain space X fails to be a retract of X. For example, let X denote the complex projective space of complex dimension $n > 1$ and A a linear subspace of X of complex dimension r with $0 < r < n$. Then A is not a retract of X. To prove this fact, let us assume that there is a retraction $r : X \supset A$. Then we obtain a ring monomorphism $r^* : H^*(A) \to H^*(X)$ of the cohomology rings with integral coefficients. Let α and ξ denote generators of the free cyclic groups $H^2(A)$ and $H^2(X)$ respectively. Since r^* is a monomorphism, there is a non-zero integer k such that $r^*(\alpha) = k\xi$. Since $n > r$, we have $\alpha^n = 0$. Since r^* preserves multiplication, we obtain

$$k^n \xi^n = r^*(\alpha^n) = 0.$$

This contradicts the fact that ξ^n is a generator of the free cyclic group $H^{2n}(X)$.

Finally, let us give an important example of retract in the form of the following

Proposition 4.2. *If (X, A) is a (finitely) triangulable pair,* [E–S; p. 60], *then the closed subspace*

$$L = (X \times 0) \cup (A \times I)$$

of the product space $M = X \times I$ is a retract of M.

Proof. First, let us prove the special case where X is the unit n-simplex Δ_n of the euclidean $(n + 1)$-space, [E–S; p. 55], and A is the boundary $(n-1)$-sphere of Δ_n which is empty if $n = 0$. Then a retraction $r: M \supset L$ can be constructed geometrically as follows. Since $I \subset R$, it follows that M is a subspace of $\Delta_n \times R$. Then we define r to be the central projection of M onto L from the point $(c, 2)$ of $\Delta_n \times R$, where c denotes the centroid of Δ_n. This proves the special case.

For a finitely triangulable pair (X, A), we may assume that X is a finite simplicial polyhedron and A is a subpolyhedron. Since a retract of a retract is also a retract, one can easily prove the proposition by induction on the number of simplexes in X but not in A and by the aid of the special case proved above. ∎

A strengthened form of (4.2) will be given in § 10. Besides, (4.2) can also be generalized to some non-triangulable pairs; see Ex.O at the end of the chapter.

5. Combined maps

Frequently, a function is constructed by prescribing it on pieces of its domain. The purpose of this section is to give sufficient conditions, for the continuity of functions so constructed.

Let $\{ X_\mu \mid \mu \in M \}$ be a given system of subspaces of a space X, indexed by the elements of a set M, such that the union of all subspaces X_μ, $\mu \in M$, is the whole space X. Let $D_{\mu\nu} = X_\mu \cap X_\nu$ for each pair of indices μ, ν in M.

For any given map $g: X \to Y$ of X into a space Y, the partial maps $g_\mu = g \mid X_\mu$ are well-defined and satisfy the relation $g_\mu \mid D_{\mu\nu} = g_\nu \mid D_{\mu\nu}$ for every pair of indices μ,ν in M. Hence, our problem in this section is to study the inverse of this process described as follows.

Let us assume that, for each index $\mu \in M$, there is given a map $f_\mu: X_\mu \to Y$ such that

$$f_\mu \mid D_{\mu\nu} = f_\nu \mid D_{\mu\nu}$$

for each pair of indices μ and ν in M. Then we may define a function $f: X \to Y$ by taking

$$f \mid X_\mu = f_\mu, \quad (\mu \in M).$$

We are concerned with the problem whether or not f is continuous.

Proposition 5.1. *If M is finite and all the subspaces X_μ, $\mu \in M$, are closed in X, then the combined function f is continuous.*

Proof. Let F be any closed set in Y. Then it follows from the continuity of f_μ, $\mu \in M$, that $f_\mu^{-1}(F)$ is a closed set of X_μ. Since X_μ is closed in X, this implies that $f_\mu^{-1}(F)$ is a closed set of X. Since M is finite,

$$f^{-1}(F) = \bigcup_{\mu \in M} f_\mu^{-1}(F)$$

is a closed set of X. Hence f is continuous. ∎

A frequent application of this proposition is as follows. Let $f,g:X \times I \to Y$ be two given maps such that $f(x,1) = g(x,0)$ for every $x \in X$. Then we may define a function $h:X \times I \to Y$ by taking:

$$h(x,t) = \begin{cases} f(x,2t), & (x \in X, 0 \leqslant t \leqslant \frac{1}{2}); \\ g(x,2t\text{-}1), & (x \in X, \frac{1}{2} \leqslant t \leqslant 1). \end{cases}$$

An application of (5.1) shows that h is continuous. This uniquely defined map h will be called the *sum* of the maps f and g denoted by $f + g$. This operation is obviously not commutative.

Proposition 5.2. *If x is an interior point of some subspace X_μ, then the combined function f is continuous at x.*

Proof. Let U be an open neighborhood of the point $f(x) = f_\mu(x)$ in Y. It follows from the continuity of f_μ that there is an open neighborhood V of x in X_μ such that $f_\mu(V) \subset U$. Call

$$W = V \cap \operatorname{Int}(X_\mu).$$

Then W is an open neighborhood of x in X and $f(W) = f_\mu(W) \subset U$. Hence f is continuous at x. ∎

Corollary 5.3. *If all the subspaces X_μ, $\mu \in M$, are open in X, then the combined function f is continuous.*

Finally, as a generalization of (5.1), we have the following

Proposition 5.4. *If the system $\{ X_\mu \mid \mu \in M \}$ forms a locally-finite closed covering of X, then the combined function f is continuous.*

Proof. Let x be an arbitrary point of X. It suffices to prove that f is continuous at x. Since the system $\{ X_\mu \mid \mu \in M \}$ is locally finite, there exists a neighborhood X_0 of x in X which meets only a finite number of the subspaces $\{ X_\mu \mid \mu \in M \}$. Since X_μ is closed in X, it follows that $X_0 \cap X_\mu$ is closed in X_0. An application of (5.1) proves that the restriction $f_0 = f \mid X_0$ is continous. So, we may enlarge the system $\{ X_\mu \mid \mu \in M \}$ by adjoining X_0. The new system usually fails to be a closed covering of X. However, since x is an interior point of X_0, (5.2) implies that f is continuous at x. ∎

6. Topological identification

Suppose that X is a space in which an equivalence relation is given. Then the points of X are divided into disjoint equivalence classes. Let us denote by Z the set of all these equivalence classes and by $[x]$ the equivalence class which contains the point x in X. The assignment $x \to [x]$ defines a function

$$p: X \longrightarrow Z$$

of X onto Z called the *natural projection*.

The set Z can be topologized as follows: a set W in Z is called open iff the inverse image $p^{-1}(W)$ is an open set in X. This topology of Z is called the *identification topology* determined by p or the *quotient topology* defined by the given equivalence relation in X. Having been topologized in this way, Z is said to be the *quotient space* obtained from X by *topological identification*.

This method of constructing new spaces from old is of extreme importance in combinatory topology. Most well-known spaces can be obtained from simpler spaces by topological identification. For example, the n-sphere is obtained from the n-cell by identifying the boundary to a single point and the real projective n-space is obtained from the n-sphere by identifying the antipodal points.

As another example of topological identification, let us consider a given group π of homeomorphisms of a space X. The group π defines an equivalence relation in X as follows: two points a, b in X are equivalent iff $b = \xi(a)$ for some $\xi \in \pi$. The quotient space thus obtained is called the *orbit space* and is denoted by X/π.

More examples of topological identification will be found in the exercises at the end of the chapter.

Proposition 6.1. *Let $f: Z \to Y$ be a given function of the quotient space Z into a space Y. If the composed function $g = fp: X \to Y$ is continuous, then so is f.*

Proof. Let V be any open set in Y. It suffices to show that the inverse image $W = f^{-1}(V)$ is an open set in Z. Since $g = fp$, we have

$$g^{-1}(V) = p^{-1}[f^{-1}(V)] = p^{-1}(W).$$

The continuity of g implies that $g^{-1}(V)$ is an open set in X. Hence, by definition of the identification topology, W is open. ▌

For further information about quotient spaces, see [K; pp. 94–100].

7. The adjunction space

As an important application of the process of topological identification, let us consider the construction of the *adjunction space* of a diagram

$$X \supset A \xrightarrow{g} Y$$

where g is a map defined on a closed subspace A of a space X into another space Y.

For this purpose, consider the disjoint union

$$W = X \cup Y$$

of the spaces X and Y. W is a space with its topology defined as follows: a set $V \subset W$ is open iff $V \cap X$ and $V \cap Y$ are open sets of the spaces X and Y respectively. This space W will be called the *topological sum* of the spaces X and Y.

If we identify each $x \in A$ with $g(x) \in Y$, W becomes a space Z which will be called the *adjunction space* obtained by *adjoining* X to Y by means of the map $g:A \to Y$. In more detail, Z is the quotient space obtained from W by means of topological identification as follows. Two points $x \in X$ and $y \in Y$ are said to be equivalent iff $x \in A$ and $g(x) = y$; two points x and x' of A are equivalent iff $g(x) = g(x')$. Letting, of course, each point be equivalent to itself, we have an equivalence relation in W; we take Z to be the quotient space of W with respect to this equivalence relation and $p: W \to Z$ the natural projection.

One can easily verify that the natural projection p maps Y homeomorphically onto a subspace $p(Y)$ of Z. By means of this imbedding, Y will be considered as a closed subspace of Z. Obviously, $Y = Z$ iff $A = X$. Furthermore, we have the following

Proposition 7.1. *Y is a retract of Z iff the map $g:A \to Y$ has an extension over X.*

Proof. *Necessity.* Let $r:Z \supset Y$ be a retraction. Using the natural projection $p: W \to Z$, we define a map $f : X \to Y$ by taking

$$f(x) = rp(x), \quad (x \in X \subset W).$$

If $x \in A$, then we have $f(x) = rp(x) = rg(x) = g(x)$. Hence f is an extension of g over X.

Sufficiency. Let $f:X \to Y$ be a map with $f \mid A = g$. Define a function $r:Z \to Y$ as follows. Let $z \in Z$. If $z \in Y$, we define $r(z) = z$. If $z \notin Y$, then there is a unique point $x \in X \setminus A$ such that $p(x) = z$. In this case, we define $r(z) = f(x)$. Let $h = rp$. Then $h \mid X = f$ and $h \mid Y$ is the identity map. According to (5.1), h is continuous. By (6.1), this implies that r is continuous. Since $r \mid Y$ is the identity map, r is a retraction. ∎

Thus the extension problem for the map $g:A \to Y$ over X is equivalent to a retraction problem which is a special case of the extension problem.

Next, let us note another interesting property of the adjunction space. The natural projection $p: W \to Z$ maps X into Z and A into Y. Hence it defines a map

$$p:(X, A) \to (Z, Y).$$

This map p is evidently a *relative homeomorphism*, that is to say, it carries $X \setminus A$ homeomorphically onto $Z \setminus Y$.

For examples of adjunction spaces, let us take $X = E^n$ and $A = S^{n-1}$. Consider a map $g:S^{n-1} \to Y$ and the adjunction space Z obtained by adjoining E^n to Y by means of g. Then the pair (Z, Y) is called a *relative n-cell*. In particular, if $Y = P^{n-1}$ is the real projective $(n-1)$-space and if $g:S^{n-1} \to P^{n-1}$ is the map obtained by identifying antipodal points, then the adjunction space Z can be identified with the real projective n-space P^n. Hence, the pair (P^n, P^{n-1}) is a relative n-cell. Similarly, (CP^n, CP^{n-1}) is a relative $2n$-cell, where CP^n denotes the complex projective n-space which is of real dimension $2n$.

8. Homotopy problem and classification problem

A family of maps $h_t:X \to Y$ $(0 \leqslant t \leqslant 1)$, indexed by the real numbers $t \in I$, is called a *homotopy* if the function $H:X \times I \to Y$, defined by

$$H(x, t) = h_t(x), \quad (x \in X, t \in I)$$

is continuous. h_0 and h_1 are called the *initial map* and the *terminal map* of the homotopy h_t. In the sequel, the homotopy h_t and the map H will be considered essentially as the same thing and we shall use whichever appears more convenient for the particular purpose at hand.

Two maps $f:X \to Y$ and $g:X \to Y$ are said to be *homotopic* (notation: $f \simeq g$), if there exists a homotopy, $h_t:X \to Y$, $(0 \leqslant t \leqslant 1)$, such that $h_0 = f$ and $h_1 = g$. In this case, h_t is called a *homotopy connecting* f and g, and is denoted by

$$h_t:f \simeq g.$$

Intuitively, f and g are homotopic iff each can be changed continuously into the other.

Given two maps $f, g:X \to Y$, it is not always true that f and g are homotopic. For example, if $X = Y$ is the n-sphere S^n, f the identity map on S^n, and g a constant map, then it follows from the Homotopy Axiom of homology theory that f and g are not homotopic, [E–S; p. 11].

On the other hand, if $f, g:X \to S^n$ are two maps such that for any $x \in X$ the points $f(x)$ and $g(x)$ are never antipodal, then f and g are homotopic. In fact, there is always defined uniquely the minor arc of the great circle joining $f(x)$ and $g(x)$ and we may define $h_t(x)$ as the point dividing this arc in the ratio t:1-t. As a consequence, if a map $f:X \to S^n$ leaves a point free from the image $f(X)$, then f is homotopic to a constant map.

The *homotopy problem* is to determine whether or not two given maps $f, g:X \to Y$ are homotopic. This problem is actually a special case of the extension problem of § 2. In fact, consider the product space $X \times I$ and its subspace

$$M = (X \times 0) \cup (X \times 1)$$

that is to say, M consists of the bottom and the top of $X \times I$. Define a map $\phi:M \to Y$ by setting

$$\phi(x, 0) = f(x), \quad \phi(x,1) = g(x)$$

for each $x \in X$. Then every extension $H: X \times I \to Y$ of ϕ is a homotopy connecting f and g and vice versa. Thus f and g are homotopic iff ϕ has an extension over $X \times I$.

The homotopy axiom of the homology and cohomology theories gives necessary conditions for two maps to be homotopic. More precisely, if $f, g: X \to Y$ are homotopic maps, then they induce the same homomorphisms on the homology groups and the cohomology groups; that is, $f_* = g_*$ and $f^* = g^*$.

These necessary conditions can be used to prove that two particular given maps $f, g: X \to Y$ fail to be homotopic. For example, let X and Y be two oriented closed n-dimensional manifolds. In the homology theory with integral coefficients, let $\alpha \in H_n(X)$ and $\beta \in H_n(Y)$ denote the generators determined by the orientations of X and Y. The *degree* of any given map $f: X \to Y$ is defined to be the integer $\deg(f)$ such that $f_*(\alpha) = \deg(f) \cdot \beta$. Then two maps $f, g: X \to Y$ are not homotopic if $\deg(f) \neq \deg(g)$.

Some special cases of homotopies are of importance. Let us suppose that X is a subspace of Y. Then a homotopy $h_t: X \to Y$, $(0 \leqslant t \leqslant 1)$, is said to be a *deformation of X in Y* if h_0 is the inclusion map $i: X \subset Y$. Furthermore, if h_1 is a constant map, then the deformation h_t is called a *contraction of X in Y*. They are simply called *deformations* and *contractions* of X if $Y = X$.

If a contraction of X (in Y) exists, X is said to be *contractible* (in Y). If $x_0 \in X$ and a contraction $h_t: X \to Y$, $(0 \leqslant t \leqslant 1)$, exists such that $h_t(x_0) = x_0$ for each $t \in I$, then X is said to be *contractible to the point x_0 in Y*. For example, every proper subspace of the n-sphere S^n is contractible to a point in S^n.

Now, for any two given spaces X and Y, let us denote the totality of maps of X into Y by

$$\Omega = Y^X.$$

Proposition 8.1. *The relation $\simeq$ between the maps Ω is an equivalence relation.*

Proof. We have to show that the relation $\simeq$ is reflexive, symmetric, and transitive. (1) To prove that it is reflexive, let us define for any $f \in \Omega$ a homotopy $f_t: X \to Y$ $(0 \leqslant t \leqslant 1)$ by taking $f_t = f$ for every $t \in I$. Then $f_t: f \simeq f$. (2) To prove the symmetry, let $h_t: f \simeq g$ be a homotopy connecting two given maps $f \in \Omega$ and $g \in \Omega$. Define a homotopy $k_t: X \to Y$ $(0 \leqslant t \leqslant 1)$ by taking $k_t = h_{1-t}$ for each $t \in I$. Then $k_t: g \simeq f$. (3) To prove the transitivity assume $\phi_t: f \simeq g$ and $\psi_t: g \simeq h$. Define a system of maps $\chi_t: X \to Y$ $(0 \leqslant t \leqslant 1)$ by taking

$$\chi_t = \begin{cases} \phi_{2t}, & (0 \leqslant t \leqslant \frac{1}{2}), \\ \psi_{2t-1}, & (\frac{1}{2} \leqslant t \leqslant 1). \end{cases}$$

By (5.1), χ_t is a homotopy. Since $\chi_0 = \phi_0 = f$ and $\chi_1 = \psi_1 = h$, we have $\chi_t: f \simeq h$. ∎

As a consequence of (8.1) the maps Ω are divided into disjoint equivalence classes, called *the homotopy classes* of these maps. Let us denote by $\pi(X; Y)$ the totality of these homotopy classes and denote by $[f]$ the homotopy class of f, that is to say, the homotopy class which contains the map $f \in \Omega$.

If $g: X \to X'$ is a map and if the maps $f_0, f_1: X' \to Y$ are homotopic, then obviously we have $f_0 g \simeq f_1 g$. This means that $\pi(X; Y)$ is a contravariant functor in X, [E–S; p. 111]. On the other hand, if $g: Y' \to Y$ and if the maps $f_0, f_1: X \to Y'$ are homotopic, then $gf_0 \simeq gf_1$. This means that $\pi(X; Y)$ is a covariant functor in Y. Consequently, *$\pi(X; Y)$ is a functor of two variables contravariant in X and covariant in Y.*

In a great many cases, it is possible to enumerate the homotopy classes $\pi(X; Y)$ whereas the set Y^X has the power of continuum. Given two spaces X and Y, the *classification problem* is to enumerate the homotopy classes $\pi(X; Y)$ of the maps Y^X and to exhibit a representative map in each homotopy class.

For example, if X is a paracompact Hausdorff space and Y is a solid space, then it is easily seen that any two maps are homotopic and hence $\pi(X; Y)$ consists of a single homotopy class; in symbols,

$$\pi(X; Y) = 0.$$

The same is true if X is any space and Y is contractible. Some interesting non-trivial examples will be given in the next chapter.

9. The homotopy extension property

In the present section, we propose to relate the concept of homotopy to the extension problem. The main result can be stated as follows: *In a great many cases, the extension problem for a given map $g: A \to Y$ over $X \supset A$ depends only on the homotopy class of g.*

Let $f: X \to Y$ be a given map and A a given subspace of X. A homotopy $h_t: A \to Y$; $(0 \leqslant t \leqslant 1)$, is called a *partial homotopy* of f if $f \mid A = h_0$; it will be simply called a *homotopy* of f in case $A = X$. Restrictions and extensions of homotopies are defined in the same manner as those for maps.

Definition 9.1. A subspace A of a space X is said to have the *homotopy extension property* (abbreviated HEP) in X with respect to a space Y, if every partial homotopy

$$h_t: A \to Y, \quad (0 \leqslant t \leqslant 1),$$

of an arbitrary map $f: X \to Y$ has an extension

$$g_t: X \to Y, \quad (0 \leqslant t \leqslant 1),$$

such that $g_0 = f$. A is said to have the *absolute homotopy extension property* (abbreviated AHEP) in X, if it has the HEP in X with respect to every space Y.

The homotopy extension property is basic to various constructions in homotopy theory. Fortunately, this important property exists in reasonably smooth spaces; in particular, we have the following two propositions.

Proposition 9.2. *If (X, A) is a (finitely) triangulable pair, then A has the AHEP in X.*

Proof. Let $f: X \to Y$ be a given map and $h_t: A \to Y$, $(0 \leqslant t \leqslant 1)$, a given partial homotopy of f. Consider the product space $M = X \times I$ and its closed subspace $L = (X \times 0) \cup (A \times I)$. Define a map $H: L \to Y$ by setting

$$H(x, t) = \begin{cases} f(x), & \text{if } x \in X, t = 0, \\ h_t(x), & \text{if } x \in A, t \in I. \end{cases}$$

According to (4.2), there is a retraction $r: M \supset L$. Define a homotopy $g_t: X \to Y$, $(0 \leqslant t \leqslant 1)$, by taking

$$g_t(x) = Hr(x, t), \quad (x \in X, t \in I).$$

Then g_t is obviously an extension of h_t such that $g_0 = f$. Since Y is arbitrary, this proves that A has the AHEP in X. ▌

A space X is said to be *binormal* if $X \times I$ is normal. Paracompact Hausdorff spaces (and hence metrizable spaces) are binormal.

Proposition 9.3. *If Y is a (finitely) triangulable space, then every closed subspace A of any binormal space X has the HEP in X with respect to Y.*

Proof. Without loss of generality, we may assume that Y is a subcomplex of the unit n-simplex Δ_n, where n depends on Y. According to a classical theorem, [E–S; p. 72], there exists an open neighborhood U of Y in Δ_n such that Y is a retract of U. Let $r: U \supset Y$ be a retraction.

Let A be any closed subspace of a given binormal space X, $f: X \to Y$ a given map, and $h_t: A \to Y$, $(0 \leqslant t \leqslant 1)$, a given partial homotopy of f. Consider the spaces $M = X \times I$ and $L = (X \times 0) \cup (A \times I)$, and define a map $H: L \to Y$ as in the proof of (9.2). By § 2, Δ_n is solid. Since L is closed in the normal space M, the map $H: L \to Y \subset \Delta_n$ has an extension $G: M \to \Delta_n$.

Let $V = G^{-1}(U)$. Then V is an open neighborhood of L in M. Since the unit interval I is compact, one can easily prove that there exists an open neighborhood W of A in X such that $W \times I$ is contained in V.

Since X is normal, it follows from Urysohn's lemma, [L_2; p. 27], that there exists a continuous real function $\chi: X \to I$ such that $\chi(X \setminus W) = 0$ and $\chi(A) = 1$. Define a map $F: M \to Y$ by taking

$$F(x,t) = r[G(x, \chi(x)t)], \quad (x \in X, t \in I).$$

Obviously F is an extension of H over M. Hence the partial homotopy h_t of f has an extension $g_t: X \to Y$, $(0 \leqslant t \leqslant 1)$, defined by $g_t(x) = F(x,t)$ for each $x \in X$ and $t \in I$. Clearly $g_0 = f$. ▌

Generalizations of (9.2) and (9.3) will be given in Ex.N and Ex.O at the end of the chapter.

Now let A be a given subspace of a space X and $g:A \to Y$ a given map. If A has the HEP in X with respect to Y, then the restricted extension problem in § 2 obviously depends only on the homotopy class of g. In this case, the restricted extension problem is equivalent to a broadened form, namely to ask *whether there is a map* $f:X \to Y$ *such that* $fh \simeq g$, *where* $h:A \subset X$ *denotes the inclusion map*.

This new broadened form is apparently weaker than the original restricted form. However, once we reflect upon the concept of homotopy, we realize that the original form of the problem was formulated in too narrow a fashion. Hereafter, when we refer to the *extension problem*, we shall mean its broadened form unless otherwise stated.

10. Relative homotopy

Let M be an abstract set and X, Y be two given spaces. Let $\{ X_\mu \}$, $\{ Y_\mu \}$ be two systems of sets indexed by $\mu \in M$ such that $X_\mu \subset X$ and $Y_\mu \subset Y$ for each index $\mu \in M$. For the sake of brevity, we shall denote by $\{ M \}$ the detailed notation $\{ X_\mu, Y_\mu \mid \mu \in M \}$ whenever there is no danger of ambiguity.

Consider the maps $f:X \to Y$ such that $f(X_\mu) \subset Y_\mu$ for each $\mu \in M$. The totality of these maps will be denoted by

$$\Omega = Y^X \{ M \} = Y^X \{ X_\mu, Y_\mu \mid \mu \in M \}.$$

For example, if M consists of a single element μ and if $X_\mu = A$, $Y_\mu = B$, then Ω is the set of all maps $f:(X, A) \to (Y, B)$, see [E–S; p. 3]. For a second example, let M contain two elements μ and ν. If $X_\mu = A$, $Y_\mu = B$ are as above and $X_\nu = \{ x_0 \}$, $Y_\nu = \{ y_0 \}$ are singletons with $x_0 \in A$, $y_0 \in B$, then Ω is the set of all maps $f:(X, A, x_0) \to (Y, B, y_0)$. It is primarily these cases that the notation is intended to cover.

Two maps f and g in Ω are said to be *homotopic relative to the system* $\{ M \}$ if there exists a homotopy $h_t:X \to Y$, $(0 \leqslant t \leqslant 1)$, such that $h_0 = f$, $h_1 = g$ and $h_t \in \Omega$ for every $t \in I$. In notation,

$$f \simeq g \text{ rel } \{ M \}.$$

In this case, h_t is called a *homotopy connecting* f *and* g *in* Ω and is denoted by

$$h_t:f \simeq g \text{ rel } \{ M \}.$$

If X_μ is a single point and $Y_\mu = f(X_\mu)$, we may identify μ with X_μ. If this holds for all $\mu \in M$, then M is identified with a subset of X. In this important case, we shall use the usual notation:

$$f \simeq g \text{ rel } M, \quad h_t:f \simeq g \text{ rel } M.$$

In particular, if M is the empty set, the notation will be simply $f \simeq g$ and $h_t : f \simeq g$ as in § 8.

As in (8.1) one can easily prove that the relation of homotopy relative to $\{ M \}$ of the maps Ω is an equivalence relation. Hence, Ω is divided into disjoint classes, called the *homotopy classes relative to* $\{ M \}$ or the *homotopy classes in* Ω. The *classification problem* of the maps Ω is to enumerate these classes in terms of known topological invariants.

Let B be a subspace of Y which intersects Y_μ for each $\mu \in M$. A map $f \in \Omega$ is said to be *deformable into* B *relative to* $\{ M \}$, if there exists a map $g : X \to B$ such that

$$f \simeq ig \text{ rel } \{ M \}$$

where i denotes the inclusion map $i : B \subset Y$. In particular, B may consist of a single point y_0 in the intersection of the subspaces Y_μ. Let 0 denote the constant map $0(X) = y_0$. Then 0 is in Ω. A map f is said to be *null-homotopic* or *inessential relative to* $\{ M \}$ if $f \simeq 0 \text{ rel } \{ M \}$, otherwise it is said to be *essential relative to* $\{ M \}$.

Now, let us define the important notion of deformation retract. A subspace A of a space X is said to be a *deformation retract* of X if there exists a deformation $h_t : X \to X$, $(0 \leqslant t \leqslant 1)$, such that h_1 is a retraction of X onto A. If the deformation h_t satisfies a further condition that $h_t(a) = a$ for each $a \in A$ and each $t \in I$, then A is said to be a *strong deformation retract* of X. In this case, h_t will be called a *retracting deformation* of X onto A. In Ex.T at the end of this chapter, we shall see that, in a great many cases, these two notions are equivalent.

If X denotes the space obtained by deleting from E^n an interior point which may be assumed to be the origin 0 without loss of generality, then S^{n-1} is a strong deformation retract of X. In fact, a retracting deformation $h_t : X \to X$, $(0 \leqslant t \leqslant 1)$, is given by

$$h_t(x) = \left(1 - t + \frac{t}{|x|}\right) x$$

for each $x \in X$ and $t \in I$, where $|x|$ denotes the distance between 0 and x. On the other hand, a point of the n-sphere S^n is a retract of S^n but not a deformation retract of S^n.

For another example of strong deformation retract, we strengthen (4.2) by the following proposition which can be proved by an easy modification of the arguments used in the proof of (4.2).

Proposition 10.1 *If* (X, A) *is a (finitely) triangulable pair, then the closed subspace* $L = (X \times 0) \cup (A \times I)$ *is a strong deformation retract of the product space* $M = X \times I$.

More examples of strong deformation retracts are given in Ex.S at the end of the chapter.

11. Homotopy equivalences

Under the general notations of the preceding section, let us consider two given maps

$$f: X \to Y, \quad g: Y \to X$$

such that $f(X_\mu) \subset Y_\mu$ and $g(Y_\mu) \subset X_\mu$ for each $\mu \in M$.

g is called a *left homotopy inverse of* f *relative to* $\{M\}$ and f is called a *right homotopy inverse of* g *relative to* $\{M\}$, if there exists a homotopy $h_t: X \to X$ $(0 \leqslant t \leqslant 1)$ such that $h_0 = gf$, h_1 is the identity map on X, and $h_t(X_\mu) \subset X_\mu$ for each $\mu \in M$ and each $t \in I$. g is called a *two-sided homotopy inverse of* f *relative to* $\{M\}$ if it is both left and right homotopy inverse of f relative to $\{M\}$.

The map $f: X \to Y$ is said to be a *homotopy equivalence relative to* $\{M\}$, if it has a two-sided homotopy inverse relative to $\{M\}$. In notation,

$$f: X \simeq Y \text{ rel } \{M\}.$$

The spaces X and Y are said to be *homotopically equivalent relative to* $\{M\}$, if there exists a homotopy equivalence $f: X \simeq Y \text{ rel } \{M\}$. In this case, we also say that X and Y are *of the same homotopy type relative to* $\{M\}$ and we shall use the notation

$$X \simeq Y \text{ rel } \{M\}.$$

The following two special cases are of frequent occurrence in the sequel.

Firstly, if $M = \square$, that is to say, if there is no system of relativity, then we simply omit the phrase "relative to $\{M\}$" in the terminology given above. For example, two spaces X and Y are said to be *homotopically equivalent*, $X \simeq Y$, if there are maps $f: X \to Y$ and $g: Y \to X$ such that the compositions gf and fg are homotopic to the identity maps on X and Y respectively.

Secondly, assume that M consists of a single element μ and that $X_\mu = A$, $Y_\mu = B$. Then a map $f: (X, A) \to (Y, B)$ is said to be a *homotopy equivalence* and is denoted by $f: (X, A) \simeq (Y, B)$ if $f: X \simeq Y \text{ rel } \{A, B\}$. The pairs (X, A) and (Y, B) are said to be *homotopically equivalent*, $(X, A) \simeq (Y, B)$, if there exists a homotopy equivalence $f: (X, A) \simeq (Y, B)$.

The notion of homotopy equivalence introduced above is justified by the fact that the extension problem as broadened in § 9 depends essentially only on the homotopy type of the pair (X, A) and that of the space Y. This result can be precisely formulated as follows.

Assume that $(X, A) \simeq (X', A')$ and $Y \simeq Y'$ and let $g: A \to Y$ be a given map. Let $h: A \subset X$ and $h': A' \subset X'$ denote the inclusion maps. By definition, there exist maps

$$\phi: (X, A) \to (X', A'), \phi': (X', A') \to (X, A), \psi: Y \to Y', \psi': Y' \to Y$$

such that composed maps $\phi'\phi$, $\phi\phi'$, $\psi'\psi$ and $\psi\psi'$ are homotopic to the corresponding identity maps. Let $\chi:A \to A'$ and $\chi':A' \to A$ denote the maps defined by the maps ϕ and ϕ' respectively, and let

$$g' = \psi g \chi' : A' \to Y'.$$

If the extension problem for g has a solution over X, i.e., if these exists a map $f:X \to Y$ such that $fh \simeq g$, then the extension problem for g' also has a solution over X'. In fact, $f' = \psi f \phi' : X' \to Y'$ satisfies the relation

$$f'h' = \psi f \phi' h' = \psi f h \chi' \simeq \psi g \chi' = g'.$$

Conversely, if there exists a map $e':X' \to Y'$ such that $e'h' \simeq g'$, then the extension problem for g has a solution over X. In fact, $e = \psi' e' \phi : X \to Y$ satisfies the relation

$$eh = \psi' e' \phi h = \psi' e' h' \chi \simeq \psi' g' \chi = \psi' \psi g \chi' \chi \simeq g.$$

Examples of homotopy equivalences: (1) If $f:X \to Y$ is a homeomorphism of X onto Y, then f^{-1} is obviously a two-sided homotopy inverse of f. Hence homeomorphs are of the same homotopy type. (2) If X is a deformation retract of Y, then the inclusion map $f:X \subset Y$ is a homotopy equivalence.

According to the homotopy axiom and the algebraic axioms, a homotopy equivalence induces isomorphisms on the homology groups and the cohomology groups. Hence, homotopically equivalent spaces have isomorphic homology groups and cohomology groups. A property of spaces is said to be a *homotopy invariant* if it is preserved by homotopy equivalences. Almost all invariants studied in algebraic topology are homotopy invariants.

The role of homotopy equivalences having been clarified, it is natural to search for two-sided homotopy inverses, if any, of a given map $f:X \to Y$ relative to a given system $\{M\}$. For this purpose, it is sometimes helpful to observe the following fact. If a given map $f:X \to Y$ has both left and right homotopy inverses relative to a given system $\{M\}$, then it has a two-sided one. In fact, if $g':Y \to X$ is a left homotopy inverse of f relative to $\{M\}$ and if $g'':Y \to X$ is a right homotopy inverse of f relative to $\{M\}$, then it is easy to verify that the composed map $g = g'fg'':Y \to X$ is a two-sided homotopy inverse of f relative to $\{M\}$.

12. The mapping cylinder

Let $f:X \to Y$ be a given map. By the process of topological identification, we shall construct a space M_f which is called the *mapping cylinder* of f.

For this purpose, let us consider the topological sum

$$W = (X \times I) \cup Y$$

of the spaces $X \times I$ and Y. If we identify $(x, 1) \in X \times I$ with $f(x) \in Y$ for every $x \in X$, we obtain a quotient space M_f, the mapping cylinder of f, and a natural projection

$$p: W \longrightarrow M_f.$$

One can easily verify that p maps Y homeomorphically onto a *closed* subspace $p(Y)$ of M_f. By means of this imbedding, Y will be considered as a closed subspace of M_f. On the other hand, the map

$$g: X \longrightarrow M_f$$

given by $g(x) = p(x, 0)$ for each $x \in X$ carries X homeomorphically onto the closed subspace $p(X \times 0)$ of M_f. By means of this topological map g, X is imbedded as a closed subspace of M_f. Thus X and Y are considered as disjoint closed subspaces of M_f and will be called the *domain* and the *range* of the mapping cylinder M_f.

Proposition 12.1. *The range Y of the mapping cylinder M_f of a map $f: X \to Y$ is a strong deformation retract of M_f.*

Proof. Define a system of functions $h_t: M_f \to M_f$, $(0 \leqslant t \leqslant 1)$, by taking

$$h_t[p(x, s)] = p(x, s + t - st)$$

$$h_t(y) = y$$

for every $x \in X$, $y \in Y$, $s \in I$ and $t \in I$. By (6.1), it is easily verified that h_t is a homotopy. It is obvious that h_0 is the identity map on M_f, h_1 is a retraction of M_f onto Y, and $h_t \mid Y$ is the identity map on Y for each $t \in I$. Hence Y is a strong deformation retract of M_f. ∎

Proposition 12.2. *For any given map $f: X \to Y$, the maps $g: X \to M_f$* and *$pf: X \to M_f$ are homotopic.*

Proof. Let $h_t: X \to M_f$, $(0 \leqslant t \leqslant 1)$, denote the homotopy defined by

$$h_t(x) = p(x, t)$$

for every $x \in X$ and $t \in I$. Then we have $h_0 = g$ and $h_1 = pf$. Hence $g \simeq pf$. ∎

Two maps $f: X \to Y$ and $f': X' \to Y'$ are said to be *homotopically equivalent* if there exist homotopy equivalences $\phi: X \simeq X'$ and $\psi: Y \simeq Y'$ such that $\psi f \simeq f'\phi$. Let $\phi': X' \simeq X$ and $\psi': Y' \simeq Y$ denote two-sided homotopy inverses of ϕ and ψ respectively. Then the three conditions

$$\psi f \simeq f'\phi, \quad f \simeq \psi' f' \phi, \quad \psi f \phi' \simeq f'$$

are mutually equivalent. As an immediate consequence of (12.1) and (12.2), we have the following theorem which justifies the introduction of the notion of mapping cylinders.

Theorem 12.3. *Every map $f: X \to Y$ is homotopically equivalent to an inclusion map, namely, $g: X \subset M_f$.*

Finally, the following special case of mapping cylinder is of importance. If Y consists of a single point v, then the map $f:X \to Y$ has to be the constant map $f(X) = v$. In this case, the mapping cylinder M_f is called *the join of X to v* or *the cone over X* and will be denoted by $k(X)$. The point v is called the *vertex* of $k(X)$.

13. A generalization of the extension problem

The extension problem as broadened in § 9 suggests a generalized problem which can be displayed diagrammatically by

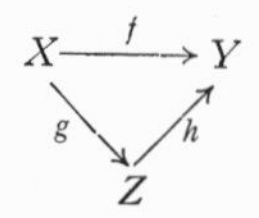

where f and g are given maps and h must be found such that $hg \simeq f$. However, this generalized problem is equivalent to the extension problem of the map $f:X \to Y$ over the mapping cylinder $M_g \supset X$ of the map $g:X \to Z$. More precisely, we have the following

Proposition 13.1. *For any two maps $f:X \to Y$ and $g:X \to Z$, the following two statements are equivalent:*

(i) There is a map $h:Z \to Y$ such that $hg \simeq f$.

(ii) There is a map $F:M_g \to Y$ such that $Fq \simeq f$, where $q:X \subset M_g$ denotes the inclusion map.

Proof. $(i) \to (ii)$. According to (12.1), there is a retraction $r:M_g \supset Z$ defined by

$$r(p(x, t)) = g(x), \quad (x \in X, t \in I).$$

Hence the map $h:Z \to Y$ has an extension $F = hr:M_g \to Y$. By (12.2), we have

$$Fq = hrq \simeq hrpg = hg \simeq f.$$

$(ii) \to (i)$. Let $h = F \mid Z$. Then, by (12.2), we have

$$hg = Fpg \simeq Fq \simeq f. \blacksquare$$

The derived algebraic problem of this general problem is as follows. In any homology theory or cohomology theory satisfying the Eilenberg-Steenrod axions, the given maps $f: X \to Y$ and $g: X \to Z$ induce the homomorphisms f_*, g_*, f^*, g^* indicated in the following diagrams:

$$\begin{array}{ccc} H_m(X) & \xrightarrow{f_*} & H_m(Y) \\ & {\scriptstyle g_*}\searrow \quad \nearrow{\scriptstyle \phi} & \\ & H_m(Z) & \end{array} \qquad \begin{array}{ccc} H^m(X) & \xleftarrow{f^*} & H^m(Y) \\ & {\scriptstyle g^*}\nwarrow \quad \swarrow{\scriptstyle \psi} & \\ & H^m(Z) & \end{array}$$

Then the derived algebraic problem is to determine whether or not there exist homomorphisms ϕ and ψ such that $\phi g_* = f_*$ and $g^*\psi = f^*$.

The existence of the homomorphisms ϕ and ψ is a necessary condition for the existence of a map $h : X \to Y$ such that $hg \simeq f$. In fact, if such an h exists, then $\phi = h_*$ and $\psi = h^*$ are solutions of the derived algebraic problem.

In many cases, this necessary condition can be used to prove the non-existence of h. For example, let us assume $X = S^m = Y$, $Z = S^n$ with $m \neq 0$ and $m \neq n$. If $f : X \to Y$ is of degree different from zero and if $g : X \to Z$ is any given map, then there exists no map $h : Z \to Y$ such that $hg \simeq f$.

14. The partial mapping cylinder

The objective of the present section is to show that the generalized problem in § 13, which is so far the broadest form of the extension problem, is equivalent to a retraction problem, which is apparently the narrowest form of the problem.

For this purpose, let us introduce the *partial mapping cylinder* $M_g(X)$ of a given map $g : A \to Y$ defined on a subspace A of a space X. Consider the topological sum

$$W = X \cup (A \times I) \cup Y$$

of the spaces X, $A \times I$ and Y. For each $a \in A \subset X$, identify a with $(a, 0) \in A \times I$, and $(a, 1)$ with $g(a) \in Y$. In this way, we obtain a quotient space $M_g(X)$ which is called the partial mapping cylinder of g over X and a natural projection

$$p : W \longrightarrow M_g(X).$$

Then one can easily see that p maps X and Y homeomorphically onto disjoint closed subspaces $p(X)$ and $p(Y)$ of $M_g(X)$ respectively. Thus, X and Y will be considered as disjoint closed subspaces of $M_g(X)$ and will be called the *domain* and the *range* of the partial mapping cylinder $M_g(X)$.

Lemma 14.1. *If the range Y of $M_g(X)$ is a retract of $M_g(X)$, then there exists a map $f : X \to Y$ such that $f \mid A \simeq g$.*

Proof. Let $r : M_g(X) \supset Y$ be a retraction and define $f : X \to Y$ by $f = r \mid X$. Consider the homotopy $h_t : A \to Y$, $(0 \leqslant t \leqslant 1)$, defined by

$$h_t(a) = r(p(a, t)), \quad (a \in A, t \in I).$$

It is obvious that $h_0 = f \mid A$ and $h_1 = g$. ∎

Lemma 14.2. *If there exist a map $f : X \to Y$ such that $f \mid A \simeq g$, then the range Y of $M_g(X)$ is a retract of $M_g(X)$.*

Proof. Since $f \mid A \simeq g$, there is a homotopy $h_t : A \to Y$ $(0 \leqslant t \leqslant 1)$ such that $h_0 = f \mid A$ and $h_1 = g$. Let $r : M_g(X) \to Y$ be defined by

$$r(z) = \begin{cases} f(x) & (\text{if } z = p(x),\ x \in X), \\ h_t(x) & (\text{if } z = p(x, t),\ (x, t) \in A \times I), \\ y & (\text{if } z = p(y),\ y \in Y). \end{cases}$$

Then it is obvious that r is a retraction. ∎

Now, consider two given maps $f: X \to Y$ and $g: X \to Z$. Since X is a subspace of the mapping cylinder M_g of g, the partial mapping cylinder $M_f(M_g)$ of f over M_g is defined. Then we have the following

Theorem 14.3. *There exists a map $h: Z \to Y$ such that $hg \simeq f$ iff the range Y of $M_f(M_g)$ is a retract of $M_f(M_g)$.*

Proof. *Sufficiency*. Assume that Y is a retract of $M_f(M_g)$. Then, by (14.1) there exists a map $F: M_g \to Y$ such that $F \mid X \simeq f$. According to (13.1), this implies the existence of a map $h: Z \to Y$ such that $hg \simeq f$.

Necessity. Assume that there is a map $h: Z \to Y$ such that $hg \simeq f$. Then, by (13.1), there is a map $F: M_g \to Y$ such that $F \mid X \simeq f$. According to (14.2), this implies that Y is a retract of $M_f(M_g)$. ∎

The construction of the partial mapping cylinder is closely related to that of the adjunction space given in § 7. In fact, for reasonable spaces X and A, (see Ex.P), the adjunction space obtained by adjoining X to Y by means of a map $g: A \to Y$ is homotopically equivalent to $M_g(X)$.

Finally, if Y consists of a single point v, then the map $g: A \to Y$ has to be the constant map $g(A) = v$. In this case, the partial mapping cylinder $M_g(X)$ is called the *partial cone* over the subspace A of the space X. Obviously,

$$M_g(X) = X \cup k(A) \subset k(X).$$

15. The deformation problem

Let us consider a given map $f: (X, A) \to (Y, B)$ and a given subspace Y_0 of Y. According to our general definition in § 10, f is said to be *deformable into* Y_0 if there exists a homotopy $f_t: (X, A) \to (Y, B)$, $(0 \leqslant t \leqslant 1)$, such that $f_0 = f$ and $f_1(X) \subset Y_0$.

The *deformation problem* of $f: (X, A) \to (Y, B)$ into Y_0 is to determine whether or not f is deformable into Y_0. Let $B_0 = B \cap Y_0$. Then this problem can be described diagrammatically by

$$\begin{array}{ccc} (X, A) & \xrightarrow{f} & (Y, B) \\ & {}_{g}\searrow \quad \nearrow_{h} & \\ & (Y_0, B_0) & \end{array}$$

where $h: (Y_0, B_0) \subset (Y, B)$ denotes the inclusion map and $g: (X, A) \to (Y_0, B_0)$ must be found such that $hg \simeq f$.

For the important special case that $A = \square$, we may always take $B = \square$ and hence $B_0 = \square$. In this case, we just omit A, B and B_0 from the notation given above. Hence, the deformation problem is actually the dual problem of the extension problem given in § 9. The derived algebraic problem and the necessary conditions of the deformability are dual to those of the extension problem in an obvious way. They can be formulated and applied similarly and hence are omitted.

Now we are in a position to prove the interesting result that the broadest extension problem in § 13 is equivalent to a deformation problem. For this purpose, let $f : X \to Y$ and $g : X \to Z$ be given maps. Let

$$W = M_f(M_g), \quad V = k(Y),$$

then $Y \subset W$ and $Y \subset V$. Define a map $\phi : (W, Y) \to (V, Y)$ by taking

$$\phi(w) = \begin{cases} w, & (\text{if } w \in Y), \\ v, & (\text{if } w \in M_g), \\ q(f(x), 1 - t), & (\text{if } w = p(x, t), x \in X, t \in I), \end{cases}$$

where v denotes the vertex of the cone $k(Y)$ and $p : M_g \cup (X \times I) \cup Y \to W$ and $q : Y \times I \to V$ denote the natural projections.

Theorem 15.1. *There exists a map $h : Z \to Y$ such that $hg \simeq f$ iff the map $\phi : (W, Y) \to (V, Y)$ is deformable into Y.*

Proof. *Necessity.* According to (14.3) there is a retraction $r : W \supset Y$. Define two homotopies $\xi_t, \eta_t : (W, Y) \to (V, Y)$, $(0 \leqslant t \leqslant 1)$, by taking

$$\xi_t(w) = \begin{cases} w, & (\text{if } w \in Y), \\ v, & (\text{if } w \in M_g), \\ q[rp(x, 1 - t + st), 1 - s], & (\text{if } w = p(x, s), x \in X, s \in I), \end{cases}$$

$$\eta_t(w) = \begin{cases} w, & (\text{if } w \in Y), \\ q[r(w), 1 - t], & (\text{if } w \in M_g), \\ q[rp(x, s), (1 - s)(1 - t)], & (\text{if } w = p(x, s), x \in X, s \in I). \end{cases}$$

Then one can easily verify that $\xi_0 = \phi$, $\xi_1 = \eta_0$ and $\eta_1 = r$. Hence ϕ is deformable into Y.

Sufficiency. There is a homotopy $\phi_t : (W, Y) \to (V, Y)$, $(0 \leqslant t \leqslant 1)$, such that $\phi_0 = \phi$ and $\phi_1(W) \subset Y$. Then we may define a retraction $r : W \supset Y$ by taking

$$r(w) = \begin{cases} w, & (\text{if } w \in Y), \\ \phi_1(w), & (\text{if } w \in M_g), \\ \phi_1 p(x, 2t), & (\text{if } w = p(x, t), x \in X, 0 \leqslant t \leqslant \frac{1}{2}), \\ \phi_{2-2t} f(x), & (\text{if } w = p(x, t), x \in X, \frac{1}{2} \leqslant t \leqslant 1). \end{cases}$$

Hence, by (14.3), there exists a map $h : Z \to Y$ such that $hg \simeq f$. ∎

On the other hand, the deformation problem is equivalent to an extension problem with side conditions. To see this, let $f : (X, A) \to (Y, B)$ be a given map and Y_0 a given subspace of Y. Consider the subspace $X \times 0$ of $X \times I$ and the map $F : X \times 0 \to Y$ defined by $F(x, 0) = f(x)$ for each $x \in X$. Then the following proposition is obvious.

Proposition 15.2. *$f : (X, A) \to (Y, B)$ is deformable into Y_0 iff F has an extension $G : X \times I \to Y$ such that $G(A \times I) \subset B$ and $G(X \times 1) \subset Y_0$.*

The special case of the deformation problem where $Y_0 = B$ is of importance. In this case, we have nice necessary conditions as given in the following proposition the proof of which is immediate and hence omitted.

Proposition 15.3. *If a map* $f : (X, A) \to (Y, B)$ *is deformable into* B, *then it induces trivial homomorphisms on homology and cohomology groups, that is to say,* $f_* = 0$ *and* $f^* = 0$.

In particular, if $(X, A) = (Y, B)$ and if f is the identity map, we have the following definition. The pair (X, A) is said to be *deformable into* A if the identity map on (X, A) is deformable into A. If this is the case, then the inclusion map $i : A \subset X$ is a homotopy equivalence. Furthermore, in case A has the HEP in X with respect to A, then the pair (X, A) is deformable into A iff A is a deformation retract of X.

16. The lifting problem

Consider a given map

$$p : E \longrightarrow B$$

of a space E *onto* a space B. Let X be a given space and $f : X \to B$ a given map. Then the *lifting problem* in the restricted form is to look for a map $g : X \to E$ such that $pg = f$. The map g is called a *lifting* of f (relative to p). This problem is dual to the restricted extension problem of § 2.

Dual to the homotopy extension property of § 9, we have the following important notion of covering homotopy property.

Definition 16.1. A map $p : E \to B$ is said to have the *covering homotopy property* (abbreviated CHP) for a space X, if, for every map $g : X \to E$ and every homotopy $f_t : X \to B$, $(0 \leqslant t \leqslant 1)$, of the map $f = pg$, there exists a homotopy $g_t : X \to E$, $(0 \leqslant t \leqslant 1)$, of g which *covers* f_t, that is to say, $pg_t = f_t$ for every $t \in I$.

The covering homotopy property is basic to many constructions in homotopy theory. Fortunately, this important property holds in a great many of cases, namely, the various fiber spaces. See Chapter III.

Now, if $p : E \to B$ has the CHP for X, then the restricted lifting problem for $f : X \to B$ obviously depends only on the homotopy class of f. In this case, it is equivalent to the broadened form, namely to *ask for a map* $g : X \to E$ *such that* $pg \simeq f$. Hereafter, when we refer to the *lifting problem*, we shall mean this broadened form unless otherwise stated.

Once the lifting problem is broadened as above, the original condition that p is onto is no longer important. In fact, we have a generalized problem displayed diagrammatically by

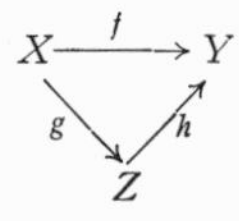

where f and h are given maps and g must be found such that $hg \simeq f$.

This general problem is dual to the generalized extension problem of § 13. If h is onto, then it is the lifting problem for f; if h is an inclusion map, then it is the deformation problem for f into the subspace Z of Y. On the other hand, this general problem is equivalent to a deformation problem. More precisely, we have the following

Theorem 16.2. *For any two maps $f: X \to Y$ and $h: Z \to Y$, the following two statements are equivalent*:

(*i*) *There exists a map $g: X \to Z$ such that $hg \simeq f$.*

(*ii*) *The map $pf: X \to M_h$ is deformable into Z, where $p: Y \subset M_h$ denotes the inclusion map.*

Proof. (*i*) $\to$ (*ii*). According to (12.2), we have $ph \simeq q$, where $q: Z \subset M_h$ denotes the inclusion map. Then, (*i*) implies that $pf \simeq phg \simeq qg$. Hence, pf is deformable into Z.

(*ii*) $\to$ (*i*). There is a homotopy $g_t: X \to M_h$, $(0 \leqslant t \leqslant 1)$, such that $g_0 = pf$ and $g_1(X) \subset Z$. Let $g: X \to Z$ denote the map defined by g_1. By (12.1), there is a retraction $r: M_h \supset Y$ such that $r \mid Z = h$. Let $f_t: X \to Y$, $(0 \leqslant t \leqslant 1)$, denote the homotopy given by $f_t = rg_t$ for each $t \in I$. Then we have $f_0 = rpf = f$ and $f_1 = rqg_1 = hg$. Hence $hg \simeq f$. ∎

17. The most general problem

In this final section of the chapter, we shall formulate the most general problem which contains essentially all problems described in this chapter as special cases.

Consider three given spaces X, Y, Z together with three given systems of subspaces $\{ X_\mu \}, \{ Y_\mu \}, \{ Z_\mu \}$ indexed by $\mu \in M$. Following our general ideas in § 10, we are concerned with only those maps and homotopies $X \to Y, X \to Z, Z \to Y$ which are relative to $\{ M \}$ in the obvious sense. Then our most general problem can be displayed diagrammatically by

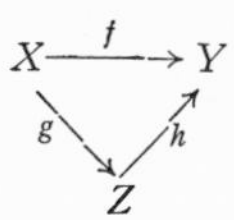

where the maps are understood to be relative to $\{ M \}$, f and one of g, h are given, and the other is to be found such that $hg \simeq f$ rel $\{ M \}$.

Nearly everything done in this chapter for the simplest extension, deformation and lifting problems admits generalizations some of which will be given in the exercises.

Although the problem is formulated in the most general form, the only cases considered in remainder of the book have 0, 1 or 2 elements as the domain M of the index μ.

EXERCISES

A. Elementary properties of retracts

1. Every retract of a Hausdorff space is closed.

2. Every subspace which consists of a single point is a retract of the containing space.

3. If A is a retract (deformation retract) of X and if B is a retract (deformation retract) of Y, then $A \times B$ is a retract (deformation retract) of $X \times Y$.

4. If X is a retract (deformation retract) of Y and if Y is a retract (deformation retract) of Z, then X is a retract (deformation retract) of Z.

B. More properties of solid spaces

1. A retract of a solid space is solid.
2. Any two maps of a binormal space into a solid space are homotopic.
3. A binormal solid space is contractible to a point.

C. AR's and ANR's

In classical algebraic topology, attention is confined to very well-behaved spaces, namely polyhedra. One class of spaces whose members retain many of the desirable properties of polyhedra is the absolute neighborhood retracts.

A subset A of a space X is said to be a *neighborhood retract* of X, if it is a retract of some open subspace U of X.

A metrizable space Y is said to be an *absolute retract* (abbreviated AR), whenever a topological image of Y as a closed subset Z_0 of any metrizable space Z is necessarily a retract of Z.

A metrizable space Y is said to be an *absolute neighborhood retract* (abbreviated ANR), whenever a topological image of Y as a closed subset Z_0 of any metrizable space Z is necessarily a neighborhood retract of Z.

Prove that every compact AR is solid.

D. Dugundji's extension of Tietze's theorem

Let X be a metrizable space and A be a closed subspace of X. Prove the following assertions.

1. *Existence of canonical coverings.* There exists of canonical covering of $X \setminus A$; by this we mean an open covering $\{ U \}$ of $X \setminus A$ satisfying the following conditions:

(CC1) $\{ U \}$ is locally finite, [K; p. 126].

(CC2) Every neighborhood of a boundary point of A in X contains infinitely many sets of $\{ U \}$.

(CC3) For each neighborhood V of $a \in A$ in X, there exists a neighborhood W of a in X, $W \subset V$, such that $U \subset V$ if $U \in \{ U \}$ and U meets W.

2. *Replacement by polyhedra.* By the aid of a canonical covering of $X \setminus A$, prove that there exists a space Y and a map $\mu : X \to Y$ with the following properties:

(RP1) The partial map $\mu \mid A$ is a homeomorphism of A onto a closed subset $\mu(A)$ of Y.

(RP2) $Y \setminus \mu(A)$ is an infinite simplicial polyhedron with Whitehead's weak topology and $\mu(X \setminus A) \subset Y \setminus \mu(A)$.

(RP3) Every neighborhood of a boundary point of $\mu(A)$ in Y contains infinitely many simplexes of the simplicial polyhedron $Y \setminus \mu(A)$.

3. *The extended Tietze's theorem.* If $f: A \to S$ is a map of A into a locally convex linear space S, then there exists an extension $F: X \to S$ of f such that $F(X)$ is contained in the convex hull of $f(A)$ in S.

Proofs are given in [Dugundji 1].

E. Kuratowski's imbedding

Let X be a given metrizable space and d a distance function which makes X a metric space. Replacing d by $d/(1 + d)$ if necessary, we may assume that d is bounded. A metric space X with a bounded distance function d is usually called a *bounded metric space.*

1. Let S denote the totality of bounded continuous real functions defined on X. Prove that S forms a Banach algebra with its norm defined by

$$| f | = \sup_{x \in X} | f(x) |$$

2. For an arbitrary point $a \in X$, consider the bounded continuous real function $f_a \in S$ defined by

$$f_a(x) = d(a, x)$$

for each $x \in X$. Prove that the correspondence $a \to f_a$ defines an isometric map $\chi: X \to S$. χ is called *Kuratowski's imbedding* of the bounded metric space X, [Kuratowski 1].

3. *Wojdyslawski's theorem.* The image $\chi(X)$ of Kuratowski's imbedding $\chi: X \to S$ is a closed subset of the convex hull Z of $\chi(X)$ in S. If X is separable, then so is Z, [Wojdyslawski 1].

F. Spaces obtained by topological identification

1. *Möbius strip.* Take a rectangle $ABCD$ and identify the side AB with the side CD so that A goes into C and B into D as shown in the following figure:

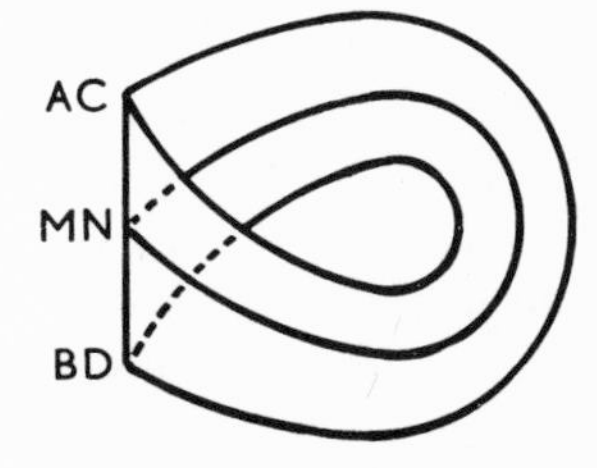

Fig. 1. On the left, there is a rectangle ABCD with M and N as mid-points of two sides. The figure on the right is obtained by pasting the two sides of the rectangle together so that A goes into C, B goes into D, and M goes into N.

The quotient space X thus obtained is called the Möbius strip. Let the construction be so carried out that the mid-points M, N of AB and CD are identified. As a consequence the line MN will go into a Jordan curve J

on the strip X. Prove that J is a deformation retract of X and that $X \setminus J$ is pathwise connected.

2. *Torus and Klein bottle.* Take the 1-sphere S^1 represented as the complex numbers z with $|z| = 1$ and consider the product space $W = S^1 \times I$. If we identify $(z, 0)$ and $(z, 1)$ for each $z \in S^1$, we obtain from W a quotient space T_2 called the *torus*. On the other hand, if we identify $(z, 0)$ and $(z^{-1}, 1)$ for each $z \in S^1$, we obtain from W a quotient space U_2 called the Klein bottle. In each case, the line $1 \times I$ will go into a Jordan curve J on the quotient space. Prove that J is a retract of the quotient space but not a deformation retract of the quotient space.

3. *Closed orientable surfaces.* Consider a plane convex region W bounded by a 4g-sided regular polygon P. Let the sides of P be denoted by

$$P = a_1 b_1 c_1 d_1 \cdots a_g b_g c_g d_g$$

in a certain positive sense. For each $i = 1, \cdots, g$, let us identify a_i with c_i^{-1} and b_i with d_i^{-1}, where c_i^{-1} denotes c_i with reverse direction. The quotient space Σ_g thus obtained is called the *closed orientable surface of genus g.* Prove that the Betti numbers of Σ_g are

$$R^0 = 1, \quad R^1 = 2g, \quad R^2 = 1.$$

4. *Closed non-orientable surfaces.* Consider a plane convex region W bounded by a $2n$-sided regular polygon P. Let the sides of P be denoted by

$$P = a_1 b_1 a_2 b_2 \cdots a_n b_n$$

in a certain positive sense. Identify a_i with b_i for each $i = 1, 2, \cdots, n$. Thus we obtain a quotient space Ω_n, which is called the *closed non-orientable surface of characteristic* $2 - n$. Prove that the Betti-numbers of Ω_n are

$$R^0 = 1, \quad R^1 = n - 1, \quad R^2 = 0.$$

G. An application of the cone k(X)

Let us use the notation of § 10. Prove that a map $f \in \Omega$ is null-homotopic relative to $\{ M \}$ iff there exists a map $F : k(X) \to Y$ of the cone $k(X)$ into Y such that $F \mid X = f$, $F(v) = y_0$, and $F[k(X_\mu)] \subset Y_\mu$ for each $\mu \in M$. An important special case is as follows: a map $f : S^n \to Y$ is homotopic to a constant map iff f has an extension $F : E^{n+1} \to Y$. Using this, prove that the identify map on S^n is not null-homotopic.

H. Extension properties

A subset A of a space X is said to have the *extension property* in X with respect to a space Y, if every map $f : A \to Y$ can be extended over X. The subset A is said to have the *neighborhood extension property* in X with respect to Y, if every map $f : A \to Y$ can be extended over some open set U of X which contains A.

Thus, a space Y is solid iff every closed subset A of any normal space X has the extension property in X with respect to Y.

The subset A is said to have the *absolute extension property* in X, if it has the extension property in X with respect to every space Y. The subset A is said to have the *absolute neighborhood extension property* in X, if it has the neighborhood extension property in X with respect to every space Y.

Prove the following two assertions:

1. A subset A of a space X is a retract of X iff it has the absolute extension property in X.

2. A subset A of a space X is a neighborhood retract of X iff it has the absolute neighborhood extension property in X.

I. Borsuk's extension theorems

By means of the results in the exercises D and E, prove the following two extension theorems:

1. A metrizable space Y is an AR iff every closed subset A of an arbitrary metrizable space X has the extension property in X with respect to Y.

2. A metrizable space Y is an ANR iff every closed subset A of an arbitrary metrizable space X has the neighborhood extension property in X with respect to Y.

As a consequence of the theorem 1, show that every convex subset of a locally convex linear space is an AR; in particular, every simplex is an AR and every euclidean space R^n is an AR.

J. Separable ANR's and compact ANR's

Prove the following two theorems:

1. For a separable metrizable space Y, the following statements are equivalent:

(*i*) Y is an ANR.

(*ii*) A topological image of Y as a closed subset Z_0 of any separable metrizable space Z is necessarily a neighborhood retract of Z.

(*iii*) Every closed subset A of an arbitrary separable metrizable space X has the neighborhood extension property in X with respect to Y.

2. For a compact metrizable space Y, the following statements are equivalent:

(*i*) Y is an ANR.

(*ii*) Y is homeomorphic with a neighborhood retract of the Hilbert parallelotope I^ω.

(*iii*) Every closed subset A of an arbitrary normal space X has the neighborhood extension property in X with respect to Y.

State and prove the analogous theorems for the AR's.

It follows from the theorem 1 that, for a separable metrizable space Y, the definitions of AR's and ANR's given in Ex.C are equivalent to Kuratowski's modifications of Borsuk's original definitions. In fact, the statement (*ii*) in theorem 1 is used as the definition of ANR's by Kuratowski as well as by Lefschetz, [L_3; p. 58].

K. Operations on ANR's

Prove the following properties of ANR's:

1. Every open subspace of an ANR is an ANR.
2. Let Y be an ANR and B a closed subspace of Y. B is an ANR iff B is a neighborhood retract of Y.
3. Let Y and Z be metrizable spaces. The product space $Y \times Z$ is an ANR iff both Y and Z are ANR's.
4. Let Y_1 and Y_2 be two closed subspaces of a metrizable space Y with $Y = Y_1 \cup Y_2$. If Y_1, Y_2, $Y_1 \cap Y_2$ are ANR's, then so is Y. [Borsuk 2].
5. Let a metrizable space Y be covered by a collection $\{ Y_\mu \mid \mu \in M \}$ of disjoint open subspaces. If Y_μ is an ANR for every $\mu \in M$, then so is Y.
6. Let a metrizable space Y be covered by a countable collection $\{ Y_n \mid n = 1, 2, \cdots \}$ of open subspaces. If Y_n is an ANR for each $n = 1, 2, \cdots$, then so is Y. [Hanner 1].

L. Locally finite simplicial polyhedra

By induction on the number of simplexes and with the aid of the property 4 in Ex.K, prove that every finite simplicial polyhedron is an ANR and hence a compact ANR. [Borsuk 2]. Then, by means of the properties 1, 5 and 6 in Ex.K, prove that every locally finite simplicial polyhedron is an ANR. [Liao 1; Hanner 1].

M. The general homotopy extension theorem

The notion of the HEP given in § 8 can be localized and apparently generalized as follows. A subspace A of a space X is said to have the *neighborhood homotopy extension property* (abbreviated NHEP) in X with respect to a space Y, if every partial homotopy

$$h_t : A \to Y, \quad (0 \leqslant t \leqslant 1),$$

of an arbitrary map $f : X \to Y$ has an extension

$$g_t : V \to Y, \quad (0 \leqslant t \leqslant 1),$$

over some open neighborhood V of A in X such that $g_0 = f \mid V$.

However, if A is a closed subspace of a normal space X, then the NHEP is equivalent to the HEP. In other words, A has the HEP in X with respect to Y iff A has the NHEP in X with respect to Y. By the aid of a Urysohn's characteristic function [L_2; p. 27], prove this *general homotopy extension theorem*.

N. Borsuk's homotopy extension theorems

By means of Borsuk's extension theorem in Ex.I and the general homotopy extension theorem in Ex.M, prove the following theorems:

1. If Y is an ANR, then every closed subspace A of an arbitrary metrizable space X has the HEP in X with respect to Y. [H–W; p. 86].

2. If Y is a compact ANR, then every closed subspace A of an arbitrary binormal space X has the HEP in X with respect to Y.

As an application of these theorems, prove that every contractible ANR is an AR.

O. Closed ANR subspaces in an ANR

Let X be an ANR, A a closed subspace of X, and T denote the subspace $(X \times 0) \cup (A \times I)$ of the product space $X \times I$. Prove that the following statements are equivalent, [Hu 3]:

(1) A has the AHEP in X.

(2) T is a retract of $X \times I$.

(3) T is an ANR.

(4) A is an ANR.

As a well-known special case, every subpolyhedron A of a locally finite simplicial polyhedron X has the AHEP in X.

P. Relation between partial mapping cylinder and adjunction space

Let $g : A \to Y$ be a given map defined on a given subspace A of a space X into a space Y. Consider the partial mapping cylinder $M_g(X)$ and the adjunction space Z obtained by adjoining X to Y by means of g. There is a natural map

$$\tau : M_g(X) \to Z$$

defined by

$$\tau(u) = \begin{cases} q(w), & (\text{if } u = p(w), w \in X \cup Y), \\ qg(a), & (\text{if } u = p(a, t), a \in A, t \in I), \end{cases}$$

where $p : X \cup (A \times I) \cup Y \to M_g(X)$ and $q : X \cup Y \to Z$ denote the natural projections. By means of these natural projections, Y can be considered as a subspace of both $M_g(X)$ and Z. Hence, τ is actually a map $(M_g(X), Y)$ into (Z, Y) such that $\tau \mid Y$ is the identity map on Y.

Using the equivalence of (2) and (4) in the preceding exercise, prove that the map

$$\tau : (M_g(X), Y) \to (Z, Y)$$

is a homotopy equivalence if X is an ANR and A is a closed ANR subspace of X.

Q. Local contractibility of ANR's

A space X is said to be *locally contractible* if, for each $x \in X$ and every open neighborhood U of x, there is an open neighborhood $V \subset U$ of x which is contractible to the point x in U.

By means of Kuratowski's imbedding, prove that every ANR is locally contractible. As a partial converse of this result, [Borsuk 2] proved that every locally contractible compact metrizable space of finite dimension is an ANR.

R. Dominating polyhedra of ANR's

A space X is said to be *dominated* by another space D if there are two maps

$$f : X \to D, \quad g : D \to X$$

such that the composed map gf is homotopic to the identity map on X. In this case, D is called a *dominating space* of X.

Prove the following assertions:

1. Every ANR has a dominating simplicial polyhedron.
2. Every separable ANR has a dominating locally finite simplicial polyhedron.
3. Every compact ANR has a dominating finite simplicial polyhedron.

S. Some strong deformation retracts

Prove the following assertions:

1. If A is a strong deformation retract of a compact space X, Y is a Hausdorff space, and $f : (X, A) \to (Y, B)$ is a relative homeomorphism, then B is a strong deformation retract of Y. [Spanier 1; p. 208].
2. In the relative n-cell (Y, B) obtained by adjoining E^n to B by means of a map $g : S^{n-1} \to B$ defined on the boundary $(n-1)$-sphere S^{n-1} of E^n, the subspace B of Y is a strong deformation retract of the space obtained from Y by deleting an interior point of E^n.
3. In the product space $S^m \times S^n$, the subspace $S^m \vee S^n = (S^m \times t_0) \cup (s_0 \times S^n)$, where $s_0 \in S^m$ and $t_0 \in S^n$ are given points, is a strong deformation retract of the space obtained from $S^m \times S^n$ by deleting one point which is not in $S^m \vee S^n$.
4. In the product space $S^l \times S^m \times S^n$, the subspace $T = (S^l \times S^m \times t_0) \cup (S^l \times s_0 \times S^n) \cup (r_0 \times S^m \times S^n)$, where $r_0 \in S^l$, $s_0 \in S^m$, $t_0 \in S^n$, is a strong deformation retract of the space obtained from $S^l \times S^m \times S^n$ by deleting one point which is not in T.
5. In the product space $(S^l \vee S^m) \times S^n$, the subspace $S^l \vee S^m \vee S^n$ is a strong deformation retract of the space obtained from $(S^l \vee S^m) \times S^n$ by deleting a point of $S^l \times S^n$ not in $S^l \vee S^n$ and a point of $S^m \times S^n$ not in

$S^m \vee S^n$. Generalize this result to the one-point union of a finite number of spheres.

6. If A is a closed subspace of a metrizable space X and if both A and X are AR's, then A is a strong deformation retract of X.

7. If A is a strong deformation retract of a metrizable space X, then

$$T = (X \times 0) \cup (A \times I) \cup (X \times 1)$$

is a strong deformation retract of the product space $X \times I$. [Hu 2].

T. Relations between different notions of deformation retracts

Assume that X and A are ANR's and that A is a closed subspace of X. Prove that the following statements are equivalent, [Fox 1]:

1. The subspace A is a strong deformation retract of X.
2. The subspace A is a deformation retract of X.
3. There exists a homotopy $h_t : (X, A) \to (X, A)$, $(0 \leqslant t \leqslant 1)$, such that h_0 is the identity map on X and $h_1(X) \subset A$.

U. Expansion of the relativity of a null homotopy

Let X, Y be given spaces, $A \subset X$, $B \subset Y$ given subspaces, and $x_0 \in A$, $y_0 \in B$ given points. Consider the maps

$$\Omega = Y^X \{ A, B; x_0, y_0 \}.$$

According to § 10, Ω consists of the totality of the maps $f : X \to Y$ such that $f(A) \subset B$ and $f(x_0) = y_0$.

Let $f \in \Omega$ and let 0 denote the constant map $0(X) = y_0$. Prove that $f \simeq 0 \text{ rel } \{ A, B \}$ implies $f \simeq 0 \text{ rel } \{ A, B; x_0, y_0 \}$ provided that the following side conditions are satisfied:

(1) The subspace $A \cup k(x_0)$ of the cone $k(A)$ has the HEP in $k(A)$ with respect to B.

(2) The subspace $X \cup k(A)$ of the cone $k(X)$ has the HEP in $k(X)$ with respect to Y.

These side conditions are satisfied, for example, if X is a finite simplicial polyhedron and A is a closed subpolyhedron of X. For more general cases see N and O.

V. A general expansion theorem

Let X be a metrizable space and

$$N \subset M \subset A \subset X$$

be closed subspaces. Let Y be a space and B a subspace of Y not necessarily closed. Establish the following

Theorem. *Let $f, g : X \to Y$ be two given maps such that*

$$f(A) \subset B, \quad f \mid M = g \mid M, \quad g(A) \subset B.$$

Then $f \simeq g \operatorname{rel} \{ A, B; N \}$ *implies* $f \simeq g \operatorname{rel} \{ A, B; M \}$ *provided that the following conditions are satisfied.* [Hu 4]:

(1) N *is a deformation retract of* M.

(2) *The subspace* $P = (A \times 0) \cup (M \times I) \cup (A \times 1)$ *has the* HEP *in* $A \times I$ *with respect to* B.

(3) *The subspace* $Q = (X \times 0) \cup (A \times I) \cup (X \times 1)$ *has the* HEP *in* $X \times I$ *with respect to* Y.

The conditions (2) and (3) are satisfied, for example, if X is a finite simplicial polyhedron and A, M are closed subpolyhedra of X.

CHAPTER II

SOME SPECIAL CASES OF THE MAIN PROBLEMS

1. Introduction

In the first chapter, we have described the main problems in homotopy theory. However, the examples and their solutions given there are essentially trivial cases. The present chapter takes up the first nontrivial cases. Answers will be obtained by using homology and cohomology groups.

By means of the elementary properties of the exponential map $p : R \to S^1$ in § 2, we begin with the classification problem of the maps $S^1 \to S^1$ in § 3. Then, the classification problem of the maps of S^1 into an arbitrarily given space X is treated in § 4 and § 5 by studying the fundamental group. In this connection, the classical relation between the fundamental group and the first homology group is given in § 6. Dually, the classification problem of the maps $X \to S^1$ is studied in § 7 by means of the Bruschlinsky group. Finally, we take up the higher dimensional sphere S^n. The Hopf theorems concerning the maps $S^n \to S^n$ and $K^n \to S^n$ are proved in § 8. Dual to the Hopf theorem is the Hurewicz theorem, which is stated in § 9; the proof of this important proposition is deferred until Chapter V, since it seems to depend in an essential way on the elementary properties of the homotopy groups.

2. The exponential map p: R → S¹

Represent the 1-sphere S^1 as the unit circle in the space K of all complex numbers, that is to say

$$S^1 = \{ z \in K \mid | z | = 1 \}.$$

Therefore, S^1 is a compact abelian topological group with the usual multiplication of complex numbers as group operation.

Consider the map $p : R \to S^1$ defined on the space R of real numbers by the formula

$$p(x) = \exp(2\pi xi) = e^{2\pi xi}, \quad (x \in R),$$

where e denotes the base of natural logarithms and i the unit of imaginary numbers. This map p will be called *the exponential map* of R onto S^1. The continuity of p and the following proposition are obvious.

Proposition 2.1. *The exponential map p is a homomorphism, that is, $p(x + y) = p(x)p(y)$ for any x and y in R. The kernel $p^{-1}(1)$ is the subgroup Z of all integers.*

Since $p(x) = \cos(2\pi x) + i \sin(2\pi x)$, it is easy to see that p is one-to-one on the open interval $(-\frac{1}{2}, \frac{1}{2})$ of R. By a standard theorem in general topology, p maps every closed subinterval of $(-\frac{1}{2}, \frac{1}{2})$ homeomorphically and hence p is a homeomorphism on the open interval itself. Since p is a homomorphism, this implies that p is a homeomorphism on each translate of $(-\frac{1}{2}, \frac{1}{2})$. Therefore, for every open interval (a, b) with $b - a \leqslant 1$, p carries (a, b) homeomorphically onto an open subspace of S^1. This and (2.1) imply the following

Proposition 2.2. *For every proper connected subspace U of S^1, p carries every component of $p^{-1}(U)$ homeomorphically onto U.*

By a *path* in a space X, we mean a map

$$\sigma : I \to X$$

of the unit interval I into X. The points $\sigma(0)$ and $\sigma(1)$ are called respectively the *initial point* and the *terminal point* of the path σ. They are said to be *connected* by σ. If $\sigma(0) = \sigma(1)$, then σ is called a *loop* in X with $\sigma(0) = \sigma(1)$ as its *basic point.*

Proposition 2.3. (The covering path property). *For every path $\sigma : I \to S^1$ and every point $x_0 \in R$ such that $p(x_0) = \sigma(0)$, there exists a unique path $\tau : I \to R$ such that $\tau(0) = x_0$ and $p\tau = \sigma$.*

Proof. By the continuity of σ and the compactness of I, there exists a partition

$$0 = t_0 < t_1 < \cdots < t_{n-1} < t_n = 1$$

of I such that the image of every closed subinterval $I_i = [t_i, t_{i+1}], 0 \leqslant i < n$, under σ is a proper connected subspace U_i of S^1.

We shall establish the proposition by proving inductively the following assertion A_i: There exists a unique map τ_i defined on the closed interval $[0, t_i]$ into R such that

$$\tau_i(0) = x_0, \quad p\tau_i = \sigma \mid [0, t_i].$$

A_0 is obvious. Assume $0 \leqslant i < n$. It remains to prove that A_i implies A_{i+1}.

For this purpose, let us assume A_i. By the choice of the partition, $U_i = \sigma(I_i)$ is a proper connected subspace of S^1. Let V_i denote the component of $p^{-1}(U_i)$ which contains the point $\tau_i(t_i)$. By (2.2), the partial map $p_i = p \mid V_i$ is a homeomorphism of V_i onto U_i and hence the inverse $p_i^{-1} : U_i \to V_i$ is defined. Define a map $\tau_{i+1} : [0, t_{i+1}] \to R$ by taking

$$\tau_{i+1}(t) = \begin{cases} \tau_i(t), & (\text{if } 0 \leqslant t \leqslant t_i), \\ p_i^{-1}\sigma(t), & (\text{if } t_i \leqslant t \leqslant t_{i+1}). \end{cases}$$

This proves the existence of τ_{i+1}. Since V_i is a component of $p^{-1}(U_i)$ and p_i is a homeomorphism, it follows easily that τ_{i+1} is also unique. ∎

If the reader is familiar with the logarithmic function and its analytic continuations, he can easily see that the unique covering path $\tau : I \to R$ in (2.3) is given by

$$\tau(0) = x_0, \quad \tau(t) = \frac{1}{2\pi i} \log_e \sigma(t), \quad t \in I.$$

Proposition 2.4. (The covering homotopy property). *For every map* $f: X \to R$ *of a space* X *into* R *and every homotopy* $h_t: X \to S^1$, $(0 \leqslant t \leqslant 1)$, *of the map* $h = pf$, *there exists a unique homotopy* $f_t : X \to R$, $(0 \leqslant t \leqslant 1)$, *of* f *such that* $pf_t = h_t$ *for every* $t \in I$.

Proof. For each point $x \in X$, the partial homotopy $h_t \mid x$ gives a path in S^1. Define $f_t : X \to R$, $(0 \leqslant t \leqslant 1)$, by taking $f_t \mid x$ to be the unique path in R which starts from $f(x)$ and covers $h_t \mid x$. Let

$$H : X \times I \to S^1, \quad F : X \times I \to R$$

be defined by $H(x, t) = h_t(x)$ and $F(x, t) = f_t(x)$. For the existence part of the proof, it remains to establish the continuity of F.

For this purpose, let $x_0 \in X$. By the continuity of H and the compactness of I, there exist a neighborhood M of x_0 in X and a partition

$$0 = t_0 < t_1 < \cdots < t_{n-1} < t_n = 1, \quad I_i = [t_i, t_{i+1}]$$

of I such that the image of $M \times I_i$ under H is contained in a proper connected subspace of S^1. By an inductive construction similar to that used in the proof of (2.3), we can define a map $G : M \times I \to R$ such that

$$G(x, 0) = f(x), \quad pG(x, t) = H(x, t)$$

for every $x \in M$ and $t \in I$. By the uniqueness part of (2.3), this implies that $G = F \mid M \times I$. Since M is a neighborhood of x_0 in X, it follows that F is continuous at (x_0, t) for every $t \in I$. Since x_0 is arbitrary, this completes the proof. ▮

These properties of the exponential map p generalize to a very important class of spaces, namely, the covering spaces. See Chapter III.

3. Classification of the maps $S^1 \to S^1$

Let Λ denote the set of all maps of S^1 into itself. According to (I; § 8), these maps are divided into disjoint homotopy classes and the set of these homotopy classes is denoted by $\pi(S^1; S^1)$. In this section we will see that the members of $\pi(S^1; S^1)$ correspond in a natural way to the integers. Use will be made of the homology group $H_1(S^1)$, but the reader will realize that this may be avoided by adopting some other definition of degree.

Before going further, let us first note that the classification problem of the maps $f: S^1 \to S^1$ is equivalent to that of the maps $\phi : I \to S^1$ with $\phi(0) = \phi(1)$. In fact, the exponential map $p : I \to S^1$ of § 2 is an identification map, (I; § 6), and hence the assignment $f \to \phi = fp$ gives a one-to-one correspondence preserving homotopy.

Now, let us consider the homology group $H_1(S^1)$ which is isomorphic to the additive group Z of integers. According to (I; § 8), the degree of a given map $f: S^1 \to S^1$ is the integer $\deg(f)$ uniquely determined by the relation

$$f_*(c) = \deg(f) \cdot c$$

for every $c \in H_1(S^1)$. Obviously, deg(f) depends only on the homotopy class $[f]$ of f. Hence we may define the degree of a homotopy class $\alpha \in \pi(S^1; S^1)$ by taking $\deg(\alpha) = \deg(f)$ where $f \in \alpha$. Thus we obtain a function

$$\deg : \pi(S^1; S^1) \to Z.$$

We are going to show that deg maps $\pi(S^1; S^1)$ onto Z is a one-to-one fashion.

The exponential map $p : R \to S^1$ maps the open interval (0,1) homeomorphically onto the open subspace $W = S^1 \setminus 1$. Let $q : W \to (0,1)$ denote the unique homeomorphism such that pq is the inclusion map $W \subset S^1$.

To prove that deg is onto, let n be any integer and let

$$\phi_n : S^1 \to S^1$$

denote the map defined by $\phi_n(z) = z^n$ for every $z \in S^1$. One can easily see that ϕ_n is also given by the following formula

$$\phi_n(z) = \begin{cases} p\xi_n q(z), & (\text{if } z \in S^1 \setminus 1), \\ 1, & (\text{if } z = 1 \in S^1), \end{cases}$$

where $\xi_n : I \to R$ denotes the map defined by $\xi_n(t) = nt$ for every $t \in I$. The following lemma is obvious and implies that deg is onto.

Lemma 3.1. $\deg(\phi_n) = n$.

To prove that deg is one-to-one, it suffices to establish the following

Lemma 3.2. *If the degree of a given map $f : S^1 \to S^1$ is n, then $f \simeq \phi_n$.*

Proof. Pick $\theta \in R$ such that $p(\theta) = f(1)$. Define a homotopy $f_t : S^1 \to S^1$, $(0 \leqslant t \leqslant 1)$, by taking

$$f_t(z) = f(z)/p(\theta t), \quad (z \in S^1, t \in I).$$

Set $g = f_1$. Then we have $f \simeq g$ and $g(1) = 1$.

Let $\sigma = gp : I \to S^1$. Since $\sigma(0) = 1 = p(0)$, it follows from (2.3) that there exists a unique path $\tau : I \to R$ such that $\tau(0) = 0$ and $p\tau = \sigma$. Since $p\tau(1) = \sigma(1) = 1$, $\tau(1)$ is an integer, say $\tau(1) = m$. Define a homotopy $\eta_t : I \to R$, $(0 \leqslant t \leqslant 1)$, by taking

$$\eta_t(s) = \tau(s) + t\xi_m(s) - t\tau(s), \quad (s \in I, t \in I).$$

Then $\eta_0 = \tau$, $\eta_1 = \xi_m$ and $\eta_t(0) = 0$, $\eta_t(1) = m$ for every $t \in I$.

Finally, define a homotopy $g_t : S^1 \to S^1$, $(0 \leqslant t \leqslant 1)$, by taking

$$g_t(z) = \begin{cases} p\eta_t q(z), & (\text{if } z \in S^1 \setminus 1), \\ 1, & (\text{if } z = 1 \in S^1). \end{cases}$$

Then $g_0 = g$ and $g_1 = \phi_m$. This implies $f \simeq g \simeq \phi_m$. It follows that $\deg(f) = \deg(\phi_m)$. Since $\deg(f) = n$ and $\deg(\phi_m) = m$, we obtain $m = n$ and hence $f \simeq \phi_n$. ∎

Thus we have proved the following

Theorem 3.3. (Classification theorem). *The homotopy classes* $\pi(S^1; S^1)$ *are in a one-to-one correspondence with the integers* Z *under the function* deg: $\pi(S^1; S^1) \to Z$. *The homotopy class of degree* n *is represented by the map* $\phi_n : S^1 \to S^1$.

To strengthen (3.3), let us make use of the fact that S^1 is a topological abelian group under the usual multiplication of complex numbers. The maps Λ form a ring with addition and multiplication defined as follows: If $f, g : S^1 \to S^1$ are any maps, then $f + g$ and fg are the maps in Λ defined by

$$(f + g)(z) = f(z)g(z), \quad (fg)(z) = f[g(z)]$$

for every $z \in S^1$. Since the homotopy classes $[f + g]$ and $[fg]$ obviously depend only on the classes $[f]$ and $[g]$, this ring structure of Λ induces a ring structure in $\pi(S^1; S^1)$. Now, (3.3) can be strengthened by the following

Corollary 3.4. *The function* deg *is an isomorphism of the ring* $\pi(S^1; S^1)$ *onto the ring* Z *of integers.*

Proof. Let $f, g \in \Lambda$. It suffices to prove that

$$\deg(f + g) = \deg(f) + \deg(g), \quad \deg(fg) = \deg(f)\deg(g).$$

Let $\deg(f) = m$ and $\deg(g) = n$. Then $f \simeq \phi_m$ and $g \simeq \phi_n$. It is obvious from the definitions that

$$\phi_m + \phi_n = \phi_{m+n}, \quad \phi_m\phi_n = \phi_{mn}.$$

This implies the corollary. ∎

4. The fundamental group

Let X be a space and let $x_0 \in X$; let Ω denote the set of all maps of S^1 into X such that the point 1 of S^1 is mapped at the point x_0 of X.

For any map $f : S^1 \to X$ in Ω, the composition $fp : I \to X$ of f with the exponential map p of § 2 is a loop in X with its basic point at x_0. This correspondence is obviously one-to-one. Hence, by identifying f with fp, we may consider Ω as the set of all loops in X with given basic point x_0. In symbols, we have

$$\Omega = \{ f : I \to X \mid f(0) = x_0 = f(1) \}.$$

We may define a *multipltication* in Ω as follows. For any two loops $f, g \in \Omega$, their *product* $f \cdot g$ is the loop defined by

$$(f \cdot g)(t) = \begin{cases} f(2t), & (\text{if } 0 \leqslant t \leqslant \frac{1}{2}), \\ g(2t - 1), & (\text{if } \frac{1}{2} \leqslant t \leqslant 1). \end{cases}$$

Intuitively speaking, $f \cdot g$ is the loop traced by moving along the loops f and g in succession.

Two loops $f, g \in \Omega$ are said to be *equivalent*, (in symbols: $f \sim g$), if there exists a homotopy $h_t : I \to X$, $(0 \leqslant t \leqslant 1)$, such that $h_0 = f$, $h_1 = g$ and $h_t(0) = x_0 = h_t(1)$ for every t. This relation $\sim$ is reflexive, symmetric and transitive; hence the loops Ω are divided into disjoint equivalence classes. Let $\pi_1(X, x_0)$ denote the set of these classes and $[f]$ denote the class containing $f \in \Omega$. The loop f is said to be a *representative* for the class $[f]$.

Let d denote the degenerate loop $d(I) = x_0$. For any loop $f \in \Omega$, let f^{-1} denote the *reverse* of f defined by $f^{-1}(t) = f(1 - t)$ for each $t \in I$. Then clearly $(f^{-1})^{-1} = f$.

The following elementary properties can be easily verified:

(1) If $f \sim f'$ and $g \sim g'$, then $f \cdot g \sim f' \cdot g'$.

(2) $(f \cdot g) \cdot h \sim f \cdot (g \cdot h)$.

(3) $f \cdot d \sim f \sim d \cdot f$.

(4) $f \cdot f^{-1} \sim d \sim f^{-1} \cdot f$.

According to property (1), the class $[f \cdot g]$ depends only on the classes $[f]$ and $[g]$. Hence we may define a multiplication in $\pi_1(X, x_0)$ by taking

$$[f]\,[g] = [f \cdot g].$$

By (2), (3) and (4), this multiplication makes $\pi_1(X, x_0)$ a group which is called the *Poincaré group* or the *fundamental group of X at x_0*. The class $e = [d]$ is the neutral element of $\pi_1(X, x_0)$, and the class $[f^{-1}]$ is the inverse element of $[f]$.

For example, take X to be the S^1 of § 2 and x_0 the point 1 of S^1. Then Ω becomes a subset of the set Λ of all maps of S^1 into itself. The inclusion $i : \Omega \subset \Lambda$ induces a function

$$i_* : \pi_1(S^1, 1) \to \pi(S^1; S^1).$$

By (3.2), i_* is onto. On the other hand, i_* is also one-to-one. To prove this, let $f, g \in \Omega$ be any two maps such that $f \simeq g$. Then there is a homotopy $h_t : S^1 \to S^1$, $(0 \leqslant t \leqslant 1)$, such that $h_0 = f$ and $h_1 = g$. Define a homotopy $k_t : S^1 \to S^1$, $(0 \leqslant t \leqslant 1)$, by taking $k_t(z) = h_t(z)/h_t(1)$ for each $z \in S^1$ and $t \in I$. Then $k_0 = f$, $k_1 = g$ and $k_t(1) = 1$ for every $t \in I$. Hence, $f \simeq g$ rel 1. This proves that i_* is one-to-one. Finally, it is not difficult to see that $\phi_{mp} \cdot \phi_{np}$ is equivalent to ϕ_{m+np}. Therefore, *$\pi_1(S^1, 1)$ is a free cyclic group generated by the class $[p]$ of the exponential map $p : I \to S^1$.*

The definition of the fundamental group $\pi_1(X, x_0)$ is by no means constructive and the problem of computing the group effectively is quite difficult. In a finite simplicial complex, an effective rule can be given for writing down generators and relations (finite in number) for the group (see Ex. A at the end of the chapter). This reduces the problem for finite simplicial complexes to solving the word problem of group theory. For many of the simpler complexes, this has been done.

Now let x_1 be another point of X and let $\sigma : I \to X$ be a path connecting x_0 to x_1. For each $i = 0, 1$, let Ω_i denote the set of all loops in X with basic point at x_i. The path σ defines a transformation $\sigma_\# : \Omega_1 \to \Omega_0$ as follows. Let $f \in \Omega_1$ and $\tau : I \to X$ denote the reverse of σ defined by $\tau(t) = \sigma(1 - t)$ for each $t \in I$. Then $\sigma_\#(f)$ is the loop $\sigma \cdot f \cdot \tau$ defined by

$$(\sigma\cdot f\cdot\tau)(t) = \begin{cases} \sigma(3t). & (\text{if } 0 \leqslant t \leqslant \frac{1}{3}). \\ f(3t-1). & (\text{if } \frac{1}{3} \leqslant t \leqslant \frac{2}{3}). \\ \tau(3t-2). & (\text{if } \frac{2}{3} \leqslant t \leqslant 1). \end{cases}$$

It is straightforward to verify the following three properties:

(5) If $f \sim g$, then $\sigma\cdot f\cdot\tau \sim \sigma\cdot g\cdot\tau$.

(6) $\sigma\cdot(f\cdot g)\cdot\tau \sim (\sigma\cdot f\cdot\tau)\cdot(\sigma\cdot g\cdot\tau)$.

(7) $\tau\cdot(\sigma\cdot f\cdot\tau)\cdot\sigma \sim f$.

According to (5), the class $[\sigma\cdot f\cdot\tau]$ depends only on the class $[f]$. Hence $\sigma_{\#}$ induces a transformation

$$\sigma_* : \pi_1(X, x_1) \to \pi_1(X, x_0).$$

The property (6) implies that σ_* is a homomorphism. Similarly, the reverse τ of σ determines a homomorphism

$$\tau_* : \pi_1(X, x_0) \to \pi_1(X, x_1).$$

It follows from (7) that the composition $\tau_*\sigma_*$ is the identity automorphism on $\pi_1(X, x_1)$. Since the roles of σ and τ are interchangeable, this implies that σ_* is an isomorphism and $\tau_* = (\sigma_*)^{-1}$. Thus we have proved the following

Proposition 4.1. *Every path $\sigma : I \to X$ induces an isomorphism $\sigma_* : \pi_1(X, x_1) \approx \pi_1(X, x_0)$, where $x_0 = \sigma(0)$ and $x_1 = \sigma(1)$.*

A space X is said to be *pathwise connected* if every pair of points of X can be connected by a path in X. If X is pathwise connected, then it follows from (4.1) that the fundamental groups $\pi_1(X, x_0)$ at various basic points $x_0 \in X$ are all isomorphic. As abstract groups, these may be considered as ths same group which will be denoted by $\pi_1(X)$ and called the (abstract) *fundamental group* of the pathwise connected space X.

Any two paths $\sigma, \tau : I \to X$ are said to be *equivalent,* (in symbols, $\sigma \sim \tau$), if $\sigma(0) = \tau(0)$, $\sigma(1) = \tau(1)$, and $\sigma \simeq \tau$ relative to the extremities of I. Then one can easily verify the following

Proposition 4.2. *If two paths $\sigma, \tau : I \to X$ are equivalent, then $\sigma_* = \tau_*$.*

If the path $\sigma : I \to X$ is a loop with basic point $x_0 \in X$, then σ represents an element ξ in $\pi_1(X, x_0)$. By the definition of σ_*, one can easily see that

$$\sigma_*(w) = \xi w \xi^{-1}$$

for every w in $\pi_1(X, x_0)$. Hence σ_* is the inner automorphism of $\pi_1(X, x_0)$ determined by the element $\xi = [\sigma]$.

Finally, let $\phi : (X, x_0) \to (Y, y_0)$ be a given map. For every loop f in X with x_0 as basic point, the composition ϕf is a loop in Y with y_0 as basic point. The following properties are obvious:

(8) If $f \sim g$, then $\phi f \sim \phi g$.

(9) If $h = f\cdot g$, then $\phi h = \phi f\cdot\phi g$.

(10) If two maps $\phi, \psi : (X, x_0) \to (Y, y_0)$ are homotopic, then $\phi f \sim \psi f$.

By (8), the map $\phi : (X, x_0) \to (Y, y_0)$ induces a transformation

$$\phi_* : \pi_1 (X, x_0) \to \pi_1 (Y, y_0).$$

The property (9) implies that ϕ_* is a homomorphism which will be called the *induced homomorphism* of ϕ (on the fundamental groups). Thus the operations $(X, x_0) \to \pi_1(X, x_0)$ and $\phi \to \phi_*$ form a covariant functor for the categories involved, [E-S ; p. 111].

5. Simply connected spaces

A pathwise connected space X is said to be *simply connected* provided that every pair of paths $\sigma, \tau : I \to X$ with $\sigma(0) = \tau(0)$ and $\sigma(1) = \tau(1)$ are homotopic with end points held fixed.

Proposition 5.1. *A pathwise connected nonvacuous space X is simply connected iff $\pi_1(X) = 0$.*

Proof. *Necessity.* Let $x_0 \in X$ and let $f : I \to X$ be any loop with x_0 as basic point. By the definition of simple connectedness, f is equivalent to the degenerate loop d at x_0. This implies that $\pi_1(X, x_0) = 0$ and hence $\pi_1(X) = 0$.

Sufficiency. Assume that $\pi_1(X) = 0$ and let $\sigma, \tau : I \to X$ be two paths such that $\sigma(0) = x_0 = \tau(0)$ and $\sigma(1) = x_1 = \tau(1)$. Then we obtain a loop $f : I \to X$ with basic point x_0 defined by $f = \sigma \cdot \tau^{-1}$, that is to say,

$$f(t) = \begin{cases} \sigma(2t), & (\text{if } 0 \leqslant t \leqslant \tfrac{1}{2}), \\ \tau(2 - 2t), & (\text{if } \tfrac{1}{2} \leqslant t \leqslant 1). \end{cases}$$

Since $\pi_1(X) = 0$, f is equivalent to the degenerate loop at x_0. It follows that there exists a map $F : I^2 \to X$ such that

$$F(0, t) = x_0, \quad F(1, t) = x_0, \quad F(t, 0) = f(t), \quad F(t, 1) = x_0$$

for every $t \in I$. Now, it is not difficult to construct a homotopy $\xi_t : I \to X$, $(0 \leqslant t \leqslant 1)$, such that $\xi_0 = \sigma$, $\xi_1 = \tau$, and $\xi_t(0) = x_0$, $\xi_t(1) = x_1$ for every $t \in I$. For this purpose, let us first consider the following geometric situation in the unit square I^2 in the euclidean plane R^2.

Let q denote the point $(\frac{1}{2}, 0)$ of I^2. Then the line L_t through q with angle of inclination $(1 - t)\pi$ meets the boundary of I^2 in q and another point p_t if $0 < t < 1$. Let $p_0 = (0,0)$ and $p_1 = (1,0)$. Define a homotopy $h_t : I \to I^2$, $(0 \leqslant t \leqslant 1)$, by taking $h_t(s)$ to be the point which divides the segment $p_t q$ in the ratio $s : (1 - s)$.

Now, let $\xi_t = F h_t$ for each $t \in I$. Then one verifies immediately $\xi_0 = \sigma$, $\xi_1 = \tau$ and $\xi_t(0) = x_0$, $\xi_t(1) = F(q) = x_1$ for every $t \in I$. Hence $\sigma \sim \tau$. ∎

It follows immediately from (5.1) that every contractible space is simply connected. For examples of non-contractible simply connected spaces, we have the following

Proposition 5.2. *The n-sphere S^n is simply connected iff $n > 1$.*

Proof. The 0-sphere S^0 is not simply connected since it is not pathwise connected. The 1-sphere S^1 is not simply connected since $\pi_1(S^1) \approx Z$. It remains to prove that S^n, $n > 1$, is simply connected. For this purpose, we shall prove that every loop $f: I \to S^n$ at a basic point $x_0 \in S^n$ is equivalent to the degenerate loop d at x_0. By the method of simplicial approximation [E-S; p. 64], f is equivalent to a loop $g: I \to S^n$ which is a simplicial map on some triangulations of I and S^n. Since $n > 1$, $g(I)$ is a proper subspace of S^n and hence is contractible in S^n to the point x_0. This implies $g \sim d$ and hence $f \sim d$. ∎

A space X is said to be *locally pathwise connected at a point* $x_0 \in X$ if, for every open neighborhood U of x_0 in X, there exists an open neighborhood $V \subset U$ of x_0 such that every pair of points in V can be connected by a path in U. If X is locally pathwise connected at every point of X, then it is said to be *locally pathwise connected.* For example, triangulable spaces are locally pathwise connected.

An important property of spaces which are both simply connected and locally pathwise connected is that the restricted lifting problem with respect to the exponential map $p: R \to S^1$ in § 2 has a unique solution. More generally, we have the following

Proposition 5.3. (Covering map property). *Let X be a connected and locally pathwise connected space, and let x_0 be a given point in X. For every map $f: X \to S^1$ and every point $r_0 \in R$ such that $p(r_0) = f(x_0)$ and $f_*\pi_1(X, x_0) = 0$, there exists a unique map $g: X \to R$ such that $g(x_0) = r_0$* and $pg = f$.

Proof. We shall first construct $g: X \to R$ as follows. Let x be an arbitrary point in X, then there exists a path $\pi: I \to X$ such that $\pi(0) = x_0$ and $\pi(1) = x$. Let $\sigma: I \to S^1$ denote the path defined by $\sigma = f\pi$. Then $\sigma(0) = f(x_0)$ and $\sigma(1) = f(x)$. According to (2.3), there is a unique path $\tau: I \to R$ such that $\tau(0) = r_0$ and $p\tau = \sigma$. Since $f_*\pi_1(X, x_0) = 0$, it is easy to see that different choices of the path π in X give rise to equivalent paths $\sigma = f\pi$ in S^1. Hence, by (2.4), one can readily prove that $\tau(1)$ does not depend on the choice of π and we may define $g(x) = \tau(1)$. Clearly we have $pg = f$ and $g(x_0) = r_0$. For the existence proof, it remains to prove that g is continuous.

Let x_1 be any point in X. To show the continuity of g, it suffices to prove that g is continuous at x_1. To this end, we shall prove that g coincides with a map in some neighborhood of x_1 in X.

Let $z_1 = f(x_1)$ and $W = S^1 \setminus z_1 e^{\pi i}$. Let J denote the component of $p^{-1}(W)$ which contains $r_1 = g(x_1)$. Then J is an open interval of length 1 and p maps J homeomorphically onto W. Let $q: W \to J$ denote the homeomorphism such that $pq(z) = z$ for every $z \in W$. Let $U = f^{-1}(W)$ and define a map $h: U \to R$ by taking $h(x) = qf(x)$ for every $x \in U$. Obviously we have $h(x_1) = r_1$ and $ph = f \mid U$. We are going to prove that g coincides with h in an open neighborhood of x_1 in X.

Since X is locally pathwise connected and U is an open neighborhood of x_1 in X, there exists an open neighborhood $V \subset U$ of x_1 in X such that every point of V can be connected to x_1 by a path in U. We shall prove that $g(x) = h(x)$ for every $x \in V$. For this purpose, let $x \in V$ be given. Then there is a path $\eta : I \to U \subset X$ such that $\eta(0) = x_1$ and $\eta(1) = x$. Since X is pathwise connected, there is a path $\xi : I \to X$ such that $\xi(0) = x_0$ and $\xi(1) = x_1$. Let $\pi = \xi \cdot \eta : I \to X$ be the path defined in the obvious way. Set $\rho = f\xi$ and $\sigma = f\pi$. Denote by $\theta : I \to R$ the unique path such that $\theta(0) = r_0$ and $p\theta = \rho$. By the very construction of g, we have $\theta(1) = g(x_1) = r_1$. On the other hand, we also have $h(x_1) = r_1$. Hence, we may define a path $\tau : I \to R$ by taking

$$\tau(t) = \begin{cases} \theta(2t), & (\text{if } 0 \leqslant t \leqslant \frac{1}{2}), \\ h\eta(2t - 1), & (\text{if } \frac{1}{2} \leqslant t \leqslant 1). \end{cases}$$

Then $\tau(1) = h(x)$. On the other hand, since $p\theta = \rho = f\xi$ and $ph\eta = f\eta$, we obtain $p\tau = \sigma$. Further, we also have $\tau(0) = \theta(0) = r_0$. Hence, it follows from the construction of g that $\tau(1) = g(x)$. This proves that $g(x) = h(x)$ whenever x is in V. Since V is an open neighborhood of x_1 in X and h is continuous, it follows that g is continuous at x_1. This completes the existence proof.

To prove the uniqueness, let g and g' be any two maps of X into R such that

$$pg = f = pg', \qquad g(x_0) = r_0 = g'(x_0).$$

We are going to prove that $g = g'$. Let x be an arbitrary point in X. It suffices to prove that $g(x) = g'(x)$. There is a path $\pi : I \to X$ connecting x_0 to x. Let $\sigma = f\pi$, $\tau = g\pi$ and $\tau' = g'\pi$. Then we have

$$p\tau = \sigma = p\tau', \qquad \tau(0) = r_0 = \tau'(0).$$

According to (2.3), we get $\tau = \tau'$. In particular,

$$g(x) = \tau(1) = \tau'(1) = g'(x). \blacksquare$$

6. Relation between $\pi_1(X,x_0)$ and $H_1(X)$

In the present section, we shall give a detailed exposition of the classical result on the relation between the fundamental group $\pi_1(X, x_0)$ of a pathwise connected space X at any given basic point x_0 and the 1-dimensional integral singular homology group $H_1(X)$ of X. We shall assume that the reader is familiar with the singular homology theory as given in [E–S; Chapter VII] and [Eilenberg 2].

First of all, let us construct a *natural homomorphism*

$$h_* : \pi_1(X, x_0) \to H_1(X)$$

described as follows. Let α be any element of $\pi_1(X, x_0)$. Choose a representative loop

$$f : S^1 \to X, \quad f(1) = x_0,$$

for α. This map f induces a homomorphism $f_* : H_1(S^1) \to H_1(X)$. Let ι denote the generator of the free cyclic group $H_1(S^1)$ corresponding to the counter-clockwise orientation of S^1. By the homotopy axiom, the element $f_*(\iota)$ in $H_1(X)$ depends only on the class α and, therefore, we define h_* by taking $h_*(\alpha) = f_*(\iota)$. By an elementary property of induced homomorphisms in homology theory, [E–S; p. 36], it is easy to see that h_* is a homomorphism.

Let us consider the *unit n-simplex* Δ_n in the $(n+1)$-space R^{n+1} which consists of the points $(t_0, t_1, \cdots, t_n)$ such that $t_0 + t_1 + \cdots + t_n = 1$ and $t_i \geqslant 0$ for each $i = 0, 1, \cdots, n$. Let $j : I \to \Delta_1$ denote the homeomorphism defined by $j(t) = (1 - t, t)$ for each $t \in I$ and let $k : \Delta_1 \to S^1$ denote the map defined by $k(t_0, t_1) = p(t_1)$ for every point (t_0, t_1) of Δ_1, where p stands for the exponential map in § 2. Then we have $kj = p$. As a singular 1-simplex in X, $T = fk : \Delta_1 \to X$ is in the group $C_1(X)$ of integral singular 1-chains. Since $\partial T = 0$, T is a singular 1-cycle. It is not difficult to verify that T represents the homology class $h_*(\alpha)$.

Since $H_1(X)$ is abelian, the commutator subgroup Comm $\{\pi_1(X, x_0)\}$ of $\pi_1(X, x_0)$ must be contained in the kernel of h_*. In fact, we have the following classical theorem.

Theorem 6.1. *If X is pathwise connected, then the natural homomorphism h_* maps $\pi_1(X, x_0)$ onto $H_1(X)$ with the commutator subgroup of $\pi_1(X, x_0)$ as its kernel. Hence $H_1(X)$ is isomorphic with the group $\pi_1(X, x_0)$ made abelian.*

To prove this theorem, let us first carry out a preliminary reduction. Consider the singular complex $S(X)$ of X, [E–S; p. 186], and denote by $S_1(X)$ the subcomplex of $S(X)$ defined as follows. A singular simplex $T : \Delta_q \to X$ is in $S_1(X)$ iff T sends the vertices of Δ_q into the point x_0. We shall call $S_1(X)$ the *first Eilenberg subcomplex* of $S(X)$. Since X is pathwise connected, it follows from Eilenberg's reduction theorem, [Eilenberg 2; p. 440], that the inclusion cellular map

$$\lambda : S_1 = S_1(X) \to S(X)$$

induces an isomorphism

$$\lambda_* : H_1(S_1) \approx H_1(X).$$

On the other hand, let α be any element of $\pi_1(X, x_0)$ and choose a representative loop $f : S^1 \to X$ for α. Then the singular simplex $T = fk : \Delta_1 \to X$ is a 1-cycle in S_1 and hence determines a homology class $\mu(\alpha) \in H_1(S_1)$ which does not depend on the choice of f. The operation $\alpha \to \mu(\alpha)$ defines a homomorphism

$$\mu : \pi_1(X, x_0) \to H_1(S_1).$$

Obviously we have the relation $h_* = \lambda_* \mu$. Hence the theorem reduces to the following

Lemma 6.2. *The homomorphism μ maps $\pi_1(X, x_0)$ onto $H_1(S_1)$ with the commutator subgroup of $\pi_1(X, x_0)$ as its kernel.*

Proof. To prove that μ is onto, let β be an arbitrary element of $H_1(S_1)$. Choose a 1-cycle z of S_1 which represents β. Since $z \in C_1(S_1)$, it can be written in the form

$$z = \sum_{i=1}^{n} a_i T_i$$

where $T_1, \cdots, T_n$ are 1-simplexes in S_1 and $a_1, \cdots, a_n$ are integers. Since T_i maps the vertices of Δ_1 into x_0, it is a 1-cycle in S_1 and hence represents an element β_i of $H_1(S_1)$. Then we have

$$\beta = \sum_{i=1}^{n} a_i \beta_i.$$

On the other hand, the loop $T_i j : I \to X$ represents an element α_i of $\pi_1(X, x_0)$. According to the definition of μ, we have $\mu(\alpha_i) = \beta_i$. Let

$$\alpha = \alpha_1^{a_1} \alpha_2^{a_2} \cdots \alpha_n^{a_n} \in \pi_1(X, x_0).$$

Since μ is a homomorphism, it follows that

$$\mu(\alpha) = \sum_{i=1}^{n} a_i \mu(\alpha_i) = \sum_{i=1}^{n} a_i \beta_i = \beta.$$

This implies that μ maps $\pi_1(X, x_0)$ onto $H_1(S_1)$.

To study the kernel of μ, let us first note that, because of the commutativity of $H_1(S_1)$, the kernel of μ contains the commutator subgroup $\mathrm{Comm}\{\pi_1(X, x_0)\}$ of $\pi_1(X, x_0)$. The quotient group

$$\pi_1{}^*(X, x_0) = \pi_1(X, x_0)/\mathrm{Comm}\{\pi_1(X, x_0)\}$$

is commutative and is known as *the group* $\pi_1(X, x_0)$ *made abelian*. We shall use the additive notation in $\pi_1{}^*(X, x_0)$ and consider the natural projection

$$\nu : \pi_1(X, x_0) \to \pi_1{}^*(X, x_0).$$

Since the kernel of μ contains $\mathrm{Comm}\{\pi_1(X, x_0)\}$, μ induces a unique homomorphism

$$\mu^* : \pi_1{}^*(X, x_0) \to H_1(S_1)$$

such that $\mu = \mu^* \nu$. To prove that the kernel of μ is $\mathrm{Comm}\{\pi_1(X, x_0)\}$, it suffices to show that μ^* is a monomorphism.

Indeed, we will produce a left inverse $\varkappa^*$ for μ^*. Since $\pi_1{}^*(X, x_0)$ is abelian and $C_1(S_1)$ is the free abelian group generated by the 1-simplexes of S_1, we may define a homomorphism

$$\varkappa : C_1(S_1) \to \pi_1{}^*(X, x_0)$$

described as follows. Let T be any 1-simplex of S_1. Then the loop $Tj : I \to X$ represents an element $[Tj]$ of $\pi_1(X, x_0)$ and $\varkappa$ is defined by taking

$$\varkappa(T) = \nu([Tj]).$$

We assert that $\varkappa$ maps $B_1(S_1)$ into the zero element of $\pi_1{}^*(X, x_0)$. To prove this, let T be an arbitrary 2-simplex in S_1. Then T is a map $T : \Delta_2 \to X$

which sends the vertices of Δ_2 into x_0. Let $T^{(0)}$, $T^{(1)}$, $T^{(2)}$ denote the 1-dimensional faces of T; then we have

$$\begin{aligned}\varkappa(\partial T) &= \varkappa(T^{(0)} - T^{(1)} + T^{(2)}) = \varkappa(T^{(2)} + T^{(0)} - T^{(1)}) \\ &= \varkappa(T^{(2)}) + \varkappa(T^{(0)}) - \varkappa(T^{(1)}) \\ &= \nu([T^{(2)}j]) + \nu([T^{(0)}j]) - \nu([T^{(1)}j]) \\ &= \nu([T^{(2)}j]\,[T^{(0)}j]\,[T^{(1)}j]^{-1}) = 0\end{aligned}$$

because of the relation $[T^{(1)}j] = [T^{(2)}j]\,[T^{(0)}j]$ since T is defined throughout Δ_2. Since $B_1(S_1)$ is generated by all ∂T with T running over the 2-simplexes of S_1, this proves our assertion. Thus $\varkappa$ induces a homomorphism

$$\varkappa^* : H_1(S_1) \to \pi_1^*(X, x_0).$$

Finally, $\varkappa^*\mu^*$ is the identity. To prove this, let α^* be any element of $\pi_1^*(X, x_0)$. Choose $\alpha \in \pi_1(X, x_0)$ with $\nu(\alpha) = \alpha^*$. Let $f: S^1 \to X$ be a representative loop for α. Then the singular simplex $T = fk : \Delta_1 \to X$ represents the class $\mu(\alpha)$ and hence we have

$$\begin{aligned}\varkappa^*\mu^*(\alpha^*) &= \varkappa^*\mu^*\nu(\alpha) = \varkappa^*\mu(\alpha) = \nu([Tj]) \\ &= \nu([fkj]) = \nu([fp]) = \nu(\alpha) = \alpha^*,\end{aligned}$$

where p denotes the exponential map. This implies that μ^* is a monomorphism. ∎

An important consequence of (6.1) is the fact that, for any pathwise connected space X, the fundamental group $\pi_1(X)$ completely determines the integral singular homology group $H_1(X)$.

7. The Bruschlinsky group

Let (X, A) be an arbitrary pair consisting of a space X and a subspace A of X which may be empty. Let us consider the set

$$W = \{ f : (X, A) \to (S^1, 1) \}$$

of all maps f of the pair (X, A) into the pair $(S^1, 1)$, where S^1 denotes the unit circle in the complex plane as in § 2.

Since S^1 is an abelian group under multiplication of complex numbers, we may define an *addition* in W as follows. If f, g are any maps in W, then $f + g$ is the map in W defined by

$$(f + g)(x) = f(x)\, g(x)$$

for every $x \in X$. In this way, W becomes an abelian group.

The homotopy class (relative to A) of the map $f + g$ depends only on that of f and that of g. Hence the set $\pi^1(X, A)$ of all homotopy classes of the maps of W forms an abelian group under the addition defined as follows: For any α, β in $\pi^1(X, A)$, we have

$$\alpha + \beta = [f + g], \quad f \in \alpha, \quad g \in \beta.$$

This abelian group $\pi^1(X, A)$ will be called the *Bruschlinsky group of the pair* (X, A). If A is empty, then it will be denoted by $\pi^1(X)$ and called the *Bruschlinsky group of the space* X. See [Bruschlinsky 1].

For example, we have

$$\pi^1(S^1) \approx \pi^1(S^1, 1) \approx \pi_1(S^1, 1) \approx Z.$$

To determine the structure of $\pi^1(X, A)$, let us assume throughout the remainder of the section that the pair (X, A) is triangulable. Thus, we assume that X is a finite simplicial complex, [E–S; p. 56], and that A is a subcomplex of X. Under these assumptions, let us consider the integral cohomology group $H^1(X, A)$ of the pair (X, A) and construct a *natural homomorphism*

$$h^* : \pi^1(X, A) \to H^1(X, A)$$

as follows. Let ι denote the generator of the free cyclic group $H^1(S^1, 1)$ determined by the counter-clockwise orientation of S^1. For an arbitrary element $\alpha \in \pi^1(X, A)$, pick a map $f : (X, A) \to (S^1, 1)$ which represents α. f induces a homomorphism

$$f^* : H^1(S^1, 1) \to H^1(X, A)$$

which depends only on α according to the homotopy axiom of cohomology theory. Hence we may define h^* by taking $h^*(\alpha) = f^*(\iota)$. It remains to verify that h^* is a homomorphism.

Let $\alpha \in \pi^1(X, A)$. By means of the homotopy extension property (I; § 9), one can easily see that α contains a map $f : (X, A) \to (S^1, 1)$ which sends all vertices of X into the point 1 of S^1. Such a map f determines an integral cochain $c^1(f) \in C^1(X, A)$ as follows. Let $\sigma = v_0 v_1$ be any 1-simplex of X and let $\lambda : I \to \sigma$ denote the linear map which sends 0 to v_0 and 1 to v_1. The composed map $f\lambda : I \to S^1$ is a loop in S^1 and hence the degree of $f\lambda$ is a well-defined integer $\deg(f\lambda)$ which depends on f and σ. Then the cochain $c^1(f)$ is defined by

$$[c^1(f)](\sigma) = \deg(f\lambda).$$

This cochain $c^1(f)$ is a cocycle. To prove this assertion, let $\tau = v_0 v_1 v_2$ be an arbitrary 2-simplex. Then we have

$$\partial\tau = \sigma_0 - \sigma_1 + \sigma_2, \quad \sigma_0 = v_1 v_2, \quad \sigma_1 = v_0 v_2, \quad \sigma_2 = v_0 v_1.$$

Let $\lambda_i : I \to \sigma_i$, $i = 0,1,2$, denote the linear maps as described above. It follows that

$$\begin{aligned}[\delta c^1(f)](\tau) &= [c^1(f)](\partial\tau) = [c^1(f)](\sigma_0) - [c^1(f)](\sigma_1) + [c^1(f)](\sigma_2) \\ &= \deg(f\lambda_0) - \deg(f\lambda_1) + \deg(f\lambda_2) = 0\end{aligned}$$

since f is defined throughout τ. This proves that $c^1(f)$ is a cocycle. By the definition of the induced homomorphism f^*, it is not difficult to see that this cocycle $c^1(f) \in Z^1(X, A)$ represents the element $f^*(\iota) \in H^1(X, A)$.

Now let β be another element of $\pi^1(X, A)$ and pick a representative map

$g:(X, A) \to (S^1, 1)$ of β which sends all vertices of X into 1. Then the map $f + g$ represents $\alpha + \beta$ and sends all vertices of X into 1. For every 1-simplex $\sigma = v_0v_1$, we have

$$\deg((f+g)\lambda) = \deg(f\lambda) + \deg(g\lambda).$$

It follows that $c^1(f+g) = c^1(f) + c^1(g)$ and hence

$$h^*(\alpha+\beta) = h^*(\alpha) + h^*(\beta).$$

This proves that h^* is a homomorphism and completes its construction.

Theorem 7.1. *The natural homomorphism h^* is an isomorphism of $\pi^1(X,A)$ onto $H^1(X,A)$.*

Proof. Let $z \in Z^1(X, A)$ be arbitrarily given. To prove that h^* is an epimorphism, it suffices to construct a map $f:(X, A) \to (S^1, 1)$ which sends all vertices of X into 1 and such that $c^1(f) = z$.

For each $n = 0, 1, \cdots, \dim X$, let K_n denote the subcomplex of X which consists of all simplexes in A together with the simplexes in $X \setminus A$ of dimension not exceeding n. We are going to construct a sequence of maps

$$f_n : K_n \to S^1, \quad n = 1, 2, \cdots, \dim X,$$

as follows. To construct f_1, let us take for each 1-simplex $\sigma = v_0v_1$ in $X \setminus A$ a loop $g_\sigma : I \to S^1$ such that

$$g_\sigma(0) = 1 = g_\sigma(1), \quad \deg(g_\sigma) = z(\sigma).$$

Let $\lambda_\sigma : I \to \sigma$ denote the linear map which sends 0 to v_0 and 1 to v_1. Then we define $f_1 : K_1 \to S^1$ by taking

$$f_1(x) = \begin{cases} 1, & (\text{if } x \in K_0), \\ g_\sigma\lambda_\sigma^{-1}(x), & (\text{if } x \in \sigma \in K_1 \setminus A). \end{cases}$$

Next, let us construct the map $f_2 : K_2 \to S^1$. Let $\tau = v_0v_1v_2$ be an arbitrary 2-simplex in $X \setminus A$ and $\sigma_0 = v_1v_2$, $\sigma_1 = v_0v_2$, $\sigma_2 = v_0v_1$. Then $f_1 \mid \partial\tau$ is a loop in S^1 of degree

$$z(\sigma_0) - z(\sigma_1) + z(\sigma_2) = z(\partial\tau) = (\delta z)(\tau) = 0$$

since z is a cocycle. By (3.3), $f_1 \mid \partial\tau$ is homotopic to the constant map. Hence it follows from the homotopy extension property that $f_1 \mid \partial\tau$ has an extension $h_\tau : \tau \to S^1$. Then we define f_2 by the following formula:

$$f_2(x) = \begin{cases} f_1(x), & (\text{if } x \in K_1), \\ h_\tau(x), & (\text{if } x \in \tau \in K_2 \setminus A). \end{cases}$$

Now we shall complete the construction of the sequence $\{f_n\}$ by induction. Let $m > 2$ and assume that $f_{m-1} : K_{m-1} \to S^1$ has already been constructed. Let θ be an arbitrary m-simplex in $X \setminus A$, and v_0 be a vertex of θ. Since $m > 2$, the $(m - 1)$-sphere $\partial\theta$ is simply connected by (5.2). Then, according to (5.3), there is a unique map $j_\theta : \partial\theta \to R$ such that

$$j_\theta(v_0) = 0, \quad pj_\theta = f_{m-1} \mid \partial\theta$$

where $p : R \to S^1$ denotes the exponential map in § 2. Since R is solid, the map j_θ has an extension $k_\theta : \theta \to R$. Then we define f_m by setting

$$f_m(x) = \begin{cases} f_{m-1}(x), & (\text{if } x \in K_{m-1}), \\ pk_\theta(x), & (\text{if } x \in \theta \in K_m \setminus A). \end{cases}$$

This completes the construction of the sequence $\{ f_n \}$.

Since $f_n \mid K_{n-1} = f_{n-1}$, we obtain a map $f : X \to S^1$ such that $f \mid K_n = f_n$. In particular, $f(K_0) = 1$ and $f \mid K_1 = f_1$. By the construction of f_1, this implies that $c^1(f) = z$. Hence, h^* is an epimorphism.

To prove that h^* is a monomorphism, let α be any element of $\pi^1(X, A)$ with $h^*(\alpha) = 0$. Pick a representative map $f : (X, A) \to (S^1, 1)$ of α which sends all vertices of X to 1. Then the cocycle $c^1(f)$ is a coboundary of X modulo A and hence there is a cochain $c^0 \in C^0(X, A)$ such that $c^1(f) = \delta c^0$.

For each $n = 1,2,\cdots, \dim X + 1$, let J_n denote the subspace of $X \times I$ defined by

$$J_n = (X \times 0) \cup (K_{n-1} \times I) \cup (X \times 1).$$

To prove that f is homotopic to the constant map $0(X) = 1$ relative to A, let us construct a sequence of maps

$$F_n : J_n \to S^1, \quad n = 1,2,\cdots, \dim X + 1$$

as follows. To construct F_1, let us take for each vertex v in $X \setminus A$ a loop $\xi_v : I \to S^1$ such that

$$\xi_v(0) = 1 = \xi_v(1), \quad \deg(\xi_v) = c^0(v).$$

Then we define $F_1 : J_1 \to S^1$ by taking

$$F_1(x, t) = \begin{cases} 1, & (\text{if } x \in A \text{ or if } t = 0), \\ \xi_v(t), & (\text{if } x = v), \\ f(x), & (\text{if } t = 1). \end{cases}$$

Next, let us construct the map $F_2 : J_2 \to S^1$. Let $\sigma = v_0 v_1$, be any 1-simplex in $X \setminus A$. Then the partial map $F_1 \mid \partial(\sigma \times I)$ on the boundary $\partial(\sigma \times I)$ of $\sigma \times I$ is a loop in S^1 of degree

$$c^0(v_1) - [c^1(f)](\sigma) - c^0(v_0) = (\delta c^0)(\sigma) - [c^1(f)](\sigma) = 0.$$

Hence $F_1 \mid \partial(\sigma \times I)$ has an extension $\eta_\sigma : \sigma \times I \to S^1$. We define F_2 by the following formula:

$$F_2(y) = \begin{cases} F_1(y), & (\text{if } y \in J_1), \\ \eta_\sigma(y), & (\text{if } y \in \sigma \times I). \end{cases}$$

Now we shall complete the construction of the sequence $\{ F_n \}$ by induction. Let $m > 2$ and assume that $F_{m-1} : J_{m-1} \to S^1$ has already been constructed. Let τ be any $(m-1)$-simplex in $X \setminus A$. Since $m > 2$, the $(m-1)$-sphere $\partial(\tau \times I)$ is simply connected. Hence, it follows just as before, that $F_{m-1}|\partial(\tau \times I)$ has an extension $\zeta_\tau : \tau \times I \to S^1$. Then we define $F_m : J_m \to S^1$ by setting

$$F_m(y) = \begin{cases} F_{m-1}(y), & (\text{if } y \in J_{m-1}). \\ \zeta_\tau(y), & (\text{if } y \in \tau \times I). \end{cases}$$

This completes the construction of the sequence $\{F_n\}$.

Since $F_n \mid J_{n-1} = F_{n-1}$, we obtain a map $F : X \times I \to S^1$ such that $F \mid J_n = F_n$. In particular, we have $F(x, 0) = 1$, $F(x, 1) = f(x)$, $F(a, t) = 1$ for every $x \in X$, $a \in A$ and $t \in I$. This implies that f is homotopic to the constant map $0(X) = 1$ relative to A. Hence $\alpha = 0$ and h^* is a monomorphism. ∎

Since $H^1(X, A)$ is effectively computable, the theorem (7.1) solves the classification problem for the maps $(X, A) \to (S^1, 1)$. In particular, if we take $A = \square$, then it gives a solution of the classification problem for the maps $X \to S^1$. On the other hand, it also solves the homotopy problem and the extension problem in the form of the following corollaries.

Corollary 7.2. *Two maps $f, g : (X, A) \to (S^1, 1)$ are homotopic (relative to A) iff $f^*(\iota) = g^*(\iota)$.*

This corollary is an immediate consequence of (7.1). In particular, let us take $A = \square$. The inclusion map $j : S^1 \subset (S^1, 1)$ induces an isomorphism $j^* : H^1(S^1, 1) \approx H^1(S^1)$ and $\varkappa = j^*(\iota)$ is the generator of the free cyclic group $H^1(S^1)$ determined by the counter-clockwise orientation of S^1. Then (7.2) gives the following corollary as a special case.

Corollary 7.3. *Two maps $f, g : X \to S^1$ are homotopic iff $f^*(\varkappa) = g^*(\varkappa)$.*

Corollary 7.4. *A map $f : A \to S^1$ can be extended over X iff the element $f^*(\varkappa)$ of $H^1(A)$ is contained in the image of the homomorphism $i^* : H^1(X) \to H^1(A)$ induced by the inclusion map $i : A \subset X$. In fact, if α is an element of $H^1(X)$ such that $i^*(\alpha) = f^*(\varkappa)$, then f has an extension $g : X \to S^1$ with $g^*(\varkappa) = \alpha$.*

Proof. The necessity of the condition is obvious. For the sufficiency, it suffices to establish the second assertion. By (7.1), there exists a map $k : X \to S^1$ such that $(jk)^*(\iota) = \alpha$. Then we have

$$(jki)^*(\iota) = i^*(jk)^*(\iota) = i^*(\alpha) = f^*(\varkappa) = f^*j^*(\iota) = (jf)^*(\iota).$$

By (7.2), this implies that $jki \simeq jf$ and hence $ki \simeq f$. According to the homotopy extension property, there exists an extension $g : X \to S^1$ of f such that $g \simeq k$. Then we have

$$g^*(\varkappa) = k^*(\varkappa) = k^*j^*(\iota) = (jk)^*(\iota) = \alpha. \ ∎$$

Following the definition of triangulable pairs in [E–S; p.60], we have assumed above that X is a *finite* simplicial complex. However, the proof of

(7.1) is so arranged that it extends to the case that X is an *infinite* simplicial complex and A is any subcomplex of X provided that Whitehead's weak topology is used in X. Hence all results in this section are true for the *infinitely* triangulable pairs (X, A) defined in the obvious way. On the other hand, these results can be extended to more general pairs (X, A) by using various Čech cohomology theories. Some of these generalizations will be given as exercises at the end of the chapter.

8. The Hopf theorems

With some dimensional restrictions on the triangulable pair (X, A), the nice results of the preceding section for S^1 can be generalized to higher spheres. For this purpose, let us first consider some preliminaries.

For every $n \geqslant 1$, let S^n denote the boundary n-sphere of the unit $(n+1)$-simplex $\Delta = \Delta_{n+1}$. For each $m = 0, 1, \cdots, n+1$, let $\Delta^{(m)}$ denote the m-th (n-dimensional) face of Δ. Then S^n is the union of the n-simplexes $\Delta^{(0)}, \Delta^{(1)}, \cdots, \Delta^{(n+1)}$. The n-dimensional cocycle ϕ of S^n defined by $\phi(\Delta^{(0)}) = 1$ and $\phi(\Delta^{(m)}) = 0$, $m = 1, \cdots, n+1$, represents a generator $\varkappa$ of the free cyclic cohomology group $H^n(S^n)$.

For any map $f: X \to S^n$, the element $f^*(\varkappa)$ of $H^n(X)$ depends only on the homotopy class $[f]$ of f and will be called the *degree* of the map f or of the class $[f]$. In particular, if $X = S^n$, then the element $f^*(\varkappa)$ of $H^n(S^n)$ determines a unique integer $\deg(f)$ such that $f^*(\varkappa) = \deg(f)\varkappa$. In this case, it is this integer $\deg(f)$ which is traditionally known as the *degree* of the map $f: S^n \to S^n$. If $n = 1$, this definition of $\deg(f)$ is obviously equivalent to the one given in § 3.

Let v_0 denote the leading vertex of S^n. For each $q = 0, 1, \cdots, n+1$, let $\Gamma^{(q)}$ denote the union of all $\Delta^{(m)}$ with $m \neq q$. Then the inclusion maps

$$\xi: S^n \subset (S^n, v_0), \quad \eta_q: (S^n, v_0) \subset (S^n, \Gamma^{(q)})$$

induce isomorphisms ξ^*, η_q^* on the cohomology groups of positive dimensions. Let

$$\iota = \xi^{*-1}(\varkappa), \qquad \lambda_q = \eta_q{}^{*-1}(\iota).$$

Then ι and λ_q are generators of the free cyclic groups $H^n(S^n, v_0)$ and $H^n(S^n, \Gamma^{(q)})$ respectively.

Next, let us consider the unit n-simplex Δ_n and its boundary $(n-1)$-sphere $\partial\Delta_n$. The n-cocycle ψ of Δ_n given by $\psi(\Delta_n) = 1$ represents a generator μ of the free cyclic group $H^n(\Delta_n, \partial\Delta_n)$. For any map $f: (\Delta_n, \partial\Delta_n) \to (S^n, v_0)$, the element $f^*(\iota)$ determines a unique integer $\deg(f)$ such that $f^*(\iota) = \deg(f) \cdot \mu$ This integer $\deg(f)$ will be called the *degree* of the map f.

For each $q = 0, 1, \cdots, n+1$, let $\zeta_q: (\Delta_n, \partial\Delta_n) \to (S^n, \Gamma^{(q)})$ denote the map defined by the order-preserving one-to-one assignment of vertices of Δ_n into $\Delta^{(q)}$. Then it is not difficult to verify that

$$\zeta_q^*(\lambda_q) = (-1)^q\mu \tag{8.1}$$

Now, let $f: S^n \to S^n$ be a map which sends the $(n-1)$-dimensional skeleton of S^n into v_0. Then, by (8.1) and a theorem in cohomology theory, [E-S; p. 37, Theorem 14.6c], one can easily deduce the following relation

$$\deg(f) = \sum_{q=0}^{n+1} (-1)^q \deg(f\zeta_q). \tag{8.2}$$

Finally, let us prove the following

Lemma 8.3. *For every integer m, there exists a map $f_m : (\Delta_n, \partial\Delta_n) \to (S^n, v_0)$ with $\deg(f_m) = m$.*

Proof. If $m = 0$, then the constant map $f_0(\Delta_n) = v_0$ is obviously of degree 0. Next, assume $m > 0$. Take a sufficiently fine simplicial subdivision K of Δ_n so that we may pick m mutually disjoint closed n-simplexes $\sigma_1, \cdots, \sigma_m$ contained in the interior of Δ_n. Let the vertices u_{ij} of σ_i be ordered in such a way that the orientation of

$$\sigma_i = (u_{i1}, u_{i2}, \cdots, u_{in+1})$$

agrees with that of Δ_n. We now define f_m to be the unique simplicial map of K into S^n which sends u_{ij} into the vertex v_j of S^n for each $i = 1, \cdots, m$ and each $j = 1, \cdots, n+1$ and sends every other vertex of K into v_0. Then it is easily seen that $f_m(\partial\Delta_n) = v_0$ and $\deg(f_m) = m$. Finally, let λ denote a linear homeomorphism of Δ_n which interchanges a pair of vertices of Δ_n and leaves other vertices fixed. Then $\deg(f_m\lambda) = -m$. ∎

An organization of the Hopf theorems is as follows.

Theorem $\mathbf{H}^n$. (Homotopy). *If a map $f: S^n \to S^n$ has degree $\deg(f) = 0$, then f is homotopic to the constant map $0(S^n) = v_0$.*

Theorem $\mathbf{E}^n$. (Extension). *Let (X, A) be a triangulable pair with $\dim(X \setminus A) \leqslant n+1$ and $f: A \to S^n$ be a given map. If there exists an element $\alpha \in H^n(X)$ such that $i^*(\alpha) = f^*(\varkappa)$, where $i: A \subset X$, then f admits an extension $g: X \to S^n$ such that $g^*(\varkappa) = \alpha$.*

Theorem $\mathbf{C}^n$. (Classification). *If X is a triangulable space with $\dim X \leqslant n$, then the assignment $f \to f^*(\varkappa)$ sets up a one-to-one correspondence between the homotopy classes of the maps $f: X \to S^n$ and the elements of the cohomology group $H^n(X)$.*

The converse of $\mathbf{E}^n$ is trivial: if f admits an extension $g: X \to S^n$, then there exists an element $\alpha \in H^n(X)$ such that $i^*(\alpha) = f^*(\varkappa)$. In fact, we have $\alpha = g^*(\varkappa)$.

Since $\mathbf{H}^1$ is a special case of (3.2), these theorems can be established inductively by proving

$$\mathbf{H}^n \Rightarrow \mathbf{E}^n, \qquad \mathbf{E}^n \Rightarrow \mathbf{H}^{n+1}, \qquad \mathbf{E}^n \Rightarrow \mathbf{C}^n$$

for every $n \geqslant 1$. Also, it worth noting that $\mathbf{H}^n$ is a special case of $\mathbf{C}^n$, $\mathbf{C}^1$ is a special case of (7.1), and $\mathbf{E}^1$ is a special case of (7.4).

Proof of $H^n \Rightarrow E^n$. Let X be a simplicial complex and A a subcomplex of X. We may assume that f sends the $(n-1)$-dimensional skeleton of A into the point v_0 of S^n, for otherwise we may replace f by a homotopic map which satisfies this condition by applying the method of simplicial approximation and observing that every proper subspace of S^n is contractible in S^n.

Under these assumptions, we may define an n-cochain $c^n(f)$ of A as follows. Let σ be any n-simplex of A and $\lambda_\sigma : \Delta_n \to \sigma$ denote the linear homeomorphism which preserves the order of vertices. Then $f\lambda_\sigma$ is a map of $(\Delta_n, \partial\Delta_n)$ into (S^n, v_0) and hence we may define the cochain $c^n(f)$ by taking

$$[c^n(f)](\sigma) = \deg(f\lambda_\sigma)$$

for every n-simplex σ of A. By (8.2), one can easily prove that $c^n(f)$ is a cocycle. By the definition of the induced homomorphism f^*, it is not difficult to see that $c^n(f)$ represents the cohomology class $f^*(\varkappa) \in H^n(A)$.

Assume that α is an element of $H^n(X)$ such that $i^*(\alpha) = f^*(\varkappa)$. Then it follows that there exists an n-cocycle z^n of X which represents α and satisfies the relation

$$z^n(\sigma) = [c^n(f)](\sigma)$$

for every n-simplex σ of A. Let K denote the union of A and the n-dimensional skeleton of X. We shall define a map $k : K \to S^n$ as follows. Let σ be any n-simplex of K which is not in A. By (8.3), there is a map

$$f_\sigma : (\Delta_n, \partial\Delta_n) \to (S^n, v_0)$$

with $\deg(f_\sigma) = z^n(\sigma)$. Let $\lambda_\sigma : \Delta_n \to \sigma$ denote the linear homeomorphism which preserves the order of vertices. Then we define k by setting

$$k(x) = \begin{cases} f(x), & (\text{if } x \in A), \\ f_\sigma\lambda_\sigma^{-1}(x), & (\text{if } x \in \sigma \in K \setminus A). \end{cases}$$

Finally, we are going to construct an extension $g : X \to S^n$ of k as follows. Let τ be any $(n+1)$-simplex of $X \setminus A$ and $\lambda_\tau : \Delta_{n+1} \to \tau$ the linear homeomorphism which preserves the order of vertices. Denote $k_\tau = k\lambda_\tau \mid S^n$. By (8.2), we have

$$\deg(k_\tau) = \sum_{i=0}^{n} (-1)^i z^n(\tau^{(i)}) = z^n(\partial\tau) = \partial z^n(\tau) = 0.$$

According to Theorem $\mathbf{H}^n$, this implies that k_τ is homotopic to the constant map 0. Then it follows from the homotopy extension property that k_τ has an extension $g_\tau : \Delta_{n+1} \to S^n$. Then we define $g : X \to S^n$ by taking

$$g(x) = \begin{cases} k(x) & (\text{if } x \in K), \\ g_\tau\lambda_\tau^{-1}(x), & (\text{if } x \in \tau \in X \setminus A). \end{cases}$$

Since g is obviously an extension of f, it remains to verify that $g^*(\varkappa) = \alpha$. According to the construction of g, it can be seen that g maps the $(n-1)$-dimensional skeleton of X into v_0 and that $c^n(g) = z^n$. Since $c^n(g)$ represents $g^*(\varkappa)$ and z^n represents α, this implies $g^*(\varkappa) = \alpha$. ∎

Proof of $E^n \Rightarrow H^{n+1}$. Let $f: S^{n+1} \to S^{n+1}$ be a given map with $\deg(f) = 0$. We are going to prove that f is homotopic to a constant map.

Using the method of simplicial approximation, we may assume that f is a simplicial map of a triangulation K of S^{n+1} into a triangulation J of S^{n+1}. Pick an $(n+1)$-simplex $\sigma = (u_0, u_1, \cdots, u_{n+1})$ of J such that the cocycle ϕ defined by $\phi(\sigma) = 1$ and $\phi(\tau) = 0$ for every $(n+1)$-simplex τ of J other than σ represents the generator $\varkappa$ of $H^{n+1}(S^{n+1})$.

Let M denote the subcomplex of K consisting of the closed $(n+1)$-simplexes of K which are mapped into σ by the simplicial map f. We may assume that no two of these $(n+1)$-simplexes have a common vertex, for otherwise we could replace K and J by their second barycentric subdivisions K'' and J'', and take for σ an $(n+1)$-simplex of J'' none of whose vertices is also a vertex of J. Under these assumptions, we have $H^q(M) = 0$ for every $q > 0$.

The simplicial map $f: K \to J$ induces an $(n+1)$-cocycle $\psi = f^{\#}(\phi)$ of K which represents the element

$$f^*(\varkappa) = \deg(f) \cdot \varkappa = 0$$

of $H^{n+1}(K)$. Hence there exists an n-cochain c of K such that $\psi = \delta c$.

The simplicial inclusion map $p: M \subset K$ induces the following two cochains of M

$$c_1 = p^{\#}(c), \qquad \psi_1 = p^{\#}(\psi) = p^{\#}f^{\#}(\phi) = (fp)^{\#}(\phi).$$

The relation $\psi = \delta c$ in K implies that $\psi_1 = \delta c_1$ in M. Next, consider the leading n-face $\sigma^{(0)} = (u_1, \cdots, u_{n+1})$ of the $(n+1)$-simplex σ and let ϕ_0 denote the n-cochain of J defined by $\phi_0(\sigma^{(0)}) = 1$ and $\phi_0(\tau) = 0$ for other n-simplexes τ of J. Then the simplicial map $fp: M \to J$ induces an n-cochain $c_0 = (fp)^{\#}(\phi_0)$ of M. Since ϕ is the coboundary of ϕ_0 on the subcomplex σ of J and fp maps M into σ, it follows that $\psi_1 = \delta c_0$. Then $\delta(c_1 - c_0) = \psi_1 - \psi_1 = 0$ and hence $c_1 - c_0$ is an n-cocycle of M. Since $H^n(M) = 0$, this implies that $c_1 - c_0$ is a coboundary of M.

Let M^n denote the n-dimensional skeleton of the complex M. Then f defines a simplicial map χ of M^n into the n-sphere $\partial\sigma$. By the construction given above, it is clear that c_0 is a cocycle of M^n and represents the degree of the map χ. Since $c_1 - c_0$ is a coboundary, the degree of χ is also represented by the cocycle c_1 of M^n.

Let $N = (K \setminus M) \cup M^n$ and consider the inclusion map $q: N \subset K$. Since $\delta c(\tau) = \psi(\tau) = 0$ for any $(n+1)$-simplex τ which is not in M, it follows that $c_2 = q^{\#}(c)$ is a cocycle of the complex N. Since c_2 is an extension of c_1, we may apply Theorem $\mathbf{E^n}$ to conclude that χ has an extension $\rho: N \to \partial\sigma$. Define a map $g: S^{n+1} \to S^{n+1}$ by taking

$$g(x) = \begin{cases} f(x) & (\text{if } x \in M), \\ \rho(x), & (\text{if } x \in N). \end{cases}$$

Since f maps N into the space $W = (J \setminus \sigma) \cup \partial\sigma$ which is solid, the maps $f \mid N$ and ρ are homotopic in W modulo M^n. This implies $f \simeq g$. Next, since the image of g is contained in σ, it follows that g is homotopic to a constant map. Since S^{n+1} is pathwise connected, this implies that $f \simeq g \simeq 0$. ∎

Proof of $\mathbf{E}^n \Rightarrow \mathbf{C}^n$. Let X be a simplicial complex with dim $X \leqslant n$. First, let α be any element of $H^n(X)$; we shall find a map $f: X \to S^n$ such that $f^*(\varkappa) = \alpha$. Let A denote the $(n-1)$-dimensional skeleton of X and $i: A \subset X$ the inclusion map. Then we have $H^n(A) = 0$ and $i^*(\alpha) = 0$. Hence, by $\mathbf{E}^n$, the constant map $k(A) = v_0$ has an extension $f: X \to S^n$ such that $f^*(\varkappa) = \alpha$.

Next, assume that $f, g: X \to S^n$ are two maps such that $f^*(\varkappa) = g^*(\varkappa)$. We are going to prove $f \simeq g$. For this purpose, let us study some elementary properties of the product space $X \times I$. Obviously, $X \times I$ is triangulable and $\dim(X \times I) \leqslant n + 1$. Let $p: X \times I \to X$ and $q_i: X \to X \times I$, $(i = 0,1)$, denote the maps defined by

$$p(x, t) = x, \qquad q_0(x) = (x, 0), \qquad q_1(x) = (x, 1).$$

Since pq_i is the identity map on X, we obtain $q_i^*p^*(\alpha) = \alpha$ for every $\alpha \in H^n(X)$. Next, consider the subspaces

$$X_0 = X \times 0, \quad X_1 = X \times 1, \quad A = X_0 \cup X_1$$

of $X \times I$. Then, according to an elementary theorem [E-S; p. 33], $H^n(A)$ is the direct sum of $H^n(X_0)$ and $H^n(X_1)$. Hence the elements of $H^n(A)$ can be represented by the pairs (β, γ) of elements of $H^n(X)$. Let $r: A \subset X \times I$ denote the inclusion map. Then we have

$$r^*p^*(\alpha) = (\alpha, \alpha).$$

Define a map $h: A \to S^n$ by taking

$$h(x, t) = \begin{cases} f(x), & (\text{if } x \in X \text{ and } t = 0), \\ g(x), & (\text{if } x \in X \text{ and } t = 1). \end{cases}$$

Then, $h^*(\varkappa) = (\alpha, \alpha) \in H^n(A)$. According to $\mathbf{E}^n$, there exists an extension $H: X \times I \to S^n$ such that $H^*(\varkappa) = q_0^*(\alpha)$. This proves $f \simeq g$. ∎

Thus we have proved the celebrated Hopf theorems $\mathbf{H}^n$, $\mathbf{E}^n$, $\mathbf{C}^n$ for every $n = 1, 2, \cdots$. In particular, if we take $X = S^n$ in Theorem $\mathbf{C}^n$, we obtain the following

Corollary 8.4. *The homotopy classes of the maps $f: S^n \to S^n$ are in a one-to-one correspondence with the integers. The correspondence is given by the assignment $f \to \deg(f)$.*

Finally, the remarks given at the end of § 7 are also true in the present circumstances.

9. The Hurewicz theorem

The dual of the Hopf classification theorem is the Hurewicz theorem, the statement of which is the main purpose of the present section. To this aim, let us define the important notion of n-connected spaces.

Let $n \geqslant 0$ be a given integer. A space X is said to be *n-connected* if, for every triangulable space T of dimension $\leqslant n$, any two maps $f, g : T \to X$ are homotopic. So, X is 0-connected iff it is pathwise connected, and X is 1-connected iff it is simply connected. Using the extension property, one can easily prove that an $(n - 1)$-connected space X is n-connected iff every map $f : S^n \to X$ is homotopic to a constant map.

Now, let X be a given space and consider the maps

$$f : S^n \to X,$$

where S^n denotes the boundary n-sphere of the unit $(n + 1)$-simplex $\Delta = \Delta_{n+1}$. Since we have fully studied the case $n = 1$ in § 4 and § 6, we will assume that $n \geqslant 2$. The fundamental n-cycle

$$z = \sum_{i=0}^{n+1} (-1)^i \Delta^{(i)}$$

of S^n represents a generator ι of the free cyclic group $H_n(S^n)$.

For each map $f : S^n \to X$, the element $f_*(\iota)$ of $H_n(X)$ depends only on the homotopy class of f and will be called the *degree* of f. If $X = S^n$, then one can easily verify that $f_*(\iota) = \deg (f) \cdot \iota$, where $\deg(f)$ denotes the integer defined in § 8.

Theorem 9.1. (*Hurewicz Theorem*). *If X is an $(n - 1)$-connected space with $n \geqslant 2$, then the assignment $f \to f_*(\iota)$ sets up a one-to-one correspondence between the homotopy classes of the maps $f : S^n \to X$ and the elements of the singular homology group $H_n(X)$.*

As an immediate consequence, we have the following important

Corollary 9.2. *A simply connected space X is n-connected iff $H_m(X) = 0$ for every $m = 2, \cdots, n$.*

In all existing proofs of the Hurewicz theorem, one must use the group structure and a few elementary properties of the homotopy group. In the sequel, we shall prove a much more general theorem. See (V; §4) and (X; §8).

EXERCISES

A. The fundamental group of a connected simplicial complex

Let X be a connected simplicial complex. Denote the vertices of X by $v_0, v_1, \cdots, v_m$. A *broken line* joining a vertex to another is a path which consists of a finite number of 1-simplexes. Since X is connected, we can join v_0 to v_i by a broken line λ_i. Suppose that these broken lines λ_i, $(i = 0, 1, \cdots, m)$, have been chosen such that λ_0 consists of only a single point v_0. Let λ_i^{-1} denote the reverse of λ_i. To each oriented 1-simplex $v_i v_j$ of X, consider the loop

$$f_{ij} = \lambda_i \cdot (v_i v_j) \cdot \lambda_j^{-1}$$

which represents an element $\alpha_{ij} \in \pi_1(X, v_0)$. Prove that these elements α_{ij} form a system of generators of $\pi_1(X, v_0)$ with the set of relations described as follows. First, for the 1-simplex $v_j v_i$, we have trivial relation

$$\alpha_{ji} = (\alpha_{ij})^{-1}.$$

Second, for each 1-simplex $v_i v_j$, the loop f_{ij} is a product of the 1-simplexes $v_\xi v_\eta$ of X, say

$$f_{ij} = P_{ij}\,(v_\xi v_\eta).$$

Then we have the relation

$$\alpha_{ij} = P_{ij}\,(\alpha_{\xi\eta}),$$

where $P_{ij}\,(\alpha_{\xi\eta})$ is the product obtained by replacing $v_\xi v_\eta$ with the corresponding element $\alpha_{\xi\eta}$ in $\pi_1(X, v_0)$. Third, for each 2-simplex $\sigma = v_i v_j v_k$ of X, we have the relation

$$\alpha_{ij}\alpha_{jk}\alpha_{ki} = 1.$$

This gives an effective method for computing the fundamental group $\pi_1(X)$ of X. For examples, prove the following assertions:

1. The fundamental group $\pi_1(X)$ depends only on the 2-dimensional skeleton X^2 of X. More precisely, the inclusion map $i : X^2 \subset X$ induces an isomorphism $i_* : \pi_1(X^2, v_0) \approx \pi_1(X, v_0)$.

2. The fundamental group of a closed orientable surface of genus g is the abstract group generated by $2g$ elements $\alpha_i, \beta_i, (i = 1, \cdots, g)$, with a single relation

$$\alpha_1\beta_1\alpha_1^{-1}\beta_1^{-1}\alpha_2\beta_2\alpha_2^{-1}\beta_2^{-1}\cdots\alpha_g\beta_g\alpha_g^{-1}\beta_g^{-1} = 1.$$

Hence the fundamental group of the (2-dimensional) torus is the free abelian group with two generators. If $g > 1$, then the fundamental group is non-abelian.

3. The fundamental group of a closed non-orientable surface of genus g is the abstract group generated by g elements $\alpha_1, \cdots, \alpha_g$ with a single relation

$$\alpha_1\alpha_1\alpha_2\alpha_2\cdots\alpha_g\alpha_g = 1.$$

Hence the fundamental group of the projective plane is the cyclic group of order two. If $g > 1$, then the group is non-abelian.

4. If a connected simplicial complex X is the union of two connected subcomplexes A and B with connected intersection $D = A \cap B$, then $\pi_1(X)$ is the quotient group of the free product $\pi_1(A) \bigcirc \pi_1(B)$, [S–T; p. 30], obtained by identifying, for each element $\delta \in \pi_1(D)$, the element $\xi_*(\delta) \in \pi_1(A)$ with the element $\eta_*(\delta) \in \pi_1(B)$, where $\xi : D \subset A$ and $\eta : D \subset B$ denote the inclusion maps. In particular, if D is simply connected, then we have

$$\pi_1(X) \approx \pi_1(A) \bigcirc \pi_1(B).$$

5. If X is a connected 1-dimensional simplicial complex, then $\pi_1(X)$ is a free group. In particular, if X is the space which consists of two circles intersecting at a single point, then $\pi_1(X)$ is the free group on two generators.

B. The bridge theorems

Assume that (1) X is a normal space, (2) Y is a connected ANR, and (3) either X or Y is compact. Let $f: X \to Y$ be a given map and α be a finite open covering of X. A map

$$g_\alpha : N_\alpha \to Y$$

of the geometric nerve N_α of α into Y will be called a *bridge map* for f if $g_\alpha h_\alpha \simeq f$ for every canonical map $h_\alpha : X \to N_\alpha$, [$L_3$; p. 40]. If such a bridge map g_α exists, α is said to be a *bridge* for the given map f. Prove the following assertions, [Hu 5]:

1. Every map $f: X \to Y$ has a bridge.
2. Every refinement of a bridge is also a bridge.
3. If α, β are two bridges for a given map $f: X \to Y$, where X is compact, and if $g_\alpha : N_\alpha \to Y$, $g_\beta : N_\beta \to Y$ are bridge maps, then there exists a common refinement γ of α and β such that $g_\alpha p_{\gamma\alpha} \simeq g_\beta p_{\gamma\beta}$, where $p_{\gamma\alpha} : N_\gamma \to N_\alpha$ and $p_{\gamma\beta} : N_\gamma \to N_\beta$ are arbitrary simplicial projections.

Study the analogous assertions for a map $f: X \to Y$ of a paracompact Hausdorff space X into a connected ANR Y by considering the locally finite open coverings of X.

C. Maps of compact spaces into spheres

The bridge theorems in Ex. B furnish us with a link between a map on a compact Hausdorf space and maps of simplicial complexes. Therefore, if one uses the Čech cohomology groups and the notion of dimension both defined by means of finite open coverings, one can extend the results in § 7 and § 8 to compact Hausdorff spaces. Establish the following theorems:

1. Hopf extension theorem. Let (X, A) be a compact Hausdorff pair with $\dim X \leqslant n + 1$, u a generator of the free cyclic group $H^n(S^n)$, and $f: A \to S^n$ a given map. If there exists an element $\alpha \in H^n(X)$ such that $i^*(\alpha) = f^*(u)$, where $i: A \subset X$, then there exists an extension $g: X \to S^n$ such that $g^*(u) = \alpha$.
2. Hopf classification theorem. If X is a compact Hausdorff space with $\dim X \leqslant n$ and if u is a generator of $H^n(S^n)$, then the assignment $f \to f^*(u)$ sets up a one-to-one correspondence between the homotopy classes of the maps $f: X \to S^n$ and the elements of the Čech cohomology group $H^n(X)$.

In the case $n = 1$, the conditions on the dimension of X may be removed. Furthermore, the assignment $f \to f^*(u)$, in this case, is an isomorphism

$$h^* : \pi^1(X) \approx H^1(X).$$

For further generalizations to non-compact spaces, see [Dowker 1].

D. The degree of a suspended maps

Consider the n-sphere S^n as the equator of an $(n + 1)$-sphere S^{n+1} with north hemisphere E_+^{n+1} and south hemisphere E_-^{n+1}. By Tietze's extension theorem, every map $f: S^n \to S^n$ has an extension $f^*: S^{n+1} \to S^{n+1}$ such that $f^*(E_+^{n+1}) \subset E_+^{n+1}$ and $f^*(E_-^{n+1}) \subset E_-^{n+1}$, which is called a *suspended map* of f. Prove

1. $\deg(f^*) = \deg(f)$.
2. (8.3) can be obtained inductively by the construction of suspended maps of a map $f: S^1 \to S^1$ of degree m.

CHAPTER III

FIBER SPACES

1. Introduction

The concept of *fiber space* is crucial in homotopy theory: it usually appears in the application of homotopy theory to geometric problems; it is a powerful weapon in the computation of the homotopy groups of various spaces; and it plays a key role in the axiomatization of homotopy theory. Thus, before turning to the discussion of the homotopy groups themselves, it is appropriate to develop certain properties of fiber spaces in some detail.

Already we have seen one important example, namely the exponential map $p : R \to S^1$ of Chapter II: in the language of the present chapter, we should say that R is a fiber space (indeed a *covering space*) over S^1 with projection p. As to the usefulness of this particular example, recall that upon it was based the classification of maps of a space X into S^1 of § 7 of Chapter II.

Historically, there were a number of examples, and definitions of fiber spaces were abstracted from these in a variety of ways; but in each case it was possible to prove a so-called *covering homotopy theorem.* It remained for J.-P. Serre in 1950 to single this theorem out as the crucial property, and to base on it his study of the singular homology of fiber spaces. The influence of this study on homotopy theory has been profound, and it now seems quite clear that his is the proper definition; hence we adopt it and consider some of its immediate consequences in §§ 2–3.

On the other hand, the classical examples of fiber spaces belong to a much more narrow class, in that all of them have a *local product structure.* This concept, together with an important example, the Hopf fiberings of spheres, is considered in §§ 4–6.

After considering certain mappings of fiber spaces (§§ 7–8) we develop (for later use, especially in Chapter IV) certain properties of spaces of paths (§§ 9–14). In particular, it will appear that they are fiber spaces in the sense of Serre mentioned above, but they do not have a local product structure.

Finally (§§ 15–17) we study the case of discrete fibers and in particular the classical covering spaces.

2. Covering homotopy property

Throughout the present section, we shall consider a given map

$$p : E \to B$$

of a space E called the *total space* into a space B called the *base space.*

Let X be a given space, $f : X \to B$ a given map, and $f_t : X \to B$, $(0 \leqslant t \leqslant 1)$, a given homotopy of f. A map $f^* : X \to E$ is said to *cover* f (relative to p) if $pf^* = f$. Similarly, a homotopy $f_t^* : X \to E$, $(0 \leqslant t \leqslant 1)$, of f^* is said to *cover* the homotopy f_t (relative to p) if $pf_t^* = f_t$ for each $t \in I$; f_t^* is called a *covering homotopy* of f_t.

The map $p: E \to B$ is said to have the *covering homotopy property* (abbreviated CHP) for the space X if, for every map $f^* : X \to E$ and every homotopy $f_t : X \to B$, $(0 \leqslant t \leqslant 1)$, of the map $f = pf^* : X \to B$, there exists a homotopy $f_t^* : X \to E$, $(0 \leqslant t \leqslant 1)$, of f^* which covers the homotopy f_t. The map $p : E \to B$ is said to have the *absolute covering homotopy property* (abbreviated ACHP) if it has the CHP for every space X. The map $p : E \to B$ is said to have the *polyhedral covering homotopy property* (abbreviated PCHP) if it has the CHP for every triangulable space X.

The map $p : E \to B$ is said to have the *covering homotopy extension property* (abbreviated CHEP) for the space X relative to a given subspace A of X if, for every map $f^* : X \to E$ and every homotopy $f_t : X \to B$, $(0 \leqslant t \leqslant 1)$, of the map $f = pf^* : X \to B$, every homotopy $g_t^* : A \to E$, $(0 \leqslant t \leqslant 1)$, of $g^* = f^* | A$ which covers $f_t | A$ can be extended to a homotopy $f_t^* : X \to E$, $(0 \leqslant t \leqslant 1)$, of f^* covering f_t. The map $p : E \to B$ is said to have *absolute covering homotopy extension property* (abbreviated ACHEP) if it has the CHEP for every space X relative to every closed subspace A of X. The map $p : E \to B$ is said to have the *polyhedral covering homotopy extension property* (abbreviated PCHEP) if, for every triangulable pair (X, A), it has the CHEP for X relative to A.

If E is the topological product $B \times D$ and if $p : E \to B$ is the natural projection, then p obviously has the ACHP. In fact, let X be a given space, $f^* : X \to E$ a given map, and $f_t : X \to B$, $(0 \leqslant t \leqslant 1)$, a given homotopy of the map $f = pf^*$. Then, f_t has a covering homotopy $f_t^* : X \to E$, $(0 \leqslant t \leqslant 1)$, of f^* defined by

$$f_t^*(x) = (f_t(x), qf^*(x))$$

for every $x \in X$ and $t \in I$, where $q : E \to D$ denotes the other natural projection.

3. Definition of fiber space

A map $p : E \to B$ is said to be a *fibering* if it has the PCHP. In this case, the space E will be called a *fiber space over the base space B with projection p : $E \to B$*. For each point $b \in B$, the subspace $p^{-1}(b)$ of E is called the *fiber over b*.

Remark. We do not assume that $p(E) = B$ and hence the fiber $p^{-1}(b)$ may be vacuous for some point b of B. However, applying CHP for a point, one can easily prove that $p(E)$ is a union of path-components of B. Therefore, if B is pathwise connected, we must have $p(E) = B$ unless E itself is empty.

From time to time, it will be useful to know that, for a map $p : E \to B$, the PCHP is equivalent to certain other covering properties. Precisely, we have the following

Theorem 3.1. *For an arbitrary map $p : E \to B$, the following statements are equivalent:*

(i) $p : E \to B$ *is a fibering*, i.e. *has the PCHP.*

(ii) *For each* $m \geqslant 0$, $p : E \to B$ *has the CHP for the m-simplex* Δ_m.

(iii) *For each* $m \geqslant 0$, $p : E \to B$ *has the CHEP for the m-simplex* Δ_m *relative to its boundary* $(m - 1)$*-sphere* S^{m-1}.

(iv) $p : E \to B$ *has the PCHEP.*

(v) *If* (X, A) *is a triangulable pair such that A is a strong deformation retract of X and if* $f : X \to B$ *and* $g^* : A \to E$ *are maps such that* $pg^* = f \mid A$, *then* g^* *has an extension* $f^* : X \to E$ *such that* $pf^* = f$.

Proof. (i) → (ii) is obvious.

(ii) → (iii). If $m = 0$, then Δ_m is a single point and its boundary sphere S^{m-1} is empty. Hence, the implication (ii) → (iii) is obvious in this case.

For the general case $m > 0$, we construct a homeomorphism h of $\Delta_m \times I$ onto itself which carries $(\Delta_m \times 0) \cup (S^{m-1} \times I)$ homeomorphically onto $\Delta_m \times 0$. For this purpose, let us first construct a homeomorphism h_0 of $(\Delta_m \times 0) \cup (S^{m-1} \times I)$ onto $\Delta_m \times 0$ as follows: For an arbitrary point $w = (x_0, \cdots, x_m ; t)$ of $(\Delta_m \times 0) \cup (S^{m-1} \times I)$, we define $h_0(w)$ to be the point $(y_0, \cdots, y_m ; 0)$ of $\Delta_m \times 0$, where

$$y_i = \begin{cases} \frac{1}{2}\left(x_i + \dfrac{1}{m+1}\right), & (\text{if } w \in \Delta_m \times 0), \\ \frac{1}{2}\left(x_i + \dfrac{1}{m+1}\right) + \frac{1}{2}\left(x_i - \dfrac{1}{m+1}\right)t, & (\text{if } w \in S^{m-1} \times I). \end{cases}$$

It follows that $h_0(x_0, \cdots, x_m; 1) = (x_0, \cdots, x_m; 0)$ for each $(x_0, \cdots, x_m) \in S^{m-1}$. Similarly, we can define a homeomorphism h_1 of $(\Delta_m \times 1) \cup (S^{m-1} \times I)$ onto $\Delta_m \times 1$ such that $h_1(x_0, \cdots, x_m; 0) = (x_0, \cdots, x_m; 1)$ for each $(x_0, \cdots, x_m) \in S^{m-1}$. Next, we define a homeomorphism h_2 of the boundary $\partial(\Delta_m \times I)$ of $\Delta_m \times I$ onto itself by taking

$$h_2(w) = \begin{cases} h_0(w), & \text{if } w \in (\Delta_m \times 0) \cup (S^{m-1} \times I), \\ h_1^{-1}(w), & \text{if } w \in \Delta_m \times 1. \end{cases}$$

Finally, the homeomorphism h_2 of $\partial(\Delta_m \times I)$ may be extended to a homeomorphism h of $\Delta_m \times I$ by radial extension from the center $(c, \frac{1}{2})$ of $\Delta_m \times I$, where c denotes the center of Δ_m. Since h is an extension of h_0, it carries $(\Delta_m \times 0) \cup (S^{m-1} \times I)$ onto $\Delta_m \times 0$. The following picture

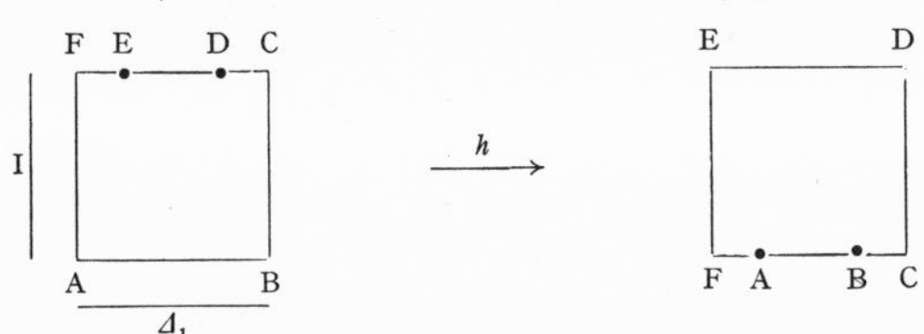

illustrates the homeomorphism h for the case $m = 1$.

Now, let $f^* : \Delta_m \to E$ be a map, $f_t : \Delta_m \to B$, $(0 \leqslant t \leqslant 1)$, a homotopy of the map $f = pf^*$, and $g_t^* : S^{m-1} \to E$, $(0 \leqslant t \leqslant 1)$, a homotopy of $g^* = f^* \mid S^{m-1}$ which covers $f_t \mid S^{m-1}$. The homotopy f_t and the partial homotopy g_t^* give rise to maps

$$F : \Delta_m \times I \to B, \quad G^* : (\Delta_m \times 0) \cup (S^{m-1} \times I) \to E$$

in the obvious way. Let

$$\Phi = Fh^{-1} : \Delta_m \times I \to B, \quad \Psi^* = G^* h_0^{-1} : \Delta_m \times 0 \to E.$$

Then we have $p\Psi^* = \Phi \mid \Delta_m \times 0$. By (ii), Ψ^* has an extension $\Phi^* : \Delta_m \times I \to E$ such that $p\Phi^* = \Phi$. Define a homotopy $f_t^* : \Delta_m \to E$, $(0 \leqslant t \leqslant 1)$, by taking

$$f_t^*(x) = \Phi^* h(x, t), \qquad (x \in \Delta_m, t \in I).$$

Then it is easy to verify that $f_0^* = f^*$, $f_t^* \mid S^{m-1} = g_t^*$ and $pf_t^* = f_t$ for every $t \in I$.

(iii) $\to$ (iv). Let (X, A) be a triangulable pair, $f^* : X \to E$ a map, $f_t : X \to B$, $(0 \leqslant t \leqslant 1)$, a homotopy of the map $f = pf^*$, and $g_t^* : A \to E$, $(0 \leqslant t \leqslant 1)$, a homotopy of $g^* = f^* \mid A$ which covers $f_t \mid A$. We are going to extend g_t^* to a homotopy $f_t^* : X \to E$, $(0 \leqslant t \leqslant 1)$, of f^* covering f_t. We may assume that X is a simplicial complex and A is a subcomplex of X. For each $m \geqslant 0$, let X^m denote the m-dimensional skeleton of X. Let K_m denote $A \cup X^m$. By successive application of (iii), one can construct for each integer $m = 0, 1, 2, \cdots$ a homotopy $h_t^m : K_m \to E$, $(0 \leqslant t \leqslant 1)$, such that

$$h_0^m = f^* \mid K_m, \qquad ph_t^m = f_t \mid K_m,$$

$$h_t^m \mid A = g_t^*, \qquad h_t^{m+1} \mid K_m = h_t^m.$$

Then the required homotopy f_t^* is defined by taking

$$f_t^* \mid K_m = h_t^m$$

for each $m = 0, 1, 2, \cdots$

(iv) $\to$ (v). Since A is a strong deformation retract of X, there exists a homotopy $h_t : X \to X$, $(0 \leqslant t \leqslant 1)$, such that h_0 is a retraction of X onto A, h_1 is the identity map on X, and $h_t(a) = a$ for every $a \in A$ and $t \in I$. Define a map $f^{\#} : X \to E$ and a homotopy $f_t : X \to B$, $(0 \leqslant t \leqslant 1)$, by taking $f^{\#} = g^* h_0$ and $f_t = fh_t$ for each $t \in I$. Then $f_0 = pf^{\#}$. Since $f_t(a) = f(a)$ for each $a \in A$ and $t \in I$, we may define a partial covering homotopy $g_t^* : A \to E$, $(0 \leqslant t \leqslant 1)$, of $f^{\#}$ by setting $g_t^* = g^*$ for every $t \in I$. According to (iv), g_t^* has an extension $f_t^* : X \to E$, $(0 \leqslant t \leqslant 1)$, such that $pf_t^* = f_t$ for every $t \in I$. Let $f^* = f_1^*$. Then we have $f^* \mid A = g^*$ and $pf^* = f_1 = fh_1 = f$.

(v) $\to$ (i). This implication follows inmmediately from the fact that $X \times 0$ is a strong deformation retract of $X \times I$. ∎

As noted at the end of § 2, the product space $E = B \times D$ is obviously a fiber space over B relative to the natural projection $p : E \to B$.

4. Bundle spaces

A map $p : E \to B$ is said to have the *bundle property* (abbreviated BP) if there exists a space D such that, for each $b \in B$, there is an open neighborhood U of b in B together with a homeomorphism

$$\phi_U : U \times D \to p^{-1}(U)$$

of $U \times D$ onto $p^{-1}(U)$ satisfying the condition

$$(DF) \qquad p\phi_U(u, d) = u, \quad (u \in U, d \in D).$$

In this case, the space E is called a *bundle space* over the *base space* B relative to the *projection* $p : E \to B$. The space D will be called a *director space*. The open sets U and the homeomorphisms ϕ_U will be called the *decomposing neighborhoods* and the *decomposing functions* respectively.

As an immediate consequence of the definition, we have $p(E) = B$ except in the trivial case where E is empty.

The main idea of this definition is that a bundle space is a space with a local product structure over every point of the base space. In particular, the product space $E = B \times D$ is a bundle space over B relative to the natural projection $p : E \to B$ with D as director space.

In this definition of bundle spaces, we have essentially followed that of Ehresmann and Feldbau, [S; p. 18]. For relations with coordinate bundles in the sense of Steenrod, see [S; pp. 18–20]. Many examples can be found in [S].

Theorem 4.1. *Every bundle space E over B relative to $p : E \to B$ is a fiber space over B relative to p.*

Since CHP is trivial for product spaces, the idea of the following proof is to reduce the construction to local ones where the local product structure is available.

Proof. Let X be a (compact) triangulable space, $f^* : X \to E$ a given map, and $f_t : X \to B$, $(0 \leqslant t \leqslant 1)$, any homotopy of the map $f = pf^* : X \to B$. It suffices to construct a homotopy $f_t^* : X \to E$, $(0 \leqslant t \leqslant 1)$, of f^* which covers f_t.

Pick a collection $\omega = \{ U \}$ of decomposing neighborhoods U which covers the base space B and define a map $F : X \times I \to B$ by taking

$$F(x, t) = f_t(x), \quad (x \in X, t \in I).$$

The collection $\{ F^{-1}(U) \mid U \in \omega \}$ of open sets of $X \times I$ forms an open covering of $X \times I$. Since $X \times I$ is compact, $\{ F^{-1}(U) \mid U \in \omega \}$ has a refinement of the form $\{ W_\lambda \times I_\mu \}$, where $\{ W_\lambda \}$ is a finite open covering of X and $\{ I_\mu \} = \{ I_1, \cdots, I_r \}$ is a finite sequence of open subintervals of I which covers I. We may assume that I_μ meets only $I_{\mu-1}$ and $I_{\mu+1}$ for each $\mu = 2, \cdots, r-1$. Choose numbers

$$0 = t_0 < t_1 < \cdots < t_r = 1$$

so that t_μ is in the intersection $I_\mu \cap I_{\mu+1}$. We shall assume, inductively, that the covering homotopy f_t^* has already been defined for all $t \leqslant t_\mu$ where $\mu \geqslant 0$ is a given integer less than r. We proceed to extend f_t^* over the closed subinterval $[t_\mu, t_{\mu+1}]$ of I.

Taking a sufficiently fine triangulation of X, we may assume that X is a simplicial complex such that every closed simplex of X is contained in some W_λ of the finite open covering $\{W_\lambda\}$ constructed above. Hence, for each closed simplex σ, we may choose a decomposing neighborhood $U_\sigma \in \omega$ such that

$$f_t(x) \in U_\sigma, \quad (x \in \sigma, t_\mu \leqslant t \leqslant t_{\mu+1}).$$

Let $\psi_{U_\sigma} : U_\sigma \times D \to D$ denote the natural projection.

Let σ be a vertex of X. Then f_t^* can be defined by taking

$$f_t^*(\sigma) = \phi_{U_\sigma}[f_t(\sigma), \psi_{U_\sigma}\phi_{U_\sigma}^{-1} f_{t_\mu}^*(\sigma)], \quad (t_\mu \leqslant t \leqslant t_{\mu+1}).$$

Thus f_t^* is defined on the zero-dimensional skeleton X^0 of X for each $t \in [t_\mu, t_{\mu+1}]$. We shall assume that f_t^* has already been defined on the $(n-1)$-dimensional skeleton X^{n-1} for each $t \in [t_\mu, t_{\mu+1}]$, where $n > 0$ is a given integer. We proceed to extend f_t^* over the n-dimensional skeleton X^n of X.

Let σ be any closed n-simplex of X. Then f_t^* has been defined on the boundary $\partial\sigma$ for every $t \in [t_\mu, t_{\mu+1}]$. Let $M = \sigma \times [t_\mu, t_{\mu+1}]$ and consider the subspace

$$N = (\sigma \times t_\mu) \cup (\partial\sigma \times [t_\mu, t_{\mu+1}])$$

of M. According to (I; 4.2), these exists a retraction $\rho : M \supset N$. Let $\theta : N \to E$ denote the map defined by $\theta(x, t) = f_t^*(x)$. Then we may extend f_t^* over σ by setting

$$f_t^*(x) = \phi_{U_\sigma}[f_t(x), \psi_{U_\sigma}\phi_{U_\sigma}^{-1}\theta\rho(x, t)]$$

for every $x \in \sigma$ and $t \in [t_\mu, t_{\mu+1}]$. This completes the construction of f_t^*. ∎

The preceding theorem is a corollary of the following general theorem which is known as *the covering homotopy theorem of bundle spaces.*

Theorem 4.2. *If a map $p : E \to B$ has the BP, then it has the CHP for every paracompact Hausdorff space.*

For a proof of (4.2), see [Huebsch 1] and also [S; p. 50].

5. Hopf fiberings of spheres

Among the early examples of bundle spaces were the three fiberings of spheres

$$p : S^{2n-1} \to S^n, \quad n = 2, 4, 8$$

discovered by Hopf [2] in 1935. We shall examine the first of these (the case $n = 2$) in detail here, and show in § 6 that it may be applied to the classification problem for the maps $f : X \to S^2$, where X denotes a triangulable space of dimension not more than 3.

To construct the fibering for $n = 2$, let us represent S^3 as the unit sphere in the space C^2 of two complex variables, that is to say, S^3 consists of the points (z_1, z_2) in C^2 such that

$$z_1\bar{z}_1 + z_2\bar{z}_2 = 1.$$

Let S^2 be represented as the complex projective line, that is to say, as pairs $[z_1, z_2]$ of complex numbers, not both zero, with equivalence relation $[z_1, z_2] \sim [\lambda z_1, \lambda z_2]$ where $\lambda \neq 0$. Then the Hopf map $p : S^3 \to S^2$ is defined by $p(z_1, z_2) = [z_1, z_2]$ for each $(z_1, z_2) \in S^3$. The continuity of p is obvious. Since any pair $[z_1, z_2]$ can be normalized by dividing by $(z_1\bar{z}_1 + z_2\bar{z}_2)^{\frac{1}{2}}$, p maps S^3 onto S^2.

To prove that S^3 is a bundle space over S^2 relative to p, let us represent S^1 as the set of all complex numbers λ with $|\lambda| = 1$. Consider the points $a = [1, 0]$ and $b = [0, 1]$ of S^2 and the open sets

$$U = S^2 \setminus a, \quad V = S^2 \setminus b.$$

Then U and V cover S^2. Every point in U can be represented by a pair $[z, 1]$. Hence we may define a map ϕ_U of $U \times S^1$ into S^3 by taking

$$\phi_U([z, 1], \lambda) = \left(\frac{\lambda z}{\sqrt{z\bar{z} + 1}}, \frac{\lambda}{\sqrt{z\bar{z} + 1}}\right)$$

for each $[z, 1] \in U$ and each $\lambda \in S^1$. One can easily verify that ϕ_U maps $U \times S^1$ homeomorphically onto $p^{-1}(U)$ and that $p\phi_U(u, d) = u$ for each $u \in U$ and $d \in D$. Hence ϕ_U is a decomposing function. Similarly, we can construct a decomposing function ϕ_V. This completes the proof that S^3 is a bundle space over S^2 relative to the Hopf map p.

If $(z_1, z_2) \in S^3$, then one verifies immediately that the fiber $p^{-1}[z_1, z_2]$ consists of all the points $(\lambda z_1, \lambda z_2)$ with $\lambda \in S^1$. Hence the fibers are just great circles of S^3. In this way the 3-sphere is decomposed into a family of great circles with the 2-sphere as a quotient space.

The Hopf fiberings $p : S^7 \to S^4$ and $p : S^{15} \to S^8$ are constructed in an analogous fashion from the quaternions and the Cayley numbers respectively; a concise and clear description may be found in [S; pp. 105–110]. In these fiberings, the fibers are 3-spheres and 7-spheres respectively.

The Hopf maps are all essential; in fact, this is a consequence of the following

Proposition 5.1. *If a sphere S^n is a fiber space over a base space B which contains more than one point, then the projection $p : S^n \to B$ is an essential map.*

Proof. Assume that p were inessential. Then there exists a homotopy $h_t : S^n \to B$ $(0 \leqslant t \leqslant 1)$ such that $h_0 = p$ and $h_1(S^n)$ is a single point b_0 of B. Let i denote the identity map on S^n; then we have $pi = p$. According to the CHP, there exists a homotopy $k_t : S^n \to S^n$, $(0 \leqslant t \leqslant 1)$, such that $k_0 = i$ and $pk_t = h_t$ for every $t \in I$. Now, k_1 maps S^n into the fiber $p^{-1}(b_0)$ which is a proper subset of S^n since B contains more than one point. Hence, according to (I; § 8), k is inessential. This implies that the identity map i is homotopic to a constant map. This is impossible by (I; § 8). ∎

It is interesting to observe that the Hopf maps are *algebraically trivial*, that is to say, their induced homomorphisms on the homology groups and the cohomology groups are all trivial. Historically, these Hopf maps were the first examples of essential maps which were algebraically trivial. The existence of these maps shows that induced homomorphisms are not by themselves sufficient to classify all maps.

6. Algebraically trivial maps $X \to S^2$

In the present section, we shall show that, by the aid of the Hopf map $p: S^3 \to S^2$, one can solve the classification problem of the algebraically trivial maps $f: X \to S^2$ of a 3-dimensional triangulable space X into S^2.

Let X be any given triangulable space. For an arbitrary map $F: X \to S^3$, it is obvious that the composed map $f = pF: X \to S^2$ is algebraically trivial and that the homotopy class of f depends only on that of F.

Proposition 6.1. *For any given triangulable space X, the assignment $F \to f = pF$ sets up a one-to-one correspondence between the homotopy classes of the maps $F: X \to S^3$ and those of the algebraically trivial maps $f: X \to S^2$.*

As an immediate consequence of (6.1) and the Hopf classification theorem $\mathbf{C}^3$ in (II; § 8), we have the following

Theorem 6.2. *The homotopy classes of the algebraically trivial maps $f: X \to S^2$ of a 3-dimensional triangulable space X into the 2-sphere S^2 are in a one-to-one correspondence with the elements of the integral cohomology group $H^3(X)$. For any $\alpha \in H^3(X)$, the homotopy class which corresponds to α contains the map $f = pF: X \to S^2$, where $F: X \to S^3$ is a map with α as its degree.*

In particular, if $X = S^3$, then every map $f: X \to S^2$ is algebraically trivial. Hence, we have the following

Corollary 6.3. *The homotopy classes of the maps $f: S^3 \to S^2$ are in a one-to-one correspondence with the integers. For any integer n, the homotopy class which corresponds to n is represented by the composition $f = pF: S^3 \to S^2$ of the Hopf map $p: S^3 \to S^2$ and a map $F: S^3 \to S^3$ with* $\deg(F) = n$.

The proof of (6.1) consists of the following two lemmas.

Lemma 6.4. *If a map $f: X \to S^2$ is algebraically trivial, then there exists a map $F: X \to S^3$ such that $pF = f$.*

Proof. Since S^3 is a bundle space over S^2 relative to the Hopf map $p: S^3 \to S^2$, we may choose a collection $\omega = \{U\}$ of decomposing neighborhoods U which covers S^2. Taking a sufficiently fine triangulation of X, we may assume that X is a simplicial complex such that the given map f carries every closed simplex σ of X into some decomposing neighborhood $U_\sigma \in \omega$.

Let X^m denote the m-dimensional skeleton of X, and let $f_m = f \mid X^m$. For each $m = 2, 3, \cdots$, we shall construct a map $F_m: X^m \to S^3$ such that

$$pF_m = f_m, \quad F_{m+1} \mid X^m = F_m.$$

First, let us construct F_2. Since the given map f is algebraically trivial, so is f_2. According to the Hopf classification theorem $\mathbf{C}^2$ of (II; § 8), f_2 is homotopic to a constant map which can be lifted. Hence it follows from the CHP that there is a map $F_2: X^2 \to S^3$ such that $pF_2 = f_2$.

Next, assume that $n > 2$ and that F_m has been constructed for every

$m < n$. Let σ be an arbitrary closed n-simplex of X and choose a decomposing neighborhood $U_\sigma \in \omega$ which contains $f(\sigma)$. Let

$$\phi_\sigma : U_\sigma \times S^1 \to p^{-1}(U_\sigma), \quad \psi_\sigma : U_\sigma \times S^1 \to S^1$$

denote the decomposing function and the natural projection respectively. Define a map $\xi_\sigma : \partial\sigma \to S^1$ by taking

$$\xi_\sigma(x) = \psi_\sigma \phi_\sigma^{-1} F_{n-1}(x)$$

for every $x \in \partial\sigma$. By (II; § 7), ξ_σ has an extension $\eta_\sigma : \sigma \to S^1$. Then we define a map $F_n : X^n \to S^3$ by taking

$$F_n(x) = \begin{cases} F_{n-1}(x), & (\text{if } x \in X^{n-1}), \\ \phi_\sigma[f_n(x), \eta_\sigma(x)], & (\text{if } x \in \sigma \in X^n). \end{cases}$$

Obviously we have $pF_n = f_n$ and $F_n \mid X^{n-1} = F_{n-1}$. This completes the inductive construction of the maps F_m, $m = 2, 3, \cdots$.

Finally, define a map $F : X \to S^3$ by taking $F \mid X^m = F_m$ for every $m \geqslant 2$. Then we have $pF = f$. ∎

Lemma 6.5. *If $F, G : X \to S^3$ are two maps such that $pF \simeq pG$, then $F \simeq G$.*

Proof. Since $pF \simeq pG$, it follows from the PCHP that F is homotopic to a map F' such that $pF' = pG$. Hence we may simply assume that

$$pF = pG.$$

Consider S^3 as the group of quaternions q with $q\bar{q} = 1$. Then the fiber which contains the quaternion 1 is a subgroup S^1 of S^3 and the other fibers are cosets of S^1 in S^3. In fact, in the usual representation

$$\begin{aligned} q &= x_1 + x_2 i + x_3 j + x_4 ij \\ &= x_1 + x_2 i + (x_3 + x_4 i) j \\ &= z_1 + z_2 j \end{aligned}$$

where $z_1 = x_1 + x_2 i$ and $z_2 = x_3 + x_4 i$, the multiplication is based on the rules $j^2 = -1$ and $zj = j\bar{z}$. Then S^1 is the subgroup defined by $z_2 = 0$ and the right cosets of S^1 are the fibers of the Hopf map $p : S^3 \to S^2$.

Define a map H : $X \to S^3$ by taking

$$H(x) = F(x) \cdot [G(x)]^{-1}$$

for every $x \in X$. Since $pF = pG$, $F(x)$ and $G(x)$ are contained in a coset of S^1 and hence H carries X into a proper subspace S^1 of S^3. Then it follows that there exists a homotopy $H_t : X \to S^3$, $(0 \leqslant t \leqslant 1)$, such that $H_0 = H$ and $H_1(X) = 1$. Define a homotopy $J_t : X \to S^3$, $(0 \leqslant t \leqslant 1)$, by taking

$$J_t(x) = H_t(x) \cdot G(x)$$

for each $x \in X$ and $t \in I$. Then we have $J_0 = F$ and $J_1 = G$. Hence $F \simeq G$. ∎

Since the proof of (6.4) and (6.5) is based on special properties of S^1 and S^3, there are no analogous results for the maps $X \to S^4$ and $X \to S^8$.

7. Liftings and cross-sections

Let E be a fiber space over a base space B with projection $p : E \to B$ and let $f : X \to B$ be a map of a space X into B. By a *lifting* of f in E, we mean a map $g : X \to E$ such that $pg = f$. As a special case of (I; § 16), the *lifting problem* for the given map f is to determine whether or not f has a lifting in E.

For example, if $p : S^3 \to S^2$ is the Hopf map and X is a triangulable, space, then (6.4) and (6.5) solve the lifting problem for any map $f : X \to S^2$, namely, f has a lifting in S^3 iff f is algebraically trivial.

If $p : E \to B$ has the CHP for X, then the lifting problem for a map $f : X \to B$ is equivalent to a broadened lifting problem, namely, to find a map $g : X \to E$ such that $pg \simeq f$. According to the definition of fiber spaces, this is always the case if X is triangulable. By (4.2), this is also true if E is a bundle space over B relative to p and X is a paracompact Hausdorff space.

If X is a subspace of B and $f : X \to B$ is the inclusion map, then the notion of a lifting for f reduces to that of a cross-section over X. A *cross-section* in E over a subspace X of B is a map $\varkappa : X \to E$ such that

$$p\varkappa(x) = x, \quad (x \in X).$$

Thus, a map $\varkappa : X \to E$ is a cross-section iff the image $\varkappa(x)$ of an arbitrary point $x \in X$ is contained in the fiber over x.

It is easy to verify that every cross-section $\varkappa : X \to E$ maps X homeomorphically onto $\varkappa(X)$ with $p \mid \varkappa(X)$ as its inverse. Therefore, a cross-section $\varkappa : X \to E$ is considered intuitively as lifting the subspace X of the base space B up into E.

If E is a bundle space over B relative to $p : E \to B$ and $U \subset B$ is a decomposing neighborhood with decomposing function and natural projection

$$\phi_U : U \times D \to p^{-1}(U), \quad \psi_U : U \times D \to D,$$

then, for any point e in $p^{-1}(U)$, there is a cross-section $\varkappa_e : U \to E$ given by

$$\varkappa_e(u) = \phi_U(u, d), \quad d = \psi_U \phi_U^{-1}(e)$$

for each $u \in U$. If $u = p(e)$, then $\varkappa_e(u) = e$. Thus, in bundle spaces, local cross-sections always exist.

However, the existence of a global cross-section, i.e. a cross-section over the whole base space B, is a rather strong condition on the structure of the fiber space. In fact, if a global cross-section $\varkappa : B \to E$ exists, then, in any homology theory satisfying the Eilenberg–Steenrod axioms, the projection $p : E \to B$ and the cross-section $\varkappa : B \to E$ induce for each m the homomorphisms

$$p_* : H_m(E) \to H_m(B), \quad \varkappa_* : H_m(B) \to H_m(E).$$

Since $p\varkappa$ is the identity on B, $p_*\varkappa_*$ must be the identity on $H_m(B)$. It follows that $\varkappa_*$ is a monomorphism, that p_* is an epimorphism, and that $H_m(E)$ decomposes into the direct sum

$$H_m(E) = \text{Kernel } p_* + \text{Image } \varkappa_*.$$

An immediate consequence of this necessary condition is that the Hopf fiberings in § 5 do not have global cross-sections.

Dually, one can deduce necessary conditions for the existence of a global cross-section in terms of cohomology:

$$H^m(E) = \text{Image } p^* + \text{Kernel } \varkappa^*.$$

These conditions (both those on the groups $H_m(E)$ and those on the groups $H^m(E)$) resemble those for retracts. This is no accident: if $\varkappa$ is a cross-section, then the image of $\varkappa$ is a homeomorph of B and is a retract of E.

Let $\varkappa : X \to E$ be a given cross-section. Corresponding to the extension problem of maps in (I; § 2), we have the *extension problem of cross-sections* to determine if $\varkappa$ can be extended over the whole base space, that is to say, whether or not there exists a cross-section $\varkappa^* : B \to E$ such that $\varkappa^* \mid X = \varkappa$. This extension problem of cross-sections is a generalization of that of maps in (I; § 2). Indeed, if $E = B \times D$ where D is a given space, then E is a bundle space over B with projection $p : E \to B$ defined by $p(b, d) = b$. For a given map $f : X \to D$ on a subspace X of B, we have a cross-section $\varkappa_f : X \to E$ defined by

$$\varkappa_f(x) = (x, f(x))$$

for each $x \in X$. Then it is obvious that f has an extension over B iff the cross-section $\varkappa_f$ can be extended throughout B.

Similar to the classification problem for maps in (I; § 8) is the *classification problem of cross-sections*. Let K denote the set of all cross-sections $\varkappa : B \to E$. Introduce an equivalence relation $\sim$ in K as follows: For any two cross-sections $f, g \in K$, $f \sim g$ iff there exists a homotopy $h_t : B \to E$, $(0 \leqslant t \leqslant 1)$, such that $h_0 = f$, $h_1 = g$, and $h_t \in K$ for every $t \in I$. Then the classification problem of cross-sections is to enumerate the classes of K divided by this equivalence relation $\sim$. An argument similar to that used above for the extension problem shows that this classification problem of cross-sections is a generalization of that of maps in (I; § 8).

8. Fiber maps and induced fiber spaces

Let $p : E \to B$ and $p' : E' \to B'$ be any two fiberings. A map $F : E \to E'$ is said to be a *fiber map* if it carries fibers into fibers. Precisely, F is a fiber map iff, for every point b in B, there exists a point b' in B' such that F carries $p^{-1}(b)$ into $p'^{-1}(b')$.

Now let $F : E \to E'$ be a given fiber map. Then F induces a function $f : B \to B'$ defined by

$$f(b) = p'Fp^{-1}(b)$$

for every $b \in B$. For any arbitrary set U in B', we have

$$f^{-1}(U) = pF^{-1}p'^{-1}(U).$$

Hence f is continuous if p is either open or closed. If this is the case, f is called the *induced map* of the fiber map F. In particular, if E is a bundle space over B relative to p, then p is obviously open and so f is continuous. The following rectangle is commutative:

$$\begin{array}{ccc} E & \xrightarrow{F} & E' \\ \downarrow{\scriptstyle p} & & \downarrow{\scriptstyle p'} \\ B & \xrightarrow{f} & B' \end{array}$$

Some special cases of fiber maps are important. First, let us take $p' : E' \to B'$ to be the *trivial fibering over B*, that is to say, $B' = B$, $E' = B$, and p' is the identity map. In this case, the projection $p : E \to B$ is a fiber map and its induced map is the identity map on B. Second, let us take $p : E \to B$ to be the trivial fibering over a space X, that is to say, $B = X$, $E = X$, and p is the identity. In this case, every map $F : X \to E'$ is a fiber map with $f = p'F : X \to B'$ as induced map. This suggests the following extension of the lifting problem in § 7.

Let $p : E \to B$ and $p' : E' \to B'$ be two fiberings. By a *lifting* of a given map $f : B \to B'$, we mean a fiber map $F : E \to E'$ which induces f. Hence, a map $F : E \to E'$ is a lifting of $f : B \to B'$ iff $p'F = fp$. The *lifting problem* for $f : B \to B'$ is to determine whether or not f has a lifting $F : E \to E'$. As we have seen in the special case of cross-sections, the answer to this problem is not always affirmative.

However, for a given fibering $p' : E' \to B'$ and a given map $f : B \to B'$ of a given space B into B', we can construct a fibering $p : E \to B$ together with a lifting $F : E \to E'$ of f as follows.

Let E denote the subspace of $B \times E'$ given by

$$E = \{ (b, e') \in B \times E' \mid f(b) = p'(e') \}$$

and let $p : E \to B$ denote the natural projection defined by $p(b, e') = b$. Let $F : E \to E'$ denote the map defined by $F(b, e') = e'$. We are going to prove that $p : E \to B$ is a fibering and that F is a lifting of f.

By the preceding construction, we have $fp = p'F$. Hence it remains to prove that $p : E \to B$ is a fibering.

Let $\phi : X \to E$ be a given map of a triangulable space X into E and $h_t : X \to B$, $(0 \leqslant t \leqslant 1)$, be a homotopy of the map $\psi = \phi p$. Let

$$\xi = F\phi : X \to E', \quad k_t = fh_t : X \to B', (0 \leqslant t \leqslant 1).$$

Then k_t is a homotopy of the map

$$k_0 = fh_0 = fp\phi = p'F\phi = p'\xi$$

Since $p' : E' \to B'$ is a fibering, there exists a homotopy $k_t^* : X \to E'$, $(0 \leqslant t \leqslant 1)$, of ξ which covers k_t. Define a homotopy $h_t^* : X \to E$, $(0 \leqslant t \leqslant 1)$, by taking

$$h_t^*(x) = (h_t(x), k_t^*(x))$$

for every $x \in X$ and $t \in I$. This definition of h_t^* is justified by the relation $p'k_t^* = k_t = fh_t$. Since

$$h_0^*(x) = (h_0(x), k_0^*(x)) = (p\phi(x), F\phi(x)) = \phi(x)$$

for every $x \in X$, h_t^* is a homotopy of ϕ. Since h_t^* obviously covers h_t, this completes the proof that $p : E \to B$ is a fibering and F is a lifting of f.

The fibering $p : E \to B$ constructed above is said to be *induced by f*; the lifting $F : E \to E'$ of f will also be said to be *induced by f*. Note that, for each $b \in B$, F maps the fiber $p^{-1}(b)$ homeomorphically onto the fiber $p'^{-1}(b')$, where $b' = f(b)$. Finally, it is straightforward to verify that, if E' is a bundle space over B', then so is E over B.

Finally, if $\varkappa' : B' \to E'$ is a cross-section, then the map $\varkappa : B \to E$ defined by $\varkappa(b) = (b, \varkappa' f(b))$ is also a cross-section. We shall call $\varkappa$ the *induced cross-section* of $\varkappa'$ by f.

The following special case will be used in the sequel. If B is a subspace of B' and $f : B \to B'$ is the inclusion map, then E can be identified with $p'^{-1}(B)$ and p with $p' \mid B$ in an obvious way. In this case, the induced fibering $p : E \to B$ will be called the *restriction* of $p' : E' \to B'$ on B. Hence we have the following

Proposition 8.1. *If E is a fiber space over a base space B with projection $p : E \to B$ and if A is any subspace of B, then $p^{-1}(A)$ is a fiber space over A with $p \mid p^{-1}(A)$ as projection.*

9. Mapping spaces

Let X and Y be arbitrarily given spaces and denote by

$$\Omega = Y^X$$

the totality of maps of X into Y. There are various ways of topologizing Ω, but we will be concerned only with the *compact-open topology*, [Fox 3]. It is also called the *k-topology*, [Arens 2], and the *topology of compact convergence*, [B; III] and [K].

For any two sets $K \subset X$ and $W \subset Y$, let $M(K, W)$ denote the subset of Ω defined by

$$M(K, W) = \{ f \in \Omega \mid f(K) \subset W \}.$$

$M(K, W)$ will be called a *subbasic set* of Ω if K is compact and W is open. The compact-open topology of Ω is defined by selecting as a subbasis for the open sets of Ω the totality of the subbasic sets $M(K, W)$ of Ω. According to the usual definition of a subbasis, every subbasic set is open in Ω and every open set of Ω is the union of a collection of the finite intersections of subbasic sets.

Throughout the present book, mapping spaces are understood to be topologized by their compact-open topologies, unless otherwise stated.

Lemma 9.1. *If X is a Hausdorff space and $\{ U \}$ is a subbasis for the open sets of Y, then the totality of the sets $M(K, U)$, for K a compact subset of X and $U \in \{ U \}$, constitutes a subbasis for the compact-open topology of Ω* [Jackson 2].

Proof. It suffices to show that if K is a compact subset of X and W an open subset of Y, and if $f \in M(K, W)$, then there exist compact subsets $K_1, \cdots, K_m$ of X and members $U_1, \cdots, U_m$ in $\{ U \}$ such that

$$f \in M(K_1, U_1) \cap \cdots \cap M(K_m, U_m) \subset M(K, W).$$

Let $x \in K$. Since $f(x) \in W$, there are a finite number of sets in $\{ U \}$, say $U_1^x, \cdots, U_{n_x}^x$, such that

$$f(x) \in U_1^x \cap \cdots \cap U_{n_x}^x \subset W.$$

Since f is continuous, there is a neighborhood G_x of x in X such that

$$f(G_x) \subset U_1^x \cap \cdots \cap U_{n_x}^x.$$

As a compact Hausdorff space, K is regular. So there is an open neighborhood H_x of x in K such that the closure $K_x = \overline{H}_x$ is contained in G_x.

The collection $\{ H_x \mid x \in K \}$ is an open covering of the compact space K, and hence there are a finite number of points in K, say $x_1, \cdots, x_q$, such that

$$K = H_{x_1} \cup \cdots \cup H_{x_q}.$$

In the subscripts and superscripts involved above, we shall simply replace x_j by j, $j = 1, \cdots, q$.

Now the sets $K_1, \cdots, K_q$ are compact. Moreover,

$$f(K_j) \subset f(G_j) \subset U_1^j \cap \cdots \cap U_{n_j}^j \subset W, \ (j = 1, \cdots, q).$$

Hence we have

$$f \in \bigcap_{j=1}^{q} [\bigcap_{i=1}^{n_j} M(K_j, U_j^i)].$$

Suppose that $g \in \Omega$ is contained in the set on the right-hand side of the preceding formula. If $x \in K$, then x is in some H_j and hence is in K_j. Therefore

$$g(x) \in U_1^j \cap \cdots \cap U_{n_j}^j \subset W.$$

Thus $g \in M(K, W)$, and so

$$f \in \bigcap_{j=1}^{q} [\bigcap_{i=1}^{n_j} M(K_j, U_i^j)] \subset M(K, W). \ \blacksquare$$

Next, there is a natural function

$$\omega : \Omega \times X \to Y$$

defined by $\omega(f, x) = f(x)$ for each $f \in \Omega$ and each $x \in X$. ω will be called the *evaluation* of the mapping space Ω.

Proposition 9.2. *If X is a locally compact regular space, then the evaluation ω of Ω is continuous.*

Proof. Let $f \in \Omega$, $x \in X$, and an open set W of Y which contains $f(x)$ be arbitrarily given. Since f is continuous, the inverse image $f^{-1}(W)$ is an open set containing x. Since X is regular and locally compact, there exists an open neighborhood V of x such that the closure $\overline{V}$ is compact and is contained in $f^{-1}(W)$. Then $U = M(\overline{V}, W)$ is a subbasic open set of Ω which contains f. $U = M(\overline{V}, W)$ implies that ω maps $U \times V$ into W. This proves the continuity of ω. $\blacksquare$

For the necessity of the local compactness of X in (9.2), see Ex. J at the end of the chapter.

For each point $y \in Y$, let $j(y)$ denote the constant map in Ω which maps X into the single point y. The assignment $y \to j(y)$ defines a function

$$j : Y \to \Omega$$

called the *natural injection* of Y into Ω. It can be easily verified that j maps Y homeomorphically onto a subspace $j(Y)$ of Ω. Furthermore, if Y is a Hausdorff space, then $j(Y)$ is closed in Ω.

For any given point $a \in X$, let

$$p_a : \Omega \to Y$$

denote the function defined by $p_a(f) = f(a)$ for each $f \in \Omega$. Clearly, p_a is a map and sends the subspace $j(Y)$ of Ω onto Y. We shall call p_a the *projection* of Ω onto Y determined by the given point $a \in X$.

From the definitions given above, we observe that $p_a j$ is the identity map on Y and jp_a is a retraction of Ω onto its subspace $j(Y)$. Hence we have the following

Proposition 9.3. *The natural injection j maps Y homeomorphically onto a retract $j(Y)$ of Ω.*

Next, let us consider three given spaces T, X, Y and the mapping spaces

$$\Omega = Y^X, \quad \Phi = Y^{X \times T}, \quad \Psi = \Omega^T.$$

For each map $\phi : X \times T \to Y$ in Φ, define a function $\theta(\phi) : T \to \Omega$ by taking

$$[\theta(\phi)(t)](x) = \phi(x, t), \quad (t \in T, x \in X).$$

$\theta(\phi)$ is said to be *associated with* ϕ.

Proposition 9.4. *The associated function $\theta(\phi)$ of $\phi \in \Phi$ is continuous.*

Proof. Let $\psi = \theta(\phi)$ and let $U = M(K, W)$ be any subbasic open set in Ω. It suffices to prove that $\psi^{-1}(U)$ is an open set of T. Let t_0 be any given point in $\psi^{-1}(U)$. By definition, we have

$$K \times t_0 \subset \phi^{-1}(W)$$

The continuity of ϕ implies that $\phi^{-1}(W)$ is an open set of $X \times T$. Hence $\phi^{-1}(W)$ is the union of a collection of open sets of the form $G_\mu \times H_\mu$, where G_μ and H_μ are open sets of X and T respectively. Since K is compact, $K \times t_0$ is contained in the union of a finite number of these open sets, say

$$G_1 \times H_1, G_2 \times H_2, \cdots, G_n \times H_n$$

with $t_0 \in H_i$ for each $i = 1, 2, \cdots, n$. Then

$$H = H_1 \cap H_2 \cap \cdots \cap H_n$$

is an open set of T containing t_0 and is contained in $\psi^{-1}(U)$. Therefore, $\psi^{-1}(U)$ is an open set of T. ∎

As a consequence of (9.4) the assignment $\phi \to \theta(\phi)$ defines a function

$$\theta : \Phi \to \Psi;$$

when continuous, this function will be called the *association map*; that this is usually the case is shown by the following

Proposition 9.5. *If T is a Hausdorff space, then $\theta : \Phi \to \Psi$ is continuous.*

Proof. Since T is a Hausdorff space and the totality of the subbasic sets $M(K, W)$ constitutes a subbasis of Ω, it follows from (9.1) that the subsets

$$M[L, M(K, W)] = \{ \psi \in \Psi \mid \psi(L) \subset M(K, W) \}$$

form a subbasis for Ψ, where L runs through the compact subsets of T, K runs through the compact subsets of X, and W runs through the open subsets of Y. It follows clearly from the definition of θ that

$$\theta^{-1} \{ M[L, M(K, W)] \} = M(K \times L, W).$$

Since $K \times L$ is compact, $M(K \times L, W)$ is open in Φ. Since $\{ M[L, M(K, W)] \}$ is a subbasis of Ψ, it follows that θ is continuous. ∎

It is obvious from the definition that θ carries Φ into Ψ in a one-to-one fashion. In general, θ is not onto. However, we have the following

Proposition 9.6. *The evaluation ω of Ω is continuous iff, for every space T, the function $\theta : \Phi \to \Psi$ is onto.*

Proof. *Necessity.* Let $\psi \in \Psi$ and consider the map $\chi : X \times T \to \Omega \times X$ defined by

$$\chi(x, t) = (\psi(t), x), \quad (x \in X, t \in T).$$

Let $\phi = \omega\chi \in \Phi$. Since

$$[\theta(\phi)(t)](x) = \phi(x, t) = \omega\chi(x, t) = \omega[\psi(t), x] = [\psi(t)](x)$$

for every $t \in T$ and $x \in X$, we have $\theta(\phi) = \psi$. Hence θ is onto.

Sufficiency. Assume that the condition holds. In particular, select $T = \Omega$ and take $\psi \in \Psi$ to be the identity map on Ω. Then there is a map $\phi \in \Phi$ with $\theta(\phi) = \psi$. Since

$$\phi(x, f) = [\psi(f)](x) = f(x) = \omega(f, x)$$

for every $x \in X$ and $f \in \Omega$, the continuity of ϕ implies that of ω. ∎

As an immediate consequence of (9.2) and (9.6), we have the following

Corollary 9.7. *If X is a locally compact regular space, then the function $\theta : \Phi \to \Psi$ sends Φ onto Ψ in a one-to-one fashion.*

Proposition 9.8. *If X and T are Hausdorff spaces, then the association map $\theta : \Phi \to \Psi$ is a homeomorphism of Φ onto a subspace of Ψ* [Jackson 2].

Proof. Since θ is one-to-one and continuous by (9.5), it remains to prove that θ^{-1} is continuous on $\theta(\Phi) \subset \Psi$. For this purpose, it suffices to prove that, if J is a compact subspace of $X \times T$ and W is an open set in Y, then the image $\theta[M(J, W)]$ is an open set of $\theta(\Phi)$.

Let $\psi \in \theta[M(J, W)]$ be arbitrarily given. Choose a $\phi \in M(J, W)$ with $\theta(\phi) = \psi$. Let J_X and J_T denote the projections of J in X and T respectively. For each point $z = (x, t) \in J$, choose an open neighborhood U_z of x in J_X and an open neighborhood V_z of t in J_T, such that

$$\phi(U_z \times V_z) \subset W.$$

As compact Hausdorff spaces, J_X and J_T are regular. Hence we may shrink U_z and V_z a little bit so that

$$\phi(K_z \times L_z) \subset W,$$

where K_z denotes the closure of U_z in J_X and L_z that of V_z in J_T.

The collection $\{ (U_z \times V_z) \cap J \mid z \in J \}$ is an open covering of the compact space J. Hence there is a finite number of points in J, say $z_1, \cdots, z_n$, such that

$$J \subset (U_{z_1} \times V_{z_1}) \cup \cdots \cup (U_{z_n} \times V_{z_n}).$$

For the subscripts in the notations of the various sets involved above, we shall simply replace z_i by i, $i = 1, \cdots, n$.

Now the sets K_i and L_i, $i = 1, \cdots, n$, are compact. Moreover,

$$[\psi(L_i)]\,(K_i) = \phi(K_i \times L_i) \subset W$$

for each $i = 1, \cdots, n$. Hence

$$\psi \in \theta(\Phi) \cap \{ \bigcap_{i=1}^{n} M[L_i, M(K_i, W)] \}.$$

Suppose that $\chi \in \Psi$ is contained in the set on the right-hand side of the preceding formula. Since $\chi \in \theta(\Phi)$, there is a $\xi \in \Phi$ with $\chi = \theta(\xi)$. If $z = (x, t) \in J$, then z is contained in some $U_i \times V_i$ and hence in $K_i \times L_i$. Since χ is in $M[L_i, M(K_i, W)]$ and since $x \in K_i$, $t \in L_i$, we have

$$\xi(z) = \xi(x, t) = [\chi(t)]\,(x) \in W.$$

This proves that $\xi(J) \subset W$. Therefore, $\xi \in M(J, W)$ and hence $\chi = \theta(\xi) \in \theta[M(J, W)]$. Thus, we obtain

$$\psi \in \theta(\Phi) \cap \{ \bigcap_{i=1}^{n} M[L_i, M(K_i, W)] \} \subset \theta[M(J, W)].$$

This proves that $\theta[M(J, W)]$ is an open set of $\theta(\Phi)$. ∎

As an immediate consequence of (9.7) and (9.8), we obtain the following

Theorem 9.9. *Let X and T be Hausdorff spaces and Y any space. If X is locally compact, then the association map θ is a homeomorphism of the mapping space $\Phi = Y^{X \times T}$ onto the mapping space $\Psi = (Y^X)^T$.*

Hereafter, when the assumptions of (9.9) are satisfied, the two mapping spaces will be identified by means of the homeomorphism θ. In symbols, we have

$$Y^{X \times T} = (Y^X)^T.$$

This will be called *the exponential law* of mapping spaces.

Now let us go back to (9.3) and look for a sufficient condition that $j(Y)$ be a strong deformation retract of Ω. Such a condition is given by the following

Proposition 9.10. *If X is a locally compact regular space and contractible to a point $a \in X$, then $j(Y)$ is a strong deformation retract of Ω.*

Proof. By (9.2) the evaluation $\omega : \Omega \times X \to Y$ is continuous. Since X is contractible to the point $a \in X$, there exists a map

$$h : X \times I \to X$$

such that
$$h(x, 0) = x, h(x, 1) = a, \quad (x \in X);$$
$$h(a, t) = a, \quad (t \in I).$$

Define a map $\phi : X \times \Omega \times I \to Y$ by taking

$$\phi(x, f, t) = \omega[f, h(x, t)], \quad (x \in X, f \in \Omega, t \in I).$$

According to (9.4), the associated function

$$\psi = \theta(\phi) : \Omega \times I \to \Omega$$

is continuous. Define a homotopy $\chi_t : \Omega \to \Omega$, $(0 \leqslant t \leqslant 1)$, by setting

$$\chi_t(f) = \psi(f, t) \quad (f \in \Omega, t \in I).$$

Then it is easily verified that χ_0 is the identity map, $\chi_1 = jp_a$, and $\chi_t(f) = f$ for every $f \in j(Y)$ and $t \in I$. ∎

Since it is easily verified that

$$\chi_t[p_a^{-1}(y)] \subset p_a^{-1}(y), \quad (y \in Y, t \in I),$$

we have proved the following

Proposition 9.11. *If X is a locally compact regular space and contractible to a point $a \in X$, then, for each $y \in Y$, the subspace $p_a^{-1}(y)$ is contractible to the point $j(y)$.*

10. The spaces of paths

Let Y denote a given space. By a *path* in Y we mean a map

$$\sigma : I \to Y$$

of the unit interval $I = [0, 1]$ into Y. The points $\sigma(0)$ and $\sigma(1)$ are called respectively the *initial point* and the *terminal point* of the path σ, and are said to be *connected by the path* σ.

The relation that two points a and b can be connected by a path in Y is obviously symmetric, reflexive and transitive; hence the points of Y are divided into disjoint classes, called the *path-components* of Y. We shall denote by C_y the path-component of Y which contains the point $y \in Y$. Y is said to be *pathwise connected* if it consists of a single path-component and hence every pair of points in Y can be connected by a path.

The totality of paths in Y forms a space

$$\Omega = Y^I$$

with the compact-open topology defined in the previous section. This space Ω together with certain of its subspaces is of fundamental importance in homotopy theory as well as in the functional topology of Morse, [M_1 and M_2].

By a *generalized triad* $(Y; A, B)$, we mean a space Y together with two subspaces A and B. $(Y; A, B)$ is said to be a *triad* if the intersection $A \cap B$ is non-empty. For a given generalized triad $(Y; A, B)$, let us denote by

$$[Y; A, B]$$

the subspace of Ω which consists of the paths σ in Y such that $\sigma(0) \in A$ and $\sigma(1) \in B$. The following particular cases are of importance.

If $A = Y$ and B consists of a single point $y \in Y$, then we shall denote the subspace $[Y; Y, y]$ of Ω simply by Ω_y. It is called the *space of paths in Y with a given terminal point y*.

A *loop* in Y is a path $\sigma : I \to Y$ such that $\sigma(0) = \sigma(1)$. The point $\sigma(0) = \sigma(1)$ will be called the *basic point* of the loop σ. The set of all loops in Y forms a subspace Λ of Ω which will be called the *space of loops* in Y.

If both A and B of a given generalized triad $(Y; A, B)$ consist of the same single point $y \in Y$, then we shall denote the subspace $[Y; y, y]$ of Ω simply by Λ_y. It is obviously a subspace of Λ and will be called the *space of loops in Y with a given basic point y*. Clearly we have

$$\Lambda_y = \Lambda \cap \Omega_y.$$

The projections $p_0, p_1 : \Omega \to Y$ determined by the points 0, 1 as in the preceding section shall be called the *initial projection* and the *terminal projection* respectively. If Y is a Hausdorff space, then the continuity of p_0 and p_1 implies that the subspaces $\Lambda, \Lambda_y, \Omega_y$ of Ω are all closed.

Let δ_y denote the degenerate loop $\delta_y(I) = y$. By (9.11), we have the following

Proposition 10.1. *Ω_y is contractible to the point δ_y.*

Intuitively, a contraction of Ω_y to the point δ_y is obtained by pushing all paths of Ω_y along themselves simultaneously to the terminal point y.

11. The space of loops

Let Y denote a given space and $y \in Y$ a given point. The space Λ_y of loops in Y with y as basic point has a central role in the study of homotopy groups; therefore, we shall study its important properties in the present section.

There is a natural multiplication defined in Λ_y as follows. For any pair of loops $f, g \in \Lambda_y$, the *product* $fg \in \Lambda_y$ is the loop in Y defined by

$$(fg)(t) = \begin{cases} f(2t), & (0 \leqslant t \leqslant \frac{1}{2}), \\ g(2t-1), & (\frac{1}{2} \leqslant t \leqslant 1). \end{cases}$$

Intuitively speaking, fg is the loop in Y which travels along the loop f with double speed while t is in the first half of I; then fg travels along the loop g with double speed while t is in the second half of I.

The correspondence $(f, g) \to \mu(f, g) = fg$ defines a function

$$\mu : \Lambda_y \times \Lambda_y \to \Lambda_y$$

called the *multiplication function* in Λ_y.

Proposition 11.1. *The natural multiplication in Λ_y is continuous; by this we mean that the multiplication function μ is continuous.*

Proof. Let $U = M(K, W) \cap \Lambda_y$ be an arbitrary non-empty subbasic open set of Λ_y. It suffices to show that the inverse image $\mu^{-1}(U)$ is an open set of $\Lambda_y \times \Lambda_y$. Let $\phi, \psi : I \to I$ denote the maps defined by

$$\phi(t) = \tfrac{1}{2}t, \quad \psi(t) = \tfrac{1}{2}(t + 1)$$

for each $t \in I$. Let $A = \phi^{-1}(K)$ and $B = \psi^{-1}(K)$; then A and B are compact subspaces of I. Consider the subbasic open sets

$$F = M(A, W) \cap \Lambda_y, \quad G = M(B, W) \cap \Lambda_y$$

of Λ_y. For any pair of loops $f, g \in \Lambda_y$, it is easy to see that $fg \in M(K, W)$ iff $f \in M(A, W)$ and $g \in M(B, W)$. Hence $\mu^{-1}(U) = F \times G$. This proves that $\mu^{-1}(U)$ is open. ∎

Corollary 11.2. *Every loop $f \in \Lambda_y$ determines two maps*

$$L_f, R_f : \Lambda_y \to \Lambda_y$$

defined by $L_f(g) = fg$ and $R_f(g) = gf$ for every $g \in \Lambda_y$.

Proposition 11.3. *If $\delta \in \Lambda_y$ denotes the degenerate loop $\delta(I) = y$, then L_δ and R_δ are deformations of Λ_y. In fact, there exists a homotopy $h_t : \Lambda_y \to \Lambda_y$, $(0 \leqslant t \leqslant 1)$, such that $h_0 = L_\delta$, h_1 is the identity map, and $h_t(\delta) = \delta$ for each $t \in I$; and analogously for R_δ.*

Proof. According to (9.2), the evaluation $\omega : \Lambda_y \times I \to Y$ is continuous. Hence we may define a map

$$\phi : I \times \Lambda_y \times I \to Y$$

by taking

$$\phi(s, f, t) = \begin{cases} f\left(\dfrac{2s}{1 + t}\right), & \left(f \in \Lambda_y, t \in I, 0 \leqslant s \leqslant \dfrac{1 + t}{2}\right), \\ y, & \left(f \in \Lambda_y, t \in I, \dfrac{1 + t}{2} \leqslant s \leqslant 1\right). \end{cases}$$

By (9.4), the associated function $\psi = \theta(\phi)$ is continuous. It is easily seen that ψ maps $\Lambda_y \times I$ into the subspace Λ_y of Ω and, by definition, ψ is given by

$$[\psi(f, t)](s) = \phi(s, f, t), \quad (f \in \Lambda_y, t \in I, s \in I).$$

Hence we may define a homotopy $k_t : \Lambda_y \to \Lambda_y$, $(0 \leqslant t \leqslant 1)$, by taking

$$k_t(f) = \psi(f, t), \quad (f \in \Lambda_y, 0 \leqslant t \leqslant 1).$$

Then it is easy to verify that $k_0 = R_\delta$, k_1 is the identity map, and $k_t(\delta) = \delta$ for each $t \in I$. This implies that R_δ is a deformation of Λ_y. Similarly, we can prove that L_δ is a deformation of Λ_y. ∎

These properties of Λ_y show that it belongs to the class of H-spaces, which are defined as follows.

By a *continuous multiplication* in a space X, we mean a map $\mu : X \times X \to X$. Let μ be a given continuous multiplication in X. For any pair of points a and b of X, $\mu(a, b)$ is customarily denoted by ab and is called the *product* of a and b. For any given point $a \in X$, the correspondences $x \to ax$ and $x \to xa$ determine respectively the maps

$$L_a : X \to X, \quad R_a : X \to X$$

called *the left* and *the right translation* of X by a. A point $a \in X$ is said to be a *left homotopy unit* if a is an idempotent, that is to say, $aa = a$, and if L_a is homotopic with the identity map relative to a. Similarly, an idempotent $a \in X$ is said to be a *right homotopy unit* if R_a is homotopic with the identity map relative to a. An idempotent $a \in X$ is called *a homotopy unit* if it is both left and right homotopy unit. By an *H-space*, we mean a space X with a given continuous multiplication μ which has a homotopy unit.

According to the propositions (11.1) and (11.3), *Λ_y is an H-space under its natural multiplication with the degenerate loop δ as a homotopy unit.* As another important example, any topological group is obviously an H-space under its group multiplication with the neutral element as a homotopy unit.

An important property of H-spaces which will be used in the sequel is expressed by the following

Proposition 11.4. *If X is an H-space with x_0 as a homotopy unit, then the fundamental group $\pi_1(X, x_0)$ is abelian.*

Proof. Let α, β be any two elements of $\pi_1(X, x_0)$ and let $f, g : I \to X$ be representative loops of α, β respectively. Since x_0 is an idempotent, we may define a loop $h : I \to X$ at x_0 by taking

$$h(t) = f(t)g(t)$$

for each $t \in I$. This loop h represents an element γ of $\pi_1(X, x_0)$ which obviously depends only on α and β. It suffices to prove that $\gamma = \alpha\beta$ and $\gamma = \beta\alpha$.

To prove $\gamma = \alpha\beta$, we may assume that the representive loops f, g have been so chosen that $f(t) = x_0$ for each $t \geqslant \frac{1}{2}$ and $g(t) = x_0$ for each $t \leqslant \frac{1}{2}$. It follows that

$$h(t) = \begin{cases} f(t) \cdot x_0, & (\text{if } 0 \leqslant t \leqslant \frac{1}{2}), \\ x_0 \cdot g(t), & (\text{if } \frac{1}{2} \leqslant t \leqslant 1). \end{cases}$$

Since x_0 is a homotopy unit of X, there exist two homotopies $\phi_s, \psi_s : X \to X$, $(0 \leqslant s \leqslant 1)$, such that $\phi_0(x) = x_0 x, \phi_1(x) = x, \psi_0(x) = x x_0, \psi_1(x) = x$ for each $x \in X$ and $\phi_s(x_0) = x_0 = \psi_s(x_0)$ for each $s \in I$. So, we may define a homotopy $h_s : I \to X$, $(0 \leqslant s \leqslant 1)$, by taking

$$h_s(t) = \begin{cases} \psi_s f(t), & (\text{if } 0 \leqslant t \leqslant \frac{1}{2}), \\ \phi_s g(t) & (\text{if } \frac{1}{2} \leqslant t \leqslant 1). \end{cases}$$

Then we have $h_0 = h$ and $h_s(0) = x_0 = h_s(1)$ for each $s \in I$. Since h_1 satisfies the relation

$$h_1(t) = \begin{cases} f(t), & (\text{if } 0 \leqslant t \leqslant \frac{1}{2}), \\ g(t) & (\text{if } \frac{1}{2} \leqslant t \leqslant 1), \end{cases}$$

it obviously represents the element $\alpha\beta$. This implies that $\gamma = \alpha\beta$.

Similarly, we can prove $\gamma = \beta\alpha$ by choosing f and g such that $f(t) = x_0$ for each $t \leqslant \frac{1}{2}$ and $g(t) = x_0$ for each $t \geqslant \frac{1}{2}$. ∎

In the preceding proof, we have also proved the following

Corollary 11.5. *Under the hypothesis of (11.4), if two elements $\alpha, \beta \in \pi_1(X, x_0)$ are represented by the loops $f, g : I \to X$ at x_0, then the element $\alpha\beta$ is represented by the functional product $h : I \to X$ of f, g defined by*

$$h(t) = f(t)\, g(t), \quad (t \in I).$$

As we shall see later, this assertion also holds for the higher homotopy groups.

Now let us come back to the space Λ_y of loops and consider its path-components. By (9.9), one can easily see that the path-components of Λ_y are exactly the equivalence classes of the loops of Λ_y as defined in (II, § 4). Hence we have the following

Proposition 11.6. *Under the natural multiplication of Λ_y, the path-components of Λ_y form a group which is essentially the fundamental group $\pi_1(Y, y)$.*

12. The path lifting property

In the present section, we are going to define, for any given map $p : E \to B$ the *path lifting property* (abbreviated PLP) which was recently introduced by Hurewicz and Curtis, and to prove that it is equivalent to the absolute covering homotopy property (abbreviated ACHP) defined in § 2.

Roughly speaking, a map $p : E \to B$ is said to have the PLP if, for each $e \in E$ and each path $f : I \to B$ with $f(0) = p(e)$, there exists a path $g : I \to E$ such that $g(0) = e$, $pg = f$, and that g depends continuously on e and f. For a precise definition, let Z denote the subspace of the product space $E \times B^I$ defined by $\quad Z = \{ (e, f) \in E \times B^I \mid p(e) = f(0) \}$.

Define a map

$$q : E^I \to Z$$

by taking $q(g) = (g(0), pg)$ for each $g : I \to E$ in E^I. Then $p : E \to B$ is said to have the PLP if there exists a map

$$r : Z \to E^I$$

such that qr is the identity map on Z. In this case, r is clearly a homeomorphism of Z onto a retract of E^I.

Proposition 12.1. *A map $p : E \to B$ has the PLP iff it has the ACHP.*

Proof. PLP $\Rightarrow$ ACHP. Let $g : X \to E$ be any given map of a space X into E and $f_t : X \to B$, $(0 \leqslant t \leqslant 1)$, a given homotopy of the map $f = pg$. According to (9.4), the homotopy f_t gives a map $h : X \to B^I$ defined by

$$[h(x)]\,(t) = f_t(x), \quad (x \in X, t \in I).$$

Let $k : X \to Z$ denote the map defined by

$$k(x) = (g(x), h(x)), \quad (x \in X).$$

Then the composition rk is a map of X into E^I. According to (9.9), we may define a homotopy $g_t : X \to E$, $(0 \leqslant t \leqslant 1)$, by taking

$$g_t(x) = [rk(x)](t), \quad (x \in X, t \in I).$$

One can easily verify that $g_0 = g$ and $pg_t = f_t$ for each $t \in I$. This implies ACHP.

ACHP $\Rightarrow$ PLP. Let $\eta : Z \to E$ denote the natural projection defined by $\eta(e, f) = e$ for each $(e, f) \in Z$. Define a homotopy $\xi_t : Z \to B$, $(0 \leqslant t \leqslant 1)$, by taking

$$\xi_t(e, f) = f(t), \quad ((e, f) \in Z, t \in I).$$

Then we have $\xi_0 = p\eta$. According to ACHP, there exists a homotopy $\eta_t : Z \to E$, $(0 \leqslant t \leqslant 1)$, such that $\eta_0 = \eta$ and $p\eta_t = \xi_t$ for each $t \in I$. According to (9.4), η_t gives a map $r : Z \to E^I$ defined by

$$[r(z)](t) = \eta_t(z), \quad (z \in Z, t \in I).$$

Then it is easily verified that $qr(z) = z$ for each $z \in Z$. Hence, $p : E \to B$ has the PLP. ∎

Corollary 12.2. *If a map $p : E \to B$ has the PLP, then E is a fiber space over B relative to p.*

13. The fibering theorem for mapping spaces

Let X be a locally compact ANR and A be a closed subspace of X which is also an ANR. On the other hand, let Y be an arbitrary space.

Let us consider the mapping spaces

$$E = Y^X, \quad B = Y^A.$$

There is a *natural map*

$$p : E \to B$$

defined by $p(f) = f \mid A$ for every map $f : X \to Y$. We are going to prove that E is a fiber space over B with p as projection; in fact, we have the following stronger result.

Theorem 13.1. *The map $p : E \to B$ has the PLP.*

Proof. Let us consider the subspace Z of $E \times B^I$ and the map $q : E^I \to Z$ described in the preceding section and try to construct a map $r : Z \to E^I$ such that qr is the identity map.

Consider the closed subspace $T = (X \times 0) \cup (A \times I)$ of $X \times I$. By (I; Ex. O), T is a retract of $X \times I$. Let $\rho : X \times I \supset T$ denote a retraction.

Consider the mapping space Y^T. For each $g : T \to Y$ in Y^T, define two maps $e : X \to Y$ and $f : I \to Y^A$ as follows:

$$e(x) = g(x, 0), \quad (x \in X);$$

$$[f(t)](a) = g(a, t), \quad (t \in I, a \in A).$$

The continuity of f follows from (9.4). Hence, the assignment $g \to \theta(g) = (e, f)$ determines a function

$$\theta : Y^T \to Z.$$

By the methods used in establishing the exponential law of mapping spaces in § 9, one can prove that θ is a homeomorphism of Y^T onto Z.

Using the inverse of θ, we can define a map $r : Z \to E^I$ by taking

$$\{ [r(z)](t) \} (x) = [\theta^{-1}(z)]\rho(x, t)$$

for every $t \in I$ and $x \in X$. Since ρ is a retraction, it follows immediately that $qr(z) = z$ for each z in Z. ∎

Note. The condition that X is locally compact is inessential. In fact, the theorem holds without this condition; for (9.7) may be replaced by Ex. L at the end of the chapter.

Now let $\{ A_\mu \mid \mu \in M \}$ and $\{ Y_\mu \mid \mu \in M \}$ be given families of subspaces of A and Y respectively, both indexed by the elements of a set M. Consider the subspace B' of B which consists of the maps $g \in B$ such that $g(A_\mu) \subset Y_\mu$ for each $\mu \in M$. Let $E' = p^{-1}(B')$. Then E' is the subspace of E consisting of the maps $f : X \to Y$ such that $f(A_\mu) \subset Y_\mu$ for every $\mu \in M$. By (8.1) and (12.2), we have the following

Corollary 13.2. *The map $p' : E' \to B'$ is a fibering.*

In fact, it is also obvious that $p' : E' \to B'$ has the PLP. So, this generalizes (13.1).

For important examples, let us take X to be the unit interval I. First, take A to be the single point 1 of I. In this case, the space E is the space Ω of all paths in Y, and the space B can be identified with Y in an obvious way. Furthermore, the natural projection is essentially the terminal projection $p_1 : \Omega \to Y$. Hence $p_1 : \Omega \to Y$ has the PLP by (13.1). Let Z be a subspace of Y, then

$$p_1^{-1}(Z) = [Y\,;\, Y, Z].$$

Therefore, by (13.2), we have the following

Corollary 13.3. *For any subspace Z of a given space Y, the space $[Y; Y, Z]$ is a fiber space over Z relative to the terminal projection. Similarly, the space $[Y; Z, Y]$ is a fiber space over Z relative to the initial projection.*

Next, let $X = I$ and take A to be the subspace of I which consists the boundary points 0 and 1 of I. Let M be an index set containing a single element μ. Define A_μ and Y_μ to be the single points $1 \in A$ and $y \in Y$, where y is a given point. In this case, the space E' in (13.2) becomes the space Ω_y of all paths in Y with y as terminal point, and the space B' can be identified with the path-component C_y of Y in an obvious way. Furthermore, the natural projection p' is essentially the initial projection p_0. Hence, we have the following important

Corollary 13.4. *The space Ω_y is a fiber space over Y relative to the initial projection. The fiber over the given point $y \in Y$ is the space Λ_y of all loops in Y with y as basic point.*

Let Z be a subspace of Y. Then the following corollary is an immediate consequence of (13.4) and (8.1).

Corollary 13.5. *For every subspace Z of a space Y and any given point y in Y, the space $[Y; Z, y]$ is a fiber space over Z relative to the initial projection.*

14. The induced maps in mapping spaces

Let X, Y, Z be arbitrarily given spaces and $\phi : Y \to Z$ a given map. For each $f \in Y^X$, the composition ϕf is in Z^X. The assignment $f \to \phi f$ defines a function

$$\phi^X : Y^X \to Z^X$$

which is obviously continuous and will be called the *induced map* of ϕ on Y^X into Z^X. Let C denote the category composed of all spaces and all maps, [E–S; p. 110]. The operation $Y \to Y^X$ and $\phi \to \phi^X$, where X is a given space, defines a covariant functor of C into itself, [E–S; p. 111].

Furthermore, if we consider the natural injection j and the projection p_a determined by a given point $a \in X$, the commutativity relations

$$j\phi = \phi^X j, \quad \phi p_a = p_a \phi^X$$

obviously hold in the following rectangles of maps:

Throughout the remainder of the present section, we are concerned with the important special case that $X = I$.

Consider a given map $\phi : Y \to Z$. Let y be a given point of Y and denote $z = \phi(y) \in Z$. As above, we shall use the following notation:

$$\Omega_y = [Y; Y, y], \quad \Lambda_y = [Y; y, y];$$
$$\Omega_z = [Z; Z, z], \quad \Lambda_z = [Z; z, z].$$

Then the induced map ϕ^I obviously sends Ω_y into Ω_z and Λ_y into Λ_z. More generally, ϕ^I maps $[Y; A, B]$ into $[Z; \phi(A), \phi(B)]$ for any given subspaces A and B of Y.

According to § 11, there are natural multiplications defined both in Λ_y and in Λ_z. It follows from the definition that

(14.1) $$\phi^I(fg) = \phi^I(f)\phi^I(g)$$

for any $f \in \Lambda_y$ and $g \in \Lambda_y$. Since ϕ^I is a map, it sends the path-components of Λ_y into those of Λ_z. Hence, (14.1) shows that ϕ^I induces a homomorphism of the group of path-components of Λ_y into that of Λ_z which is the induced homomorphism ϕ_* of $\pi_1(Y, y)$ into $\pi_1(Z, z)$.

Now let Y, Z be given spaces, $A \subset Y$, $B \subset Z$ given subspaces, and $y_0 \in A$, $z_0 \in B$ given points. Let

$$U = [Y; Y, y_0], \quad C = [Y; A, y_0],$$
$$V = [Z; Z, z_0], \quad D = [Z; B, z_0].$$

and denote by $u_0 \in C$ and $v_0 \in D$ the degenerate loops $u_0(I) = y_0$ and $v_0(I) = z_0$ respectively. Let

$$p : U \to Y, \quad q : V \to Z$$

denote the initial projections. We are going to establish a *covering map theorem* which will be important in the next chapter.

Theorem 14.2. *If $\psi : U \to Z$ is a map such that $\psi(C) \subset B$ and $\psi(u_0) = z_0$, then there exists a map $\psi^* : U \to V$ such that $q\psi^* = \psi, \psi^*(C) \subset D$, and $\psi^*(u_0) = v_0$.*

Proof. According to (9.2), the evaluation $\omega: U \times I \to Y$ is continuous. Define a map $\sigma : I \times I \times U \to Y$ by setting

$$\sigma(s, t, f) = \omega(f, s + t - st), \quad (s \in I, t \in I, f \in U).$$

By (9.3), the associated function $\chi : I \times U \to U$ defined by

$$[\chi(t, f)](s) = \sigma(s, t, f), \quad (t \in I, f \in U, s \in I),$$

is continuous. Let $\xi = \psi\chi : I \times U \to Z$. By (9.3), the associated function $\psi^* : U \to V$ defined by

$$[\psi^*(f)](t) = \xi(t, f), \quad (t \in I, f \in U),$$

is continuous. It remains to verify the relations $q\psi^* = \psi$, $\psi^*(C) \subset D$ and $\psi^*(u_0) = v_0$.

From the definition of χ, one can easily see that $\chi(0, f) = f$ for every $f \in U$. Hence we have

$$q\psi^*(f) = [\psi^*(f)](0) = \psi\chi(0, f) = \psi(f)$$

for every $f \in U$. This proves that $q\psi^* = \psi$. $\psi^*(C) \subset D$ is an immediate consequence of $q\psi^* = \psi$ and $\psi(C) \subset D$. To check $\psi^*(u_0)$, $= v_0$ we first note that $\chi(t, u_0) = u_0$ for every $t \in I$. Then we have

$$[\psi^*(u_0)](t) = \xi(t, u_0) = \psi\chi(t, u_0) = \psi(u_0) = z_0.$$

for every $t \in I$. This implies $\psi^*(u_0) = v_0$. ∎

If we removed the condition $\psi^*(u_0) = v_0$ from the conclusion, then (14.2) would be easier to prove. By (13.1), the map $q : V \to Z$ has the PLP and hence also the ACHP. Since U is contractible, an application of the CHP for U gives a lifting $\psi^* : U \to V$ of the map $\psi : U \to Z$. That $\psi^*(C) \subset D$ is obvious but the condition $\psi^*(u_0) = v_0$ does not necessarily hold. The lifting ψ^* constructed in the proof of (14.2) may be called the *canonical lifting* of ψ.

15. Fiberings with discrete fibers

Motivated by the results concerning the exponential map $p : R \to S^1$ in Chapter II, we are going to establish similar results for fiber spaces *with discrete fibers*, i.e. in which all fibers are discrete. Let E be a given fiber space over B relative to $p : E \to B$ with discrete fibers.

Lemma 15.1. *For each path $\sigma : I \to B$ joining b_0 to b_1 and for each $e_0 \in p^{-1}(b_0)$, there exists one and only one* (covering) *path $\sigma^* : I \to E$ such that $\sigma^*(0) = e_0$ and $p\sigma^* = \sigma$.*

Proof. The path σ may be considered as a homotopy of the partial map $\sigma \mid 0$; hence, the existence of σ^* is an immediate consequence of the covering homotopy property.

To prove the uniqueness, let $\sigma^*, \sigma^\# : I \to E$ be any two paths in E such that $p\sigma^* = \sigma = p\sigma^\#$ and $\sigma^*(0) = e_0 = \sigma^\#(0)$. Let $s \in I$ be arbitrarily given. It remains to show that $\sigma^*(s) = \sigma^\#(s)$. For this purpose, let us define a map $g : I \to E$ by taking

$$g(t) = \begin{cases} \sigma^*(s - 2st), & (\text{if } 0 \leqslant t \leqslant \frac{1}{2}), \\ \sigma^\#(2st - s), & (\text{if } \frac{1}{2} \leqslant t \leqslant 1). \end{cases}$$

Then the map $f = pg : I \to B$ has a homotopy $f_r : I \to B$, $(0 \leqslant r \leqslant 1)$, defined by

$$f_r(t) = \begin{cases} \sigma(s - 2st + 2rst), & (\text{if } 0 \leqslant t \leqslant \frac{1}{2}), \\ \sigma(2rs + 2st - s - 2rst), & (\text{if } \frac{1}{2} \leqslant t \leqslant 1). \end{cases}$$

Since $f_r(0) = \sigma(s) = f_r(1)$ for every $r \in I$, it follows from (iv) of (3.1) that g has a homotopy $g_r : I \to E$, $(0 \leqslant r \leqslant 1)$, such that

$$pg_r = f_r, \quad g_r(0) = \sigma^*(s), \quad g_r(1) = \sigma^\#(s)$$

for every $r \in I$. Since $f_1(I) = \sigma(s)$, $pg_1 = f_1$ implies that the connected set $g_1(I)$ is contained in the fiber over $\sigma(s)$ which is discrete. Therefore, $g_1(I)$ must be a single point. This implies that $\sigma^*(s) = \sigma^\#(s)$. ∎

We recall that two paths $\sigma, \tau : I \to B$ joining b_0 to b_1 are equivalent, $\sigma \sim \tau$, if they are homotopic with the end points held fixed. The paths in B joining b_0 to b_1 are thus divided into disjoint equivalence classes which are actually the path-components of the space $[B ; b_0, b_1]$.

Lemma 15.2. *The terminal point $\sigma^*(1)$ of the covering path $\sigma^* : I \to E$ in (15.1) depends only on $e_0 \in p^{-1}(b_0)$ and the equivalence class of the path $\sigma : I \to B$.*

Proof. Assume that $\sigma, \tau : I \to B$ are equivalent paths joining b_0 to b_1 and that $\sigma^*, \tau^* : I \to E$ are the covering paths with common initial point $e_0 \in p^{-1}(b_0)$. It suffices to prove that $\sigma^*(1) = \tau^*(1)$.

Since $\sigma \sim \tau$, there exists a homotopy $h_t : I \to B$, $(0 \leqslant t \leqslant 1)$, such that $h_0 = \sigma$, $h_1 = \tau$ and $h_t(0) = b_0$, $h_t(1) = b_1$ for each $t \in I$. According to (iv) of (3.1), there exists a homotopy $h_t^* : I \to E$, $(0 \leqslant t \leqslant 1)$, such that $h_0^* = \sigma^*$, $ph_t^* = h_t$, $h_t^*(0) = e_0$, and $h_t^*(1) = \sigma^*(1)$ for every $t \in I$. Since $h_1^*(0) = e_0$ and $ph_1^* = h_1 = \tau$, it follows from the uniqueness part of (15.1) that $h_1^* = \tau^*$. Hence we have

$$\tau^*(1) = h_1^*(1) = \sigma^*(1). \; ∎$$

Lemma 15.3. *For each $b_0 \in B$ and each $e_0 \in p^{-1}(b_0)$, the projection $p : (E, e_0) \to (B, b_0)$ induces a monomorphism*

$$p_* : \pi_1(E, e_0) \to \pi_1(B, b_0).$$

Proof. Assume that $\alpha \in \pi_1(E, e_0)$ and $p_*(\alpha) = 1$. Let $g : I \to E$ be a representative loop of α. Since the loop $f = pg : I \to B$ represents the element $p_*(\alpha) = 1$, there exists a homotopy $f_t : I \to B$, $(0 \leqslant t \leqslant 1)$, such that $f_0 = f$, $f_1(I) = b_0$ and $f_t(0) = b_0 = f_t(1)$ for each $t \in I$. According to (iv) of (3.1), there exists a homotopy $g_t : I \to E$, $(0 \leqslant t \leqslant 1)$, such that $g_0 = g$, $pg_t = f_t$

and $g_t(0) = e_0 = g_t(1)$ for each $t \in I$. Since $f_1(I) = b_0$ and $pg_1 = f_1$, the connected set $g_1(I)$ is contained in the fiber $p^{-1}(b_0)$ which is discrete. Hence, $g_1(I)$ must be a single point. This implies that $g_1(I) = e_0$ and $\alpha = 1$. ∎

Therefore, $\pi_1(E, e_0)$ is isomorphic to a subgroup $p_*[\pi_1(E, e_0)]$ of $\pi_1(B, b_0)$ which obviously depends on the choice of e_0. In general, there are no relations among these subgroups $p_*[\pi_1(E, e_0)]$ for various choices of $e_0 \in p^{-1}(b_0)$ unless they can be joined by paths in E. Hence, let us assume that E is pathwise connected. It follows, of course, that B must also be pathwise connected.

Now let e_1 be another point in $p^{-1}(b_0)$. Since E is pathwise connected, there exists a path $\sigma : I \to E$ joining e_0 to e_1. According to (II; 4.1), σ determines an isomorphism

$$\sigma_* : \pi_1(E, e_1) \approx \pi_1(E, e_0).$$

On the other hand, $p\sigma : I \to B$ is a loop at b_0 and hence represents an element w of $\pi_1(B, b_0)$. By the definitions of p_* and σ_*, one can easily see that

$$p_*\sigma_*(\alpha) = w \cdot p_*(\alpha) \cdot w^{-1}$$

for each $\alpha \in \pi_1(E, e_1)$. This implies that $p_*[\pi_1(E, e_1)]$ is the transform $w^{-1} \cdot p_*[\pi_1(E, e_0)] \cdot w$ of $p_*[\pi_1(E, e_0)]$.

Conversely, let w be an arbitrary element of $\pi_1(B, b_0)$. Pick a representative loop $\tau : I \to B$ for w. By (15.1) there exists a path $\sigma : I \to E$ such that $\sigma(0) = e_0$ and $p\sigma = \tau$. By (15.2), the point $e_1 = \sigma(1)$ in $p^{-1}(b_0)$ depends only on the element w. One can also easily see that $e_1 = e_0$ iff w is in the subgroup $p_*[\pi_1(E, e_0)]$. Hence, we have proved the following

Theorem 15.4. *If E is a pathwise connected fiber space over B relative to $p : E \to B$ with discrete fibers, then, for each $b_0 \in B$, the images $p_*[\pi_1(E, e_0)]$ of the induced monomorphisms*

$$p_* : \pi_1(E, e_0) \to \pi_1(B, b_0)$$

for all $e_0 \in p^{-1}(b_0)$ constitute a class of conjugate subgroups of $\pi_1(B, b_0)$. Furthermore, for a fixed $e_0 \in p^{-1}(b_0)$, there is a natural one-to-one correspondence between the points of $p^{-1}(b_0)$ and the right cosets of $p_*[\pi_1(E, e_0)]$ in $\pi_1(B, b_0)$ with e_0 corresponding to the subgroup $p_*[\pi_1(E, e_0)]$.

This class $\chi(E ; b_0)$ of conjugate subgroups $\{ p_*[\pi_1(E, e_0)] \mid e_0 \in p^{-1}(b_0) \}$ of $\pi_1(B, b_0)$ will be called the *characteristic class* of E at b_0. Each group in $\chi(E ; b_0)$ is isomorphic to the fundamental group $\pi_1(E)$ of E. $\chi(E ; b_0)$ will consist of a single group iff $p_*[\pi_1(E, e_0)]$ is an invariant subgroup of $\pi_1(B, b_0)$ for some and hence every $e_0 \in p^{-1}(b_0)$. If $\chi(E, b_0)$ consists of a single group, then so does $\chi(E, b_1)$ for any other $b_1 \in B$. In this case, we say that the fibering (E, p, B) is *regular*.

For a fixed $e_0 \in p^{-1}(b_0)$, the natural one-to-one correspondence is defined as follows: The right coset of $p_*[\pi_1(E, e_0)]$ in $\pi_1(B, b_0)$ corresponding to $e_1 \in p^{-1}(b_0)$ is the one which contains the element $w \in \pi_1(B, b_0)$ represented by the loop $p\sigma : I \to B$, where $\sigma : I \to E$ is any path joining e_0 to e_1.

Note. All the assertions as well as their proofs obviously hold for fiber spaces with totally pathwise disconnected fibers. Here, a space is said to be *totally pathwise disconnected* if it has no pathwise connected subspaces except single points.

16. Covering spaces

Throughout the present section, let B be a given connected and locally pathwise connected space.

A connected space E is said to be a *covering space* over B relative to a map $p : E \to B$ if the following conditions are satisfied:

(CS1) p maps E onto B.

(CS2) For each $b \in B$, there is a connected open neighborhood V of b in B such that each component of $p^{-1}(V)$ is open in E and is mapped homeomorphically onto V by p.

As immediate consequences of (CS2), the following two conditions are also satisfied:

(CS3) E is locally pathwise connected.

(CS4) $p : E \to B$ is an open map.

Examples. According to (II; 2.2), the real line R is a covering space over S^1 relative to the exponential map $p : R \to S^1$. Next, let us consider the unit n-sphere S^n in the euclidean $(n + 1)$-space R^{n+1}. If we identify the antipodal points of S^n, we obtain the real projective n-space P^n with natural projection $p : S^n \to P^n$. It is easy to verify that S^n is a covering space over P^n relative to p and the fibers are the pairs of antipodal points in S^n.

Covering spaces are the earliest examples of fiber spaces. In fact, we have the following theorem which is an easy consequence of (CS2) and the connectedness of B.

Theorem 16.1. *Every covering space E over B relative to $p : E \to B$ is a bundle space over B relative to p with discrete fibers.*

For generalizations and converses of (16.1), see Ex. P at the end of the chapter.

Now, let us consider the lifting problem for a map $f : X \to B$ of a connected and locally pathwise connected space X into the base space B of a covering space E relative to a projection $p : E \to B$. Pick $x_0 \in X$ and denote $b_0 = f(x_0)$. Let e_0 be a point of E with $p(e_0) = b_0$. The following theorem is a generalization of (II; 5.3).

Theorem 16.2. (The lifting theorem). *There exists a unique map $g : X \to E$ such that $g(x_0) = e_0$ and $pg = f$ iff the image of $f_* : \pi_1(X, x_0) \to \pi_1(B, b_0)$ is contained in that of $p_* : \pi_1(E, e_0) \to \pi_1 (B, b_0)$.*

Proof. The necessity of the condition is obvious. In fact, if a map $g : (X, x_0) \to (E, e_0)$ exists such that $pg = f$, then we have $p_*g_* = f_*$ and hence the image of f_* is contained in that of p_*. It remains to establish the sufficiency of the condition. The remainder of the proof is motivated by that of the special case (II; 5.3).

To construct the map g, let x be an arbitrary point in X. Then there exists a path $\pi : I \to X$ with $\pi(0) = x_0$ and $\pi(1) = x$. The composed map $\sigma = f\pi : I \to B$ is a path in B with $\sigma(0) = f(x_0)$ and $\sigma(1) = f(x)$. According to (15.1), there exists a unique path $\tau : I \to E$ such that $\tau(0) = e_0$ and $p\tau = \sigma$.

We assert that the point $\tau(1)$ of E does not depend on the choice of the path $\pi : I \to X$. To prove this, let $\pi' : I \to X$ be another path in X joining x_0 to x and let $\tau' : I \to E$ denote the unique path such that $\tau'(0) = e_0$ and $p\tau' = f\pi'$. We are going to prove that $\tau(1) = \tau'(1)$. The loop $\lambda = \pi \cdot \pi'^{-1} : I \to X$ represents an element $\alpha \in \pi_1(X, x_0)$ and so the loop $f\lambda : I \to B$ represents the element $f_*(\alpha)$ of $\pi_1(B, b_0)$. According to our condition, this element is contained in the subgroup $p_*\pi_1(E, e_0)$. Hence there exists a loop $\mu : I \to E$ at e_0 such that $p\mu = f\lambda$. Since $\lambda = \pi \cdot \pi'^{-1}$, it follows from the uniqueness of the path in (15.1) that $\tau(t) = \mu(\frac{1}{2}t)$ and $\tau'(t) = \mu(1 - \frac{1}{2}t)$ for each $t \in I$. In particular, $\tau(1) = \mu(\frac{1}{2}) = \tau'(1)$. This proves our assertion.

Because of the preceding assertion, we may define a function $g : X \to E$ by setting $g(x) = \tau(1)$ for each $x \in X$. By the construction of the point $\tau(1)$, it is obvious that $g(x_0) = e_0$ and $pg = f$.

By the same method as used in the proof of (II; 5.3), one can establish the continuity and the uniqueness of the function $g : X \to E$. ∎

If we omit the condition $g(x_0) = e_0$ in (16.2), we obtain the result that there exists a map $g : X \to E$ with $pg = f$ iff f_* maps $\pi_1(X, x_0)$ into a group of the characteristic class $\chi\ (E, b_0)$.

Now, let E be a covering space over B relative to $p : E \to B$, E' a covering space over B' relative to $p' : E' \to B'$, and $f : B \to B'$ a given map. Let $e_0 \in E$, $b_0 \in B$, $e'_0 \in E'$, and $b_0' \in B'$ be given points such that

$$p(e_0) = b_0, \quad p'(e'_0) = b'_0, \quad f(b_0) = b'_0.$$

The maps p, p' and f induce homomorphisms indicated in the following diagram:

$$\begin{array}{ccc} \pi_1(E, e_0) & \xrightarrow{g_*} & \pi_1(E', e'_0) \\ \downarrow{\scriptstyle p_*} & & \downarrow{\scriptstyle p'_*} \\ \pi_1(B, b_0) & \xrightarrow{f_*} & \pi_1(B', b'_0) \end{array}$$

Theorem 16.3. (The fiber map theorem). *There exists a unique map $g : E \to E'$ such that $g(e_0) = e'_0$ and $p'g = fp$ iff f_* carries the image of p_* into that of p'_*.*

Since X can be considered as the trivial covering space over itself, (16.2) is a special case of (16.3). However, (16.2) also implies (16.3) by considering the map $fp : E \to B'$.

This fiber map theorem has quite a few important consequences to which we devote the remainder of this section.

First, let us assume that $B = B'$ and that f is the identity map on B. Then we obtain the following

Theorem 16.4. (The covering theorem). *Let E and E' be two covering spaces over the same space B relative to $p : E \to B$ and $p' : E' \to B$ respectively, and let $b_0 \in B$. If $e_0 \in E$ and $e'_0 \in E'$ are such that*

$$p(e_0) = b_0 = p'(e'_0), \quad p_*[\pi_1(E, e_0)] \subset p'_*[\pi_1(E', e'_0)],$$

then there exists a unique map $g : E \to E'$ such that

$$g(e_0) = e'_0, \qquad p'g = p.$$

Furthermore, E is a covering space over E' relative to g.

Proof. The first assertion is a special case of (16.3). Hence, it remains to prove that E is a covering space over E' relative to g.

For this purpose, let us first verify (CS1). Let $e \in E$, $b = p(e)$ and $e' = g(e)$. Choose a connected open neighborhood V of b in B such that (CS2) holds for both covering spaces E and E' over B. Let W denote the component of $p^{-1}(V)$ containing e and W' that of $p'^{-1}(V)$ containing e'. Then the restrictions

$$q = p \mid W, \qquad q' = p' \mid W'$$

are homeomorphisms onto V. Since $p'g = p$, we have $g \mid W = q'^{-1}q$. Hence g maps W homeomorphically onto W'. This implies that $g(E)$ is both open and closed in the connected space E'; so g maps E onto E'.

Next, let us verify (CS2). Let $e' \in E'$ and $b = p'(e')$. Choose a connected open neighborhood V of b in B such that (CS2) holds for both covering spaces E and E' over B. Let W' denote the component of $p'^{-1}(V)$ which contains e'. Then W' is a connected open neighborhood of e' in E' and $g^{-1}(W')$ is the union of a set of components of $p^{-1}(V)$. It follows that every component of $g^{-1}(W')$ is open in E and is mapped homeomorphically onto W' by g. ▮

In particular, if E is simply connected, then (16.4) implies that E is a universal covering space over B relative to $p : E \to B$. Here, a covering space E over B relative to $p : E \to B$ is said to be *universal* if, for every covering space E' over B relative to $p' : E' \to B$, there exists a map $g : E \to E'$ such that $p'g = p$ and that E is a covering space over E' relative to g.

Next, let us define the notion of equivalent covering spaces as follows. Two covering spaces E and E' over a same base space B relative to projections $p : E \to B$ and $p' : E' \to B$ are said to be *equivalent* if there exists a homeomorphism $g : E \to E'$ of E onto E' such that $p'g = p$. By (16.2) and (15.4), the characteristic classes $\chi(E, b_0)$ and $\chi(E', b_0)$ are defined for every $b_0 \in B$.

Theorem 16.5. (The equivalence theorem). *For any given point $b_0 \in B$, two covering spaces E and E' over B relative to projections $p : E \to B$ and $p' : E' \to B$ are equivalent iff $\chi(E, b_0) = \chi(E', b_0)$.*

Proof. *Necessity.* Let $g : E \to E'$ be a homeomorphism of E onto E' such that $p'g = p$. Let $e_0 \in p^{-1}(b_0)$ and $e'_0 = g(e_0)$. Then g induces an isomorphism g_* of $\pi_1(E, e_0)$ onto $\pi_1(E', e'_0)$. On the other hand, $p'g = p$ gives $p'_* g_* = p_*$. This implies $p_*[\pi_1(E, e_0)] = p'_*[\pi_1(E', e'_0)]$ and hence $\chi(E, b_0) = \chi(E', b_0)$.

Sufficiency. Assume that $\chi(E, b_0) = \chi(E', b_0)$. Then there are points $e_0 \in E$ and $e'_0 \in E'$ such that

$$p(e_0) = b_0 = p'(e'_0), \quad p_*[\pi_1(E, e_0)] = p'_*[\pi_1(E', e'_0)].$$

According to (16.3), there exists a map $g : E \to E'$ such that $g(e_0) = e'_0$ and $p'g = p$. Similarly, there exists a map $h : E' \to E$ such that $h(e'_0) = e_0$ and $ph = p'$. Now, consider the composed map $hg : E \to E$. Since $hg(e_0) = e_0$ and $phg = p'g = p$, it follows from the uniqueness part of (16.4) that hg is the identity map on E. Similarly, gh is the identity map on E'. Hence g is a homeomorphism and the covering spaces E, E' are equivalent. ∎

In particular, any two simply connected covering spaces over the same base space are equivalent.

Next, let us consider the regular covering spaces. A covering space E over a base space B relative to $p : E \to B$ is said to be *regular* if the fibering $p : E \to B$ is regular in the sense of § 15. Hence, E is a regular covering space over B iff for every $b_0 \in B$, $\chi(E, b_0)$ consists of a single invariant subgroup of $\pi_1(B, b_0)$.

Now, let E be a given regular covering space over B relative to $p : E \to B$ and let b_0 be a given point in B. Since $\chi(E, b_0)$ is an invariant subgroup of $\pi_1(B, b_0)$, the quotient group

$$W = \pi_1(B, b_0) \,/\, \chi(E, b_0)$$

is well-defined. One can easily verify that, as an abstract group, W does not depend on the choice of the basic point $b_0 \in B$. Then, we have the following

Theorem 16.6. *The group W operates on the right of the regular covering space E. More precisely, to each element w of the group W and each point e of the space E, there corresponds a unique point ew of E such that*

$$(ew_1)w_2 = e(w_1w_2)$$

and, for each $w \in W$, the correspondence $e \to ew$ defines a homeomorphism w^* of E onto itself. Furthermore, the operation has the following two properties:

(i) *$p(ew) = p(e)$ for every $e \in E$ and $w \in W$.*

(ii) *For a given $e \in E$, $ew = e$ implies $w = 1$.*

In words, the condition (ii) is equivalent to saying that W operates freely on E.

Proof. Pick a point $e_0 \in p^{-1}(b_0)$. According to (15.4), there is a natural one-to-one correspondence $\nu : W \to p^{-1}(b_0)$ of W onto $p^{-1}(b_0)$ with $\nu(1) = e_0$.

For an arbitrary element $w \in W$, let $e_1 = \nu(w)$. Since E is a regular covering space over B, we have

$$p_*[\pi_1(E, e_0)] = p_*[\pi_1(E, e_1)].$$

Hence, just as in the proof of (16.5), there is a unique homeomorphism w^* of E onto itself such that $w^*(e_0) = e_1$ and $pw^* = p$. By the construction of ν and w^*, one can easily verify that

$$(w_1w_2)^* = w_2{}^*w_1{}^*, \qquad (w_1, w_2 \in W).$$

If the homeomorphism w^* of E admits a fixed point $e \in E$, then it follows from the uniqueness part of (16.3) that w^* must be the identity map on E. This implies that $e_1 = w^*(e_0) = e_0$ and hence we have $w = \nu^{-1}(e_0) = 1$.

Then, the theorem follows immediately if we set $ew = w^*(e)$ for every $e \in E$ and $w \in W$. ∎

In particular, if E is a simply connected covering space over a space B, then the fundamental group $\pi_1(B)$ operates freely on E.

In the classical theory, the homeomorphisms w^* in (16.6) are referred to as the *covering transformations* (Deckbewegungen) of the regular covering space E.

17. Construction of covering spaces

A space B is said to be *locally simply connected* if, for every point $b \in B$ and every open neighborhood V of b in B, there exists an open neighborhood $U \subset V$ of b in B such that, for any two points u_0 and u_1 in U, every pair of paths in U joining u_0 to u_1 are homotopic in V with end points held fixed. Obviously, every locally contractible space is locally simply connected and hence so is every simplicial complex. For our purpose in the present section, a slightly weaker condition is enough: we can replace the open neighborhood V by the space B itself. In this case, the space B is said to be *semi-locally simply connected.* Observe that every locally simply connected space is semi-locally simply connected and so is every simply connected space.

Throughout the present section, we assume that B is a given space which is connected, locally pathwise connected, and semi-locally simply connected. Let b_0 be a given point in B.

Theorem 17.1. (The existence theorem). *For every subgroup G of the fundamental group $\pi_1(B, b_0)$, there exist a covering space E over B relative to a projection $p: E \to B$ and a point $e_0 \in p^{-1}(b_0)$ such that G is exactly the image of the induced homomorphism* $p_*: \pi_1(E, e_0) \to \pi_1(B, b_0)$.

Proof. Let us consider the space of paths

$$\Omega = [B; b_0, B].$$

As in § 10, we denote by $p_1: \Omega \to B$ the terminal projection defined by $p_1(\sigma) = \sigma(1)$ for each $\sigma \in \Omega$. We shall define a new equivalence relation in Ω as follows. Two paths $\sigma, \tau \in \Omega$ are said to be *equivalent modulo G*, (in symbols: $\sigma \sim \tau$ mod G), if $p_1(\sigma) = p_1(\tau)$ and the element of $\pi_1(B, b_0)$ represented by the loop $\sigma \cdot \tau^{-1}$ is in G. Let E denote the quotient space of Ω defined by this equivalence relation. Then the points of E are equivalence classes (mod G) of the paths Ω. We shall denote by $[\sigma]$ the class which contains the path $\sigma \in \Omega$. Define a map

$$p: E \to B$$

by taking $p[\sigma] = p_1(\sigma)$ for every $[\sigma] \in E$. The continuity of p follows from (I; 12.1). Since B is pathwise connected, it follows that p maps E onto B.

We shall construct a convenient basis for the open sets of E as follows. For a given $e \in E$ and a given open neighborhood U of $p(e)$ in B, choose a path

$\sigma \in e$ and denote by $N(e, U)$ the subset of E which consists of the classes each containing a path of the form $\tau : I \to B$ such that $\tau(t) = \sigma(2t)$ for every $t \leqslant \frac{1}{2}$ and $\tau(t) \in U$ for every $t \geqslant \frac{1}{2}$. It is obvious that $N(e, U)$ does not depend on the choice of the representative path σ from the class $e \in E$. Since B is locally pathwise connected and semi-locally simply connected, it is straightforward to prove that $N(e, U)$ is open in E. Then it follows easily that the collection $\{N(e, U)\}$ for all $e \in E$ and all open neighborhoods U of $p(e)$ in B constitutes a basis for the open sets of E. An immediate consequence of this result is that $p : E \to B$ is open.

To prove that E is a covering space over B relative to p, let us establish (CS1) and (CS2). Since p maps E onto B, (CS1) is satisfied. To prove ($CS2$), let b be an arbitrarily given point of B. According to our assumption on B, there exists a pathwise connected open neighborhood U of b in B such that, for any two points u_0 and u_1 in U, every pair of paths in U joining u_0 to u_1 are homotopic with end points held fixed. Then it suffices to prove that $p^{-1}(U)$ is the disjoint union of the open sets $\{ N(e, U) \mid e \in p^{-1}(b) \}$ in E each of which is mapped homeomorphically onto U by p.

For this purpose, we shall first prove that p maps $N(e, U)$ homeomorphically onto U for each e in $p^{-1}(b)$. By the definition of $N(e, U)$, it is obvious that p maps $N(e, U)$ into U. Since U is pathwise connected, there is a path $\theta : I \to U$ joining b to u. Choose a path $\sigma \in \Omega$ with $[\sigma] = e$ and consider the path $\tau = \sigma \cdot \theta$. Clearly $[\tau] \in N(e, U)$ and $p[\tau] = u$. This proves that p maps $N(e, U)$ onto U. Let us assume that e_1 and e_2 are any two points in $N(e, U)$ such that $p(e_1) = p(e_2)$. There are two paths $\tau_i \in \Omega$, $(i = 1,2)$, such that $[\tau_i] = e_i$, $\tau_i(t) = \sigma(2t)$ whenever $t \leqslant \frac{1}{2}$, and $\tau_i(t) \in U$ whenever $t \geqslant \frac{1}{2}$. Since $p(e_1) = p(e_2)$, we have $\tau_1(1) = \tau_2(1)$. Call this common terminal point $v \in U$. Denote by $\xi_i : I \to U$ the path defined for each $i = 1, 2$ by

$$\text{(i)} \qquad \xi_i(t) = \tau_i\left(\frac{1 + t}{2}\right), \quad (t \in I).$$

Then ξ_1 and ξ_2 are two paths in U joining b to v. According to the choice of U, ξ_1 and ξ_2 are homotopic in B with end points held fixed. This implies that τ_1 and τ_2 are homotopic with end points held fixed and hence $e_1 = e_2$. This proves that p maps $N(e, U)$ in a one-to-one fashion. Since p is both open and continuous, we conclude that p maps $N(e, U)$ homeomorphically onto U.

Next, we are going to prove that the open sets $N(e, U)$, $e \in p^{-1}(b)$, are disjoint. Assume that $N(e_1, U)$ and $N(e_2, U)$ have a common point x. Choose paths $\sigma_i \in \Omega$, $(i = 1, 2)$, with $[\sigma_i] = e_i$. Then there are paths $\tau_i \in \Omega$, such that $[\tau_i] = x$, $\tau_i(t) = \sigma_i(2t)$ if $t \leqslant \frac{1}{2}$, and $\tau_i(t) \in U$ if $t \geqslant \frac{1}{2}$. Let $\xi_i : I \to U$, $(i = 1, 2)$, denote the path defined by (i). Then ξ_1 and ξ_2 are two paths in U joining b to $p(x)$. By the choice of U, ξ_1 and ξ_2 are homotopic in B with end points held fixed. This implies that

$$\tau_1 \cdot \tau_2^{-1} \sim \sigma_1 \cdot \xi_1 \cdot \xi_2^{-1} \cdot \sigma_2^{-1} \sim \sigma_1 \cdot \sigma_2^{-1}.$$

Since $[\tau_1] = x = [\tau_2]$, $\tau_1 \cdot \tau_2^{-1}$ represents an element of G and so does $\sigma_1 \cdot \sigma_2^{-1}$. Hence $e_1 = e_2$. This proves that the open sets $N(e, U)$, $e \in p^{-1}(b)$, are disjoint.

Now, we shall prove that $p^{-1}(U)$ is the union of the collection $\{ N(e, U) \mid e \in p^{-1}(b) \}$. Since p maps $N(e, U)$ into U, $N(e, U)$ is contained in $p^{-1}(U)$. Let $x \in p^{-1}(U)$ be arbitrarily given. We have to prove that there is some $e \in p^{-1}(b)$ such that $N(e, U)$ contains x. For this purpose, choose a path $\tau \in \Omega$ with $[\tau] = x$ and let $u = p(x) = \tau(1)$. Since U is pathwise connected, there exists a path $\theta : I \to U$ joining b to u. Let $\sigma = \tau \cdot \theta^{-1}$ and $\xi = \sigma \cdot \theta$. Set $e = [\sigma]$. Since $\sigma(1) = \theta(0) = b$, we have $e \in p^{-1}(b)$. By an easy homotopy, one can prove that ξ and τ are homotopic with end points held fixed. It follows that $x = [\tau] = [\xi] \in N(e, U)$. Hence, $p^{-1}(U)$ is the union of the collection $\{ N(e, U) \mid e \in p^{-1}(b) \}$. This completes the proof of (CS2).

As a continuous image of a pathwise connected space Ω, E is pathwise connected. This completes the proof that E is a covering space over B relative to p.

Let $\delta \in \Omega$ denote the degenerate path at b_0, that is to say, $\delta(I) = b_0$. Denote $e_0 = [\delta] \in E$. Then $p(e_0) = b_0$ and p induces a monomorphism

$$p_* : \pi_1(E, e_0) \to \pi_1(B, b_0).$$

It remains to prove that the image of p_* is exactly the given subgroup G of $\pi_1(B, b_0)$.

For this purpose, let us first prove an assertion as follows. Let $\sigma : I \to B$ be any path with $\sigma(0) = b_0$. According to (15.1), σ has a unique covering path $\sigma^* : I \to E$ with $\sigma^*(0) = e_0$. We assert that, for each $t \in I$, $\sigma^*(t)$ is the class which contains the path $\sigma_t : I \to B$ defined by $\sigma_t(s) = \sigma(st)$ for each $s \in I$. This follows immediately from the fact that the assignment $t \to \sigma_t$ defines a path $\tau : I \to \Omega$ according to (9.10) and that E is a quotient space of Ω.

Now, let us prove that the image of p_* is exactly G. Let α be any element in $\pi_1(E, e_0)$. Choose a representative loop $\sigma^* : I \to E$ for α, then the loop $\sigma = p\sigma^*$ represents the element $p_*(\alpha)$ of $\pi_1(B, b_0)$. According to the foregoing assertion, we have

$$[\sigma] = \sigma^*(1) = e_0 = [\delta].$$

This implies that $p_*(\alpha) \in G$. Conversely, let β be any element of G. Choose a representative loop $\sigma : I \to B$ for β. By (15.1), σ has a unique covering path $\sigma^* : I \to E$ with $\sigma^*(0) = e_0$. According to the foregoing assertion, we have $\sigma^*(1) = [\sigma]$. Since σ represents $\beta \in G$, this implies that $\sigma^*(1) = e_0$. Hence σ^* represents an element α of $\pi_1(E, e_0)$. This implies that $\beta = p_*(\alpha)$. Thus, we have proved that G is exactly the image of p_*. ∎

In particular, if G is the trivial subgroup of $\pi_1(B, b_0)$ consisting of the neutral element 1 of $\pi_1(B, b_0)$, then E is a simply connected covering space over B. According to (16.4) this implies that E is a universal covering space over B. By (15.3) it follows that every universal covering space of B is simply connected. Finally, by (16.5) we conclude that every pair of universal covering spaces over B are equivalent. Hence, E is essentially the only universal covering space over B. Hereafter, E will be referred to as *the universal covering space of* B.

Combining (16.5) and (17.1), we obtain the following classification of covering spaces.

Theorem 17.2. (The classification theorem). *For any connected, locally pathwise connected, and semi-locally simply connected space B, the equivalence classes of the covering spaces over B are in a one-to-one correspondence with the conjugate classes of subgroups of the fundamental group $\pi_1(B)$.*

The following simple examples are given to illustrate the preceding results.

(*i*) Covering spaces of S^1. The real line R is the universal covering space over S^1 relative to the exponential map $p : R \to S^1$. Since $\pi_1(S^1)$ is abelian, it follows that every covering space over S^1 is regular. Since $\pi_1(S^1)$ is free cyclic, the non-trivial subgroups of $\pi_1(S^1)$ are the free cyclic subgroups G_n, $(n = 1, 2, \cdots)$, where G_n is of index n in $\pi_1(S^1)$. The covering space which corresponds to G_n is S^1 itself relative to the projection $p_n : S^1 \to S^1$ defined by $p_n(z) = z^n$ for each $z \in S^1$. These are essentially all the covering spaces over S^1. This example reveals the fact that homeomorphic covering spaces over the same base space are not necessarily equivalent.

(*ii*) Covering spaces of S^n and P^n with $n > 1$. Since S^n is simply connected, every covering space over S^n is equivalent to the trivial covering space by which we mean the covering space S^n over itself relative to the identity map. On the other hand, S^n is the universal covering space over P^n relative to the natural projection $p : S^n \to P^n$. Since the fibers of this covering space are the pairs of antipodal points in S^n, it follows from (15.4) that $\pi_1(P^n)$ is the cyclic group of order 2. Hence, by (17.2), the universal covering space is essentially the only non-trivial covering space over P^n.

(*iii*) Covering spaces of closed surfaces. Let M denote a closed surface other than S^2 and P^2. Then, by (II; Ex. A), $\pi_1(M)$ is an infinite group. It follows from (15.4) that the universal covering space E of M is not compact. Hence, E is a simply connected infinite 2-manifold. In fact, it is a classical theorem that E is homeomorphic to the euclidean 2-space, [V; p. 153].

As an application of the preceding results, we have the following

Theorem 17.3. *Assume that X is a connected triangulable space and E is the universal covering space over B relative to $p : E \to B$. Then, for any $x_0 \in X$, $b_0 \in B$ and $e_0 \in p^{-1}(b_0)$, the assignment $g \to f = pg$ sets up a one-to-one correspondence between the homotopy classes of the maps $g : (X, x_0) \to (E, e_0)$ and those of the maps $f : (X, x_0) \to (B, b_0)$ with f_* sending $\pi_1(X, x_0)$ into the neutral element of $\pi_1(B, b_0)$.*

Proof. Let us use the usual notation $f_* = 0$ to denote the fact that f_* sends $\pi_1(X, x_0)$ into the neutral element of $\pi_1(B, b_0)$.

If $f = pg$ for some $g : (X, x_0) \to (E, e_0)$, obviously we have $f_* = p_* g_* = 0$. Hence, the assignment $g \to f = pg$ defines a function ϕ of the homotopy classes of the maps $g : (X, x_0) \to (E, e_0)$ into those of the map $f : (X, x_0) \to (B, b_0)$ with $f_* = 0$.

Let $f : (X, x_0) \to (B, b_0)$ be a map with $f_* = 0$. According to (16.2) there exists a unique map $g : (X, x_0) \to (E, e_0)$ such that $pg = f$. Hence ϕ is onto.

Let $g, g' : (X, x_0) \to (E, e_0)$ be any two maps such that $pg \simeq pg'$ *rel* x_0. Then there exists a homotopy $f_t : (X, x_0) \to (B, b_0)$, $0 \leqslant t \leqslant 1$, such that $f_0 = pg$ and $f_1 = pg'$. According to (iv) of (3.1), there exists a homotopy $g_t : (X, x_0) \to (E, e_0)$, $0 \leqslant t \leqslant 1$, such that $g_0 = g$ and $pg_t = f_t$ for every $t \in I$. Since $g_1(x_0) = e_0 = g'(x_0)$ and $pg_1 = f_1 = pg'$, it follows from the uniqueness part of (16.2) that $g_1 = g'$. Hence $g \simeq g'$ *rel* x_0. This proves that ϕ is one-to-one. ▮

In particular, if X is simply connected, then the assignment $g \to f = pg$ sets up a one-to-one correspondence between the homotopy classes of the maps $g : (X, x_0) \to (E, e_0)$ and those of the maps $f : (X, x_0) \to (B, b_0)$. For example, let $B = P^2$, $E = S^2$, and let $p : S^2 \to P^2$ denote the natural projection. If X is either S^2 or S^3, then it follows that the homotopy classes of the maps $f : (X, x_0) \to (B, b_0)$ are in a one-to-one correspondence with the set Z of all integers.

Note. In (17.3), the condition that X is triangulable can be replaced by the weaker condition that X is locally connected. In fact, in the proof of (17.3), one may use Ex. P instead of (3.1).

EXERCISES

A. Sliced fiber spaces

Let $p : E \to B$ be a given map. By a *slicing structure* for p, we mean a collection $S = \{ \omega, \phi_U \}$ of the following entities:

(1) a system $\omega = \{ U \}$ of open sets of B which covers B, called the *slicing neighborhoods.*

(2) a system of maps $\{ \phi_U \mid U \in \omega \}$ indexed by the slicing neighborhoods, called the *slicing functions,* where each ϕ_U, is defined on the subspace $U \times p^{-1}(U)$ of the product space $B \times E$ with images in E in such a way that the following two conditions are satisfied:

(SF1) $$p\phi_U(b, x) = b, \quad (b \in U, x \in p^{-1}(U));$$

(SF2) $$\phi_U(p(x), x) = x, \quad (x \in p^{-1}(U)).$$

If a slicing structure $S = \{ \omega, \phi_U \}$ for p exists, we say that $p : E \to B$ has the *slicing structure property* (abbreviated SSP).

We shall use the abbreviation Para CHP to stand for the covering homotopy property for all paracompact Hausdorff spaces. Prove the following implications, [Hu 8; Huebsch 1]:

$$\text{BP} \Rightarrow \text{SSP} \Rightarrow \text{Para CHP}$$

Hence, $p : E \to B$ is a fibering if it has the SSP. In this case, E is called a *sliced fiber space* over B relative to p, and in particular, every bundle space is a sliced fiber space.

Let E be a sliced fiber space over B relative to $p : E \to B$. Prove the following assertions:

1. The projection $p : E \to B$ is open, that is to say, it maps open sets onto open sets.

2. If two points a and b of B can be connected by a path in B, then the fibers $p^{-1}(a)$ and $p^{-1}(b)$ are homotopically equivalent.

3. E is a bundle space over B relative to p iff the following two conditions are satisfied, [Griffin 1]:

(GC1) There exists a space D which is homeomorphic with every fiber $p^{-1}(b)$, $b \in B$.

(GC2) For each point $b \in B$, there exist a slicing neighborhood U which contains b and a slicing function $\phi_U : U \times p^{-1}(U) \to E$ such that, for every pair of points u and v in U, we always have

$$\phi_U[u, \phi_U(v, x)] = x, \quad (x \in p^{-1}(u)).$$

4. If E and B are metrizable then $p : E \to B$ has the ACHP and hence the PLP, [Curtis 1].

Now assume that E and B are spaces such that $B \times E$ is a paracompact Hausdorff space and B is an ANR. Let $p : E \to B$ be a given map. Prove that E is a sliced fiber space over B relative to p iff, for any map $g : X \to E$ of a paracompact Hausdorff space X into E and any homotopy $f_t : X \to B$, $(0 \leqslant t \leqslant 1)$, of the map $f = pg$, there exists a homotopy $g_t : X \to E$, $(0 \leqslant t \leqslant 1)$, of g which covers f_t and is stationary with f_t, [Fox 2]. For the definition of the stationary property, see [S ; p. 50].

A sliced fiber space E over B relative to p is said to have a *unified slicing function* if there exists a slicing structure $S = \{\omega, \phi_U\}$ for $p : E \to B$ such that, for any two slicing neighborhoods U and V in ω, we always have $\phi_U = \phi_V$ on the intersection of $U \times p^{-1}(U)$ and $V \times p^{-1}(V)$. For such a slicing structure S, we may define a *unified slicing function* ϕ on the union W of the subspaces $U \times p^{-1}(U)$ for all $U \in \omega$ by taking $\phi = \phi_U$ on each $U \times p^{-1}(U)$. The fiber spaces defined by Hurewicz and Steenrod [1] and Fox [2] are those with unified slicing functions. Prove that every metrizable sliced fiber space over an ANR has a unified slicing function, [H; p. 170].

B. Local path lifting property

A map $p : E \to B$ is said to have the *local path lifting property* (abbreviated LPLP) if, for each $b \in B$, there exists an open neighborhood U of b in B such that the map

$$q : W \to U, \quad W = p^{-1}(U), \quad q = p \mid W$$

has the PLP. Prove

1. SSP $\Rightarrow$ LPLP $\Rightarrow$ Para CHP.

2. If B is a paracompact Hausdorff space, then LPLP implies PLP. [Hurewicz 2].

C. Relations between various notions of fiber space

Let us consider a given map $p : E \to B$ having one of the following properties:

PCHP = CHP for polyhedra, (§ 2),
Para CHP = CHP for paracompact Hausdorff spaces, (Ex. A),
ACHP = CHP for arbitrary spaces, (§ 2),
PLP = path lifting property, (§ 12),
LPLP = local path lifting property, (Ex. B),
SSP = slicing structure property, (Ex. A),
BP = bundle property, (§ 4),

where the abbreviation CHP stands for the covering homotopy property, (§ 2). The first four properties are of global nature while the last three are apparently local properties.

A number of implications among these properties are known and can be summarized by the following diagram

$$\begin{array}{ccccccccc}
\text{BP}\{\text{para}\} & \Rightarrow & \text{SSP}\{\text{para}\} & \Rightarrow & \text{LPLP}\{\text{para}\} & \Rightarrow & \text{PLP} & \Longleftrightarrow & \text{ACHP} \\
\Downarrow & & \Downarrow & & \Downarrow & & & & \Downarrow \\
\text{BP} & \Longrightarrow & \text{SSP} & \Longrightarrow & \text{LPLP} & & \Longrightarrow & & \text{Para CHP} \\
 & & & & & & & & \Downarrow \\
 & & & & & & & & \text{PCHP}
\end{array}$$

where the attached symbol { para } means that the base space B is assumed to be a paracompact Hausdorff space. Check if all of these implications have been established in the text and in the preceding two exercises.

D. Homogeneous spaces

Let E be a topological group and let F be a closed subgroup of E. Define an equivalence relation in E as follows: two elements a, b of E are said to equivalent iff there is an element $f \in F$ such that $af = b$. Thus the elements of E are divided into disjoint equivalence classes called the *left cosets* of F in E. The left coset containing $a \in E$ is obviously the closed set aF of E. According to (I; § 12), we obtain a quotient space $B = E/F$ whose elements are left cosets of F in E and a natural projection $p : E \to B$ which maps $a \in E$ onto the left coset $aF \in B$. $B = E/F$ is called the *quotient space of E by F*; it will be called simply a *homogeneous space*. Prove the following assertions:

1. $B = E/F$ is a Hausdorff space.

2. The natural projection p is an open map.

3. E operates transitively on B under the homeomorphisms $e : B \to B, e \in E$, defined by
$$e(b) = p[e \cdot p^{-1}(b)], \quad (b \in B).$$

4. E is a bundle space over B relative to p iff there is a *local cross-section of B in E*; by this we mean a cross-section $\varkappa : V \to E$ defined on an open neighborhood V of the point $b_0 = F$ in B.

5. If E is a Lie group, then a local cross-section of B in E exists and hence E is a bundle space over B relative to p, [Che; p. 110].

E. Spheres as homogeneous spaces

Let Q denote one of the three fields of real numbers, complex numbers, or quaternions. Consider the right vector space Q^n whose elements are ordered sets $x = (x_1, \cdots, x_n)$ of n elements of Q. The *inner product* xy of x and y in Q^n is defined by

$$xy = \bar{x}_1 y_1 + \cdots + \bar{x}_n y_n,$$

where $\bar{x}_i$ denotes the conjugate of x_i. The topological group G_n of all linear transformations in Q^n which preserve the inner product is called the *orthogonal*, *unitary* or *symplectic group* according as the scalars are real, complex or quaternionic. It is a compact Lie group. Let S denote the unit sphere in Q_n; then S is the sphere of dimension $n - 1$, $2n - 1$ or $4n - 1$ according as the scalars are real, complex or quaternionic. Prove the following assertions:

1. G_n operates transitively on S.

2. G_n is a bundle space over S relative to the projection $p : G_n \to S$ defined by $p(f) = f(x_0)$ for every $f \in G_n$, where $x_0 = (1, 0, \cdots, 0) \in S$.

3. Let G_{n-1} denote the subgroup of G_n leaving x_0 fixed. Then the fibers $p^{-1}(x)$ are the left cosets of G_{n-1} in G_n and hence S can be considered as the homogeneous space G_n/G_{n-1}.

F. Fiberings of spheres over projective spaces

Consider the non-zero elements of the vector space Q^n in the preceding exercise. Define an equivalence relation in $X = Q^n \setminus 0$ as follows: for any two elements x and y in X, $x \sim y$ iff there is a q in Q such that $xq = y$. According to (I, § 12), we obtain a quotient space M called the *projective space associated with* Q^n and a natural projection $\pi : X \to M$. Let S denote the unit sphere in Q^n and $p = \pi \mid S$. Prove the following assertions:

1. S is a bundle space over M relative to p.

2. The fibers $p^{-1}(b)$, $b \in M$, are great spheres of dimension 0, 1 or 3 according as the scalars Q are real, complex or quaternionic.

3. The group G_n operates transitively on M in some natural way; and hence M can be identified with the quotient space of G_n by its subgroup of the elements leaving a given point of M fixed.

G. Stiefel manifolds

A k-frame, v^k, in R^n is an ordered set of k linearly independent vectors. Let $V'_{n,k}$ denote the set of all k-frames in R^n. Let L_n denote the full linear group, and let $L_{n,k}$ be the subgroup of L_n leaving fixed each vector of a given frame $v_0{}^k$. Then we may identify

$$V'_{n,k} = L_n / L_{n,k}$$

and hence $V'_{n,k}$ becomes a homogeneous space called the *Stiefel manifold of k-frames in n-space*. Let $V_{n,k}$ denote the subspace of $V'_{n,k}$ consisting of the *orthogonal k-frames*. Prove the following assertions:

1. The orthogonal group O_n operates transitively on $V_{n,k}$. If O_{n-k} is represented as the subgroup of O_n leaving fixed a given orthogonal k-frame $v_0{}^k$, then one may identify $V_{n,k} = O_n / O_{n-k}$.

2. If $k < n$, the rotation group R_n operates transitively on $V_{n,k}$ and hence one may identify $$V_{n,k} = R_n / R_{n-k}, \quad k < n.$$

3. $V_{n,k}$ may be interpreted as the space of all orthogonal $(k - 1)$-frames tangent to S^{n-1}. In particular, $V_{n,1} = S^{n-1}$ and $V_{n,2}$ is the space of all unit tangent vectors on S^{n-1}.

4. $V_{n,k}$ can be identified with the space of all orthogonal maps S^{k-1} into S^{n-1}.

5. Let $v_0{}^n$ be a fixed orthogonal n-frame in R^n and let $v_0{}^k$ denote the first k vectors of $v_0{}^n$. Let O_{n-k} be the subgroup of O_n leaving $v_0{}^k$ fixed. Then we obtain a chain of Stiegel manifolds and projections

$$O_n = V_{n,n} \to V_{n,n-1} \to \cdots \to V_{n,2} \to V_{n,1} = S^{n-1}.$$

Each projection or any composition of them is a bundle map, that is to say, the projection of a bundle space over its base space.

H. Grassmann manifolds

Let $M_{n,k}$ denote the set of all k-dimensional linear subspaces (k-planes through the origin) of R^n. The orthogonal group O_n operates transitively on $M_{n,k}$. If R^k is a fixed k-plane through the origin and R^{n-k} its orthogonal complement, the subgroup of O_n mapping R^k onto itself splits up into the direct product $O_k \times O_{n-k}$ of two orthogonal subgroups the first of which leaves R^{n-k} pointwise fixed and the second leaves R^k pointwise fixed. Hence we may identify $$M_{n,k} = O_n / (O_k \times O_{n-k}).$$

Thus $M_{n,k}$ becomes a homogeneous space called the *Grassmann manifold of k-planes in n-space.* Prove the following assertions:

1. The natural projection $O_n \to M_{n,k}$ maps the rotation group R_n onto $M_{n,k}$. Let R_k and R_{n-k} denote the rotation subgroups of O_k, O_{n-k}. Define

$$\tilde{M}_{n,k} = R_n / (R_k \times R_{n-k}).$$

Then $\tilde{M}_{n,k}$ is called the *manifold of oriented k-planes in n-space.* $\tilde{M}_{n,k}$ is a covering space over $M_{n,k}$ relative to the natural projection with 0-spheres as fibers.

2. $V_{n,k}$ is a bundle space over $M_{n,k}$ relative to the natural projection induced by the inclusion $O_{n-k} \subset O_k \times O_{n-k}$. The fibers are homeomorphic to O_k.

3. If $k < n$, $V_{n,k}$ is a bundle space over $\tilde{M}_{n,k}$ with fibers homeomorphic to R_k.

4. The correspondence between any k-plane and its orthogonal $(n - k)$ plane sets up a homeomorphism $M_{n,k} \leftrightarrow M_{n,n-k}$.

5. $M_{n,1}$ is essentially the $(n - 1)$- dimensional real projective space P^{n-1} and $\tilde{M}_{n,1}$, the $(n - 1)$-sphere S^{n-1}.

I. Elementary properties of mapping spaces

Let Ω denote the mapping space Y^X with the compact-open topology. Prove the following assertions:

1. If Y is a T_0-, T_1-, T_2-, or regular space, then so is Ω. Conversely, if Ω is a T_0-, T_1-, T_2-, or regular space, then so is Y.

2. If X is a locally compact Hausdorff space and if X, Y are both separable (being separable means having a countable basis), then so is Ω. On the other hand, if Ω is separable, then so is Y.

3. Assume that X is a compact metrizable space and Y is a metrizable space. Then Ω is an ANR iff Y is such; similarly, Ω is an AR iff Y is such.

J. Admissible topologies

A great variety of topologies may be introduced into the set $\Omega = Y^X$ making it a space. We shall denote by Ω_τ the space obtained by topologizing Ω with a topology τ. A topology τ of Ω is said to be *admissible* if the evaluation

$$\omega : \Omega_\tau \times X \to Y$$

is a map. Thus, by (9.1), the compact-open topology of Ω is admissible provided that X is a locally compact regular space. Prove the following assertions, [Arens 1]:

1. Any admissible topology of Ω contains every open set in the compact-open topology of Ω.

2. If X is a completely regular space and Y is a T_1-space containing a nondegenerate path, then a necessary and sufficient condition for the compact-open topology to be admissible is the local compactness of X.

K. The topology of uniform convergence

Let Y be a metrizable space and let d be a given bounded metric consistent with the topology of Y. There is a natural metric d^* defined on $\Omega = Y^X$ by

$$d^*(f, g) = \sup_{x \in X} d[f(x), g(x)]$$

for each pair of map $f, g \in \Omega$. The metric d^* determines a topology of Ω called the *d^*-topology*, or the *topology of uniform convergence with respect to d*. Prove the following assertions:

1. The d^*-topology of Ω is admissible.

2. If X is compact, then the compact-open topology of Ω coincides with the d^*-topology induced by any given bounded metric d on Y. The word "bounded" in this assertion might have been omitted.

3. If X is a completely regular Hausdorff space and Y a metrizable space containing a nondegenerate path, then a necessary and sufficient condition for the compact-open topology of Ω to coincide with the d^*-topology induced by a given bounded metric d on Y is the compactness of X [Jackson 1].

4. The d^*-topology of Ω depends not only on the topologies of X and Y but also on the metric d chosen for Y. Construct a few examples.

L. Maps on topological products

Consider three given spaces T, X, Y and the mapping spaces:

$$\Omega = Y^X, \Phi = Y^{X \times T}, \Psi = \Omega^T.$$

In § 9, we have defined the association $\theta : \Phi \to \Psi$. Prove the following assertions:

1. If both X and T satisfy the first countability axiom, then θ sends Φ onto Ψ, [Fox 3].

2. If X and T are Hausdorff spaces satisfying the first countability axiom, then θ is a homeomorphism of Φ onto Ψ and hence the exponential law holds in this case.

$$Y^{X\times T} = (Y^X)^T$$

M. Borsuk's fibering theorem

Let X be a compact metrizable space, A a closed subspace of X, and Y a compact ANR. Consider the mapping space $E = Y^X$ and the subspace B of the mapping space Y^A consisting of the maps $g: A \to Y$ which can be extended over X. Prove that E is a sliced fiber space over B relative to the natural projection $p: E \to B$ defined by $p(f) = f \mid A$ for every $f \in E$. By Exercises C and I, show that this fibering has a unified slicing function. [H; p. 173].

N. Change of the boundary sets

There are changes of the boundary sets A and B without changing the homotopy type of the space $[X;$ A, B$]$ of paths. Because of symmetry, we may study only the changes of the terminal set B.

Let B_1 and B_2 be two subspaces of X. A *deformation of B_1 into B_2* in X is a homotopy $h_t: B_1 \to X$, $(0 \leqslant t \leqslant 1)$, such that h_0 is the inclusion map and $h_1(B_1) \subset B_2$. Such a deformation h_t induces a map $h^{\#}: [X; A, B_1] \to [X; A, B_2]$ described as follows: for each $f \in [X; A, B_1]$, $g = h^{\#}(f)$ is given by

$$g(t) = \begin{cases} f(2t), & (0 \leqslant t \leqslant \frac{1}{2}), \\ h_{2t-1} f(1), & (\frac{1}{2} \leqslant t \leqslant 1). \end{cases}$$

B_1 is said to be a *deformation homeomorph of B_2 in X* if there exists a homotopy $h_t: B_1 \to X$, $(0 \leqslant t \leqslant 1)$, such that h_0 is the inclusion map and h_1 is a homeomorphism of B_1 onto B_2.

Prove that the spaces $[X; A, B_1]$ and $[X; A, B_2]$ are homotopically equivalent if B_1 is either a strong deformation retract of B_2 or a deformation homeomorph of B_2. Consequently, if X is pathwise connected, then the homotopy type of the space $[X; a, b]$ is independent of the choice of the points a and b. In particular, the spaces $[X; a, b]$, Λ_a and Λ_b are homotopically equivalent.

O. The space of curves

In our definition of the space of paths, the domain I is the same for all paths. In his studies on Pontrjagin products, J. C. Moore, has found that it is more convenient to allow different domains for different paths. To avoid possible ambiguity, these will be called curves.

Precisely, let Y be a given space. Then, a *curve* in Y consists of a real number $a \geqslant 0$ and a map f of the closed interval $[0, a]$ into Y. The points $f(0)$ and $f(a)$ are called the *initial point* and the *terminal point* of f respectively. A curve $f : [0, a] \to Y$ is said to be *closed* if $f(0) = f(a)$; in this case, $f(0)$ is called the *basic point* of the closed curve f.

Now, let Γ denote the set of all curves in Y. To topologize Γ, let us consider the subspace J of the real line R consisting of the real numbers $a \geqslant 0$ and the space $\Omega = Y^I$ of all paths in Y. Define a function $\phi : \Gamma \to J \times \Omega$ by taking $\phi(f) = (a, \sigma_f)$ for every curve $f : [0, a] \to Y$, where $\sigma_f : I \to Y$ is the path given by $\sigma_f(t) = f(at)$ for each $t \in I$. This function ϕ carries Γ onto a subspace $\phi(\Gamma)$ of $J \times \Omega$ in a one-to-one fashion. We topologize Γ in such a way that ϕ becomes a homeomorphism. Prove that $(J \times \Omega) \setminus \phi(\Gamma)$ is the subspace $0 \times [\Omega \setminus j(Y)]$ where $j : Y \to \Omega$ denotes the natural injection $j : Y \to \Omega$ of § 9.

Since $I = [0, 1]$, every path in Y is also a curve in Y. The topology of Γ defined above permits us to consider Ω as a subspace of Γ. Construct a homotopy $h_t : \Gamma \to \Gamma$, $(0 \leqslant t \leqslant 1)$, such that the following conditions are satisfied:

(*i*) h_0 is the identity map on Γ.

(*ii*) h_1 is a retraction of Γ onto Ω.

(*iii*) $h_t(f) = f$ for every $f \in \Omega$ and $t \in I$.

(*iv*) For every $f \in \Gamma$, the initial point and the terminal point of the curve $h_t(f)$ are the same as those of f for each $t \in I$.

If $f : [0, a] \to Y$ and $g : [0, b] \to Y$ are two curves in Y such that $f(a) = g(0)$, then we may define a product $f \cdot g : [0, a + b] \to Y$ by taking

$$[f \cdot g](t) = \begin{cases} f(t), & (\text{if } 0 \leqslant t \leqslant a), \\ g(t - a), & (\text{if } a \leqslant t \leqslant a + b). \end{cases}$$

This multiplication $(f, g) \to f \cdot g$ of curves is continuous and associative whenever it is defined.

Now, let y be a given point in Y and let Θ_y denote the subspace of Γ consisting of the closed curves with y as the basic point. Then the homotopy h_t shows that the space of loops Λ_y is a deformation retract of Θ_y. On the other hand, the multiplication $(f, g) \to f \cdot g$ is defined for every pair $f, g \in \Theta_y$ with the trivial curve $e : [0, 0] \to y$ as a two-sided unit. Hence, if Y is a Hausdorff space, then Θ_y is a *mob* with e as a two-sided unit in the sense of Wallace [1].

P. Generalized covering spaces

A space E is called a *generalized covering space* over a space B relative to a map $p : E \to B$ if the following conditions are satisfied:

(GCS1) p maps E onto B.

(GCS2) For each $b \in B$, there exists an open neighborhood U of b in B such that $p^{-1}(U)$ can be represented as a disjoint union of open sets in E each of which is mapped homeomorphically onto U by p.

Prove that every bundle space E over B relative to $p : E \to B$ with discrete fibers is a generalized covering space over B relative to p.

Let E be a generalized covering space over B relative to $p : E \to B$. Prove the following assertions:

1. E is a sliced fiber space over B relative to p with discrete fibers.

2. If B is connected, then E is a bundle space over B relative to p with discrete fibers.

3. If E is connected and B is locally connected, then E is a covering space over B relative to p.

4. Let $q : X \to Y$ be a given map of a space X into a space Y. If $g : X \to E$ is a map and $f_t : Y \to B$, $(0 \leqslant t \leqslant 1)$, a homotopy with $f_0 q = pg$, then there exists a unique homotopy $g_t : X \to E$, $(0 \leqslant t \leqslant 1)$, of g such that $f_t q = pg_t$ for every $t \in I$ and that g_t is stationary with f_t. In particular, $p : E \to B$ has the ACHP. This assertion is also true for sliced fiber spaces with totally pathwise disconnected fibers, [Griffin 1].

Q. Local homeomorphisms

A map $p : E \to B$ of a space E onto a space B is called a *local homeomorphism* of E onto B if every point x of E has an open neighborhood which is mapped homeomorphically by p onto an open neighborhood of $p(x)$ in B.

Prove that, if $p : E \to B$ is a local homeomorphism of a regular Hausdorff space E onto a space B such that $p^{-1}(b)$ is finite for every $b \in B$, then E is a generalized covering space over B relative to p.

R. The covering spaces of the torus

As an exercise to determine all equivalence classes of the covering spaces over a given space B, let us study to case where B is the torus $S^1 \times S^1$.

According to (II; Ex. A), the fundamental group $\pi_1(B)$ is a free abelian group with two free generators a and b. Among the subgroups of $\pi_1(B)$ there is a doubly indexed system

$$G_{m,n}, \quad (m, n = 0, 1, 2, \cdots),$$

where $G_{m,n}$ is the subgroup generated by a^m and b^n. Hence, $G_{o,o} = 0$ and $G_{1,1} = \pi_1(B)$.

Prove the following assertions:

1. Every covering space over B is regular.

2. Corresponding to $G_{o,o}$, we have the universal covering space $E = R^2 = R \times R$ over $B = S^1 \times S^1$ relative to the projection $p_{o,o} : E \to B$ defined by $p_{o,o}(x, y) = (px, py)$, where $p : R \to S^1$ denotes the exponential map of (II; § 2).

3. Corresponding to $G_{o,n}$, $n > 0$, we have the covering space $E = R \times S^1$ over $B = S^1 \times S^1$ relative to the projection $p_{o,n} : E \to B$ defined by $p_{o,n}(x, z) = (px, z^n)$ for each $x \in R$ and $z \in S^1$. Similarly, we may get the covering space corresponding to $G_{m,o}$, $m > 0$.

4. Corresponding to $G_{m,n}$, $m > 0$, $n > 0$, we have the covering space $E = S^1 \times S^1$ over $B = S^1 \times S^1$ relative to the projection $p_{m,n} : E \to B$ defined by $p_{m,n}(u, v) = (u^m, v^n)$ for each $(u, v) \in E$.

S. Maps of the torus into the projective plane

Consider the torus T and the projective plane P. Pick arbitrary basic points $t_0 \in T$ and $p_0 \in P$, and study the maps $f : (T, t_0) \to (P, p_0)$.

The fundamental group $\pi_1(T, t_0)$ is a free abelian group with two free generators a and b, while $\pi_1(P, p_0)$ is a group of order 2 generated by c. Hence, there are four possible homomorphisms

$$h_{m,n} : \pi_1(T, t_0) \to \pi_1(P, p_0),$$

with m, n running over 0 and 1, defined by

$$h_{m,n}(a) = mc, \quad h_{m,n}(b) = nc.$$

By considering T as the unit square with the opposite sides identified, construct for each (m, n) a map

$$f_{m,n} : (T, t_0) \to (P, p_0)$$

such that $(f_{m,n})_* = h_{m,n}$. Hence, the homotopy classes of the maps $f : (T, t_0) \to (P, p_0)$ are divided into four disjoint collections $C_{m,n}$ such that $f \in C_{m,n}$ iff $f_* = h_{m,n}$.

Next, prove that, for each (m, n), the homotopy classes in the collection $C_{m,n}$ are in a one-to-one correspondence with the homotopy classes of the maps $(S^2, s_0) \to (P, p_0)$ and hence with the integers.

T. Maps of a surface into a surface

Let X and Y be any two closed surfaces other than the sphere and the projective plane. Study the classification problem of the maps $f: X \to Y$ as follows.

Pick arbitrary basic points $x_0 \in X$ and $y_0 \in Y$. Two homomorphisms

$$h, k : \pi_1(X, x_0) \to \pi_1(Y, y_0)$$

are said to be *equivalent* (notation : $h \simeq k$) if there exists an element $b \in \pi_1(Y, y)$ such that

$$k(a) = b^{-1} \cdot h(a) \cdot b$$

holds for every $a \in \pi_1(X, x_0)$. Thus the homomorphisms $h : \pi_1(X, x_0) \to \pi_1(Y, y_0)$ are divided into disjoint equivalence classes. Let C denote the set of these equivalence classes.

Prove that the homotopy classes of the maps $f : X \to Y$ are in a one-to-one correspondence with the equivalence classes C by establishing the following assertions:

1. Every map $f : X \to Y$ is homotopic to a map $g : X \to Y$ such that $g(x_0) = y_0$.
2. For any two maps $f, g : X \to Y$ satisfying $f(x_0) = y_0 = g(x_0)$, $f \simeq g$ implies $f_* \simeq g_*$.
3. For any $h \in C$, there exists a map $f: X \to Y$ such that $f(x_0) = y_0$ and $f_* = h$.
4. If $f : X \to Y$ is a map with $f(x_0) = y_0$ and $h \in C$ is equivalent to f_*, then there exists a map $g : X \to Y$ such that $g(x_0) = y_0$, $g \simeq f$ and $g_* = h$.
5. If $f, g : X \to Y$ are two maps such that $f(x_0) = y_0 = g(x_0)$ and $f_* = g_*$, then $f \simeq g$.

CHAPTER IV

HOMOTOPY GROUPS

1. Introduction

The basic problem which led to the discovery of "homotopy groups" was to classify homotopically the maps of an n-sphere S^n into a given space X. In the case $n = 1$, this was facilitated by pinning down a base point to obtain a group structure as in (II; § 4). The same trick was found to work in higher dimensions; in fact, if we pinch the equator of S^n to a point, we obtain two n-spheres with one point in common. If $n > 1$, there is a rotation of S^n which gives a homotopy interchanging the two hemispheres. This implies the striking feature that the group is abelian.

The relation between homotopy and homology groups and the existence of relative homology groups $H_n(X, A)$ led quickly to the relative homotopy groups $\pi_n(X, A, x_0)$, giving a system highly analogous to homology theory. But it differs in several ways: $\pi_0(X, x_0)$ and $\pi_1(X, A, x_0)$ are not ordinarily groups; $\pi_1(X, x_0)$ and $\pi_2(X, A, x_0)$ are not usually abelian; and the excision property for homology does not hold for homotopy.

A very important fact in the minds of those involved in the development of the theory was the result of Hopf in 1930 that $\pi_3(S^2)$ is infinite. This showed that the groups express some very deep topological properties of spaces. A second important fact is that the definition of π_n is not effectively computable; and there was no definition available which led immediately to effective computations as in the case of homology groups of complexes. Successful calculations in special cases came slowly. It is only recently that methods have been found which apply to a reasonably broad variety of cases. These are the subject of much current research.

The objectives of the present chapter are to define the groups and related homomorphisms, to establish their main general properties, and to show that certain of these properties are characteristic.

2. Absolute homotopy groups

Let (X, x_0) be a given pair consisting of a space X and a point x_0 in X. Let $\pi_0(X, x_0)$ denote the set of all path-components of X. The path-component of X which contains x_0 will be called the *neutral element* of $\pi_0(X, x_0)$ and will be denoted by 0. As in (II; § 4), we shall denote by $\pi_1(X, x_0)$ the fundamental group of X at x_0.

For each integer $n > 1$, the definition of the n-th (absolute) *homotopy group* $\pi_n(X, x_0)$ is strictly analogous to that of the fundamental group. We replace the unit interval I by the *n-cube* I^n, i.e. the topological product of n copies of I.

Every point $t \in I^n$ is represented by n real numbers $t = (t_1, \cdots, t_n)$, $t_i \in I$, $(i = 1, 2, \cdots, n)$, called the *coordinates* of t. The number t_i is called the i-th coordinate of t. An $(n-1)$-*face* of I^n is obtained by setting some coordinate t_i to be 0 or 1. The union of the $(n-1)$-faces forms the *boundary* ∂I^n of I^n; it is topologically equivalent to the unit $(n-1)$-sphere S^{n-1}.

Consider the set $F^n = F^n(X, x_0)$ of all maps

$$f : (I^n, \partial I^n) \to (X, x_0).$$

These maps are divided into *homotopy classes* (relative to ∂I^n). We shall denote by $\pi_n(X, x_0)$ the totality of these homotopy classes. We shall also denote by $[f]$ the class which contains the map f and by 0 the class which contains the unique constant map $d_0(I^n) = x_0$. Topologize F^n by means of the compact-open topology as in (III; § 9); then $\pi_n(X, x_0)$ becomes the set of all path-components of the space F^n.

We may define an *addition* (usually non-commutative) in F^n as follows. For any two maps f, g in F^n, their *sum* $f + g$ is the map defined by

$$(f+g)(t) = \begin{cases} f(2t_1, t_2, \cdots, t_n), & \text{if } 0 \leqslant t_1 \leqslant \frac{1}{2}, \\ g(2t_1 - 1, t_2, \cdots t_n), & \text{if } \frac{1}{2} \leqslant t_1 \leqslant 1, \end{cases}$$

for every point $t = (t_1, \cdots, t_n)$ in I^n. Obviously, $f + g$ is in F^n.

The homotopy class $[f + g]$ clearly depends only on the classes $[f]$ and $[g]$. Hence we may define an addition in $\pi_n(X, x_0)$ by taking

$$[f] + [g] = [f + g].$$

Just as in (II; § 4) for the fundamental group, one can easily verify that this addition makes $\pi_n(X, x_0)$ a group which will be called the *n-th homotopy group* of X at x_0. The class 0 is the group-theoretic neutral element of $\pi_n(X, x_0)$, and the inverse element of $[f]$ is the class $[f\theta]$, where $\theta : I^n \to I^n$ denotes the map defined by

$$\theta(t) = (1 - t_1, t_2, \cdots, t_n)$$

for every $t = (t_1, t_2, \cdots, t_n)$ in I^n.

If the boundary ∂I^n of I^n is identified to a point, we get a quotient space which is topologically equivalent to an n-sphere S^n with a given basic point $s_0 \in S^n$. It follows that one might equally well define an element of $\pi_n(X, x_0)$ as a homotopy class (relative to s_0) of the maps $f : (S^n, s_0) \to (X, x_0)$. Since the two halves of I^n, defined by the conditions $t_1 \leqslant \frac{1}{2}$ and $t_1 \geqslant \frac{1}{2}$ respectively, correspond to two hemispheres of S^n, one can clearly see how to define the group operation of $\pi_n(X, x_0)$ from this point of view. For details, see [Hu 4]. Since, when $n > 1$, there exists a rotation of S^n which leaves s_0 fixed and interchanges the two hemispheres, this definition suggests the following striking property of $\pi_n(X, x_0)$ which however will be proved in a different way.

Proposition 2.1. *For every $n > 1, \pi_n(X, x_0)$ is an abelian group.*

Proof. As in (III; § 11), one can prove that F^{n-1} is an H-space with the constant map $d_0 \in F^{n-1}$ as a two-sided homotopy unit. Hence, by (III; 9.9), we have
we have

$$X^{I^n} = X^{I^{n-1} \times I} = (X^{I^{n-1}})^I.$$

Then one can easily see that $\pi_n(X, x_0)$ and $\pi_1(F^{n-1}, d_0)$ are essentially the same group. Hence $\pi_n(X, x_0)$ is abelian. ∎

In the preceding proof, we have incidentally obtained an interesting result:

$$\pi_n(X, x_0) = \pi_1(F^{n-1}, d_0).$$

Hence every homotopy group of a space can be expressed as the fundamental group of some other space. This relation can be used to define higher homotopy groups in terms of fundamental groups; indeed, Hurewicz used this definition when he introduced these groups in 1935.

One can say more: let p be any positive integer less than n and let $q = n - p$. By (III; 9.9), we have

$$X^{I^n} = X^{I^p \times I^q} = (X^{I^p})^{I^q}.$$

By means of this relation, it is easy to deduce the following result:

$$\pi_n(X, x_0) = \pi_q(F^p, d_0), \quad p + q = n,$$

where d_0 denotes the constant map $d_0(I^q) = x_0$. In particular, when $p = 1$, F^p becomes the space W of loops in X with basic point x_0 and d_0 the degenerate loop at x_0. Thus we have proved the following proposition which will be used in the sequel.

Proposition 2.2. $\pi_n(X, x_0) = \pi_{n-1}(W, d_0)$.

Finally, the following proposition is an immediate consequence of the fact that I^n is pathwise connected.

Proposition 2.3. *If X_0 denotes the path-component of X containing x_0, then*

$$\pi_n(X_0, x_0) = \pi_n(X, x_0), \quad n > 0.$$

Examples. If X is a contractible space, then $\pi_n(X, x_0) = 0$ for every $n \geqslant 0$. Next, by (II; 7.1) we have

$$\pi_0(S^1, 1) = 0, \quad \pi_1(S^1, 1) \approx Z,$$
$$\pi_n(S^1, 1) = 0, \quad \text{if } n > 1.$$

On the other hand, by means of (II; § 8), one can prove that

$$\pi_m(S^n, s_0) = 0, \quad (\text{if } m < n),$$
$$\pi_n(S^n, s_0) \approx Z.$$

Finally, by (III; 6.4), we deduce the result

$$\pi_3(S^2, s_0) \approx Z.$$

3. Relative homotopy groups

The objective of the present section is to generalize the notion of homotopy groups in § 2 by defining the relative homotopy groups $\pi_n(X, A, x_0)$.

By a *triplet* (X, A, x_0), we mean a space X, a nonvacuous subspace A of X, and a point x_0 in A. If x_0 is the only point of A, then the triplet (X, A, x_0) will be simply denoted by (X, x_0) and may be considered as a pair consisting of a space X and a point x_0 in X.

Let $n > 0$ and define the *n-th relative homotopy set* $\pi_n(X, A, x_0)$ as follows.

Consider again the n-cube I^n. The *initial* $(n - 1)$-face of I^n defined by $t_n = 0$ will be identified with I^{n-1} hereafter. The union of all remaining $(n - 1)$-faces of I^n is denoted by J^{n-1}. Then we have

$$\partial I^n = I^{n-1} \cup J^{n-1}, \quad \partial I^{n-1} = I^{n-1} \cap J^{n-1}.$$

By a map $f : (I^n, I^{n-1}, J^{n-1}) \to (X, A, x_0)$, we mean a continuous function from I^n to X which carries I^{n-1} into A and J^{n-1} into x_0. In particular, it sends ∂I^n into A and ∂I^{n-1} into x_0. We denote by $F^n = F^n(X, A, x_0)$ the set of all such maps. These maps are divided into *homotopy classes* (relative to the system $\{ I^{n-1}, A ; J^{n-1}, x_0 \}$). We shall denote by $\pi_n(X, A, x_0)$ the totality of these homotopy classes. We shall also denote by $[f]$ the class which contains the map f and by 0 the class which contains the constant map $d_0(I^n) = x_0$. Topologize F^n by means of the compact-open topology; then $\pi_n(X, A, x_0)$ becomes the set of all path-components of the space F^n.

If $n > 1$, we may define an *addition* (usually non-commutative) in F^n. For any two maps f, g in F^n, their *sum* $f + g \in F^n$ is defined by the formula given in § 2 for the absolute case. The homotopy class $[f + g]$ depends only on the classes $[f]$ and $[g]$ and hence we may define an addition in $\pi_n(A,X,x_0)$ by taking

$$[f] + [g] = [f + g].$$

As in § 2, one can verify that this addition makes $\pi_n(X, A, x_0)$ a group which will be called the *n-th relative homotopy group* of X modulo A at x_0. The class 0 is the group-theoretic neutral element of $\pi_n(X, A, x_0)$, and the inverse element of $[f]$ is the class $[f\theta]$, where $\theta : I^n \to I^n$ denotes the map defined in § 2.

If x_0 is the only point of A, then we have

$$F^n(X, A, x_0) = F^n(X, x_0).$$

Hence, in this case, $\pi_n(X, A, x_0)$ reduces to the absolute homotopy group $\pi_n(X, x_0)$ defined in § 2.

If J^{n-1} is pinched to a point s_0, (I^n, I^{n-1}, J^{n-1}) becomes a configuration topologically equivalent to the triplet (E^n, S^{n-1}, s_0) consisting of the unit n-cell E^n, its boundary $(n - 1)$-sphere S^{n-1}, and a reference point $s_0 \in S^{n-1}$. It follows that one might equally well define an element of $\pi_n(X, A, x_0)$ as a homotopy class (relative to the system $\{ S^{n-1}, A ; s_0, x_0 \}$) of the maps of (E^n, S^{n-1}, s_0) into (X, A, x_0). Since, when $n > 2$, there exists a rotation of

E^n which leaves s_0 fixed and interchanges the two halves of E^n, we see that $\pi_n(X, A, x_0)$ is abelian for every $n > 2$. This commutativity property is also an immediate consequence of the next proposition. $\pi_2(X, A, x_0)$ is in general non-abelian.

Next, let us introduce the notion of the *derived triplet* of a given triplet which will be frequently used in the sequel. Let

$$T = (X, A, x_0)$$

be a given triplet. Consider the space of paths

$$X' = [X ; X, x_0]$$

and the initial projection $p : X' \to X$ as defined in (III; § 10). Let

$$A' = p^{-1}(A) = [X ; A, x_0] \subset X'$$

and denote by $x'_0 \in A'$ the degenerate loop $x'_0(I) = x_0$. Thus we obtain a triplet

$$T' = (X', A', x'_0),$$

called the *derived triplet* of T, and a map

$$p : (X', A', x'_0) \to (X, A, x_0),$$

called the *derived projection* over (X, A, x_0).

Proposition 3.1. *For every $n > 0$, we have*

$$\pi_n(X, A, x_0) = \pi_{n-1}(A', x'_0).$$

Proof. If $n = 1$, this is obvious since each side can be considered as the set of all path-components of A'. Assume $n > 1$. Since $X^{I^n} = (X^I)^{I^{n-1}}$ by (III; 9.9), it follows that

$$F^n(X, A, x_0) = F^{n-1}(A', x'_0).$$

Hence $\pi_n(X, A, x_0)$ and $\pi_{n-1}(A', x'_0)$ coincide set-theoretically. It is also clear that the group structures are the same for any $n > 1$. ∎

Another consequence of (3.1) is that every relative homotopy group can be expressed as an absolute homotopy group and hence as a fundamental group.

Since ∂I^n is pathwise connected for $n > 1$, the following proposition is obvious.

Proposition 3.2. *If X_0 denotes the path-component of X containing x_0 and A_0 that of A, then*

$$\pi_n(X, A, x_0) = \pi_n(X_0, A_0, x_0), \quad n > 1.$$

Finally, the following proposition will be used in the sequel.

Proposition 3.3. *If $\alpha \in \pi_n(X, A, x_0)$ is represented by a map $f \in F^n(X, A, x_0)$ such that $f(I^n) \subset A$, then $\alpha = 0$.*

Proof. Since $f \in F^n(X, A, x_0)$ and $f(I^n) \subset A$, we may define a homotopy $f_t \in F^n(X, A, x_0)$, $0 \leqslant t \leqslant 1$, by taking

$$f_t(t_1, \cdots, t_{n-1}, t_n) = f(t_1, \cdots, t_{n-1}, t + t_n - tt_n).$$

Then we have $f_0 = f$ and $f_1(I^n) = x_0$. Hence $\alpha = 0$. ▮

As an application of (3.3), let us prove the following

Proposition 3.4. *If (X, A, x_0) is a triplet such that (X, A) is a relative n-cell, then $\pi_m(X, A, x_0) = 0$ for every m satisfying $0 < m < n$.*

Proof. By the definition of relative n-cells in (I; § 7), X is the adjunction space obtained by adjoining E^n to A by means of a map $g : S^{n-1} \to A$ defined on the boundary $(n-1)$-sphere S^{n-1} of E^n. Let e_0 denote an interior point of E^n. Then, by (I; Ex. S), A is a strong deformation retract of $X \setminus e_0$.

Let $\alpha \in \pi_m(X, A, x_0)$ with $0 < m < n$ and choose a representative map

$$f : (I^m, I^{m-1}, J^{m-1}) \to (X, A, x_0)$$

for α. Applying (II; Ex. C), we can easily prove that we may free e_0 from the image of f by means of a suitable homotopy of f relative to $\{ I^{m-1}, A ; J^{m-1}, x_0 \}$. Since A is a strong deformation retract of $X \setminus e_0$, this implies that there exists a homotopy

$$f_t : (I^m, I^{m-1}, J^{m-1}) \to (X, A, x_0), \quad (0 \leqslant t \leqslant 1),$$

such that $f_0 = f$ and $f_1(I^m) \subset A$. By (3.3), we conclude that $\alpha = 0$. Hence $\pi_m(X, A, x_0) = 0$ for every m satisfying $0 < m < n$. ▮

4. The boundary operator

Let (X, A, x_0) be a given triplet. For every $n > 0$, we shall define a transformation

$$\partial : \pi_n(X, A, x_0) \to \pi_{n-1}(A, x_0).$$

Let α be any element of $\pi_n(X, A, x_0)$. Then, by definition, α is a homotopy class represented by a map

$$f : (I^n, I^{n-1}, J^{n-1}) \to (X, A, x_0).$$

If $n = 1$, $f(I^{n-1})$ is a point of A which determines a path-component $\beta \in \pi_{n-1}(A, x_0)$ of A. If $n > 1$, then the restriction $f \mid I^{n-1}$ is a map of $(I^{n-1}, \partial I^{n-1})$ into (A, x_0) and hence represents an element $\beta \in \pi_{n-1}(A, x_0)$.

Obviously, the element $\beta \in \pi_{n-1}(A, x_0)$ does not depend on the choice of the map f which represents the given element $\alpha \in \pi_n(X, A, x_0)$. Hence we may define the transformation ∂ by setting $\partial(\alpha) = \beta$. Hereafter, ∂ will be called the *boundary operator*.

The following two properties of ∂ are obvious from the definition.

Proposition 4.1. *The boundary operator ∂ sends the neutral element of $\pi_n(X, A, x_0)$ into that of $\pi_{n-1}(A, x_0)$.*

Proposition 4.2. *If $n > 1$, then the boundary operator ∂ is a homomorphism.*

5. Induced transformations

Let (X, A, x_0) and (Y, B, y_0) be given triplets. By a map of (X, A, x_0) into (Y, B, y_0), we mean a continuous function X to Y which carries A into B and x_0 into y_0. Consider such a map

$$f : (X, A, x_0) \to (Y, B, y_0).$$

Since f is continuous, it sends the path-components of X into those of Y. Hence, f determines an *induced transformation*

$$f_* : \pi_0(X, x_0) \to \pi_0(Y, y_0)$$

which obviously sends the neutral element of $\pi_0(X, x_0)$ into that of $\pi_0(Y, y_0)$.

Now let $n > 0$. For any map $\phi \in F^n(X, A, x_0)$, the composition $f\phi$ is in $F^n(Y, B, y_0)$ and the assignment $\phi \to f\phi$ defines a map

$$f_\# : F^n(X, A, x_0) \to F^n(Y, B, y_0).$$

The continuity of $f_\#$ implies that $f_\#$ carries the path-components of $F^n(X, A, x_0)$ into those of $F^n(Y, B, y_0)$. Hence it determines an *induced transformation*

$$f_* : \pi_n(X, A, x_0) \to \pi_n(Y, B, y_0)$$

which obviously sends the neutral element of $\pi_n(X, A, x_0)$ into that of $\pi_n(Y, B, y_0)$.

If $n = 1$, $A = x_0$, $B = y_0$, or if $n > 1$, then $\pi_n(X, A, x_0)$ and $\pi_n(Y, B, y_0)$ are groups. For any two maps ϕ, ψ in $F^n(X, A, x_0)$, one can easily see that

$$f(\phi + \psi) = f\phi + f\psi$$

in $F^n(Y, B, y_0)$. Hence it follows that f_* is a homomorphism. Thus we have established the following two properties of f_*.

Proposition 5.1. *If $n = 0$, $A = x_0$, $B = y_0$, or if $n > 0$, then the induced transformation f_* sends the neutral element of $\pi_n(X, A, x_0)$ into that of $\pi_n(Y, B, y_0)$.*

Proposition 5.2. *If $n = 1$, $A = x_0$, $B = y_0$, or if $n > 1$, then the induced transformation f_* is a homomorphism.*

In the case of (5.2), f_* will be called the *induced homomorphism.*

Now, consider the derived triplets (X', A', x'_0), (Y', B', y'_0) of the given triplets together with the derived projections

$$p : (X', A', x'_0) \to (X, A, x_0), \quad r : (Y', B', y'_0) \to (Y, B, y_0).$$

The given map $f : (X, A, x_0) \to (Y, B, y_0)$ considered as a map from X into Y induces a map

$$f^I : X^I \to Y^I$$

according to (III; § 14). Since $f(A) \subset B$ and $f(x_0) = y_0$, it follows that f^I carries X' into Y', A' into B', and x'_0 into y'_0. Hence f^I defines a map

$$f' : (X', A', x'_0) \to (Y', B', y'_0)$$

which satisfies the relation $rf' = fp$ and which will be called the *derived map* of f. Let $\bar{f} = f' \mid (A', x'_0)$. Then $\bar{f}$ induces

$$\bar{f}_* : \pi_{n-1}(A', x'_0) \to \pi_{n-1}(B', y'_0).$$

Under the identifications in (3.1) it is obvious that $f_* = \bar{f}_*$.

Finally, the boundary operator ∂ in § 4 is essentially a special case of induced transformations. In fact, for any triplet (X, A, x_0), consider the pair (A', x'_0) of the derived triplet and the restriction $q = p \mid (A', x'_0)$ of the derived projection p. Then q induces

$$q_* : \pi_{n-1}(A', x'_0) \to \pi_{n-1}(A, x_0).$$

Under the identification in (3.1), it is obvious that $\partial = q_*$.

6. The algebraic properties

The homotopy groups $\pi_n(X, A, x_0)$, the boundary operator ∂, and the induced transformations defined in the previous sections possess seven fundamental properties which will be given in this and the next few sections.

By the definition of induced transformations, the following two properties are obvious.

Property I. *If $f : (X, A, x_0) \to (X, A, x_0)$ is the identity map, then f_* is the identity transformation on $\pi_n(X, A, x_0)$ for every n.*

Property II. *If $f : (X, A, x_0) \to (Y, B, y_0)$ and $g : (Y, B, y_0) \to (Z, C, z_0)$ are maps, then for every $n \geqslant 0$ we have $(gf)_* = g_* f_*$.*

Hence, for any given n, the assignment $(X, A, x_0) \to \pi_n(X, A, x_0)$ and $f \to f_*$ defines a covariant functor.

The next property gives a relation between the boundary operator and the induced transformations. It is an obvious consequence of their definitions.

Property III. *If $f : (X, A, x_0) \to (Y, B, y_0)$ is a map and if $g : (A, x_0) \to (B, y_0)$ is the restriction of f, then the commutativity relation $\partial f_* = g_* \partial$ holds in the following rectangle for every $n > 0$:*

$$\begin{array}{ccc} \pi_n(X, A, x_0) & \xrightarrow{\partial} & \pi_{n-1}(A, x_0) \\ \downarrow f_* & & \downarrow g_* \\ \pi_n(Y, B, y_0) & \xrightarrow{\partial} & \pi_{n-1}(B, y_0) \end{array}$$

Note that Property III is also an easy consequence of Property II. To see this, let us consider the derived map

$$f' : (X', A', x'_0) \to (Y', B', y'_0)$$

of f as well as the derived projections

$$p : (X', A', x'_0) \to (X, A, x_0), \quad r : (Y', B', y'_0) \to (Y, B, y_0).$$

Let $\bar{f} = f' \mid (A', x'_0)$, $q = p \mid (A', x'_0)$ and $s = r \mid (B', y'_0)$. After the various identifications described in (3.1) and § 5, the preceding rectangle reduces to the following form:

$$\begin{array}{ccc} \pi_{n-1}(A', x'_0) & \xrightarrow{q_*} & \pi_{n-1}(A, x_0) \\ \downarrow \bar{f}_* & & \downarrow g_* \\ \pi_{n-1}(B', y'_0) & \xrightarrow{s_*} & \pi_{n-1}(B, y_0) \end{array}$$

Since f' satisfies the relation $rf' = fp$, we have $s\bar{f} = gq$. Hence, Property II implies that the commutativity relation $s_*\bar{f}_* = g_*q_*$ holds in this rectangle.

7. The exactness property

Let (X, A, x_0) be any given triplet. The inclusion maps

$$i : (A, x_0) \subset (X, x_0), \quad j : (X, x_0) \subset (X, A, x_0)$$

induce transformations i_* and j_* for each n $\geqslant 0$. Together with the boundary operators ∂, they form a beginningless sequence:

$$\cdots \xrightarrow{j_*} \pi_{n+1}(X, A, x_0) \xrightarrow{\partial} \pi_n(A, x_0) \xrightarrow{i_*} \pi_n(X. x_0) \xrightarrow{j_*} \pi_n(X, A, x_0) \xrightarrow{\partial} \cdots$$
$$\cdots \xrightarrow{j_*} \pi_1(X, A, x_0) \xrightarrow{\partial} \pi_0(A, x_0) \xrightarrow{i_*} \pi_0(X, x_0)$$

which will be called the *homotopy sequence* of the triplet (X, A, x_0) and will be denoted by $\pi(X, A, x_0)$.

Every set in $\pi(X, A, x_0)$ has a specified element called its neutral element and every transformation in $\pi(X, A, x_0)$ carries the neutral element into the neutral element. We define the *kernel* of a transformation in $\pi(X, A, x_0)$ to be the inverse image of the neutral element. Such a sequence is said to be *exact* if the kernel of each transformation coincides exactly with the image of the preceding transformation.

Property IV. *The homotopy sequence of any triplet (X, A, x_0) is exact.*

The proof breaks up into the proofs of the following six statements:

(1) $j_*i_* = 0$, (2) $\partial j_* = 0$, (3) $i_*\partial = 0$.

(4) If $\alpha \in \pi_n(X, x_0)$ and $j_*\alpha = 0$, then there exists an element $\beta \in \pi_n(A, x_0)$ such that $i_*\beta = \alpha$.

(5) If $\alpha \in \pi_n(X, A, x_0)$ and $\partial\alpha = 0$, then there exists an element $\beta \in \pi_n(X, x_0)$ such that $j_*\beta = \alpha$.

(6) If $\alpha \in \pi_{n-1}(A, x_0)$ and $i_*\alpha = 0$, then there exists an element $\beta \in \pi_n(X, A, x_0)$ such that $\partial\beta = \alpha$.

In the preceding statements, the symbol 0 denotes either the neutral element of the set involved or the transformation which sends every element into the neutral element.

Proof of (1). For each $n > 0$, let $\alpha \in \pi_n(A, x_0)$ and choose a map $f \in F^n(A, x_0)$ which represents α. Then the element $j_*i_*(\alpha)$ in $\pi_n(X, A, x_0)$ is represented by the composition $jif \in F^n(X, A, x_0)$. Since obviously $jif(I^n) \subset A$, it follows from (3.3) that $j_*i_*\alpha = 0$. Since α is arbitrary, this implies $j_*i_* = 0$. ∎

Proof of (2). For each $n > 0$, let $\alpha \in \pi_n(X, x_0)$ and choose a map $f \in F^n(X, x_0)$ which represents α. Then the element $\partial j_* \alpha$ is determined by the restriction $jf \mid I^{n-1} = f \mid I^{n-1}$. Since $f(I^{n-1}) = x_0$, we have $\partial j_* \alpha = 0$. Hence $\partial j_* = 0$. ∎

Proof of (3). For each $n > 0$, let $\alpha \in \pi_n(X, A, x_0)$ and choose a map $f \in F^n(X, A, x_0)$ which represents α. Then the element $i_* \partial \alpha$ is determined by the restriction $g = f \mid I^{n-1}$. Define a homotopy $g_t : I^{n-1} \to X$, $0 \leqslant t \leqslant 1$, by setting

$$g_t(t_1, \cdots, t_{n-1}) = f(t_1, \cdots, t_{n-1}, t).$$

Then $g_0 = g$, $g_1(I^{n-1}) = x_0$, and $g_t \in F^{n-1}(X, x_0)$ if $n > 1$. This implies $i_* \partial \alpha = 0$. Hence $i_* \partial = 0$. ∎

Proof of (4). Choose a map $f \in F^n(X, x_0)$ which represents α. The condition $j_* \alpha = 0$ implies that there exists a homotopy $f_t : I^n \to X$, $0 \leqslant t \leqslant 1$, such that $f_0 = f$, $f_1(I^n) = x_0$, and $f_t \in F^n(X, A, x_0)$ for each $t \in I$. Define a homotopy $g_t : I^n \to X$, $0 \leqslant t \leqslant 1$, by setting

$$g_t(t_1, \cdots, t_{n-1}, t_n) = \begin{cases} f_{2t_n}(t_1, \cdots, t_{n-1}, 0), & (\text{if } 0 \leqslant 2t_n \leqslant t), \\ f_t\left(t_1, \cdots, t_{n-1}, \dfrac{2t_n - t}{2 - t}\right), & (\text{if } t \leqslant 2t_n \leqslant 2). \end{cases}$$

Then $g_0 = f$, $g_1(I^n) \subset A$, and $g_t(\partial I^n) = x_0$ for every $t \in I$. Now, g_1 represents an element $\beta \in \pi_n(A, x_0)$ and the homotopy g_t proves that $i_* \beta = \alpha$. ∎

Proof of (5). First, assume that $n > 1$. Choose a map $f \in F^n(X, A, x_0)$ which represents α. Then the condition $\partial \alpha = 0$ implies that there exists a homotopy $g_t : I^{n-1} \to A$, $0 \leqslant t \leqslant 1$, such that $g_0 = f \mid I^{n-1}$, $g_1(I^{n-1}) = x_0$, and $g_t(\partial I^{n-1}) = x_0$ for every $t \in I$. Define a partial homotopy $h_t : \partial I^n \to A$, $0 \leqslant t \leqslant 1$, by setting

$$h_t(s) = \begin{cases} g_t(s), & (s \in I^{n-1}, t \in I), \\ x_0, & (s \in J^{n-1}, t \in I). \end{cases}$$

Since $h_0 = f \mid \partial I^n$, it follows from the homotopy extension property that the homotopy h_t has an extension $f_t : I^n \to X$, $0 \leqslant t \leqslant 1$, such that $f_0 = f$. Since $f_1(\partial I^n) = h_1(\partial I^n) = x_0$, f_1 represents an element β in $\pi_n(X, x_0)$. Since $f_t \in F^n(X, A, x_0)$ for $t \in I$, it follows that $j_* \beta = \alpha$. For the remaining case $n = 1$, α is represented by a path $f : I \to X$ such that $f(0) \in A$ and $f(1) = x_0$. The condition $\partial \alpha = 0$ means that $f(0)$ is contained in the same path-component of A as x_0. Hence there exists a homotopy $f_t : I \to X$, $0 \leqslant t \leqslant 1$, such that $f_0 = f$, $f_t(0) \in A$, $f_t(1) = x_0$, and $f_1(0) = x_0$. Then, f_1 represents an element $\beta \in \pi_1(X, x_0)$ and the homotopy f_t implies that $j_* \beta = \alpha$. ∎

Proof of (6). First, assume that $n > 1$. Choose a map $f \in F^{n-1}(A, x_0)$ which represents α. Then the condition $i_* \alpha$ implies that there exists a homotopy $f_t : I^{n-1} \to X$, $0 \leqslant t \leqslant 1$, such that $f_0 = f$, $f_1(I^{n-1}) = x_0$ and $f_t(\partial I^{n-1}) = x_0$ for every $t \in I$. Define a map $g : I^n \to X$ by taking

$$g(t_1, \cdots, t_{n-1}, t_n) = f_{t_n}(t_1, \cdots, t_{n-1}).$$

This map g is in $F^n(X, A, x_0)$ and represents an element α of $\pi_n(X, A, x_0)$. Since $g \mid I^{n-1} = f$, we have $\partial\beta = \alpha$. For the remaining case $n = 1$, α is a path-component of A. The condition $i_*\alpha = 0$ means that α is contained in the path-component of X which contains x_0. Pick a point x from α. Then there exists a path $f: I \to X$ such that $f(0) = x$ and $f(1) = x_0$. This path f represents an element β of $\pi_1(X, A, x_0)$. Since $f(0) \in \alpha$, we have $\partial\beta = \alpha$. ∎

8. The homotopy property

Consider any two given triplets (X, A, x_0) and (Y, B, y_0), and any two given maps

$$f, g: (X, A, x_0) \to (Y, B, y_0).$$

We recall that f and g are said to be *homotopic* (relative to $\{ A, B ; x_0, y_0 \}$) if there exists a homotopy

$$h_t: (X, A, x_0) \to (Y, B, y_0), \quad 0 \leqslant t \leqslant 1,$$

such that $h_0 = f$ and $h_1 = g$.

Property V. *If f and g are homotopic, then their induced transformations*

$$f_*, g_*: \pi_n(X, A, x_0) \to \pi_n(Y, B, y_0)$$

are equal for every n.

Proof. Let $\alpha \in \pi_n(X, A, x_0)$. It suffices to prove that $f_*\alpha = g_*\alpha$.

First, assume that $n > 0$. Choose a map $\phi \in F^n(X, A, x_0)$ which represents α. Then the elements $f_*\alpha$ and $g_*\alpha$ are represented by the compositions $f\phi$ and $g\phi$ respectively. The composition $h_t\phi$ of ϕ and a homotopy $h_t: f \simeq g$ gives a homotopy connecting $f\phi$ and $g\phi$. Hence $f_*\alpha = g_*\alpha$.

It remains to settle the case $n = 0$, $A = x_0$, and $B = y_0$. Here α is a path-component of X. Pick a point $x \in \alpha$. Then $f_*\alpha$ and $g_*\alpha$ are the path-components of Y containing the points $f(x)$ and $g(x)$ respectively. Let $h_t: f \simeq g$. Define a path $\sigma: I \to Y$ by taking $\sigma(t) = h_t(x)$ for each $t \in I$. Since σ joins $f(x)$ to $g(x)$, it follows that $f_*\alpha = g_*\alpha$. ∎

We recall that a map $f: (X, A, x_0) \to (Y, B, y_0)$ is said to be a *homotopy equivalence* if there exists a map $g: (Y, B, y_0) \to (X, A, x_0)$ such that gf and fg are homotopic with the identity maps on the triplets (X, A, x_0) and (Y, B, y_0) respectively. As an immediate consequence of the properties I, II and V, we have the following

Corollary 8.1. *If $f: (X, A, x_0) \to (Y, B. y_0)$ is a homotopy equivalence, then the induced transformation f_* sends $\pi_n(X, A, x_0)$ onto $\pi_n(Y, B, y_0)$ in a one-to-one fashion.*

The significance of (8.1) is that $\pi_n(X, A, x_0)$ depends only on the homotopy type of (X, A, x_0). In particular, if A is a strong deformation retract of X, then i_* sends $\pi_n(A, x_0)$ onto $\pi_n(X, x_0)$ in a one-to-one fashion for every $n \geqslant 0$. Hence, by Property IV, this implies that $\pi_n(X, A, x_0) = 0$ for every $n > 0$.

9. The fibering property

Consider two given triplets (X, A, x_0) and (Y, B, y_0) and a given map

$$f : (X, A, x_0) \to (Y, B, y_0).$$

We are concerned with the notion of *fibering* as defined in (III; § 3).

Property VI. *If $f : X \to Y$ is a fibering and $A = f^{-1}(B)$, then the transformation*

$$f_* : \pi_n(X, A, x_0) \to \pi_n(Y, B, y_0)$$

sends $\pi_n(X, A, x_0)$ onto $\pi_n(Y, B, y_0)$ in a one-to-one fashion for every $n > 0$.

Proof. To prove that f_* is onto, let α be an arbitrary element of $\pi_n(Y, B, y_0)$. By § 3, α is represented by a map

$$\phi : (I^n, I^{n-1}, J^{n-1}) \to (Y, B, y_0).$$

Since J^{n-1} is a strong deformation retract of I^n, it follows from (v) of (III; 3.1) that there exists a map $\psi : I^n \to X$ such that $f\psi = \phi$ and $\psi(J^{n-1}) = x_0$. Since $A = f^{-1}(B)$, $f\psi = \phi$ implies that $\psi(I^{n-1}) \subset A$ and hence we obtain a map

$$\psi : (I^n, I^{n-1}, J^{n-1}) \to (X, A, x_0).$$

This map ψ represents an element β of $\pi_n(X, A, x_0)$. Since $f\psi = \phi$, we have $f_*\beta = \alpha$. This proves that f_* is onto.

To prove that f_* is one-to-one, let α and β be elements of $\pi_n(X, A, x_0)$ such that $f_*\alpha = f_*\beta$. Choose representative maps

$$\phi, \psi : (I^n, I^{n-1}, J^{n-1}) \to (X, A, x_0)$$

for α and β respectively. Since $f_*\alpha = f_*\beta$, the composed maps $f\phi$ and $f\psi$ represent the same element of $\pi_n(Y, B, y_0)$. Hence, there exists a map

$$F : (I^n \times I, I^{n-1} \times I, J^{n-1} \times I) \to (Y, B, y_0)$$

such that $F(z, 0) = f\phi(z)$ and $F(z, 1) = f\psi(z)$ for each $z \in I^n$. Consider the closed subspace

$$T = (I^n \times 0) \cup (J^{n-1} \times I) \cup (I^n \times 1)$$

of $I^n \times I$ and define a map $G : T \to X$ by setting

$$G(z, t) = \begin{cases} \phi(z), & (z \in I^n, t = 0), \\ x_0, & (z \in J^{n-1}, t \in I), \\ \psi(z), & (z \in I^n, t = 1). \end{cases}$$

Then we have $fG = F \mid T$. Since T is clearly a strong deformation retract of $I^n \times I$, it follows from (v) of (III; 3.1) that G has an extension $G^* : I^n \times I \to X$ such that $fG^* = F$. Since F maps $I^{n-1} \times I$ into B and $A = f^{-1}(B)$, the condition $fG^* = F$ implies that $G^*(I^{n-1} \times I) \subset A$. Hence we obtain a map

$$G^* : (I^n \times I, I^{n-1} \times I, J^{n-1} \times I) \to (X, A, x_0)$$

with $G^*(z, 0) = \phi(z)$ and $G^*(z, 1) = \psi(z)$ for each $z \in I^n$. This proves that ϕ and ψ represent the same element of $\pi_n(X, A, x_0)$. Hence $\alpha = \beta$ and f_* is one-to-one. **I**

In homology theory, the corresponding fibering property is false in general. Instead of this, we have the famous *excision property* which does not hold in homotopy theory.

As a special case, let us consider the derived projection $p : (X', A', x'_0) \to (X, A, x_0)$. By Properties VI and IV, p_* and ∂ in the following diagram

$$\pi_n(X, A, x_0) \xleftarrow{p_*} \pi_n(X', A', x'_0) \xrightarrow{\partial} \pi_{n-1}(A', x'_0)$$

are both one-to-one and onto. Furthermore, if $\pi_n(X, A, x_0)$ and $\pi_{n-1}(A', x'_0)$ are identified as in (3.1) one can easily see that $p_* = \partial$. On the other hand, the identification in (3.1) may be considered as being effected by the one-to-one correspondence

$$\chi = \partial p_*^{-1} : \pi_n(X, A, x_0) \to \pi_{n-1}(A', x'_0)$$

which will be called the *natural correspondence*.

10. The triviality property

If X is a space which consists of a single point x_0, then, for each n, the constant map $f(I^n) = x_0$ is the only map of I^n into X. Hence we have the following

Property VII. *If X is a space consisting of a single point x_0, then $\pi_n(X, x_0) = 0$ for every $n \geqslant 0$.*

This property plays the role similar to that of the *dimension property* in homology theory. Since it apparently has nothing to do with the choice of the dimension n in $\pi_n(X, x_0)$, we propose to call it the *triviality property*.

11. Homotopy systems

In the preceding sections, we constructed geometrically the homotopy groups $\pi_n(X, A, x_0)$ and established seven basic properties of these groups. In next few sections, we shall show that they are characteristic; in fact, these seven properties, stated in a certain apparently weaker form, together with $\pi_0(X, x_0)$ for all pairs (X, x_0) determines all $\pi_n(X, A, x_0)$ for all triplets (X, A, x_0) up to one-to-one correspondence.

A *homotopy system*

$$\mathrm{H} = \{\pi, \partial, {}_*\}$$

consists of three functions π, ∂ and ${}_*$. The first function π assigns to each triplet (X, A, x_0) and each integer $n > 0$ an abstract set $\pi_n(X, A, x_0)$. The second function ∂ assigns to each triplet (X, A, x_0) and each integer $n > 0$ a transformation

$$\partial : \pi_n(X, A, x_0) \to \pi_{n-1}(A, x_0),$$

where, in the case of $n = 1$, $\pi_0(A, x_0)$ denotes the set of all path-components of A as in § 2. The third function $_*$ assigns to each map $f: (X, A, x_0) \to (Y, B, y_0)$ and each integer $n > 0$ a transformation

$$f_* : \pi_n(X, A, x_0) \to \pi_n(Y, B, y_0).$$

Furthermore, the system H must satisfy the following seven axioms:

Axiom I. *If $f: (X, A, x_0) \to (X, A, x_0)$ is the identity map, then f_* is the identity transformation on $\pi_n(X, A, x_0)$ for every $n > 0$.*

Axiom II. *If $f: (X, A, x_0) \to (Y, B, y_0)$ and $g: (Y, B, y_0) \to (Z, C, z_0)$ are maps, then for every $n \geqslant 0$ we have $(gf)_* = g_* f_*$.*

Axiom III. *If $f: (X, A, x_0) \to (Y, B, y_0)$ is a map and $g: (A, x_0) \to (B, y_0)$ is the restriction of f, then the commutativity relation $\partial f_* = g_* \partial$ holds in the following rectangle for every $n > 0$:*

$$\begin{array}{ccc} \pi_n(X, A, x_0) & \xrightarrow{\partial} & \pi_{n-1}(A, x_0) \\ \downarrow f_* & & \downarrow g_* \\ \pi_n(Y, B, y_0) & \xrightarrow{\partial} & \pi_{n-1}(B, y_0) \end{array}$$

where, in the case of $n = 1$, $g_* : \pi_0(A, x_0) \to \pi_0(B, y_0)$ denotes the induced transformation in § 5.

Let (X, A, x_0) be any given triplet and consider the inclusion maps

$$i: (A, x_0) \subset (X, x_0), \quad j: (X, x_0) \subset (X, A, x_0).$$

The transformations i_*, j_* and ∂ form a beginningless sequence as in § 7 which will be called the *homotopy sequence* of the triplet (X, A, x_0) in the system H.

Axiom IV. *The homotopy sequence of any triplet (X, A, x_0) is weakly exact.*

This means that, if $\pi_n(X, x_0) = 0$ for all $n \geqslant 0$, then ∂ sends $\pi_n(X, A, x_0)$ onto $\pi_{n-1}(A, x_0)$ in a one-to-one fashion for every $n > 0$.

Axiom V. *If the maps $f, g: (X, x_0) \to (Y, y_0)$ are homotopic, then $f_* = g_*$ for every $n > 0$.*

Axiom VI. *If $p: (X', A', x'_0) \to (X, A, x_0)$ is the derived projection over (X, A, x_0), then p_* sends $\pi_n(X', A', x'_0)$ onto $\pi_n(X, A, x_0)$ in a one-to-one fashion for every $n > 0$.*

Axiom VII. *If X is a space consisting of a single point x_0, then $\pi_n(X, x_0) = 0$ for every $n > 0$.*

Since the derived projection $p: X' \to X$ is a fibering by (III; 13.4), it follows that the axioms I–VII are weaker than the properties I–VII respectively. Hence, if we neglect the group operation in $\pi_n(X, A, x_0)$, the three functions $\pi, \partial, _*$ as defined in §§ 3–5 constitute a homotopy system. This proves the *existence* of homotopy systems. One can also easily construct a homotopy system by induction and without using any geometrical representation, (see Ex. A at the end of the chapter).

12. The uniqueness theorem

Two homotopy systems $H = \{\pi, \partial, _*\}$ and $H' = \{\pi', \partial', \#\}$ are said to be *equivalent* if there exists, for each triplet (X, A, x_0) and each $n > 0$, a transformation

$$h_n : \pi_n(X, A, x_0) \to \pi'_n(X, A, x_0)$$

satisfying the conditions:

($E1$) h_n sends $\pi_n(X, A, x_0)$ onto $\pi'_n(X, A, x_0)$ in a one-to-one fashion.

($E2$) For each triplet (X, A, x_0) and each $n > 0$ the commutativity relation $h_{n-1}\partial = \partial' h_n$ holds in the following rectangle:

$$\begin{array}{ccc} \pi_n(X, A, x_0) & \xrightarrow{\partial} & \pi_{n-1}(A, x_0) \\ \downarrow h_n & & \downarrow h_{n-1} \\ \pi'_n(X, A, x_0) & \xrightarrow{\partial'} & \pi'_{n-1}(A, x_0), \end{array}$$

where, in the case of $n = 1$, h_0 denotes the identity map.

($E3$) For each map $f : (X, A, x_0) \to (Y, B, y_0)$, the commutativity relation $h_n f_* = f_{\#} h_n$ holds in the following rectangle:

$$\begin{array}{ccc} \pi_n(X, A, x_0) & \xrightarrow{f_*} & \pi_n(Y, B, y_0) \\ \downarrow h_n & & \downarrow h_n \\ \pi'_n(X, A, x_0) & \xrightarrow{f\#} & \pi'_n(Y, B, y_0). \end{array}$$

A collection of transformations $h = \{h_n\}$ satisfying the conditions ($E1$) through ($E3$) is called an *equivalence* between the homotopy systems H and H' and is denoted by

$$h : H \approx H'.$$

Theorem 12.1. *Any two homotopy systems are equivalent.* [Milnor 1].

Proof. Let $H = \{\pi, \partial, _*\}$ and $H' = \{\pi', \partial', \#\}$ be any two homotopy systems. We are going to construct an equivalence $h : H \approx H'$ as follows.

Let $n \geqslant 1$ and assume that we have already constructed the transformations

$$h_m : \pi_m(X, A, x_0) \to \pi'_m(X, A, x_0)$$

for each $m < n$ and each triplet (X, A, x_0) such that the conditions ($E1$) through ($E3$) are satisfied. Let us construct h_n as follows.

Let (X, A, x_0) be any triplet. Consider the derived projection $p : (X', A', x'_0) \to (X, A, x_0)$. By Axiom VI, p_* sends $\pi_n(X', A', x'_0)$ onto $\pi_n(X, A, x_0)$ in a one-to-one fashion and analogously for $p_{\#}$. According to (III; 10.1), X' is contractible to the point x'_0. Hence, by Axioms I, II, V and VII, we obtain

$\pi_m(X', x'_0) = 0$ and $\pi'_m(X', x'_0) = 0$ for every $m \geqslant 0$. By Axiom IV, ∂ sends $\pi_n(X', A,' x'_0)$ onto $\pi_{n-1}(A', x'_0)$ in a one-to-one fashion, and analogously for ∂'. By our assumption of induction, h_{n-1} sends $\pi_{n-1}(A', x'_0)$ onto $\pi'_{n-1}(A', x'_0)$ in a one-to-one fashion. Hence we may define a transformation

$$h_n : \pi_n(X, A, x_0) \to \pi'_n(X, A, x_0)$$

by taking

$$h_n = p_{\#}\partial'^{-1}h_{n-1}\partial p_*^{-1},$$

where, in the case of $n = 1$, h_0 denotes the identity map.

Since $(E1)$ is obviously satisfied, it remains to verify $(E2)$ and $(E3)$.

To check $(E2)$, let $q = p \mid (A', x'_0)$. Then we have

$$\begin{aligned}\partial' h_n &= \partial' p_{\#}\partial'^{-1}h_{n-1}\partial p_*^{-1} = q_{\#}\partial'\partial'^{-1}h_{n-1}\partial p_*^{-1} \\ &= q_{\#}h_{n-1}\partial p_*^{-1} = h_{n-1}q_*\partial p_*^{-1} = h_{n-1}\partial\end{aligned}$$

by Axiom III. This proves that h_n satisfies $(E2)$.

To check $(E3)$, let (Y, B, y_0) be a second triplet and $f : (X, A, x_0) \to (Y, B, y_0)$ be any given map. Let

$$r : (Y', B', y'_0) \to (Y, B, y_0), \quad f' : (X', A', x'_0) \to (Y', B', y'_0)$$

denote respectively the derived projection over (Y, B, y_0) and the derived map of f. The relation $fp = rf'$ is satisfied. Set $\bar{f} = f' \mid (A', x'_0)$. Then we have

$$\begin{aligned}f_{\#}h_n &= f_{\#}p_{\#}\partial'^{-1}h_{n-1}\partial p_*^{-1} = r_{\#}f'_{\#}\partial'^{-1}h_{n-1}\partial p_*^{-1} \\ &= r_{\#}\partial'^{-1}\bar{f}_{\#}h_{n-1}\partial p_*^{-1} = r_{\#}\partial'^{-1}h_{n-1}\bar{f}_*\partial p_*^{-1} \\ &= r_{\#}\partial'^{-1}h_{n-1}\partial f'_* p_*^{-1} = r_{\#}\partial'^{-1}h_{n-1}\partial r_*^{-1}f_* = h_n f_*\end{aligned}$$

by Axioms II and III. This proves that h_n satisfies $(E3)$.

Thus we have completed the inductive construction of an equivalence $h = \{ h_n \}$ between H and H' which will be called the *natural equivalence* $h : H \approx H'$. ▮

The natural equivalence is the only possible equivalence between H and H' as will be seen in Ex. B at the end of the chapter.

The uniqueness theorem (12.1) shows that the homotopy system constructed geometrically in §§ 3–5 is essentially the only homotopy system. As a consequence of this, it follows that the set of seven properties in §§ 6–10 is equivalent to the set of seven axioms in § 11 which are apparently weaker. In fact, one can deduce the seven properties right from the axioms without using any geometrical representation of the sets $\pi_n(X, A, x_0)$. The details of the proof will be left to the reader as an exercise. See Ex. C at the end of the chapter.

13. The group structures

In the last two sections, we have shown that, apart from the group structures in the homotopy sets $\pi_n(X, A, x_0)$, the homotopy system constructed in §§ 3–5 is completely characterized by the seven axioms in § 11 and the sets $\pi_0(X, x_0)$. To complete the axiomatic approach, it remains to determine all the possible group structures that can be introduced into the essentially unique homotopy system

$$H = \{\pi, \partial, {}_*\}.$$

For this purpose, let us first consider the sets $\pi_1(X, x_0)$ in H. According to the uniqueness theorem (12.1), we may assume $\pi_1(X, x_0)$ to be the underlying set of the fundamental group of X at x_0. The product in $\pi_1(X, x_0)$ as defined in (II; § 5) will be called the *customary product* which is denoted by juxtaposition. The *reverse* of this product, denoted by a dot, is defined by

$$\alpha \cdot \beta = \beta\alpha, \quad (\alpha, \beta \in \pi_1(X, x_0)).$$

For any map $f : (X, x_0) \to (Y, y_0)$, the transformation $f_* : \pi_1(X, x_0) \to \pi_1(Y, y_0)$ is a homomorphism under the customary product as well as its reverse. These are the only group structures in all $\pi_1(X, x_0)$ such that f_* is a homomorphism for every map f. In fact, we have the following

Lemma 13.1. *There are exactly two ways of introducing a group structure into the sets $\pi_1(X, x_0)$ in such a way that f_* is a homomorphism for every map $f : (X, x_0) \to (Y, y_0)$. These two group structures are defined by the customary product and its reverse respectively.*

Proof. Assume that there is a new product, denoted by $\alpha \circ \beta$, in each $\pi_1(X, x_0)$ such that, for each pair (X, x_0), $\pi_1(X, x_0)$ is a group under this product and that f_* is a homomorphism with respect to this product for every map $f : (X, x_0) \to (Y, y_0)$. It suffices to prove that either $\alpha \circ \beta = \alpha\beta$ for any $\alpha, \beta \in \pi_1(X, x_0)$ of every (X, x_0) or $\alpha \circ \beta = \beta\alpha$ similarly.

Let Z be the space which consists of two circles, intersecting at a single point z_0. According to (II; Ex. A5), $\pi_1(Z, z_0)$ is a free group on two generators a and b. For any two elements α, β of $\pi_1(X, x_0)$, obviously there is a map

$$f : (Z, z_0) \to (X, x_0)$$

such that $f_*(a) = \alpha$ and $f_*(b) = \beta$. Since f_* is a homomorphism under the new product, we have

$$f_*(a \circ b) = \alpha \circ \beta.$$

In terms of the customary group structure of $\pi_1(Z, z_0)$, $a \circ b$ is equal to some word $w(a, b)$ of the free group. Since f_* is also a homomorphism under the customary product, we have

$$f_*(a \circ b) = f_*[w(a, b)] = w[f_*(a), f_*(b)] = w(\alpha, \beta).$$

This implies that $\alpha \circ \beta = w(\alpha, \beta)$. Thus it remains to prove that either $w(a, b) = ab$ or $w(a, b) = ba$.

For this purpose, let us first prove that the word $w(a, b)$ has the following two propeties:

$$w(a, 1) = a, \quad w(1, b) = b, \tag{1}$$

$$w[a, w(b, c)] = w[w(a, b), c], \tag{2}$$

where (2) is an identity in the free group on three generators a, b, c.

To prove (1), note that the identity element 1 of $\pi_1(Z, z_0)$ can be defined as the image of the homomorphism

$$i_* : \pi_1(z_0, z_0) \to \pi_1(Z, z_0)$$

induced by the inclusion map $i : (z_0, z_0) \subset (Z, z_0)$. It follows that the new product must have this same identity element. Hence

$$w(a, 1) = a \circ 1 = a, \quad w(1, b) = 1 \circ b = b.$$

To prove (2), choose X to be the space which consists of three circles tangent to each other at the same point x_0. Then, as in (II; Ex. A5), $\pi_1(X, x_0)$ is a free group on three generators a, b, c. Since $a \circ b = w(a, b)$ and $b \circ c = w(b, c)$, the associative law for the new product implies (2).

Finally, we shall complete the proof by showing that, if a reduced word, $w(a, b)$ in the free group on two generators satisfies the conditions (1) and (2), then either $w(a, b) = ab$ or $w(a, b) = ba$. The proof is a long but easy exercise in the manipulation of reduced words sketched as follows.

Let $w(a, b)$ be a reduced word which satisfies (1) and (2). By (1), $w(a, b) \neq a^m$. More generally, it is impossible to have

$$w(a, b) = a^{m_1} b^{n_1} \cdots a^{m_k} b^{n_k} a^{m_{k+1}}$$

with non-zero (positive or negative) integers $m_1, n_1, \cdots, m_k, n_k$, and m_{k+1}. To see this, let us assume that $w(a, b)$ were of this form with $k \geqslant 1$. Then, by (1), one can easily see that the reduced words of $(w(a, b))^{m_1}$ and $(w(b, c))^{n_1}$ would be of the same form, say,

$$(w(a, b))^{m_1} = a^{p_1} b^{q_1} \cdots a^{p_i} b^{q_i} a^{p_{i+1}},$$

$$(w(b, c))^{n_1} = b^{r_1} c^{s_1} \cdots b^{r_j} c^{s_j} b^{r_{j+1}},$$

with non-zero integral exponents. By (1), $i \geqslant 1$ and $j \geqslant 1$. Then we would have

$$w[a, w(b, c)] = a^{m_1} b^{r_1} c^{s_1} \cdots,$$

$$w[w(a, b), c] = a^{p_1} b^{q_1} a^{p_2} \cdots$$

as their reduced words. This contradicts (2).

Similarly, $w(a, b) \neq b^m$ and it is impossible to have

$$w(a, b) = b^{m_1} a^{n_1} \cdots b^{m_k} a^{n_k} b^{m_{k+1}}$$

with non-zero integral exponents.

Next, assume that $w(a, b) = a^{m_1} b^{n_1} \cdots a^{m_k} b^{n_k}$

with non-zero integral exponents. If $m_1 < 0$, then the reduced word of $w[w(a, b), c]$ begins with b^{-n_k} while that of $w[a, w(b, c)]$ begins with a^{m_1}. This contradicts (2) and hence we have $m_1 > 0$. If $n_1 < 0$, then the reduced word of $w[a, w(b, c)]$ begins with $a^{m_1}c^{-n_k}$ while that of $w[w(a, b), c]$ begins with $a^{m_1}b^{n_1}$. This contradicts (2) and so we have both $m_1 > 0$ and $n_1 > 0$. Then we have

$$w[a, w(b, c)] = a^{m_1}b^{m_1}c^{n_1}\cdots,$$

$$w[w(a, b), c] = a^{m_1}b^{n_1}\cdots a^{m_k}b^{n_k}\cdots$$

as their reduced words. By (2), it follows that $k = 1$ and hence $w(a, b) = a^{m_1}b^{n_1}$. Then, (1) implies that $m_1 = 1$ and $n_1 = 1$. Consequently, we have $w(a, b) = ab$.

Similarly, we can prove that, if

$$w(a, b) = b^{m_1}a^{n_1}\cdots b^{m_k}a^{n_k}$$

with non-zero integral exponents, then $w(a, b) = ba$. This exhausts all possible cases. ∎

Theorem 13.2. *There are exactly two ways of introducing a group structure into the sets $\pi_n(X, A, x_0)$, $n \geqslant 2$, and $\pi_1(X, x_0)$, in such a way that the transformations ∂ and f_* are homomorphisms. These two group structures are defined respectively by the customary group operation given in § 3 and its reverse.* [Milnor 1].

Proof. Let us denote the customary group operation by juxtaposition and assume that there is a new group operation, denoted by $\alpha \circ \beta$, in the sets $\pi_n(X, A, x_0)$, $n \geqslant 2$, and $\pi_1(X, x_0)$ such that the transformations ∂ and f_* are homomorphisms. We have to show that $\alpha \circ \beta$ is equal to $\alpha\beta$ or $\beta\alpha$. By (13.1), this is true if α and β are in $\pi_1(X, x_0)$. To prove the theorem by induction on n, consider the natural one-to-one correspondence

$$\chi = \partial p_*^{-1} : \pi_n(X, A, x_0) \to \pi_{n-1}(A', x'_0)$$

of § 9. Since χ is a homomorphism for both group operations, we have

$$\chi(\alpha \circ \beta) = \chi(\alpha) \circ \chi(\beta), \quad \chi(\alpha\beta) = \chi(\alpha)\,\chi(\beta)$$

for any two elements α, β in $\pi_n(X, A, x_0)$. By the inductive hypothesis, $\chi(\alpha) \circ \chi(\beta)$ is equal to $\chi(\alpha)\,\chi(\beta)$ or $\chi(\beta)\,\chi(\alpha)$. Hence $\chi(\alpha \circ \beta)$ is equal to $\chi(\alpha\beta)$ or $\chi(\beta\alpha)$. Since χ is one-to-one this implies $\alpha \circ \beta = \alpha\beta$ or $\beta\alpha$. ∎

The significance of (13.2) is that the group structure in the essentially unique homotopy system $H = \{\pi, \partial, {}_*\}$ is also essentially unique.

This completes the axiomatic approach.

14. The role of the basic point

In the notion of the homotopy groups $\pi_n(X, x_0)$ and $\pi_n(X, A, x_0)$ the basic point x_0 is explicity used in any geometrical construction of these groups. The objective of the next few sections is to study the role played by the basic point, to compare the homotopy groups with various basic points, and to free these groups from the basic point wherever it is possible.

Let us consider a given space X and two given points x_0, x_1 connected by a given path

$$\sigma : I \to X, \quad \sigma(0) = x_0, \quad \sigma(1) = x_1.$$

By the definition in § 2, we have

$$\pi_0(X, x_0) = P(X) = \pi_0(X, x_1)$$

as the set of all path-components of X. Moreover, since x_0 and x_1 are contained in the same path-component of X, the neutral element of $\pi_0(X, x_0)$ is the same as that of $\pi_0(X, x_1)$. Let us denote by

$$\sigma_0 : \pi_0(X, x_1) \to \pi_0(X, x_0)$$

the identity map on $\pi_0(X, x_1) = \pi_0(X, x_0)$.

Theorem 14.1. *For each $n > 0$, every path $\sigma : I \to X$ gives in a natural way an isomorphism*

$$\sigma_n : \pi_n(X, x_1) \approx \pi_n(X, x_0), \quad x_0 = \sigma(0), \quad x_1 = \sigma(1),$$

which depends only on the homotopy class of the path σ (relative to end points). If σ is the degenerate path $\sigma(I) = x_0$, then σ_n is the identity automorphism. If σ, τ are paths with $\tau(0) = \sigma(1)$, then $(\sigma\tau)_n = \sigma_n\tau_n$. Finally, for each path $\sigma : I \to X$ and each map $f : X \to Y$, we have a commutative rectangle

$$\begin{array}{ccc} \pi_n(X, x_1) & \xrightarrow{\sigma_n} & \pi_n(X, x_0) \\ \downarrow f_* & & \downarrow f_* \\ \pi_n(Y, y_1) & \xrightarrow{\tau_n} & \pi_n(Y, y_0), \end{array}$$

where $\tau = f\sigma$, $y_0 = f(x_0)$ and $y_1 = f(x_1)$.

As an immediate consequence of (14.1), we deduce the following

Corollary 14.2. *The fundamental group $\pi_1(X, x_0)$ acts on $\pi_n(X, x_0)$, $n \geqslant 1$, as a group of automorphisms.*

To prove (14.1), let us construct σ_n as follows. Let α be any element of $\pi_n(X, x_1)$ and choose a representative map

$$f : (I^n, \partial I^n) \to (X, x_1)$$

for α. The geometrical idea of the construction is to pull the image of ∂I^n along the path σ back to the point x_0 with the image of I^n being dragged in an arbitrary way. The map obtained after this homotopy represents an element β of $\pi_n(X, x_0)$ which depends only on α and the homotopy class of σ. Then, we define $\sigma_n(\alpha) = \beta$. The details are as follows.

First, let us prove that there exists a homotopy $f_t : I^n \to X$, $(0 \leqslant t \leqslant 1)$, of f such that $f_t(\partial I^n) = \sigma(1 - t)$ for every $t \in I$. For this purpose, define a partial homotopy $\phi_t : \partial I^n \to X$, $(0 \leqslant t \leqslant 1)$, of f by taking $\phi_t(\partial I^n) = \sigma(1 - t)$

for each $t \in I$. By (I; 9.2), ∂I^n has the AHEP in I^n. Hence the homotopy ϕ_t has an extension $f_t : I^n \to X$, $(0 \leqslant t \leqslant 1)$, such that $f_0 = f$. We call f_t a *homotopy of f along σ*.

Since f_1 maps ∂I^n into $\sigma(0) = x_0$, it represents an element β of $\pi_n(X, x_0)$. That β depends only on α and the homotopy class of σ is an obvious consequence of the following

Lemma 14.3. *If $f, g : (I^n, \partial I^n) \to (X, x_1)$ are maps homotopic relative to ∂I^n, $\sigma, \tau : I \to X$ are paths homotopic relative to end points, and $f_t, g_t : I^n \to X$, $(0 \leqslant t \leqslant 1)$, are homotopies of f along σ and of g along τ respectively, then f_1, g_1 are homotopic relative to ∂I^n.*

Proof. Define two maps $F, G : I^n \times I \to X$ by means of the homotopies f_t, g_t as usual. Consider the subspace $A = (I^n \times 0) \cup (\partial I^n \times I)$ of $I^n \times I$. By the hypothesis $f \simeq g$ and $\sigma \simeq \tau$, it follows that $F \mid A$ abd $G \mid A$ are homotopic relative to $\partial I^n \times 0$ and $\partial I^n \times 1$. Since A has the AHEP in $I^n \times I$ by (I; 9.2), there is a homotopy $F_t : I^n \times I \to X$, $(0 \leqslant t \leqslant 1)$, such that $F_0 = F$, $F_1 \mid A = G \mid A$, and $F_t(\partial I^n \times 1) = x_0$ for every $t \in I$. The map F_1 gives a homotopy $h_t : I^n \to X$, $(0 \leqslant t \leqslant 1)$, of g along τ. Since $F_t(\partial I^n \times 1) = x_0$ for each $t \in I$, it follows that f_1 and h_1 are homotopic relative to ∂I^n. It remains to prove that g_1 and h_1 are homotopic relative to ∂I^n.

Define a map $M : I^n \times I \to X$ by taking

$$M(p, q) = \begin{cases} g_{1-2q}(p), & (p \in I^n, 0 \leqslant q \leqslant \frac{1}{2}). \\ h_{2q-1}(p), & (p \in I^n, \frac{1}{2} \leqslant q \leqslant 1). \end{cases}$$

Then, for each $p \in \partial I^n$, we have $M(p, q) = M(p, 1 - q)$. Therefore we may define a partial homotopy $N_t : B \to X$, $(0 \leqslant t \leqslant 1)$, of M on the boundary $B = \partial(I^n \times I) = (\partial I^n \times I) \cup (I^n \times \partial I)$ by taking

$$N_t(p, q) = \begin{cases} M(p, q), & (p \in I^n, q \in \partial I), \\ M(p, q - tq), & (p \in \partial I^n, 0 \leqslant q \leqslant \frac{1}{2}), \\ N_t(p, 1 - q), & (p \in \partial I^n, \frac{1}{2} \leqslant q \leqslant 1). \end{cases}$$

Since B has the AHEP in $I^n \times I$, the homotopy N_t has an extension $M_t : I^n \times I \to X$, $(0 \leqslant t \leqslant 1)$, such that $M_0 = M$. The map M_1 gives a homotopy $k_t : I^n \to X$, $(0 \leqslant t \leqslant 1)$, such that $k_0 = g_1$, and $k_1 = h_1$. Since $k_t(\partial I^n) = M_1(\partial I^n \times t) = x_0$, this implies that g_1 and h_1 are homotopic relative to ∂I^n. ▮

Now let us continue the proof of (14.1). We have constructed a transformation

$$\sigma_n : \pi_n(X, x_1) \to \pi_n(X, x_0)$$

which depends only on the homotopy class of the path σ. Since the last two assertions of (14.1) are obvious, it remains to prove that σ_n is an isomorphism.

Let α, β be arbitrary elements of $\pi_n(X, x_1)$ represented by the maps $f, g : (I^n, \partial I^n) \to (X, x_1)$. Let $f_t, g_t : I^n \to X$, $(0 \leqslant t \leqslant 1)$, be homotopies along

σ of f, g respectively; then f_1 represents $\sigma_n(\alpha)$ and g_1 represents $\sigma_n(\beta)$. Define a homotopy $h_t : I^n \to X$, $(0 \leqslant t \leqslant 1)$, by taking $h_t = f_t + g_t$. Then h_0 represents $\alpha + \beta$, h_1 represents $\sigma_n(\alpha) + \sigma_n(\beta)$, and h_t is a homotopy along σ. This implies that $\sigma_n(\alpha + \beta) = \sigma_n(\alpha) + \sigma_n(\beta)$ and hence σ_n is a homomorphism.

Finally, let us prove that the homomorphism σ_n is an isomorphism. Let τ denote the reverse of σ, that is to say, $\tau(t) = \sigma(1 - t)$ for every $t \in I$. Because of symmetry, it suffices to show that σ_n is an epimorphism and τ_n is a monomorphism. For this purpose, consider the product $\sigma\tau$. Then we have $\sigma_n\tau_n = (\sigma\tau)_n$. Since τ is the reverse of σ, it is clear that $\sigma\tau$ is homotopic to the degenerate path at x_0 with end points fixed. It follows that $(\sigma\tau)_n$ is the identity automorphism on $\pi_n(X, x_0)$. By an elementary group-theoretic argument, it follows that σ_n is an epimorphism and τ_n is a monomorphism. This completes the proof of (14.1).

Thus, for a pathwise connected space X, all the groups $\pi_n(X, x_0)$ with various basic points $x_0 \in X$ are isomorphic. Hence, as an abstract group, $\pi_n(X, x_0)$ does not depend on the basic point x_0 and may be denoted simply by $\pi_n(X)$. This abstract group $\pi_n(X)$ will be called the n-th (*abstract*) *homotopy group* of the pathwise connected space X. In this terminology, we have

$$\pi_1(S^1) = Z, \quad \pi_n(S^1) = 0, \quad (n > 1).$$

$$\pi_m(S^n) = 0, \quad (m < n), \quad \pi_n(S^n) = Z.$$

$$\pi_3(S^2) = Z.$$

Now, let us consider the spaces of loops

$$W_0 = [X ; x_0, x_0], \quad W_1 = [X ; x_1, x_1]$$

as well as the degenerate loops $w_0 \in W_0$ and $w_1 \in W_1$. Then a given path $\sigma : I \to X$ which connects x_0 to x_1 induces a map $\xi : W_1 \to W_0$ defined as follows: for each $w \in W_1$, $\xi(w) \in W_0$ is the loop defined by

$$[\xi(w)](t) = \begin{cases} \sigma(3t), & (\text{if } 0 \leqslant t \leqslant \frac{1}{3}), \\ w(3t - 1), & (\text{if } \frac{1}{3} \leqslant t \leqslant \frac{2}{3}), \\ \sigma(3 - 3t), & (\text{if } \frac{2}{3} \leqslant t \leqslant 1). \end{cases}$$

Intuitively speaking, $\xi(w)$ is the loop traced by running first from x_0 to x_1 along the path σ, next around the loop w once, and then back to x_0 along the reverse of σ. On the other hand, σ also induces a path $\eta : I \to W_0$ defined as follows: for each $s \in I$, $\eta(s) \in W_0$ is the loop defined by

$$[\eta(s)](t) = \begin{cases} \sigma(3st), & (\text{if } 0 \leqslant t \leqslant \frac{1}{3}), \\ \sigma(s), & (\text{if } \frac{1}{3} \leqslant t \leqslant \frac{2}{3}), \\ \sigma(3s - 3st), & (\text{if } \frac{2}{3} \leqslant t \leqslant 1). \end{cases}$$

Then, obviously we have $\eta(0) = w_0$ and $\eta(1) = \xi(w_1)$.

Proposition 14.4. *If $\xi : (W_1, w_1) \to (W_0, \xi(w_1))$ and $\eta : I \to W_0$ are induced by the path σ, the commutativity holds in the following diagram*

$$\begin{array}{ccccc}
\pi_n(X, x_1) & & \xrightarrow{\quad\sigma_n\quad} & & \pi_n(X, x_0) \\
\downarrow \chi & & & & \downarrow \chi \\
\pi_{n-1}(W_1, w_1) & \xrightarrow{\xi_*} & \pi_{n-1}(W_0, \xi(w_1)) & \xrightarrow{\eta_{n-1}} & \pi_{n-1}(W_0, w_0)
\end{array}$$

where n is any positive integer and χ denotes the natural correspondence given by (2.2).

Proof. Let $\alpha \in \pi_n(X, x_1)$ and choose a representative map $f : (I^n, \partial I^n) \to (X, x_1)$ for α. Then it is not difficult to see that both $\sigma_n(\alpha)$ and $\chi^{-1}\eta_{n-1}\xi_*\chi(\alpha)$ are represented by the map $g : (I^n, \partial I^n) \to (X, x_0)$ defined as follows:

$$g(t) = \begin{cases} \sigma(3\theta(t)), & (\text{if } \theta(t) \leqslant \tfrac{1}{3}), \\ f(3t_1 - 1, \cdots, 3t_n - 1), & (\text{if } \theta(t) \geqslant \tfrac{1}{3}), \end{cases}$$

where $t = (t_1, \cdots, t_n)$ is an arbitrary point of I^n and $\theta(t)$ denotes the smallest of the $2n$ real numbers $t_1, \cdots, t_n, 1 - t_1, \cdots, 1 - t_n$. ∎

The property (14.4) of the operations σ_n is characteristic. To formulate this fact precisely, let us define the notion of a system of operations as follows. By a *system of operations* in a homotopy system $H = \{\pi, \partial, *\}$, we mean for each path $\sigma : I \to X$ in any space X and each integer $n \geqslant 0$ a transformation

$$\sigma_n : \pi_n(X, x_1) \to \pi_n(X, x_0), \quad x_0 = \sigma(0), x_1 = \sigma(1)$$

such that σ_0 is the identity on $\pi_0(X, x_1) = \pi_0(X, x_0)$ and that (14.4) is satisfied.

Since (14.4) implies $\sigma_n = \chi^{-1}\eta_{n-1}\xi_*\chi$, the inductive proof of the following theorem is obvious.

Theorem 14.5. *In any given homotopy system $H = \{\pi, \partial, *\}$, there exists one and only one system of operations. Furthermore, for any two given homotopy systems H and H', the natural equivalence $h : H \approx H'$ commutes with the operations in H and H'.*

Analogously, for each path $\sigma : I \to A$, one can define the operations σ_n on the relative homotopy group and deduce similar results. See Ex. D at the end of the chapter.

15. Local system of groups

The homotopy groups together with the operations σ_n constructed in the preceding section motivated the notion of a local system of groups in a space X, [Steenrod 1].

We shall say that we have a *local system of groups* $\{G_x\}$ in a space X, if the following conditions are satisfied:

(LSG1) For each point $x \in X$, there is given a group G_x.

(LSG2) For each path $\sigma : I \to X$ joining x_0 to x_1, there is given a homomorphism $\sigma_{\#} : G_{x_1} \to G_{x_0}$.

(LSG3) If σ is the degenerate path $\sigma(I) = x_0$, then $\sigma_{\#}$ is the identity automorphism on G_{x0}.

(LSG4) If two paths $\sigma, \tau : I \to X$ are equivalent, i.e., if σ, τ have the same end-points and are homotopic with end-points held fixed, then $\sigma_{\#} = \tau_{\#}$.

(LSG5) If two paths $\sigma, \tau : I \to X$ are consecutive, i.e., if $\sigma(1) = \tau(0)$, then $(\sigma\tau)_{\#} = \sigma_{\#}\tau_{\#}$.

According to (14.1), the collection of the homotopy groups $\{\pi_n(X, x_0) \mid x_0 \in X\}$, for a given space X and a given integer $n > 0$, forms a local system of groups in X. Similarly, the collection of the relative homotopy groups $\{\pi_n(X, A, x_0) \mid x_0 \in A\}$, $n > 1$, forms a local system of groups in the subspace A of X.

As an easy consequence of (LSG3)–(LSG5), we deduce as in the proof of (14.1) that every $\sigma_{\#}$ is an isomorphism. Hence, if X is pathwise connected, then all the groups G_x, $x \in X$, are isomorphic.

Since the elements of the fundamental group $\pi_1(X, x_0)$ are homotopy classes of the loops $[X ; x_0, x_0]$ with end-points held fixed, we deduce as a direct consequence of (LSG3)–(LSG5) that, for each $x_0 \in X$, $\pi_1(X, x_0)$ acts as a group of operators (or automorphisms) on G_{x0} in the sense defined as follows.

A multiplicative group H is said to *act as a group of operators* on an additive group G, (or, simply H *acts on* G), if, for every $h \in H$ and every $g \in G$, an element $hg \in G$ is defined in such a way that

$$h(g_1 + g_2) = hg_1 + hg_2, \quad h_2(h_1 g) = (h_2 h_1)g, \quad 1g = g$$

where $g, g_1, g_2 \in G$, $h, h_1, h_2 \in H$ are arbitrary elements and $1 \in H$ denotes the neutral element.

Applying this to the local system of groups $\{\pi_n(X, x_0) \mid x_0 \in X\}$, where $n > 0$, we obtain (14.2) restated as follows.

Proposition 15.1. *For each $x_0 \in X$, the fundamental group $\pi_1(X, x_0)$ acts on the n-th homotopy group $\pi_n(X, x_0)$ as a group of operators.*

In the special case $n = 1$, one can easily see that, for any two elements g and h in $\pi_1(X, x_0)$, h acts on g as follows:

(15.2) $$h(g) = hgh^{-1}.$$

Similarly, for each $x_0 \in A$, $\pi_1(A, x_0)$ acts on the n-th relative homotopy group $\pi_n(X, A, x_0)$, $n > 1$, as a group of operators.

As a consequence of these operations, let us consider two given homotopic maps

$$f, g : X \to Y.$$

Let $h_t : X \to Y$, $(0 \leqslant t \leqslant 1)$, be a homotopy such that $h_0 = f$ and $h_1 = g$. Choose a point $x_0 \in X$ and denote $y_0 = f(x_0)$ and $y_1 = g(x_0)$. Define a path $\sigma : I \to Y$ by taking

$$\sigma(t) = h_t(x_0), \quad (t \in I);$$

then $\sigma(0) = y_0$ and $\sigma(1) = y_1$. According to § 5, the maps f and g induce homomorphisms

$$f_* : \pi_n(X, x_0) \to \pi_n(Y, y_0), \quad g_* : \pi_n(X, x_0) \to \pi_n(Y, y_1)$$

for each $n \geqslant 1$. On the other hand, the path σ determines an isomorphism

$$\sigma_n : \pi_n(Y, y_1) \to \pi_n(Y, y_0).$$

Proposition 15.3. $f_* = \sigma_n g_*$.

Proof. Let $\alpha \in \pi_n(X, x_0)$ and choose a map $\phi : (I^n, \partial I^n) \to (X, x_0)$ which represents α. Define a homotopy $\psi_t : I^n \to Y$, $(0 \leqslant t \leqslant 1)$, by taking $\psi_t = h_t \phi$ for every $t \in I$. Then ψ_0 represents $f_*(\alpha)$ and ψ_1 represents $g_*(\alpha)$. Since $\psi_t(\partial I^n) = \sigma(t)$ for each $t \in I$, it follows that $f_*(\alpha) = \sigma_n g_*(\alpha)$. ∎

Corollary 15.4. *If $f, g : X \to Y$ are homotopic maps such that $f(x_0) = y_0 = g(x_0)$, then there exists an element $w \in \pi_1(Y, y_0)$ such that $f_* = w g_*$.*

As another consequence of (15.3), we have the following

Proposition 15.5. *If $f : X \to Y$ is a homotopy equivalence and if $f(x_0) = y_0$, then*

$$f_* : \pi_n(X, x_0) \to \pi_n(Y, y_0)$$

is an isomorphism for every $n > 0$.

Proof. Since f is a homotopy equivalence, there exists a map $g : Y \to X$ such that gf and fg are homotopic to the identity maps. Let $x_1 = g(y_0)$. Then g induces

$$g_* : \pi_n(Y, y_0) \to \pi_n(X, x_1).$$

Let $h_t : X \to X$, $(0 \leqslant t \leqslant 1)$, be a homotopy such that $h_0 = gf$ and h_1 is the identity map on X. Define a path $\sigma : I \to X$ by $\sigma(t) = h_t(x_0)$ for every $t \in I$. Then, by (15.3), we have

$$g_* f_* = \sigma_n.$$

Since σ_n is an isomorphism, this implies that f_* is a monomorphism and g_* is an epimorphism. Since g is also a homotopy equivalence, it follows that g_* is also a monomorphism. Hence g_* is an isomorphism and so is $f_* = g_*^{-1}\sigma_n$. ∎

Similarly, if $f : (X, A) \to (Y, B)$ is a homotopy equivalence and if $f(x_0) = y_0$, then

$$f_* : \pi_n(X, A, x_0) \to \pi_n(Y, B, y_0)$$

is an isomorphism for every $n > 1$.

16. n-Simple spaces

A local system of groups $\{ G_x \}$ in a space X is said to be *simple*, if the homomorphism $\sigma_\#$ depends only on the initial point $\sigma(0)$ and the terminal point $\sigma(1)$ of the path $\sigma : I \to X$.

Let W be a group which acts on a group G. We shall say that *W acts simply on G* if $wg = g$ for every $w \in W$ and $g \in G$.

The proofs of the following two propositions are straightforward and hence are left to the reader.

Proposition 16.1. *A local system of groups* $\{ G_x \}$ *in a space* X *is simple iff, for every* $x_0 \in X$, $\pi_1(X, x_0)$ *acts simply on* G_{x_0}.

Proposition 16.2. *A local system of groups* $\{ G_x \}$ *in a pathwise connected space* X *is simple iff there exists a point* $x_0 \in X$ *such that* $\pi_1(X, x_0)$ *acts simply on* G_{x_0}.

Corollary 16.3. *A local system of groups* $\{ G_x \}$ *in a simply connected space* X *is always simple.*

Let $n > 0$ be any given integer. A space X is said to be *n-simple* if the local system $\{ \pi_n(X, x_0) \mid x_0 \in X \}$ of the n-th homotopy groups in X is simple.

The following assertions are immediate consequences of the definition and (16.1)–(16.3).

Proposition 16.4. *A space* X *is n-simple iff, for every* $x_0 \in X$, $\pi_1(X, x_0)$ *acts simply on* $\pi_n(X, x_0)$.

Proposition 16.5. *A pathwise connected space* X *is n-simple iff there exists a point* $x_0 \in X$ *such that* $\pi_1(X, x_0)$ *acts simply on* $\pi_n(X, x_0)$.

Corollary 16.6. *A simply connected space is n-simple for every* $n > 0$.

Corollary 16.7. *A pathwise connected space* X *is n-simple if* $\pi_n(X) = 0$.

Corollary 16.8. *A pathwise connected space* X *is* 1*-simple iff* $\pi_1(X)$ *is commutative.*

Thus, the m-sphere S^m is n-simple for every $m > 0$ and $n > 0$.

Now let us consider the unit n-sphere S^n and a given point $s_0 \in S^n$. The geometrical meaning of n-simplicity is given by the following

Theorem 16.9. *A space* X *is n-simple iff, for every point,* $x_0 \in X$ *and any two maps* $f, g : S^n \to X$ *with* $f(s_0) = x_0 = g(s_0)$, $f \simeq g$ *implies* $f \simeq g$ *rel* s_0.

Proof. Assume that X is n-simple. Then, by (16.4), $\pi_1(X, x_0)$ acts simply on $\pi_n(X, x_0)$. Since $f \simeq g$, there exists a homotopy $h_t : S^n \to X$, $(0 \leqslant t \leqslant 1)$, such that $h_0 = f$ and $h_1 = g$. According to the remark given in the paragraph which precedes (2.1), the maps f and g represent elements α and β of $\pi_n(X, x_0)$ respectively. Define a path $\sigma : I \to X$ by taking $\sigma(t) = h_t(s_0)$ for each $t \in I$. Since $\sigma(0) = x_0 = \sigma(1)$, σ represents an element w of $\pi_1(X, x_0)$. By § 14 and § 15, it is easy to see that $\alpha = w\beta$. Since $\pi_1(X, x_0)$ acts simply on $\pi_n(X, x_0)$, we have $w\beta = \beta$. Hence $\alpha = \beta$. This proves that $f \simeq g$ *rel* s_0.

Next, let us assume that the condition is satisfied. Let $w \in \pi_1(X, x_0)$ and choose a loop σ which represents w. Let α be any element of $\pi_n(X, x_0)$ represented by a map $f : S^n \to X$ with $f(s_0) = x_0$. Then the element $w\alpha$ of

$\pi_n(X, x_0)$ is represented by a map $g : S^n \to X$ with $g(s_0) = x_0$ and satisfying $f \simeq g$. By our condition, this implies that $f \simeq g$ *rel* s_0. Hence $w\alpha = \alpha$. By (16.4), X is n-simple. ∎

As a consequence of (16.9), let us prove the following

Proposition 16.10. *Every pathwise connected topological group is n-simple for every $n > 0$.*

Proof. Let X be a pathwise connected topological group and x_0 its neutral element. Let $f, g : S^n \to X$ be any two homotopic maps such that $f(s_0) = x_0 = g(s_0)$. Then these exists a homotopy $h_t : S^n \to X$, $(0 \leqslant t \leqslant 1)$, with $h_0 = f$ and $h_1 = g$. Define a homotopy $k_t : S^n \to X$, $(0 \leqslant t \leqslant 1)$, by taking

$$k_t(s) = [h_t(s_0)]^{-1} \cdot [h_t(s)], \quad (s \in S^n, t \in I).$$

Then we have $k_0 = f$, $k_1 = g$ and $k_t(s_0) = x_0$ for each $t \in I$. Hence $f \simeq g$ *rel* s_0. By the sufficiency proof of (16.9), $\pi_1(X, x_0)$ operates simply on $\pi_n(X, x_0)$. By (16.2) this implies that X is n-simple. ∎

This proposition (16.10) can be generalized to the H-spaces as defined in (III; § 11). See Ex. G.

The usefulness of n-simplicity is that, for a pathwise connected n-simple space X, the abstract homotopy group $\pi_n(X)$ as defined in § 14 has a natural geometrical meaning as follows.

Let us define $\pi_n(X)$ to be the set of all homotopy classes of the maps of S^n into X. In other words, $\pi_n(X)$ is the set of all path-components of the mapping space

$$\Phi = X^{S^n}.$$

Choose an arbitrary basic point $x_0 \in X$ and consider the subspace Ψ of Φ which consists of the maps of (S^n, s_0) into (X, x_0). Then the path-components of Ψ can be considered as the elements of $\pi_n(X, x_0)$. Hence the inclusion map $\Psi \subset \Phi$ induces a transformation

$$\chi : \pi_n(X, x_0) \to \pi_n(X).$$

Lemma 16.11. *If X is pathwise connected and n-simple, then χ sends $\pi_n(X, x_0)$ onto $\pi_n(X)$ in a one-to-one fashion.*

Proof. Let $\alpha \in \pi_n(X)$ and choose a map $f : S^n \to X$ which represents α. Since X is pathwise connected, there is a path $\sigma : I \to X$ such that $\sigma(0) = x_0$ and $\sigma(1) = x_1 = f(s_0)$. By the method used in the construction of σ_n in § 14, one can show that there is a homotopy $f_t : S^n \to X$, $(0 \leqslant t \leqslant 1)$, such that $f_0 = f$ and $f_t(s_0) = \sigma(1 - t)$ for each $t \in I$. Then $f_1 \in \Psi$ and represents an element β of $\pi_n(X, x_0)$. The homotopy f_t proves that $\chi(\beta) = \alpha$. Hence χ is onto.

Next, let $\alpha, \beta \in \pi_n(X, x_0)$ be such that $\chi(\alpha) = \chi(\beta)$. Choose maps $f, g : (S^n, s_0) \to (X, x_0)$ representing α, β respectively. Since $\chi(\alpha) = \chi(\beta)$, we have $f \simeq g$. Since X is n-simple, it follows from (16.9) that $f \simeq g$ *rel* s_0. This implies that $\alpha = \beta$. Hence χ is one-to-one. ∎

By means of χ, we may define a group structure in $\pi_n(X)$ so that χ becomes an isomorphism, which will be called the *canonical isomorphism* of $\pi_n(X, x_0)$ onto $\pi_n(X)$. To justify our geometrical construction of $\pi_n(X)$ given above, it remains to show that the group structure defined in $\pi_n(X)$ by means of χ is independent of the choice of $x_0 \in X$.

The geometrical meaning of the group operation defined in $\pi_n(X)$ by means of χ is clearly as follows. Let S^n_- and S^n_+ denote the hemispheres defined by $t_n \leqslant 0$ and $t_n \geqslant 0$ respectively, where $(t_0, \cdots, t_n)$ denotes an arbitrary point of S^n. Then the basic point $s_0 = (1, 0, \cdots, 0)$ is in the equator

$$S^{n-1} = S^n_- \cap S^n_+.$$

Let α and β be arbitrarily given elements of $\pi_n(X)$. Then there exist maps $f \in \alpha$ and $g \in \beta$ such that

$$f(S^n_+) = x_0 = g(S^n_-).$$

Define a map $h : S^n \to X$ by taking

$$h(s) = \begin{cases} f(s), & (\text{if } s \in S^n_-), \\ g(s), & (\text{if } s \in S^n_+). \end{cases}$$

Then h represents the element $\alpha + \beta$ of $\pi_n(X)$ which does not depend on the choice of f from α and g from β. We call the attention of the reader to the fact that, in case $n = 1, \pi_1(X)$ is commutative by (16.8) and hence the additive notation is preferred.

Now let $x_1 \in X$. Since X is pathwise connected, there is a path $\sigma : I \to X$ such that $\sigma(0) = x_0$ and $\sigma(1) = x_1$. By a standard method, one can show that there exists a homotopy $f_t : S^n \to X$, $(0 \leqslant t \leqslant 1)$, such that $f_0 = f$ and $f_t(S^n_+) = \sigma(t)$ for each $t \in I$. Similarly, there is a homotopy $g_t : S^n \to X$, $(0 \leqslant t \leqslant 1)$, such that $g_0 = g$ and $g_t(S^n_-) = \sigma(t)$ for each $t \in I$. Define a homotopy $h_t : S^n \to X$, $(0 \leqslant t \leqslant 1)$, by setting

$$h_t(s) = \begin{cases} f_t(s), & (\text{if } s \in S^n_-), \\ g_t(s), & (\text{if } s \in S^n_+). \end{cases}$$

for each $t \in I$. Then $h_0 = h$ and $h_1 \simeq h$. Hence h_1 represents $\alpha + \beta$. Also, since

$$f_1(S^n_+) = x_1 = g_1(S_-),$$

this proves that $\alpha + \beta$ does not depend on the choice of the basic point $x_0 \in X$.

Thus, for a pathwise connected n-simple space X, we have freed the basic point from the definition of the n-th homotopy group.

We have just seen that the homotopy classes of the maps of S^n into a pathwise connected n-simple space X form a group $\pi_n(X)$ which is isomorphic with $\pi_n(X, x_0)$ for every $x_0 \in X$. If X is pathwise connected but not n-simple this is not true; in fact, the homotopy classes of the maps $S^n \to X$ are in a one-to-one correspondence with the equivalence classes in $\pi_n(X, x_0)$ under the operations of $\pi_1(X, x_0)$. The proof is left to the reader.

EXERCISES

A. Inductive construction of a homotopy system

Construct a homotopy system $H = \{\pi, \partial, *\}$ by induction as follows.

According to the definition of a homotopy system in § 11, the homotopy set $\pi_0(X, x_0)$ and the induced transformation $f_* : \pi_0(X, x_0) \to \pi_0(Y, y_0)$ are well-defined for every pair (X, x_0) and every map $f : (X, x_0) \to (Y, y_0)$. Let $n \geqslant 1$ be a given integer and assume that we have already constructed the homotopy sets $\pi_m(X, A, x_0)$ for each $1 \leqslant m < n$ and each triplet (X, A, x_0), together with the boundary operators ∂ and the induced transformations f_* on these homotopy sets, such that the seven axions in § 11 are satisfied.

To construct the homotopy set $\pi_n(X, A, x_0)$ of a given triplet (X, A, x_0), consider the derived triplet (X', A', x'_0) of (X, A, x_0) and the derived projection $p : (X', A', x'_0) \to (X, A, x_0)$. We define

$$\pi_n (X, A, x_0) = \pi_{n-1}(A', x'_0).$$

Next, let $q = p \mid (A', x'_0)$. Then, the boundary operator $\partial : \pi_n(X, A, x_0) \to \pi_{n-1}(A, x_0)$ is defined by

$$\partial = q_* : \pi_{n-1}(A', x'_0) \to \pi_{n-1}(A, x_0).$$

Finally, let $f : (X, A, x_0) \to (Y, B, y_0)$ be a given map. Then, f induces a derived map $f' : (X', A', x'_0) \to (Y', B', y'_0)$. Let $\bar{f} = f' \mid (A', x'_0)$. Then define

$$f_* = \bar{f}_* : \pi_{n-1}(A', x'_0) \to \pi_{n-1}(B', y'_0).$$

Verify the seven axions of § 11 for the system $H = \{\pi, \partial, *\}$ constructed above.

B. The equivalence theorem

Consider two given homotopy systems

$$H = \{\pi, \partial, *\}, \quad H' = \{\pi', \partial', \#\}$$

together with their natural equivalence $h = \{h_n\} : H \approx H'$ constructed in § 12.

By an *admissible transformation* $k = \{k_n\} : H \to H'$, we mean for each triplet (X, A, x_0) and each integer $n > 0$ a transformation

$$k_n : \pi_n(X, A, x_0) \to \pi'_n(X, A, x_0)$$

satisfying the conditions:

(AT1) For each triplet (X, A, x_0) and each integer $n > 0$, we have the commutativity relation $k_{n-1}\partial = \partial' k_n$, where, in case of $n = 1$, k_0 denotes the identity map.

(AT2) For each map $f : (X, A, x_0) \to (Y, B, y_0)$ and each integer $n > 0$, we have the commutativity relation $k_n f_* = k_n f_\#$.

By (E1), (E3) and (E4) of § 12, every equivalence between H and H' is an admissible transformation. Conversely, prove the following

Equivalence Theorem. *Every admissible transformation k is an equivalence between H and H'. In fact, it coincides with the natural equivalence h, that is to say, $k_n = h_n$ for every $n > 0$.*

This shows that the natural equivalence $h = \{ h_n \}$ is the only possible equivalence between any two homotopy systems. Furthermore, in order to construct geometrically the natural equivalence between two homotopy systems given by geometric definitions, it suffices to establish an admissible transformation by means of some natural geometric method.

C. Properties of the homotopy system

The uniqueness theorem (12.1) implies that the set of seven properties in §§ 6–10 is equivalent to the set of seven axioms in § 11. The following is an outline of deducing these seven basic properties together with other properties of the homotopy system right from the seven axioms. Since the properties I, II, III, V, VII are stated exactly the same as the corresponding axioms, it remains to prove the properties IV and VI.

1. If $f : (X, A, x_0) \to (Y, B, y_0)$ is a homotopy equivalence, then, for each $n > 0$, f_* sends $\pi_n(X, A, x_0)$ onto $\pi_n(Y, B, y_0)$ in a one-to-one fashion.

2. If X is contractible to the point x_0, then $\pi_n(X, x_0) = 0$ for every $n \geqslant 0$.

3. For any given triplet (X, A, x_0), consider the derived projection $p : (X', A', x'_0) \to (X, A, x_0)$. Then in the diagram

$$\pi_n(X, A, x_0) \xleftarrow{p_*} \pi_n(X', A', x'_0) \xrightarrow{\partial} \pi_{n-1}(A', x'_0),$$

both p_* and ∂ are one-to-one and onto and hence we obtain a *natural correspondence*

$$\chi = \partial p_*^{-1} : \pi_n(X, A, x_0) \to \pi_{n-1}(A', x'_0)$$

which sends $\pi_n(X, A, x_0)$ onto $\pi_{n-1}(A', x'_0)$ in a one-to-one fashion.

4. For every triplet (X, A, x_0) and every $n \geqslant 0, \pi_n(X, A, x_0)$ is non-empty. Furthermore, one can uniquely define a *neutral element* of $\pi_n(X, A, x_0)$ in such a way that χ, ∂ and the induced transformations send the neutral element into a neutral element.

5. In the homotopy sequence

$$\cdots \xrightarrow{j_*} \pi_{n+1}(X, A, x_0) \xrightarrow{\partial} \pi_n(A, x_0) \xrightarrow{i_*} \pi_n(X, x_0) \xrightarrow{j_*} \pi_n(X, A, x_0) \xrightarrow{\partial} \cdots$$
$$\cdots \xrightarrow{j_*} \pi_1(X, A, x_0) \xrightarrow{\partial} \pi_0(A, x_0) \xrightarrow{i_*} \pi_0(X, x_0)$$

of (X, A, x_0) every set has a specific neutral element. By the *kernel* of a transformation in this sequence, we mean the inverse image of the neutral element. Then prove simultaneously the following two theorems:

The exactness theorem. *The homotopy sequence of any triplet is exact*, that is to say, the kernel of every transformation in the sequence coincides with the image of the preceding transformation.

The fibering theorem. *If $f : X \to Y$ is a fibering, $A = f^{-1}(B)$ and $x_0 \in f^{-1}(y_0)$, then the induced transformation f_* carries $\pi_n(X, A, x_0)$ onto $\pi_n(Y, B, y_0)$ in a one-to-one fashion for every $n > 0$.*

These two theorems cover the properties IV and VI respectively.

D. The role of the basic point in the relative homotopy groups

Consider a given space X, a given subspace A of X, and two given points x_0, x_1 connected by a path

$$\sigma : I \to A, \quad \sigma(0) = x_0, \quad \sigma(1) = x_1.$$

For each $n > 0$, define a transformation

$$\sigma_n : \pi_n(X, A, x_1) \to \pi_n(X, A, x_0)$$

as follows. Let $\alpha \in \pi_n(X, A, x_1)$. Choose a representative map

$$f : (I^n, I^{n-1}, J^{n-1}) \to (X, A, x_1)$$

for α. Pull the image of J^{n-1} retreating along the path σ back to x_0 with the image of I^n being dragged in such a way that the image of I^{n-1} is always in A. The map obtained after this homotopy represents an element β of $\pi_n(X,A,x_0)$ which depends only on α and σ. Then, we define $\sigma_n(\alpha) = \beta$. Give the details of this geometrical construction as in § 14, and prove the following assertions:

1. For every $n \geqslant 2$, σ_n is a homomorphism.
2. For every $n > 0$, σ_n depends only on the class of the path σ.
3. If σ is the degenerate path $\sigma(I) = x_0$ in A, then σ_n is the identity transformation on $\pi_n(X, A, x_0)$ for each $n > 0$.
4. If σ, τ are consecutive paths in A, i.e. $\sigma(1) = \tau(0)$, then $(\sigma\tau)_n = \sigma_n\tau_n$ for each $n > 0$.
5. For every $n > 0$, σ_n carries $\pi_n(X, A, x_1)$ onto $\pi_n(X, A, x_0)$ in a one-to-one fashion. Hence σ_n is an isomorphism for every $n \geqslant 2$.
6. Each rectangle of the following ladder is commutative

$$\begin{array}{ccccccccc}
\cdots \xrightarrow{\partial} & \pi_n(A, x_0) & \xrightarrow{i_*} & \pi_n(X, x_0) & \xrightarrow{j_*} & \pi_n(X, A, x_0) & \xrightarrow{\partial} & \pi_{n-1}(A, x_0) & \xrightarrow{i_*} \cdots \\
& \big\downarrow \sigma_n & & \big\downarrow \sigma_n & & \big\downarrow \sigma_n & & \big\downarrow \sigma_n & \\
\cdots \xrightarrow{\partial} & \pi_n(A, x_1) & \xrightarrow{i_*} & \pi_n(X, x_1) & \xrightarrow{j_*} & \pi_n(X, A, x_1) & \xrightarrow{\partial} & \pi_{n-1}(A, x_1) & \xrightarrow{i_*} \cdots
\end{array}$$

7. For any triplet (X, A, x_0), $\pi_2(X, A, x_0)$ is a crossed $[\pi_1(A, x_0), \partial]$-module. By this, we mean that the following two conditions are satisfied for every w in $\pi_1(A, x_0)$ and α, β in $\pi_2(X, A, x_0)$:

(i) $\partial(w\alpha) = w(\partial\alpha)w^{-1}$,

(ii) $(\partial\alpha)\beta = \alpha\beta\alpha^{-1}$.

Hence $j_*\pi_2(X, x_0)$ is contained in the center of $\pi_2(X, A, x_0)$ and $i_*\pi_1(A, x_0)$ acts as a group of operators on $j_*\pi_2(X, x_0)$. See [Hi, pp. 39–41].

The significance of the assertion (5) is as follows. If A is pathwise connected and $n \geqslant 2$, then all the groups $\pi_n(X, A, x_0)$ for various basic points x_0 are isomorphic. Hence, as an abstract group, $\pi_n(X, A, x_0)$ does not depend on the basic point x_0 and may be denoted simply by $\pi_n(X, A)$. This abstract group $\pi_n(X, A)$ will be called the n-th (*abstract*) *relative homotopy group* of X modulo A. For example, we have

$$\pi_m(E^n, S^{n-1}) = 0, \quad (m < n), \quad \pi_n(E^n, S^{n-1}) = Z,$$
$$\pi_m(E^2, S^1) = 0, \quad (m > 2), \quad \pi_4(E^3, S^2) = Z.$$

Next, consider the spaces of paths

$$W_0 = [X ; A, x_0], \quad W_1 = [X ; A, x_1]$$

as well as the degenerate paths $w_0 \in W_0$ and $w_1 \in W_1$. Then a given path $\sigma : I \to A$ which connects x_0 to x_1 induces a map $\xi : W_1 \to W_0$ defined as follows: for each $w \in W_1$, $\xi(w) \in W_0$ is the path defined by

$$[\xi(w)](t) = \begin{cases} w(2t), & \text{if } 0 \leqslant t \leqslant \frac{1}{2}, \\ \sigma(2 - 2t), & \text{if } \frac{1}{2} \leqslant t \leqslant 1. \end{cases}$$

On the other hand, σ also induces a path $\eta : I \to W_0$ defined as follows: for each $s \in I$, $\eta(s) \in W_0$ is the path defined by

$$[\eta(s)](t) = \begin{cases} \sigma(s), & \text{if } 0 \leqslant t \leqslant \frac{1}{2}, \\ \sigma(2s - 2st), & \text{if } \frac{1}{2} \leqslant t \leqslant 1. \end{cases}$$

Then, obviously we have $\eta(0) = w_0$ and $\eta(1) = \xi(w_1)$. Prove that commutativity holds in the following diagram

$$\begin{array}{ccccc} \pi_n(X, A, x_1) & & \xrightarrow{\quad \sigma_n \quad} & & \pi_n(X, A, x_0) \\ \downarrow \chi & & & & \downarrow \chi \\ \pi_{n-1}(W_1, w_1) & \xrightarrow{\xi_*} & \pi_{n-1}(W_0, \xi(w_1)) & \xrightarrow{\eta_{n-1}} & \pi_{n-1}(W_0, w_0) \end{array}$$

where χ denotes the natural correspondences. Hence

$$\sigma_n = \chi^{-1}\eta_{n-1}\xi_*\chi, \quad n > 0.$$

E. Relative n-simplicity

Let $n \geqslant 2$ be a given integer. A space X is said to be *n-simple relative to a subspace* A if the local system of groups $\{\pi_n(X, A, x_0) \mid x_0 \in A\}$ in A is simple. In this case, we also say that *the pair* (X, A) *is n-simple*. Establish analogues to (16.4)–(16.7), as well as these assertions:

1. If (X, A) is 2-simple, then $\pi_2(X, A, x_0)$ is abelian for every $x_0 \in A$.

2. If, for every $x_0 \in A$, $\pi_2(X, A, x_0)$ is abelian and i_* sends $\pi_1(A, x_0)$ into the neutral element of $\pi_1(X, x_0)$ then (X, A) is 2-simple.

3. A pair (X, A) is n-simple iff, for every point $x_0 \in A$ and any two maps

$$f, g : (E^n, S^{n-1}, s_0) \to (X, A, x_0),$$

$f \simeq g \; rel \{S^{n-1}, A\}$ implies that $f \simeq g \; rel \{S^{n-1}, A ; s_0, x_0\}$.

4. If (X, A) is n-simple and if A is pathwise connected, the elements of the abstract relative homotopy group $\pi_n(X, A)$ can be considered as the homotopy classes of the maps of (E^n, S^{n-1}) into (X, A).

F. The Whitehead product

Consider a given space X and a given basic point $x_0 \in X$. Let $m \geqslant 1$ and $n \geqslant 1$ be given integers. For any two given elements

$$\alpha \in \pi_m(X, x_0), \quad \beta \in \pi_n(X, x_0),$$

we are going to construct an element $[\alpha, \beta]$ of $\pi_{m+n-1}(X, x_0)$ which will be called *the Whitehead product* of α and β.

For this purpose, let us choose representative maps

$$f : (I^m, \partial I^m) \to (X, x_0), \quad g : (I^n, \partial I^n) \to (X, x_0)$$

for α, β respectively. Since $I^{m+n} = I^m \times I^n$, we have $\partial I^{m+n} = (I^m \times \partial I^n) \cup (\partial I^m \times I^n)$. Hence we define a map $h : \partial I^{m+n} \to X$ by taking for each point (s, t) in ∂I^{m+n}

$$h(s, t) = \begin{cases} f(s), & \text{if } t \in \partial I^n, \\ g(t), & \text{if } s \in \partial I^m. \end{cases}$$

Since the point $r_0 = (0, \cdots, 0)$ of ∂I^{m+n} is in $\partial I^m \times \partial I^n$, we have $h(r_0) = x_0$. Since ∂I^{m+n} is homeomorphic to S^{m+n-1}, h represents an element γ of $\pi_{m+n-1}(X, x_0)$.

Prove that γ depends only on the elements α and β. So, we may define $[\alpha, \beta] = \gamma$.

Establish the following properties of the Whitehead products:

1. If $\alpha \in \pi_1(X, x_0)$ and $\beta \in \pi_1(X, x_0)$, then $[\alpha, \beta]$ is the commutator $\alpha\beta\alpha^{-1}\beta^{-1}$ of $\pi_1(X, x_0)$.

2. If $\alpha \in \pi_m(X, x_0)$ and $\beta \in \pi_1(X, x_0)$ with $m > 1$, then $[\alpha, \beta]$ is the element $\beta\alpha - \alpha$ of $\pi_m(X, x_0)$.

3. If $m > 1$, then the assignment $\alpha \to [\alpha, \beta]$ for a given $\beta \in \pi_n(X, x_0)$ defines a homomorphism

$$\beta_* : \pi_m(X, x_0) \to \pi_{m+n-1}(X, x_0).$$

4. If $m + n > 2$, then, for every $\alpha \in \pi_m(X, x_0)$ and $\beta \in \pi_n(X, x_0)$ we have

$$[\beta, \alpha] = (-1)^{mn}[\alpha, \beta].$$

5. If $\sigma : I \to X$ is a path joining x_0 to x_1, then, for every $\alpha \in \pi_m(X, x_1)$ and $\beta \in \pi_n(X, x_1)$, we have

$$\sigma_{m+n-1}[\alpha, \beta] = [\sigma_m(\alpha), \sigma_n(\beta)].$$

6. If $\phi : (X, x_0) \to (Y, y_0)$ is a map, then, for every $\alpha \in \pi_m(X, x_0)$ and $\beta \in \pi_n(X, x_0)$, we have $\quad \phi_*[\alpha, \beta] = [\phi_*(\alpha), \phi_*(\beta)]$.

7. For any $\alpha \in \pi_m(X, x_0)$, $\beta \in \pi_n(X, x_0)$, $r \in \pi_q(X, x_0)$, the following *Jacobi identity* holds:

$$(-1)^{mq}[[\alpha, \beta], \gamma] + (-1)^{nm}[[\beta, \gamma], \alpha] + (-1)^{qn}[[\gamma, \alpha], \beta] = 0.$$

Whitehead products may be also defined between relative homotopy groups and between $\pi_m(X, A, x_0)$ and $\pi_n\,(A, x_0)$. See [Hu 11] and [Blakers and Massey 2].

G. Homotopy groups of H-spaces

Let X be a given H-space as defined in (III; § 11) and let x_0 be a homotopy unit of X. Then the group operation in $\pi_n(X, x_0)$ is closely related to the multiplication in X as follows.

Let α, β be arbitrarily given elements of $\pi_n(X, x_0)$ with $n > 0$. Choose any representative maps

$$f, g : (I^n, \partial I^n) \to (X, x_0)$$

for α and β respectively. By means of the multiplication in X, we may define a map $h : (I^n, \partial I^n) \to (X, x_0)$ by taking

$$h(t) = f(t) \cdot g(t), \quad t \in I^n.$$

Prove that h represents the element $\alpha + \beta$ of $\pi_n(X, x_0)$. Furthermore, if X is a topological group and x_0 is the neutral element of X, then we may define a map $k : (I^n, \partial I^n) \to (X, x_0)$ by taking

$$k(t) = f(t) \cdot [g(t)]^{-1}, \quad t \in I^n.$$

Prove that k represents the element $\alpha - \beta$ of $\pi_n(X, x_0)$.

Next, let $\alpha \in \pi_m(X, x_0)$ and $\beta \in \pi_n(X, x_0)$. Choose representative maps

$$f : (I^m, \partial I^m) \to (X, x_0), \quad g : (I^n, \partial I^n) \to (X, x_0)$$

for α and β and define a map $h : I^{m+n} \to X$ by taking

$$h(s, t) = f(s) \cdot g(t), \quad s \in I^m, t \in I^n.$$

Prove that $h \mid \partial I^{m+n}$ represents $[\alpha, \beta]$. This implies that

$$[\alpha, \beta] = 0.$$

Hence, it follows from the assertion (2) of Ex. F that every pathwise connected H-space is n-simple for every $n > 0$; in particular, we deduce again that $\pi_1(X, x_0)$ is abelian.

H. Semi-simplicial complexes

A *semi-simplicial complex* K is a collection of elements $\{ \sigma \}$ called *cells* together with two functions. The first function assigns to each cell σ an integer $m \geqslant 0$ called the *dimension* of σ, $m = dim(\sigma)$; we then say that σ is an *m-cell*. The second function assigns to each m-cell σ, $(m > 0)$, of K and each integer i, $(0 \leqslant i \leqslant m)$, an $(m-1)$-cell $\partial_i \sigma$ called the *i-th face* of σ, also denoted by $\sigma^{(i)}$, subject to the condition (SSC)

$$\partial_i \partial_j \sigma = \partial_{j-1} \partial_i \sigma$$

for $m > 1$ and $0 \leqslant i < j \leqslant m$. We call ∂_i the *i-th face operator*. It may happen that $\partial_i \sigma = \partial_j \sigma$ for some $i \neq j$. Lower dimensional *faces* of σ may be defined by iterating the face operators. For any two cells σ and τ of K, we shall write $\tau < \sigma$ if either $\tau = \sigma$ or $\tau = \partial_{i_1} \cdots \partial_{i_n} \sigma$ for some set $\{ i_1, \cdots, i_n \}$ of integers with $0 \leqslant i_1 < \cdots < i_n \leqslant m$. Thus we obtain a proper reflexive partial ordering relation $<$ in K. [Eilenberg and Zilber 1].

1. A *simplicial complex* K is a set whose elements are finite subsets of a given set V, subject to the condition that if $\sigma \in K$ and τ is a non-empty subset of σ, then $\tau \in K$. Prove that, if V is partially ordered in such a way that every $\sigma \in K$ is lineally ordered, then K is a semi-simplicial complex.

2. Given a simplicial complex K, construct a semi-simplicial complex $O(K)$ as follows. The m-cells, $m \geqslant 0$, are the $(m + 1)$-tuples $(v_0, \cdots, v_m)$, repetitions allowed, of the vertices of some simplex of K; and $\partial_i(v_0, \cdots, v_m) = (v_0, \cdots, \hat{v}_i, \cdots, v_n)$, where the circumflex over v_i means that v_i is omitted.

3. Prove that the singular complex $S(X)$ of a space X, [E–S; p. 185], is a semi-simplicial complex.

4. Construct the homology and the cohomology groups of a given semi-simplicial complex K modulo a subcomplex L over a given abelian coefficient group.

I. Degeneracy operators

By a *system of degeneracy operators* in a given semi-simplicial complex K, we mean a function which assigns to each m-cell σ, $(m \geqslant 0)$, of K and each integer i, $(0 \leqslant i \leqslant m)$, an $(m + 1)$-cell $\theta_i\sigma$ of K, satisfying the conditions:

$$\theta_i\theta_j(\sigma) = \theta_{j+1}\theta_i(\sigma), \quad (i \leqslant j);$$

$$\partial_j\theta_j(\sigma) = \sigma = \partial_{j+1}\theta_j(\sigma);$$

$$\partial_i\theta_j(\sigma) = \theta_{j-1}\partial_i(\sigma), \quad (m > 0, i < j);$$

$$\partial_i\theta_j(\sigma) = \theta_j\partial_{i-1}(\sigma), \quad (m > 0, i > j + 1).$$

We call θ_i the *i-th degeneracy operator* of the system. An m-cell σ, $(m > 0)$, of K is said to be *degenerate* if $\sigma = \theta_i(\tau)$ for some $(m - 1)$-cell τ of K and some i with $0 \leqslant i \leqslant m - 1$.

1. Prove that the semi-simplicial complex $O(K)$ of a simplicial complex K admits a system of degeneracy operators defined by

$$\theta_i(v_0, \cdots, v_m) = (v_0, \cdots, v_i, v_i, v_{i+1}, \cdots, v_m).$$

2. Prove that the singular complex $S(X)$ of a space X admits a system of degeneracy operators defined by

$$\theta_i\sigma(t_0, \cdots, t_{m+1}) = \sigma(t_0, \cdots, t_{i-1}, t_i + t_{i+1}, t_{i+2}, \cdots, t_{m+1}).$$

3. Prove that the semi-simplicial complex obtained from a simplicial complex by a partial ordering of the vertices admits no system of degeneracy.

4. Prove that every semi-simplicial complex of finite dimension admits no system of degeneracy.

J. Complete semi-simplicial complexes

A semi-simplicial complex K is said to satisfy the *extension condition* if given m-cells $\sigma_0, \cdots, \sigma_{k-1}, \sigma_{k+1}, \cdots, \sigma_{m+1}$ of K such that

$$\partial_i\sigma_j = \partial_{j-1}\sigma_i, \quad (i \neq k, j \neq k, i < j),$$

for the case $m > 0$, then there exists an $(m + 1)$-cell σ such that $\partial_i\sigma = \sigma_i$ for each $i \neq k$. A semi-simplicial complex K is said to be *complete* if it satisfies the extension condition and admits a system of degeneracy operators. A complete semi-simplicial complex K with a given system of degeneracy operators is called a *Kan complex*, [Kan 1].

Prove that the singular complex $S(X)$ of a space X is complete and that the semi-simplicial complex $O(K)$ of a simplicial complex K may fail to be complete. Hence, the class of Kan complexes is rather limited.

K. Homotopy groups of Kan complexes

Let K be a given Kan complex. The 0-cells and the 1-cells of K will be called *vertices* and *edges* respectively. Two vertices v_0, v_1 of K are said to be *equivalent*, $v_0 \sim v_1$, if there exists an edge $e \in K$ with $\partial_0 e = v_1$ and $\partial_1 e = v_0$. By means of the extension property of K, prove that this relation $\sim$ is symmetric, reflexive and transitive, and hence the vertices of K are divided into disjoint equivalence classes called the *components* of K. Let $\pi_0(K)$ denote the set of all components of K.

Next, let v be a given vertex of K. We are going to define a *derived complex.*

$$K' = D(K, v)$$

as follows. The n-cells of K' are defined to be the $(n + 1)$-cells σ of K such that v is the only vertex $< \sigma$ and that $\partial_0 \sigma = \theta_0{}^n(v)$, where $\theta_0{}^n$ is the n-fold iteration of the degeneracy operator θ_0. The face operators ∂'_i and the degeneracy operators θ'_i in K' are defined by

$$\partial'_i \sigma = \partial_{i+1} \sigma, \quad \theta'_i \sigma = \theta_{i+1} \sigma.$$

Verify that K' is a Kan complex and $v' = \theta_0 v$ is a vertex of K'. Then, for every $n \geqslant 0$, we define $\pi_n(K, v)$ inductively by

$$\pi_n(K, v) = \pi_{n-1}(K', v'), \quad \pi_0(K, v) = \pi_0(K).$$

Next, let us define the group structure in $\pi_n(K, v)$ for each $n > 0$. Because of the inductive definition given above, it suffices to define a group structure in $\pi_0(K')$. Let α, β be any two components of K' and pick vertices $x \in \alpha$ and $y \in \beta$. By the definition of K', x and y are edges in K such that $\partial_0 x = v = \partial_1 x$ and $\partial_0 y = v = \partial_1 y$. By the extension property of K, there exists a 2-cell σ in K such that $\partial_2 \sigma = x$ and $\partial_0 \sigma = y$. Then $z = \partial_1 \sigma$ is a vertex in K'. Prove that the component γ of K' which contains z depends only on α and β. Define $\alpha\beta = \gamma$. Prove that $\pi_1(K')$ becomes a group under this multiplication.

Prove the following assertions:

1. $\pi_n(K, v)$ is abelian for each $n \geqslant 2$.

2. If K is the singular complex $S(X)$ of a space X and v the vertex determined by a given point $x_0 \in X$, then $\pi_n(K, v)$ and $\pi_n(X, x_0)$ are isomorphic for each n.

CHAPTER V

THE CALCULATION OF HOMOTOPY GROUPS

1. Introduction

Neither the geometrical construction nor the axiomatic approach to the homotopy groups in the previous chapter leads to effective computation of these groups. In the present chapter, we shall study a few methods which yield successful calculations of the homotopy groups in various special cases.

In the first part of the chapter, we give the celebrated Hurewicz isomorphism theorem. For every integer $n > 0$, there is a natural homomorphism h_n of $\pi_n(X, x_0)$ into the integral singular homology group $H_n(X)$. If $n \geqslant 2$ and if X is $(n - 1)$-connected, then Hurewicz's theorem states that h_n is an isomorphism. Hence the first non-zero homotopy group of a triangulable space is effectively computable.

In the second part of the chapter, we give the exact homotopy sequence of a fibering $p : E \to B$ together with a few direct sum theorems. By employing the numerous known fiberings in conjunction with this exact sequence, many homotopy groups may be computed.

In the last part of the chapter, we introduce Freudenthal's suspension together with the notion of triad homotopy groups. These suspensions are crucial in the calculation of the homotopy groups of spheres, some of which will be given in the final chapter of the book.

2. Homotopy groups of the product of two spaces

Let X, Y be two given spaces and $x_0 \in X$, $y_0 \in Y$ be given points. Consider the product $Z = X \times Y$ and the point $z_0 = (x_0, y_0)$ in Z. Let

$$p : (Z, z_0) \to (X, x_0), \quad q : (Z, z_0) \to (Y, y_0)$$

denote the *natural projections* defined by $p(z) = x$ and $q(z) = y$ for each $z = (x, y)$ in Z. On the other hand, let

$$i : (X, x_0) \to (Z, z_0), \quad j : (Y, y_0) \to (Z, z_0)$$

denote the maps (called the *injections*) defined by $i(x) = (x, y_0)$ and $j(y) = (x_0, y)$ for each $x \in X$ and each $y \in Y$. Hence pi, qj are identity maps and pj, qi are constant maps.

For each $n > 0$, the maps p, q, i, j induce the homomorphisms

$$p_* : \pi_n(Z, z_0) \to \pi_n(X, x_0), \quad q_* : \pi_n(Z, z_0) \to \pi_n(Y, y_0),$$
$$i_* : \pi_n(X, x_0) \to \pi_n(Z, z_0), \quad j_* : \pi_n(Y, y_0) \to \pi_n(Z, z_0),$$

satisfying the relations

$$p_* i_* = 1, \quad q_* j_* = 1, \quad p_* j_* = 0, \quad q_* i_* = 0.$$

Hence i_*, j_* are monomorphisms and p_*, q_* are epimorphisms.

Let us consider the direct product of the groups $\pi_n(X, x_0)$ and $\pi_n(Y, y_0)$. If $n > 1$, these groups are abelian and their direct product is also called the direct sum. We shall use the additive notation

$$\pi_n(X, x_0) + \pi_n(Y, y_0), \quad (n > 0),$$

for the direct product even in the case $n = 1$ where the groups are not necessarily abelian.

Theorem 2.1. *For every $n > 0$, we have*

$$\pi_n(Z, z_0) \approx \pi_n(X, x_0) + \pi_n(Y, y_0).$$

Proof. Define a homomorphism

$$h : \pi_n(Z, z_0) \to \pi_n(X, x_0) + \pi_n(Y, y_0)$$

by setting $h(\alpha) = (p_*(\alpha), q_*(\alpha))$ for every $\alpha \in \pi_n(Z, z_0)$. It remains to prove that h is an isomorphism.

For arbitrarily given $\alpha \in \pi_n(X, x_0)$ and $\beta \in \pi_n(Y, y_0)$, let $\gamma = i_*\alpha + j_*\beta \in \pi_n(Z, z_0)$. Then we have

$$h(\gamma) = (p_* i_* \alpha + p_* j_* \beta, \quad q_* i_* \alpha + q_* j_* \beta) = (\alpha, \beta).$$

Hence h is an epimorphism.

On the other hand, let $\delta \in \pi_n(Z, z_0)$ be any element such that $h(\delta) = 0$. Then, by definition, we have $p_*\delta = 0$ and $q_*\delta = 0$. Let $f : (I^n, \partial I^n) \to (Z, z_0)$ be any map which represents δ. Since $p_*\delta = 0$ and $q_*\delta = 0$, there exist two homotopies

$$g_t : (I^n, \partial I^n) \to (X, x_0), \quad h_t : (I^n, \partial I^n) \to (Y, y_0),$$

$(0 \leqslant t \leqslant 1)$, such that $g_0 = pf$, $g_1(I^n) = x_0$, $h_0 = qf$ and $h_1(I^n) = y_0$. Define a homotopy $f_t : I^n \to Z$, $(0 \leqslant t \leqslant 1)$, by taking

$$f_t(s) = (g_t(s), h_t(s)), \quad (s \in I^n, t \in I).$$

Then clearly we have $f_0 = f$, $f_1(I^n) = z_0$ and $f_t(\partial I^n) = z_0$ for each $t \in I$. This implies that $\delta = 0$. Hence h is also a monomorphism. ∎

The inverse of h is the isomorphism

$$h^{-1} : \pi_n(X, x_0) + \pi_n(Y, y_0) \to \pi_n(Z, z_0)$$

given by $h^{-1}(\alpha, \beta) = i_*(\alpha) + j_*(\beta)$.

By means of finite induction, (2.1) can be easily generalized in an obvious way to topological product of more than two factor spaces. In particular, the homotopy groups of the m-dimensional torus T_m, i.e. the product $T_m = S^1 \times \cdots \times S^1$ of m copies of the unit circle S^1, are as follows: $\pi_1(T_m)$ is the free abelian group with m free generators and $\pi_n(T_m) = 0$ for every $n > 1$.

3. The one-point union of two spaces

Let X, Y be two given spaces and $x_0 \in X$, $y_0 \in Y$ be given points. Consider the topological sum $W = X \cup Y$ as in (I, § 7). If we identify the point $x_0 \in X$ with the point $y_0 \in Y$, we get a quotient space U of W with a specified point u_0 which is the class consisting of x_0 and y_0. This space U will be called the *one-point union* of X and Y, and sometimes denoted by $X \vee Y$.

Using the notations of § 2, we can imbed U as the subspace $(X \times y_0) \cup (x_0 \times Y)$ of the product space Z by means of the map

$$k : (U, u_0) \to (Z, z_0)$$

defined by

$$k(u) = \begin{cases} (u, y_0), & (\text{if } u \in X), \\ (x_0, u), & (\text{if } u \in Y). \end{cases}$$

Theorem 3.1. For every $n > 1$, we have

$$\pi_n(U, u_0) \approx \pi_n(X, x_0) + \pi_n(Y, y_0) + \pi_{n+1}(Z, U, z_0).$$

Proof. The inclusion maps i', j' and the map k induce the homomorphisms

$$i'_* : \pi_n(X, x_0) \to \pi_n(U, u_0), \quad j'_* : \pi_n(Y, y_0) \to \pi_n(U, u_0)$$
$$k_* : \pi_n(U, u_0) \to \pi_n(Z, z_0).$$

Define a homomorphism $l : \pi_n(Z, z_0) \to \pi_n(U, u_0)$ by taking

$$l(\alpha) = i'_* p_*(\alpha) + j'_* q_*(\alpha), \quad \alpha \in \pi_n(Z, z_0).$$

Since obviously $ki' = i$ and $kj' = j$, it follows that

$$k_* l(\alpha) = k_*[i'_* p_*(\alpha) + j'_* q_*(\alpha)] = k_* i'_* p_*(\alpha) + k_* j'_* q_*(\alpha)$$
$$= i_* p_*(\alpha) + j_* q_*(\alpha) = h^{-1}h(\alpha) = \alpha$$

for each $\alpha \in \pi_n(Z, z_0)$. Since $n > 1$, all the groups are abelian. This implies that l is a monomorphism, k_* is an epimorphism, and $\pi_n(U, u_0)$ splits into the direct sum of the image of l and the kernel of k_*.

Since l is a monomorphism, we have

$$\text{Image } l \approx \pi_n(Z, z_0) \approx \pi_n(X, x_0) + \pi_n(Y, y_0)$$

by (2.1). To determine the kernel of k_*, consider the homotopy sequence

$$\cdots \to \pi_{n+1}(U, u_0) \xrightarrow{k_*} \pi_{n+1}(Z, z_0) \to \pi_{n+1}(Z, U, z_0) \xrightarrow{\partial} \pi_n(U, u_0) \xrightarrow{k_*} \pi_n(Z, z_0) \to$$

of the triplet (Z, U, z_0). Since k_* is an epimorphism, it follows from the exactness of the sequence that ∂ is a monomorphism. Hence we have

$$\text{Kernel } k_* = \text{Image } \partial \approx \pi_{n+1}(Z, U, z_0). \blacksquare$$

In fact, $\pi_n(U, u_0)$ decomposes into the direct sum of the images of the three monomorphisms i'_*, j'_* and ∂.

The condition $n > 1$ is used only once, namely, to assure that the image of l is a normal subgroup of $\pi_n(U, u_0)$. In general, this is not true for the case

$n = 1$. However, we have anyway the weaker assertion that $\pi_1(U, u_0)$ is an extension of $\pi_2(Z, U, z_0)$ by the direct product of $\pi_1(X, x_0)$ and $\pi_1(Y, y_0)$. If X and Y are polyhedra, we recall that $\pi_1(U, u_0)$ is the free product of $\pi_1(X, x_0)$ and $\pi_1(Y, y_0)$ by (II; Ex. A4). This is also true if X and Y are regular and locally simply connected.

As an application of (3.1), we have the following

Proposition 3.2. *For every $p > 0$, $q > 0$ and $n < p + q - 1$, we have*

$$\pi_n(S^p \vee S^q) \approx \pi_n(S^p) + \pi_n(S^q).$$

Proof. Since $(S^p \times S^q, S^p \vee S^q)$ is a relative $(p + q)$-cell in the sense of (I; § 7), it follows from (IV; 3.4) that

$$\pi_m(S^p \times S^q, S^p \vee S^q) = 0$$

for every $m < p + q$. Hence (3.2) follows from (3.1) immediately. ∎

For example, we have

$$\pi_1(S^2 \vee S^3) = 0, \quad \pi_2(S^2 \vee S^3) = Z, \quad \pi_3(S^2 \vee S^3) = Z + Z.$$

4. The natural homomorphisms from homotopy groups to homology groups

Let (X, A, x_0) be any given triplet and consider the homotopy set

$$\pi_n(X, A, x_0), \quad n \geqslant 1.$$

Let α be an arbitrary element of $\pi_n(X, A, x_0)$ and choose a map

$$\phi : (E^n, S^{n-1}, s_0) \to (X, A, x_0)$$

which represents α, where E^n denotes the unit n-cell in the euclidean n-space R^n, S^{n-1} the unit $(n - 1)$-sphere in R^n, and $s_0 = (1, 0, \cdots, 0)$. The natural coordinate system in R^n determines an orientation of R^n and hence a generator ξ_n of the free cyclic homology group $H_n(E^n, S^{n-1})$. As a map of (E^n, S^{n-1}) into (X, A), ϕ induces a homomorphism

$$\phi_* : H_n(E^n, S^{n-1}) \to H_n(X, A)$$

where $H_n(X, A)$ denotes the singular homology group with integral coefficients. According to the homotopy axiom of homology theory, ϕ_* depends only on the given element $\alpha \in \pi_n(X, A, \alpha)$. Hence the assignment $\alpha \to \phi_*(\xi_n)$ defines a transformation

$$\varkappa_n : \pi_n(X, A, x_0) \to H_n(X, A).$$

Proposition 4.1. *If either $n > 1$ or $A = x_0$, then $\varkappa_n$ is a homomorphism which will be called the natural homomorphism of $\pi_n(X, A, x_0)$ into $H_n(X, A)$.*

Proof. Let α and β be any two elements of $\pi_n(X, A, x_0)$. Then we may choose representative maps $\phi, \psi : (E^n, S^{n-1}, s_0) \to (X, A, x_0)$ such that

$\phi(t_1,\cdots,t_n) = x_0$ if $t_1 \geqslant 0$ and $\psi(t_1,\cdots,t_n) = x_0$ if $t_1 \leqslant 0$. Define a map $\chi : (E^n, S^{n-1}, s_0) \to (X, A, x_0)$ by taking

$$\chi(t_1,\cdots,t_n) = \begin{cases} \phi(t_1,\cdots,t_n), & \text{if } t_1 \leqslant 0, \\ \psi(t_1,\cdots,t_n), & \text{if } t_1 \geqslant 0. \end{cases}$$

for an arbitrary point $(t_1,\cdots,t_n)$ in E^n. Then χ clearly represents $\alpha + \beta$. By a theorem in homology theory, [E–S; p. 36], we have

$$\chi_*(\xi_n) = \phi_*(\xi_n) + \psi_*(\xi_n).$$

This implies $\varkappa_n(\alpha + \beta) = \varkappa_n(\alpha) + \varkappa_n(\beta)$ and hence $\varkappa_n$ is a homomorphism. ∎

The following proposition is obvious from the above construction of $\varkappa_n$.

Proposition 4.2. *For any map $f : (X, A, x_0) \to (Y, B, y_0)$, the following rectangle is commutative*:

$$\begin{array}{ccc} \pi_n(X, A, x_0) & \xrightarrow{f_*} & \pi_n(Y, B, y_0) \\ \downarrow{\scriptstyle \varkappa_n} & & \downarrow{\scriptstyle \varkappa_n} \\ H_n(X, A) & \xrightarrow{f_*} & H_n(Y, B) \end{array}$$

For the case $A = x_0$, we have an isomorphism

$$j_* : H_n(X) \approx H_n(X, x_0)$$

and hence we obtain a homomorphism

$$h_n = j_*^{-1}\varkappa_n : \pi_n(X, x_0) \to H_n(X)$$

which will be called the *natural homomorphism* of $\pi_n(X, x_0)$ into $H_n(X)$. Clearly, h_1 coincides with the homomorphism h_* of (II; § 6).

Since $\partial : H_{n+1}(E^{n+1}, S^n) \approx H_n(S^n)$, it follows that $\eta_n = \partial\xi_{n+1}$ is a generator of the free cyclic group $H_n(S^n)$.

If we identify the boundary S^{n-1} of E^n to a single point s_0, we obtain an n-sphere S^n and a point s_0. Hence there is a relative homeomorphism

$$p : (E^n, S^{n-1}) \to (S^n, s_0)$$

such that $p_*(\xi_n) = j_*(\eta_n)$, where

$$p_* : H_n(E^n, S^{n-1}) \approx H_n(S^n, s_0), j_* : H_n(S^n) \approx H_n(S^n, s_0)$$

are the isomorphisms induced by the map p and the inclusion map $j : S^n \subset (S^n, s_0)$.

Let $\alpha \in \pi_n(X, x_0)$ and pick a map $\phi : (E^n, S^{n-1}) \to (X, x_0)$ which represents α. Then there exists a unique map $\psi : (S^n, s_0) \to (X, x_0)$ such that $\phi = \psi p$. We may consider α as represented by ψ. As a map of S^n into X, ψ induces a homomorphism

$$\psi_* : H_n(S^n) \to H_n(X).$$

Then clearly we have $h_n(\alpha) = \psi_*(\eta_n)$. This may be used as the definition of the natural homomorphism h_n.

Then, the following proposition is obvious.

Proposition 4.3. *For any triplet* (X, A, x_0), *the following rectangle is commutative:*

$$\begin{array}{ccc} \pi_{n+1}(X, A, x_0) & \xrightarrow{\partial} & \pi_n(A, x_0) \\ \downarrow{\scriptstyle \varkappa_n} & & \downarrow{\scriptstyle h_n} \\ H_{n+1}(X, A) & \xrightarrow{\partial} & H_n(A). \end{array}$$

Now consider the following homotopy-homology ladder:

$$\begin{array}{ccccccccc} \cdots \xrightarrow{j_*} & \pi_{n+1}(X, A, x_0) & \xrightarrow{\partial} & \pi_n(A, x_0) & \xrightarrow{i_*} & \pi_n(X, x_0) & \xrightarrow{j_*} & \pi_n(X, A, x_0) & \xrightarrow{\partial} \cdots \\ & \downarrow{\scriptstyle \varkappa_n} & & \downarrow{\scriptstyle h_n} & & \downarrow{\scriptstyle h_n} & & \downarrow{\scriptstyle \varkappa_n} & \\ \cdots \xrightarrow{j_*} & H_{n+1}(X, A) & \xrightarrow{\partial} & H_n(A) & \xrightarrow{i_*} & H_n(X) & \xrightarrow{j_*} & H_n(X, A) & \xrightarrow{\partial} \cdots \end{array}$$

By (4.3), the rectangle on the left is commutative. By (4.2), the middle rectangle and the one on the right are also commutative. Hence the whole ladder is commutative.

The remainder of this section is devoted to the proof of the Hurewicz theorem for polyhedra. Generalizations will be given in the exercises at the end of the chapter and also in a later chapter.

In (II; § 9), we defined the notion of the n-connected spaces. In terms of homotopy groups, one can easily prove that, for a given integer $n \geqslant 0$, a space X is *n-connected* iff it is pathwise connected and $\pi_m(X) = 0$ for every $m \leqslant n$.

Theorem 4.4. (Hurewicz theorem). *If* X *is an* $(n-1)$*-connected finite simplicical complex with* $n > 1$, *then the natural homomorphism* h_n *is an isomorphism.*

Proof. If X is the n-sphere S^n, then the theorem is given already by the Hopf theorem in (II; § 8). By (3.2), it follows easily that the theorem holds for the case $X = S^n \vee S^n$. Then, by means of (3.1) and finite induction, one can easily prove the theorem for the case where X is the one-point union of a finite number of n-spheres.

Next, assume that $\dim X \leqslant n$. If we identify the $(n-1)$-dimensional skeleton X^{n-1} of X to a single point y_0, we obtain a quotient space Y with a natural projection

$$p : (X, X^{n-1}) \to (Y, y_0).$$

If Y is different from y_0, then it is obviously homeomorphic to the one-point union of a finite number of n-spheres with y_0 as the common point.

Pick a vertex $x_0 \in X^{n-1}$. Since X is $(n-1)$-connected, X^{n-1} must be contractible to the point x_0 in X. By an application of the homotopy extension property, it follows that there exists a homotopy $f_t : X \to X$, $(0 \leqslant t \leqslant 1)$, such that f_0 is the identity map, $f_1(X^{n-1}) = x_0$, and $f_t(x_0) = x_0$ for each $t \in I$. By (I; 6.1), the map f_1 determines a map $q : (Y, y_0) \to (X, x_0)$ such that the composed map $qp = f_1$ is homotopic to the identity map on X relative to x_0.

Now, consider the following diagram

$$\begin{array}{ccc} \pi_n(X, x_0) & \xrightarrow{h_n} & H_n(X) \\ p_* \downarrow \uparrow q_* & & p_\# \downarrow \uparrow q_\# \\ \pi_n(Y, y_0) & \xrightarrow{k_n} & H_n(Y) \end{array}$$

where h_n, k_n denote the natural homomorphisms, p_*, $p_\#$ are induced by the natural projection p, and q_*, $q_\#$ are induced by the map q. By what we have proved above, k_n is an isomorphism and $q_* p_*$, $q_\# p_\#$ are identities. By (4.2) we have $p_\# h_n = k_n p_*$ and $h_n q_* = q_\# k_n$. These imply that h_n is an isomorphism. Indeed, let $\alpha \in \pi_n(X, x_0)$ be such that $h_n(\alpha) = 0$. Then we have

$$k_n p_*(\alpha) = p_\# h_n(\alpha) = p_\#(0) = 0.$$

Since k_n and p_* are monomorphisms, this implies $\alpha = 0$. Hence h_n is a monomorphism. On the other hand, let $\beta \in H_n(X)$. Let $\alpha = q_* k_n^{-1} p_\#(\beta)$. Then we have

$$h_n(\alpha) = h_n q_* k_n^{-1} p_\#(\beta) = q_\# k_n k_n^{-1} p_\#(\beta) = q_\# p_\#(\beta) = \beta.$$

Hence h_n is also an epimorphism. This proves the theorem for the case $\dim X \leqslant n$.

Now, assume that $\dim X = n + 1$. Then the natural homomorphism

$$\varkappa_{n+1} : \pi_{n+1}(X, X^n, x_0) \to H_{n+1}(X, X^n) = C_{n+1}(X)$$

is an epimorphism. Indeed, $C_{n+1}(X)$ is the free abelian group generated by the $(n + 1)$-simplexes of X. Let s be any $(n + 1)$-simplex of X. Pick a vertex x_1 of s. Since $\partial s \subset X^n$, the inclusion map $s \subset X$ represents an element α of $\pi_{n+1}(X, X^n, x_1)$. Since X^n is pathwise connected, there exists a path $\sigma : I \to X^n$ joining x_0 to x_1. Then $\varkappa_{n+1}$ clearly sends the element $\sigma_n(\alpha)$ of $\pi_{n+1}(X, X^n, x_0)$ into the generator s of $C_{n+1}(X)$. Since s is arbitrary, this proves that $\varkappa_{n+1}$ is an epimorphism.

By (IV; 3.4), we have $\pi_n(X, X^n, x_0) = 0$. On the other hand, we have $H_n(X, X^n) = 0$. Then, in the homotopy-homology ladder of (X, X^n, x_0), we have the following diagram

$$\begin{array}{ccccccc} \pi_{n+1}(X, X^n, x_0) & \xrightarrow{\partial} & \pi_n(X^n, x_0) & \xrightarrow{i_*} & \pi_n(X, x_0) & \xrightarrow{j_*} & 0 \\ \downarrow \varkappa_{n+1} & & \downarrow k_n & & \downarrow h_n & & \\ H_{n+1}(X, X^n) & \xrightarrow{\partial} & H_n(X^n) & \xrightarrow{i_*} & H_n(X) & \xrightarrow{j_*} & 0 \end{array}$$

where $\varkappa_{n+1}$ is an epimorphism and k_n is an isomorphism. These imply that h_n is an isomorphism. In fact, let $\beta \in H_n(X)$. Since i_* carries $H_n(X^n)$ onto $H_n(X)$, there exists an element $\gamma \in H_n(X^n)$ such that $i_*(\gamma) = \beta$. Let $\alpha = i_* k_n^{-1}(\gamma) \in \pi_n(X, x_0)$. Then, we have

$$h_n(\alpha) = h_n i_* k_n^{-1}(\gamma) = i_* k_n k_n^{-1}(\gamma) = i_*(\gamma) = \beta.$$

Hence h_n is an epimorphism. On the other hand, let $\alpha \in \pi_n(X, x_0)$ be any element such that $h_n(\alpha) = 0$. Choose an element $\beta \in \pi_n(X^n, x_0)$ with $i_*(\beta) = \alpha$. Since $i_* k_n(\beta) = h_n i_*(\beta) = h_n(\alpha) = 0$, it follows from the exactness that there exists a $\gamma \in H_{n+1}(X, X^n)$ with $\partial(\gamma) = k_n(\beta)$. Since $\varkappa_{n+1}$ is an epimorphism, there is a $\delta \in \pi_{n+1}(X, X^n, x_0)$ with $\varkappa_{n+1}(\delta) = \gamma$. Then, we have

$$\begin{aligned} i_*\partial(\delta) &= i_* k_n^{-1} k_n \partial(\delta) = i_* k_n^{-1} \partial \varkappa_{n+1}(\delta) \\ &= i_* k_n^{-1} \partial(\gamma) = i_* k_n^{-1} k_n(\beta) = i_*(\beta) = \alpha. \end{aligned}$$

Hence $\alpha = i_*\partial(\delta) = 0$. This proves that h_n is also a monomorphism. So, the theorem is proved for the case $\dim X = n + 1$. Note that the argument used in this paragraph is a special case of the "five" lemma, [E–S; p. 16].

Finally, let X be any $(n - 1)$-connected finite simplicial complex. Consider the $(n + 1)$-dimensional skeleton X^{n+1} of X and the following diagram

$$\begin{array}{ccc} \pi_n(X^{n+1}, x_0) & \xrightarrow{i_*} & \pi_n(X, x_0) \\ \downarrow{\scriptstyle k_n} & & \downarrow{\scriptstyle h_n} \\ H_n(X^{n+1}) & \xrightarrow{i_\#} & H_n(X) \end{array}$$

where the natural homomorphism k_n is proved to be an isomorphism. By (IV; 3.4), it follows that i_* is an isomorphism. On the other hand, $i_\#$ is also an isomorphism. Hence, $h_n = i_\# k_n i_*^{-1}$ is an isomorphism. ∎

5. Direct sum theorems

In the present section, we shall derive three useful consequences of the exactness of the homotopy sequence.

Proposition 5.1. *If A is a retract of X and $x_0 \in A$, then*

$$\pi_n(X, x_0) \approx \pi_n(A, x_0) + \pi_n(X, A, x_0)$$

for every $n \geqslant 2$ and $i_* : \pi_n(A, x_0) \to \pi_n(X, x_0)$ is a monomorphism for every $n \geqslant 1$.

Proof. Let $r : X \supset A$ be a retraction. Since ri is the identity map on A, it follows that $r_* i_*$ is the identity automorphism on $\pi_n(A, x_0)$ for every $n \geqslant 1$. This implies that i_* is a monomorphism and r_* is an epimorphism for every $n \geqslant 1$.

If $n \geqslant 2$, then $\pi_n(X, x_0)$ is abelian. Hence it follows from $r_* i_* = 1$ that $\pi_n(X, x_0)$ decomposes into the direct sum

$$\pi_n(X, x_0) = J + K, J = \text{Image } i_*, K = \text{Kernel } r_*.$$

Since i_* is a monomorphism, we have $J \approx \pi_n(A, x_0)$. Further, by the exactness of the homotopy sequence of (X, A, x_0), it is easy to deduce that $j_* : \pi_n(X, x_0) \to \pi_n(X, A, x_0)$ is an epimorphism for every $n \geqslant 2$. By exactness, the kernel of j_* is J and so j_* sends K isomorphically onto $\pi_n(X, A, x_0)$. Hence $K \approx \pi_n(X, A, x_0)$. ∎

Consequently, if A is a retract of X, then $\pi_2(X, A, x_0)$ must be abelian.

Proposition 5.2. *If X is deformable into A relative to a point $x_0 \in A$, then*

$$\pi_n(A, x_0) \approx \pi_n(X, x_0) + \pi_{n+1}(X, A, x_0)$$

for every $n \geqslant 2$ and $i_* : \pi_n(A, x_0) \to \pi_n(X, x_0)$ is an epimorphism for every $n \geqslant 1$.

Proof. According to the hypothesis, there exists a homotopy $h_t : X \to X$, $(0 \leqslant t \leqslant 1)$, such that

$$h_0(x) = x, \quad h_1(x) \in A, \quad h_t(x_0) = x_0$$

for each $x \in X$ and $t \in I$. Define a map $h : (X, x_0) \to (A, x_0)$ by taking $h(x) = h_1(x)$ for each $x \in X$. Since $ih = h_1 \simeq h_0 \text{ rel } x_0$, it follows that i_*h_* is the identity automorphism on $\pi_n(X, x_0)$. Hence h_* is a monomorphism and i_* is an epimorphism for every $n \geqslant 1$.

If $n \geqslant 2$, then $\pi_n(A, x_0)$ is abelian. Hence it follows from $i_*h_* = 1$ that $\pi_n(A, x_0)$ decomposes into the direct sum

$$\pi_n(A, x_0) = J + K, \quad J = \text{Image } h_*, \quad K = \text{Kernel } i_*.$$

Since h_* is a monomorphism, we have $J \approx \pi_n(X, x_0)$. On the other hand, since i_* is an epimorphism for every $n \geqslant 1$, it follows from the exactness that $\partial : \pi_{n+1}(X, A, x_0) \to \pi_n(A, x_0)$ is a monomorphism. Hence

$$K = \text{Kernel } i_* = \text{Image } \partial \approx \pi_{n+1}(X, A, x_0). \quad \blacksquare$$

For the case $n = 1, \pi_1(A, x_0)$ is an extension of $\pi_2(X, A, x_0)$ by $\pi_1(X, x_0)$ if X is deformable into A relative to x_0.

Proposition 5.3. *If A is contractible in X relative to a point $x_0 \in A$, then*

$$\pi_n(X, A, x_0) \approx \pi_n(X, x_0) + \pi_{n-1}(A, x_0)$$

for every $n \geqslant 3$ and i_* sends $\pi_n(A, x_0)$ into the neutral element of $\pi_n(X, x_0)$ for every $n \geqslant 1$.

Proof. By hypothesis, there exists a homotopy $h_t : A \to X$, $(0 \leqslant t \leqslant 1)$, such that $h_0 = i$, $h_1(A) = x_0$ and $h_t(x_0) = x_0$ for each $t \in I$. This implies that $i_* = 0$ for every $n \geqslant 1$.

Now let $n \geqslant 2$. By the exactness of the homotopy sequence, $i_* = 0$ implies that j_* is a monomorphism and ∂ is an epimorphism. This proves that $\pi_n(X, A, x_0)$ is an extension of $\pi_n(X, x_0)$ by $\pi_{n-1}(A, x_0)$.

By means of the contraction h_t, we define a homomorphism $h_* : \pi_{n-1}(A, x_0) \to \pi_n(X, A, x_0)$ for each $n \geqslant 2$ as follows. Let $\alpha \in \pi_{n-1}(A, x_0)$ be represented by a map $f : (I^{n-1}, \partial I^{n-1}) \to (A, x_0)$. Define a map $g : (I^n, I^{n-1}, J^{n-1}) \to (X, A, x_0)$ by taking

$$g(t_1, \cdots, t_n) = h_{t_n} f(t_1, \cdots, t_{n-1}).$$

Then $h_*(\alpha)$ is defined to be the element represented by g. Since $g \mid I^{n-1} = f$, it follows that ∂h_* is the identity automorphism on $\pi_{n-1}(A, x_0)$. Hence h_* is a monomorphism for every $n \geqslant 2$.

If $n \geqslant 3$, then $\pi_n(X, A, x_0)$ is abelian. Hence, $\partial h_* = 1$ implies that $\pi_n(X, A, x_0)$ decomposes into the direct sum

$$\pi_n(X, A, x_0) = J + K, \quad J = \text{Image } h_*, \quad K = \text{Kernel } \partial.$$

Since h_* is a monomorphism, $J \approx \pi_{n-1}(A, x_0)$. On the other hand, we have

$$K = \text{Kernel } \partial = \text{Image } j_* \approx \pi_n(X, x_0)$$

since j_* is a monomorphism. ∎

For the case $n = 2$, the last argument of the preceding proof breaks since $\pi_2(X, A, x_0)$ is in general non-abelian. However, it can be proved that $\pi_2(X, A, x_0)$ is isomorphic to the direct product of $\pi_2(X, x_0)$ and $\pi_1(A, x_0)$.

The assumed relativity with respect to x_0 in (5.2) and (5.3) is convenient but not essential; for proofs without this assumption see [Hu 1].

6. Homotopy groups of fiber spaces

Let E be a fiber space over a base space B with projection $p : E \to B$. Choose a basic point $b_0 \in B$ such that the fiber $F = p^{-1}(b_0)$ is not empty. Call F the basic fiber and choose a basic point $e_0 \in F$. Thus, we obtain a triplet (E, F, e_0).

Since $p(F) = b_0$, the projection $p : (E, e_0) \to (B, b_0)$ defines a map $q : (E, F, e_0) \to (B, b_0)$ and $p = qj$, where $j : (E, e_0) \subset (E, F, e_0)$ denotes the inclusion map. According to (IV, § 9), q_* sends $\pi_n(E, F, e_0)$ onto $\pi_n(B, b_0)$ in a one-to-one fashion for each $n \geqslant 1$. Let

$$d_* = \partial q_*^{-1} : \pi_n(B, b_0) \to \pi_{n-1}(F, e_0), \quad (n \geqslant 1).$$

Since $p_* = q_* j_*$, we can construct from the homotopy sequence of the triplet (E, F, e_0) an exact sequence

$$\cdots \xrightarrow{p_*} \pi_{n+1}(B, b_0) \xrightarrow{d_*} \pi_n(F, e_0) \xrightarrow{i_*} \pi_n(E, e_0) \xrightarrow{p_*} \pi_n(B, b_0) \xrightarrow{d_*} \cdots$$
$$\cdots \xrightarrow{p_*} \pi_1(B, b_0) \xrightarrow{d_*} \pi_0(F, e_0) \xrightarrow{i_*} \pi_0(E, e_0)$$

which is called the *homotopy sequence of the fibering $p : E \to B$ based at e_0*.

Proposition 6.1. *If the basic fiber F is totally pathwise disconnected, then*

$$p_* : \pi_n(E, e_0) \approx \pi_n(B, b_0), \quad n \geqslant 2,$$

and p_* is a monomorphism if $n = 1$.

Proof. If F is totally pathwise disconnected, then $\pi_n(F, e_0) = 0$ for every $n \geqslant 1$. Hence the proposition is an immediate consequence of the exactness of the homotopy sequence of the fibering $p : E \to B$. ∎

Proposition 6.2. *If the fibering $p : E \to B$ admits a cross-section $\chi : B \to E$, then, for every $b_0 \in B$ and $e_0 = \chi(b_0) \in F = p^{-1}(b_0)$, we have*

$$\pi_n(E, e_0) \approx \pi_n(B, b_0) + \pi_n(F, e_0)$$

for each $n \geqslant 2$ and p_* is an epimorphism for every $n \geqslant 1$.

Proof. Since $p\chi$ is the identity map on B, it follows that $p_*\chi_*$ is the identity automorphism on $\pi_n(B, b_0)$. Hence, χ_* is a monomorphism and p_* is an epimorphism for every $n \geqslant 1$.

If $n \geqslant 2$, then $\pi_n(E, e_0)$ is abelian. Then, $p_*\chi_* = 1$ implies that $\pi_n(E, e_0)$ decomposes into the direct sum

$$\pi_n(E, e_0) = J + K, \quad J = \text{Image}\, \chi_*, \quad K = \text{Kernel}\, p_*.$$

Since χ_* is a monomorphism, we have $J \approx \pi_n(B, b_0)$. On the other hand, since p_* is an epimorphism for every $n \geqslant 1$, it follows from an exactness argument that i_* is a monomorphism for every $n \geqslant 1$. Hence

$$K = \text{Kernel}\, p_* = \text{Image}\, i_* \approx \pi_n(F, e_0). \blacksquare$$

If $n > 1$, then (6.2) is a generalization of (2.1).

As immediate consequences of (5.1)–(5.3), we have the following three propositions concerning the homotopy sequence of a fibering $p : E \to B$.

Proposition 6.3. *If F is a retract of E, then*

$$\pi_n(E, e_0) \approx \pi_n(B, b_0) + \pi_n(F, e_0)$$

for every $n \geqslant 2$ and p_* is an epimorphism for every $n \geqslant 1$.

Proposition 6.4. *If E is deformable into F, then*

$$\pi_n(F, e_0) \approx \pi_n(E, e_0) + \pi_{n+1}(B, b_0)$$

for every $n \geqslant 2$ and $p_* = 0$ for every $n \geqslant 1$.

Proposition 6.5. *If F is contractible in E, then*

$$\pi_n(B, b_0) \approx \pi_n(E, e_0) + \pi_{n-1}(F, e_0)$$

for every $n \geqslant 2$ and p_* is a monomorphism for every $n \geqslant 1$.

As an application, let us consider the Hopf fiberings

$$p : S^{2m-1} \to S^m, \quad (m = 2, 4, 8).$$

In each of these fibering, the fiber F is an $(m-1)$-sphere S^{m-1} and is contractible in S^{2m-1}. Hence, by (6.5), we have

$$\pi_n(S^m) \approx \pi_n(S^{2m-1}) + \pi_{n-1}(S^{m-1}) \tag{6.6}$$

for every $m = 2, 4, 8$ and every $n \geqslant 2$. When $m = 2$, this gives

$$\pi_n(S^2) \approx \pi_n(S^3), \quad n \geqslant 3. \tag{6.7}$$

When $m = 4$ or 8, we have

$$\pi_n(S^4) \approx \pi_{n-1}(S^3), \quad (2 \leqslant n \leqslant 6).$$

$$\pi_n(S^8) \approx \pi_{n-1}(S^7), \quad (2 \leqslant n \leqslant 14).$$

$$\pi_7(S^4) \approx Z + \pi_6(S^3).$$

$$\pi_{15}(S^8) \approx Z + \pi_{14}(S^7).$$

As another application, let us consider the fibering $p : S \to M$ in (III;

Ex. F), where S denotes the unit sphere in the space C^n of n complex variables and M the projective space associated with C^n. Then S is an $(2n - 1)$-sphere and the fiber F is a 1-sphere. If $n = 1$, M consists of a single point. For the case $n > 1$, it follows immediately from the exactness of the homotopy sequence of the fibering $p : S \to M$ that

$$\pi_1(M) = 0, \quad \pi_2(M) \approx Z,$$
$$\pi_m(M) \approx \pi_m(S^{2n-1}), \quad m > 2.$$

7. Homotopy groups of covering spaces

As a special case of (6.1), we have the following

Proposition 7.1. *If E is a covering space over a base space B relative to a projection $p : (E, e_0) \to (B, b_0)$, then*

$$p_* : \pi_n(E, e_0) \approx \pi_n(B, b_0), \quad n \geqslant 2,$$

and p_* is a monomorphism if $n = 1$.

In particular, if E is the universal covering space over B, then we have

$$\pi_1(E) = 0, \quad \pi_n(E) \approx \pi_n(B), \quad n \geqslant 2.$$

Therefore, by Hurewicz theorem (4.4 and Ex. C), we have the following

Proposition 7.2. *For every connected, locally pathwise connected and semi-locally simply connected space B, the second homotopy group $\pi_2(B)$ is isomorphic to the second singular homology group $H_2(E)$ of its universal covering space E.*

Applying (7.1) to the universal covering $p : S^m \to P^m$ of the m-sphere onto the real projective space P^m, we have

$$\pi_n(P^m) \approx \pi_n(S^m), \quad n \geqslant 2.$$

For another example, since the euclidean 2-space is the universal covering space of any closed surface M other than S^2 and P^2, we have

$$\pi_n(M) = 0, \quad n \geqslant 2.$$

Now let B be a given space which is connected, locally pathwise connected, and semi-locally simply connected. Let $b_0 \in B$ and $n > 1$. By (IV; 15.1), $\pi_1(B, b_0)$ acts on $\pi_n(B, b_0)$ as a group of operators. Using the isomorphisms p_* for the universal covering $p : E \to B$ and the covering transformations of E, these operators can be interpreted nicely as follows. Let $w \in \pi_1(B, b_0)$. By (III; 16.6), w determines a covering transformation of E which will be denoted by $h : E \to E$ for the moment. Pick $e_0 \in E$ with $p(e_0) = b_0$ and let $e_1 = h(e_0)$. Then $p(e_1) = b_0$. By the construction of h and that of the operator w on $\pi_n(B, b_0)$, one can prove that commutativity holds in the following diagram

$$\begin{array}{ccc} \pi_n(E, e_0) & \xrightarrow{h_*} & \pi_n(E, e_1) \\ \downarrow{\scriptstyle p_*} & & \downarrow{\scriptstyle p_*} \\ \pi_n(B, b_0) & \xleftarrow{w} & \pi_n(B, b_0) \end{array}$$

Since h_* and p_* are isomorphisms, we have

(7.3) $$w = p_* h_*^{-1} p_*^{-1}.$$

For example, let B denote the one-point union $S^n \vee S^1$ obtained by identifying $s_0 \in S^n$ with $1 \in S^1$, where $n \geqslant 2$. Let b_0 denote the identified point of s_0 and 1. By (II; Ex. A), we get

$$\pi_1(B, b_0) \approx Z$$

with a generator w represented by the exponential map $p : I \to S^1 \subset B$ of (II; § 2). In the product space $S^n \times R$ of S^n and the real line R, let us consider the subspace

$$E = (S^n \times Z) \cup (s_0 \times R),$$

where Z denotes the subspace of R consisting of all integers. This space E is pathwise connected and $\pi_m(E) = 0$ for every $m < n$. In particular, since $n > 1$, E is simply connected and hence n-simple. The n-th homotopy group $\pi_n(E)$ as interpreted in (IV; § 16) is a free abelian group with a countable set $\{ \alpha_i \mid i \in Z \}$ of free generators, where α_i is represented by the map $f_i : S^n \to E$ which is defined as follows:

$$f_i(s) = (s, i) \in S^n \times Z \subset E, \quad (s \in S^n).$$

For a proof of this result, see Ex. C at the end of the chapter.

Define a map $q : E \to B$ by taking

$$q(s, t) = \begin{cases} s, & (s \in S^n, t \in Z), \\ p(t), & (s = s_0, t \in R), \end{cases}$$

where $p : R \to S^1$ denotes the exponential map of (II; § 2). One can easily verify that E is a covering space over B relative to q. Hence, $\pi_m(B, b_0) = 0$ for each m with $1 < m < n$ and $\pi_n(B, b_0)$ is a free abelian group with a countable set $\{ \beta_i \mid i \in Z \}$ of free generators, where β_i corresponds to α_i in the natural way.

The generator w of $\pi_1(B, b_0)$ determines a covering transformation $h : E \to E$ given by $h(s, t) = (s, t + 1)$ for every point (s, t) of E. It follows that $h_* \alpha_i = \alpha_{i+1}$ for each $i \in Z$ and hence $w \beta_i = \beta_{i-1}$ for each $i \in Z$. This determines the operations of $\pi_1(B, b_0)$ on $\pi_n(B, b_0)$. Note that B is *not* n-simple.

8. The n-connective fiberings

The interesting property (7.1) of the universal covering space suggests the notion of n-connective fiber spaces as follows. A fibering $p : E \to B$ over a given pathwise connected space B is said to be *n-connective* if E is n-connected and

$$p_* : \pi_m(E, e_0) \approx \pi_m(B, b_0), \quad m > n,$$

where $e_0 \in E$ and $p(e_0) = b_0$, and hence

$$\pi_{n+1}(B, b_0) \approx H_{n+1}(E).$$

Thus, the identity map on B is a 0-connective fibering and the universal covering over B, if it exists, is a 1-connective fibering. If $p : E \to B$ is an n-connective fibering, then E is said to be an *n-connective fiber space* over B. The main objective of the present section is to construct an n-connective fiber space over a given pathwise connected space B.

Lemma 8.1. *If B is a given space, b_0 a given point in B, and n a given positive integer, then there exists a space X which satisfies the following three conditions:*

(i) *X contains B as a closed subspace.*

(ii) $\pi_n(X, b_0) = 0$.

(iii) *The inclusion map $i : B \subset X$ induces an isomorphism*

$$i_* : \pi_m(B, b_0) \approx \pi_m(X, b_0)$$

for every m satisfying $0 < m < n$.

Proof. Let A be a set of elements of $\pi_n(B, b_0)$ which generates $\pi_n(B, b_0)$ and consider A as a space with discrete topology. Let E^{n+1} denote the $(n + 1)$-cell bounded by the unit n-sphere S^n in the euclidean $(n + 1)$-space. Let

$$W = E^{n+1} \times A, \quad C = S^n \times A \subset W.$$

Let $s_0 = (1, 0, \cdots, 0) \in S^n$. For each $\alpha \in A$, choose a map $g_\alpha : (S^n, s_0) \to (B, b_0)$ which represents α. Then, define a map $g : C \to B$ by taking

$$g(s, \alpha) = g_\alpha(s), \quad (s, \alpha) \in C.$$

Let X denote the adjunction space obtained by adjoining W to B by means of the partial map $g : C \to B$. It remains to verify the conditions (i)–(iii).

Since C is closed in W, (i) is obvious from the definition of adjunction space in (I; § 7). To verify (ii) and (iii), let us first prove that

$$\pi_m(X, B, b_0) = 0, \quad 0 < m \leqslant n.$$

For this purpose, let ξ be an arbitrary element of $\pi_m(X, B, b_0)$ and pick a map

$$f : (I^m, I^{m-1}, J^{m-1}) \to (X, B, b_0)$$

which represents ξ. Since $m \leqslant n$, it follows from the method of simplicial approximation that, by a suitable homotopy of f, we can free an interior point e_α of $E^{n+1} \times \alpha$ from the image of f for every $\alpha \in A$. Hence there exists a homotopy

$$f_t : (I^m, I^{m-1}, J^{m-1}) \to (X, B, b_0), \quad (0 \leqslant t \leqslant 1).$$

such that $f_0 = f$ and $f_1(I^m) \subset B$. By (IV; 3.3), this implies that $\xi = 0$. Hence $\pi_m(X, B, b_0) = 0$ for every m with $0 < m \leqslant n$.

From the exactness of the homotopy sequence, we deduce immediately that

$$i_* : \pi_m(B, b_0) \to \pi_m(X, b_0)$$

is an isomorphism for $m = 1, 2, \cdots, n-1$ and is an epimorphism for $m = n$. This verifies (iii).

Since g_a represents α, it is clear that $i_*(\alpha) = 0$. Since A generates $\pi_n(B, b_0)$, it follows that i_* sends every element of $\pi_n(B, b_0)$ into 0. Since i_* is an epimorphism for $m = n$, this implies that $\pi_n(X, b_0) = 0$. This verifies (ii). ∎

Lemma 8.2. *If B is a given space, b_0 a given point in B, and n a given non-negative integer, then there exists a space X which satisfies the following three conditions:*

(i) *X contains B as a closed subspace.*

(ii) *$\pi_m(X, b_0) = 0$ for every $m > n$.*

(iii) *The inclusion map $i : B \subset X$ induces an isomorphism*

$$i_* : \pi_m(B, b_0) \approx \pi_m(X, b_0)$$

for every m satisfying $0 < m \leqslant n$.

Proof. By recurrent application of the construction given in the proof of (8.1), we obtain a sequence of spaces $X_0, X_1, \cdots, X_r, \cdots$ such that

(1) $X_0 = B$,

(2) X_r contains X_{r-1} as a closed subspace for every $r > 0$,

(3) $\pi_{n+r}(X_r, b_0) = 0$ for every $r > 0$, and

(4) the inclusion map $i_r : X_{r-1} \subset X_r$ induces an isomorphism $(i_r)_* : \pi_m(X_{r-1}, b_0) \approx \pi_m(X_r, b_0)$ for every m satisfying $0 < m < n + r$.

As a set, let X denote the union of X_r for all $r \geqslant 0$. Introduce a topology in X as follows: a set U of X is said to be open iff $U \cap X_r$ is open in X_r for every $r \geqslant 0$. Thus, we obtain a space X. It remains to verify the conditions (i)–(iii).

By (2) and the definition of the topology in X, it follows that X_r is a closed subspace of X for every $r \geqslant 0$. In particular, (i) is true since $X_0 = B$.

Next, let us prove that the inclusion map $j_r : X_r \subset X$ induces an isomorphism

$$(j_r)_* : \pi_m(X_r, b_0) \approx \pi_m(X, b_0)$$

for every m such that $0 < m < n + r$. To prove that $(j_r)_*$ is an epimorphism, let $\alpha \in \pi_m(X, b_0)$ and choose a representative map $f : (I^m, \partial I^m) \to (X, b_0)$ for α. Since $f(I^m)$ is compact, there exists an integer $q \geqslant 0$ such that $f(I^m) \subset X_{r+q}$. Hence f also represents an element β of $\pi_m(X_{r+q}, b_0)$ with $(j_{r+q})_*(\beta) = \alpha$. According to (4), the inclusion $k : X_r \subset X_{r+q}$ induces an isomorphism $k_* : \pi_m(X_r, b_0) \approx \pi_m(X_{r+q}, b_0)$. Let $\gamma = k_*^{-1}(\beta)$. Then $(j_r)_*(\gamma) = (j_{r+q})_* k_*(\gamma) = (j_{r+q})_*(\beta) = \alpha$. So, $(j_r)_*$ is an epimorphism. To prove that $(j_r)_*$ is a monomorphism, let $\xi \in \pi_m(X_r, b_0)$ be an element such that $(j_r)_*(\xi) = 0$. Choose a map $g : (I^m, \partial I^m) \to (X_r, b_0)$ representing ξ. Then $(j_r)_*(\xi) = 0$ implies that there exists a homotopy $g_t : (I^m, \partial I^m) \to (X, b_0)$, $0 \leqslant t \leqslant 1$, such that $g_0 = g$ and $g_1(I^m) = b_0$. This homotopy g_t defines a map $G : I^{m+1} \to X$. Since $G(I^{m+1})$ is compact, there is an integer $q \geqslant 0$ such that $G(I^{m+1}) \subset X_{r+q}$. Hence $k_*(\xi) = 0$, where $k : X_r \subset X_{r+q}$. Since k_*

is an isomorphism by (4), this implies that $\xi = 0$. So, $(j_r)_*$ is also a monomorphism. This completes the proof that $(j_r)_*$ is an isomorphism. By means of this, (ii) and (iii) follow readily from (3) and (4). ▮

Theorem 8.3. *For any given pathwise connected space B and any non-negative integer n, there exists an n-connective fibering $p : E \to B$.*

Proof. Let $b_0 \in B$ and choose a space X satisfying the conditions (i)–(iii) of (8.2). Thus we obtain a triplet (X, B, b_0). Let (X', B', b'_0) denote the derived triplet of (X, B, b_0) and $q : (X', B', b'_0) \to (X, B, b_0)$ the derived projection as defined in (IV; § 3). Then X' is a fiber space over X and B' is a fiber space over B relative to q. Let $F = q^{-1}(b_0)$. The homotopy sequences of the fiberings are connected by the following commutative diagram

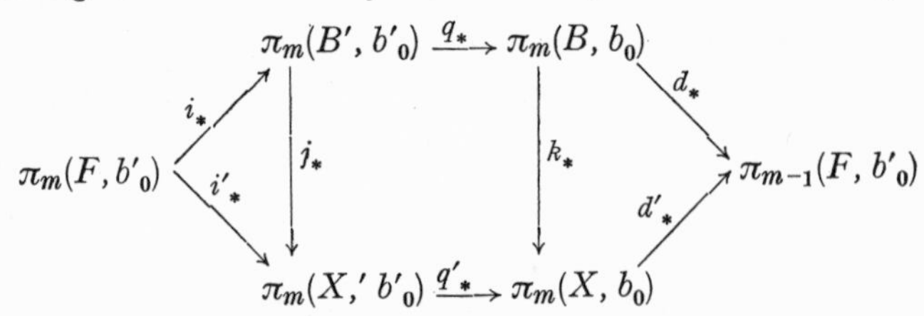

where i_*, i'_*, j_*, k_* are induced by inclusion maps, q_*, q'_* are induced by q, and d_*, d'_* are the boundary operations. Now, since X' is contractible, we have $\pi_m(X', b'_0) = 0$ for every $m \geqslant 0$. Therefore, d'_* carries $\pi_m(X, b_0)$ onto $\pi_{m-1}(F, b'_0)$ in a one-to-one fashion for every $m \geqslant 1$.

If $m > n$, then we have

$$\pi_m(F, b'_0) \approx \pi_{m+1}(X, b_0) = 0, \quad \pi_{m-1}(F, b'_0) \approx \pi_m(X, b_0) = 0.$$

Hence, q_* is an isomorphism for every $m > n$. On the other hand, if $m \leqslant n$, then k_* is an isomorphism and hence $d_* = d'_* k_*$ carries $\pi_m(B, b_0)$ onto $\pi_{m-1}(F, b'_0)$ in a one-to-one fashion. Hence, $\pi_m(B', b'_0) = 0$ for every m satisfying $0 < m < n$. Finally, since $\pi_n(F, b'_0) \approx \pi_{n+1}(X, b_0) = 0$, we obtain $\pi_n(B', b'_0) = 0$.

Now, let E denote the path-component of B' containing the point $e_0 = b'_0$ and let $p = q \mid E$. In other words, E denotes the path-component of the space of paths $[X; B, b_0]$ which contains the degenerate path $e_0(I) = b_0$, and $p : E \to B$ denotes the initial projection. Then, it follows from what we have proved for $q : B' \to B$ that $p : E \to B$ is a fibering, that E is n-connected, and that $p_* : \pi_m(E, e_0) \approx \pi_m(B, b_0)$ for every $m > n$. Hence $p : E \to B$ is n-connective. ▮

Corollary 8.4. *If B is an $(n-1)$-connected space and $p : E \to B$ is an n-connective fibering, then the fiber $F = p^{-1}(b_0)$ over any given point $b_0 \in B$ has the property:*

$$\pi_{n-1}(F, e_0) \approx \pi_n(B, b_0),$$

$$\pi_m(F, e_0) = 0, \quad m \neq n-1,$$

where e_0 is any point of F.

This corollary is an easy consequence of the homotopy sequence of the fibering $p : E \to B$ at the basic point e_0.

9. The homotopy sequence of a triple

By a *triple* (X, A, B) we mean a space X together with two nonvoid subspaces A and B such that $A \supset B$. Hence a triplet is a special case of triple where B consists of a single point. Pick a basic point $x_0 \in B$. In the homotopy system $\{\pi, \partial, *\}$, the inclusion maps

$$\bar{i} : (A, B, x_0) \subset (X, B, x_0), \quad \bar{j} : (X, B, x_0) \subset (X, A, x_0)$$

give rise to the induced transformations $\bar{i}_*$ and $\bar{j}_*$. On the other hand, the inclusion map

$$k : (A, x_0) \subset (A, B, x_0)$$

has the induced transformations k_*. Define a *boundary operator*

$$\bar{\partial} : \pi_n(X, A, x_0) \to \pi_{n-1}(A, B, x_0)$$

for every $n > 0$ by taking the composition $\bar{\partial} = k_* \partial$ of

$$\pi_n(X, A, x_0) \xrightarrow{\partial} \pi_{n-1}(A, x_0) \xrightarrow{k_*} \pi_{n-1}(A, B, x_0).$$

Thus we obtain a beginningless sequence

$$\cdots \xrightarrow{\bar{j}_*} \pi_{n+1}(X,A,x_0) \xrightarrow{\bar{\partial}} \pi_n(A,B,x_0) \xrightarrow{\bar{i}_*} \pi_n(X,B,x_0) \xrightarrow{\bar{j}_*} \pi_n(X,A,x_0) \xrightarrow{\bar{\partial}} \cdots$$
$$\cdots \xrightarrow{\bar{\partial}} \pi_1(A,B,x_0) \xrightarrow{\bar{i}_*} \pi_1(X, B, x_0) \xrightarrow{\bar{j}_*} \pi_1(X,A,x_0).$$

This will be called the *homotopy sequence of the triple* (X, A, B) at the basic point x_0.

Theorem 9.1. *The homotopy sequence of a triple is exact.*

Proof. Consider the spaces of paths

$$X' = [X; X, x_0], \quad A' = [X; A, x_0], \quad B' = [X; B, x_0]$$

with x'_0 denoting the degenerate path $x'_0(I) = x_0$. Then the initial projection $p : X' \to X$ defines a fibering

$$q : (A', B', x'_0) \to (A, B, x_0).$$

We shall also make use of the inclusion maps $i : (B', x'_0) \subset (A', x'_0)$ and $j : (A', x'_0) \subset (A', B', x'_0)$. Consider the following diagram

$$\begin{array}{ccccccccccccc}
\to \pi_n(A',x'_0) & \xrightarrow{j_*} & \pi_n(A',B',x'_0) & \xrightarrow{\partial} & \pi_{n-1}(B',x'_0) & \xrightarrow{i_*} & \pi_{n-1}(A',x'_0) & \xrightarrow{j_*} & \cdots & \xrightarrow{\partial} & \pi_0(B',x'_0) & \xrightarrow{i_*} & \pi_0(A',x'_0) \\
\uparrow \chi & & \downarrow q_* & & \uparrow \chi & & \uparrow \chi & & & & \uparrow \chi & & \uparrow \chi \\
\to \pi_{n+1}(X,A,x_0) & \xrightarrow{\bar{\partial}} & \pi_n(A,B,x_0) & \xrightarrow{\bar{i}_*} & \pi_n(X,B,x_0) & \xrightarrow{\bar{j}_*} & \pi_n(X,A,x_0) & \xrightarrow{\bar{\partial}} & \cdots & \xrightarrow{\bar{i}_*} & \pi_1(X,B,x_0) & \xrightarrow{\bar{j}_*} & \pi_1(X,A,x_0)
\end{array}$$

As the homotopy sequence of the triplet (A', B', x'_0), the top row is exact. By (IV; § 9), χ and q_* are one-to-one and onto. One can easily verify that commutativity holds in each of the rectangles. These imply that the bottom is exact. ∎

Since the part of the homotopy sequence of (X, A, B) which ends with $\pi_2(X, A, x_0)$ consists of groups and homomorphisms, the exactness of this can also be derived from the algebraic axioms and the exactness of the homotopy sequences of triplets. See [S; p. 77] and [E–S; pp. 24–28]. The proof given above is considerably simplified by the fibering axiom which does not hold in homology theory.

10. The homotopy groups of a triad

We recall that a *triad* $(X; A, B)$ consists of a space X together with two subspaces A and B such that the intersection $C = A \cap B$ is not empty. Pick a basic point $x_0 \in C$. We shall define the homotopy groups of the triad $(X; A, B)$ as follows.

Let us consider the spaces of paths

$$P = [X; B, x_0], \quad Q = [A; C, x_0]$$

and denote by σ_0 the degenerate path $\sigma_0(I) = x_0$. Since $A \subset X$ and $C \subset B$, it follows that $\sigma_0 \in Q \subset P$ and hence we obtain a triplet (P, Q, σ_0).

According to (IV; § 9), we have the natural correspondences

$$\chi : \pi_n(X, B, x_0) \approx \pi_{n-1}(P, \sigma_0), \quad \chi : \pi_n(A, C, x_0) \approx \pi_{n-1}(Q, \sigma_0)$$

for every $n > 0$. Furthermore, we define for every $n > 1$

$$\pi_n(X; A, B, x_0) = \pi_{n-1}(P, Q, \sigma_0).$$

If n $\geqslant 3$, then $\pi_n(X; A, B, x_0)$ is a group which is abelian in case $n > 3$. This group is called *the n-th homotopy group of the triad* $(X; A, B)$ at the basic point x_0. For completeness, we shall also call $\pi_n(X; A, B, x_0)$, $n \geqslant 2$, the *n-th homotopy set* of $(X; A, B)$ at x_0.

Thus, the homotopy sequence of the triplet (P, Q, σ_0) gives rise to the following exact sequence

$$\cdots \xrightarrow{j_*} \pi_{n+1}(X;A,B,x_0) \xrightarrow{\partial} \pi_n(A,C,x_0) \xrightarrow{i_*} \pi_n(X,B,x_0) \xrightarrow{j_*} \pi_n(X;A,B,x_0) \xrightarrow{\partial}$$
$$\cdots \xrightarrow{j_*} \pi_2(X;A,B,x_0) \xrightarrow{\partial} \pi_1(A, C, x_0) \xrightarrow{i_*} \pi_1(X, B, x_0)$$

which will be called *the first homotopy sequence of the triad* $(X; A, B)$ *at the basic point* x_0.

The homomorphisms

$$i_* : \pi_n(A, C, x_0) \to \pi_n(X, B, x_0), \quad n \geqslant 2,$$

are called *the excision homomorphisms*, [E–S; p. 207]. Thus the triad homotopy sets $\pi_n(X; A, B, x_0)$, $n \geqslant 2$, measure the extent by which the excision axiom fails for the relative homotopy groups.

The above definition of $\pi_n(X; A, B, x_0)$, $n \geqslant 2$, gives a geometrical representation as follows. An element of $\pi_n(X; A, B, x_0)$ is represented by a map $f: I^n \to X$ such that, for any point $t = (t_1, \cdots, t_n)$ in the boundary ∂I^n of I^n, $f(t) \in A$ if $t_{n-1} = 0$, $f(t) \in B$ if $t_n = 0$, and $f(t) = x_0$ otherwise. Precisely, the elements of $\pi_n(X; A, B, x_0)$ are the path-components of the space of these maps. See [Blakers and Massey 1].

Consider the homeomorphism $h: I^n \to I^n$ defined by

$$h(t_1, \cdots, t_{n-2}, t_{n-1}, t_n) = (t_1, \cdots, t_{n-2}, t_n, t_{n-1}).$$

Then the assignment $f \to fh$ induces an isomorphism

$$h_* : \pi_n(X; A, B, x_0) \approx \pi_n(X; B, A, x_0)$$

for every $n > 2$ and a one-to-one correspondence for $n = 2$. Hence the first homotopy sequence of $(X; B, A)$ at x_0 gives rise to the following exact sequence

$$\cdots \xrightarrow{j'_*} \pi_{n+1}(X;A,B,x_0) \xrightarrow{\partial'} \pi_n(B,C,x_0) \xrightarrow{i'_*} \pi_n(X,A,x_0) \xrightarrow{j'_*} \pi_n(X;A,B,x_0) \xrightarrow{\partial'} \cdots$$

$$\cdots \xrightarrow{j'_*} \pi_2(X; A, B, x_0) \xrightarrow{\partial'} \pi_1(B, C, x_0) \xrightarrow{i'_*} \pi_1(X, A, x_0)$$

which will be called *the second homotopy sequence of the triad* $(X; A, B)$ *at the basic point* x_0.

Another geometrical representation of $\pi_n(X; A, B, x_0)$, $n \geqslant 2$, can be described as follows. Consider the unit n-cell E^n in the euclidean n-space R^n and its boundary $(n-1)$-sphere S^{n-1}. Let E_+^{n-1} and E_-^{n-1} denote the hemispheres of S^{n-1} defined by $t_n \geqslant 0$ and $t_n \leqslant 0$ respectively. Let $s_0 = (1, 0, \cdots, 0)$. Then the elements of $\pi_n(X; A, B, x_0)$ are the path-components of the space of maps

$$f: (E^n; E_+^n, E_-^n, s_0) \to (X; A, B, x_0).$$

As an illustrative example of the triad homotopy groups, let us take $(X; A, B)$ to be the triad given by

$$X = S^2, \quad A = E_+^2, \quad B = E_-^2.$$

Then we have $C = S^1$. Take $x_0 = (1, 0, 0)$. Since A and B are both contractible in themselves, it follows from the exactness of the homotopy sequences of the triplets (A, C, x_0) and (X, B, x_0) that

$$\pi_n(A, C, x_0) \approx \pi_{n-1}(C, x_0), \quad \pi_n(X, B, x_0) \approx \pi_n(X, x_0),$$

for every $n > 0$. On the other hand,

$$\pi_n(A, C, x_0) \approx \pi_{n-1}(S^1) = 0$$

for every $n \geqslant 3$. This implies that j_* in the first homotopy sequence of $(X; A, B)$ is an isomorphism for every $n > 3$. Hence

$$\pi_n(X; A, B, x_0) \approx \pi_n(S^2), \quad n > 3.$$

Finally, it is easy to see that in the following part of the first homotopy sequence of $(X; A, B)$

$$0 \to \pi_3(X,B,x_0) \to \pi_3(X;A,B,x_0) \to \pi_2(A,C,x_0) \xrightarrow{i_*} \pi_2(X,B,x_0) \to \pi_2(X;A,B,x_0) \to 0,$$

the groups $\pi_2(A, C, x_0)$ and $\pi_2(X, B, x_0)$ are free cyclic and i_* sends a generator of $\pi_2(A, C, x_0)$ onto a generator of $\pi_2(X, B, x_0)$. Hence i_* is an isomorphism. This implies that

$$\pi_2(X; A, B, x_0) = 0,$$

$$\pi_3(X; A, B, x_0) \approx \pi_3(X, B, x_0) \approx \pi_3(S^2) \approx Z.$$

11. Freudenthal's suspension

Let $(X; A, B)$ be a given triad such that both A and B are contractible in themselves. Set $C = A \cap B$ and pick a basic point $x_0 \in C$. Consider the spaces of paths

$$U = [X; A, B], \quad V = [C; C, C]$$

and denote by σ_0 the degenerate path $\sigma_0(I) = x_0$. Since $\sigma_0 \in V \subset U$, we obtain the following exact homotopy sequence of the triplet (U, V, σ_0):

$$\cdots \xrightarrow{j_*} \pi_{n+1}(U, V, \sigma_0) \xrightarrow{\partial} \pi_n(V, \sigma_0) \xrightarrow{i_*} \pi_n(U, \sigma_0) \xrightarrow{j_*} \pi_n(U, V, \sigma_0) \xrightarrow{\partial} \cdots$$

$$\cdots \xrightarrow{j_*} \pi_1(U, V, \sigma_0) \xrightarrow{\partial} \pi_0(V, \sigma_0) \xrightarrow{i_*} \pi_0(U, \sigma_0).$$

According to (III; § 9), there is a homeomorphism $\xi : C \to V$ of C onto a subspace $\xi(C)$ of V. ξ is called the natural injection of C into V and is defined by taking $\xi(x)$ to be the degenerate path at x for every $x \in C$. By (III; 9.10), $\xi(C)$ is a strong deformation retract of V and hence we have

$$\xi_* : \pi_n(C, x_0) \approx \pi_n(V, \sigma_0), \quad n \geqslant 0.$$

On the other hand, let $W = [X; x_0, x_0]$. Since both A and B are contractible, it follows from (III; Ex. N) that the inclusion map $\eta : W \subset U$ has a homotopy inverse. This implies that

$$\eta_* : \pi_n(W, \sigma_0) \approx \pi_n(U, \sigma_0), \quad n \geqslant 0.$$

According to (IV; § 9), we have the natural correspondences

$$\chi : \pi_{n+1}(X, x_0) \approx \pi_n(W, \sigma_0), \quad n \geqslant 0.$$

Next, consider the space of paths $\Omega = [U; V, \sigma_0]$. By (III; 9.9), Ω can be considered as the space of maps $f : I^2 \to X$ such that

$$f(0, t) \in A, \quad f(1, t) \in B, \quad f(t, 0) \in C, \quad f(t, 1) = x_0$$

for every $t \in I$. Let θ_0 denote the constant map $\theta_0(I^2) = x_0$. Then, we have the natural correspondences

$$\chi : \pi_{n+1}(U, V, \sigma_0) \approx \pi_n(\Omega, \theta_0), \quad n \geqslant 0.$$

On the other hand, let $P = [X; B, x_0]$ and $Q = [A; C, x_0]$ and consider $\Lambda = [P; Q, \sigma_0]$. Then Λ can be considered as the space of maps $g: I^2 \to X$ with

$$g(0, t) \in A, \quad g(t, 0) \in B, \quad g(1, t) = x_0 = g(t, 1)$$

for every $t \in I$. Obviously $\theta_0 \in \Lambda$. According to the definition in § 10, we have

$$\pi_{n+1}(X, A, B, x_0) = \pi_n(P, Q, \sigma_0) \overset{\chi}{\approx} \pi_{n-1}(\Lambda, \theta_0), \quad n \geqslant 1.$$

We are going to prove that Ω and Λ are of the same homotopy type. Let ∂I^2 denote the boundary of I^2 and define a map $\theta : \partial I^2 \to \partial I^2$ by taking for each $t \in I$

$$\phi(t, 0) = \begin{cases} (2t, 0), & \text{if } 0 \leqslant t \leqslant \frac{1}{2} \\ (1, 2t - 1), & \text{if } \frac{1}{2} \leqslant t \leqslant 1, \end{cases}$$

$$\phi(0, t) = (0, t), \quad \phi(t, 1) = (t, 1), \quad \phi(1, t) = (1, 1).$$

On can easily see that ϕ is homotopic to the identity map on ∂I^2 and hence has an extension $\Phi : I^2 \to I^2$. For each $f \in \Omega$, the composed map $f\Phi$ is in Λ. The assignment $f \to f\Phi$ defines a map $\zeta : (\Omega, \theta_0) \to (\Lambda, \theta_0)$. By constructing another but somewhat similar map $\Psi : I^2 \to I^2$, one can prove that ζ has a homotopy inverse defined by $g \to g\Psi$ for each $g \in \Lambda$. Hence we have

$$\zeta^* : \pi_{n-1}(\Omega, \theta_0) \approx \pi_{n-1}(\Lambda, \theta_0), \quad n \geqslant 1.$$

Now let us consider the following composed transformations

$$r = \xi_*^{-1}\partial\chi^{-1}\zeta_*^{-1}\chi : \pi_{n+2}(X; A, B, x_0) \to \pi_n(C, x_0),$$

$$s = \chi^{-1}\eta_*^{-1}i_*\xi_* : \pi_n(C, x_0) \to \pi_{n+1}(X, x_0),$$

$$t = \chi^{-1}\zeta_*\chi j_*\eta_*\chi : \pi_{n+2}(X, x_0) \to \pi_{n+2}(X; A, B, x_0)$$

for each $n \geqslant 0$. If $n > 0$, then all of these are homomorphisms. The homotopy sequence of the triplet (U, V, σ_0) gives rise to the following exact sequence

$$\cdots \xrightarrow{t} \pi_{n+2}(X;A,B,x_0) \xrightarrow{r} \pi_n(C,x_0) \xrightarrow{s} \pi_{n+1}(X,x_0) \xrightarrow{t} \pi_{n+1}(X;A,B,x_0) \xrightarrow{r} \cdots$$

$$\cdots \xrightarrow{t} \pi_2(X; A, B, x_0) \xrightarrow{r} \pi_0(C, x_0) \xrightarrow{s} \pi_1(X, x_0)$$

which will be called the *suspension sequence* of the triad $(X; A, B)$. In particular, the transformations s are called the *suspensions*. Thus, the triad homotopy sets $\pi_n(X; A, B, x_0)$ measure the extent by which the suspensions s, $n > 0$, fail to be isomorphisms.

Proposition 11.1. *For any integer $m \geqslant 2$, the following three statements are equivalent:*

(i) $\pi_n(X; A, B, x_0) = 0$ *for every* $n \leqslant m$.

(ii) *The suspension* $s : \pi_n(C, x_0) \to \pi_{n+1}(X, x_0)$ *is an isomorphism for each* $n = 1, \cdots, m - 2$ *and is an epimorphism for* $n = m - 1$.

(iii) *The excision* $i_* : \pi_n(A, C, x_0) \to \pi_n(X, B, x_0)$ *is an isomorphism for each* $n = 2, \cdots, m - 1$ *and is an epimorphism for* $n = m$.

Proof. The equivalence of (i) and (ii) is a consequence of the exactness of the suspension sequence of $(X; A, B)$, and the equivalence of (i) and (iii) is a consequence of properties of first homotopy sequence of $(X; A, B)$. ∎

If X is the r-sphere with $r \geqslant 2$ and $A = E_+^r$, $B = E_-^r$, then it will be proved in the sequel that the statement (ii) is true for $m = 2r - 2$. See (XI; 2.1).

By the definition $s = \chi^{-1}\eta_*^{-1}i_*\xi_*$, it is not difficult to see that a geometrical represention of the suspension $s : \pi_n(C, x_0) \to \pi_{n+1}(X, x_0)$ can be described as follows. For any $\alpha \in \pi_n(C, x_0)$, let us choose a representative map $f: (S^n, s_0) \to (C, x_0)$. Since both A and B are contractible, f has an extension $F: (S^{n+1}, s_0) \to (X, x_0)$ such that $F(E_+^{n+1}) \subset A$ and $F(E_-^{n+1}) \subset B$. Then $s(\alpha)$ is the element of $\pi_{n+1}(X, x_0)$ represented by this map F. This beautiful geometrical representation of s is its original definition given by Freudenthal to the special triad $(S^r; E_+^r, E_-^r)$.

As a generalization of the triad $(S^r; E_+^r, E_-^r)$, let C be a given non-empty space. Consider the space X obtained by joining C to two distinct vertices a and b. Precisely, X is obtained from $C \times I$ by identifying the subsets $C \times 0$ and $C \times 1$ into single points a and b respectively. If $p : C \times I \to X$ denotes the natural projection, then C can be considered as a subspace of X by the imbedding $i : C \to X$ defined by $i(c) = p(c, \frac{1}{2})$ for each $c \in C$. Let

$$A = \{ p(c, t) \mid c \in C, 0 \leqslant t \leqslant \tfrac{1}{2} \}, \quad B = \{ p(c, t) \mid c \in C, \tfrac{1}{2} \leqslant t \leqslant 1 \}.$$

Then A and B are cones over C and hence are contractible in themselves. Further, obviously we have $C = A \cap B$. Thus we obtain a triad $(X; A, B)$ which satisfies the conditions given at the beginning of this section. In particular, if C is the $(r - 1)$-sphere S^{r-1}, then $(X; A, B)$ is topologically equivalent to the triad $(S^r; E_+^r, E_-^r)$.

Finally, let us determine the suspension sequence of the triad $(S^2; E_+^2, E_-^2)$. In this case, we have $C = S^1$; pick a basic point $s_0 \in S^1$. Since $\pi_0(S^1, s_0) = 0 = \pi_2(S^1, s_0)$, we get the following part of the suspension sequence

$$0 \to \pi_3(S^2, s_0) \xrightarrow{t} \pi_3(S^2; E_+^2, E_-^2, s_0) \xrightarrow{r} \pi_1(S^1, s_0) \xrightarrow{s} \pi_2(S^2, s_0) \to 0$$

where each of the four groups is free cyclic. We assert that t is an epimorphism. For, otherwise, the image of r would be of finite order equal to the index of $t[\pi_3(S^2, s_0)]$ in $\pi_3(S^2; E_+^2, E_-^2, s_0)$. Since $\pi_1(S^1, s_0)$ is a free group, this is impossible. Hence s, t are isomorphisms and $r = 0$. On the other hand, since $\pi_n(S^1, s_0) = 0$ for every $n \geqslant 2$, we have

$$t : \pi_n(S^2, s_0) \approx \pi_n(S^2; E_+^2, E_-^2, s_0)$$

for every $n \geqslant 4$.

EXERCISES

A. The homotopy addition theorem

Consider the unit n-simplex Δ_n of the euclidean $(n + 1)$-space R^{n+1}, [E–S; p. 55]. For every pair (X, x_0) and any integer $n > 0$, the elements of $\pi_n(X, x_0)$ may be considered as the homotopy classes of the maps of $(\Delta_n, \partial\Delta_n)$ into (X, x_0); for details of this definition see [Hu 12].

Let $n > 0$ and denote by K^{n-1} the $(n - 1)$-dimensional skeleton of the boundary n-sphere S^n of the unit $(n+1)$-simplex Δ_{n+1} in the euclidean $(n+2)$-space. Consider any given map

$$f: (S^n, K^{n-1}) \to (X, x_0).$$

Since f sends the leading vertex of S^n onto x_0, f represents an element $[f]$ of $\pi_n(X, x_0)$. On the other hand, let

$$e_i^n : \Delta_n \to \Delta_{n+1}, \quad (i = 0, \cdots, n+1),$$

denote the simplicial map defined in [E–S; p. 185]. Then, the composed map

$$fe_i^n : (\Delta_n, \partial\Delta_n) \to (X, x_0)$$

represents an element $[fe_i^n]$ of $\pi_n(X, x_0)$ for every $i = 0, 1, \cdots, n+1$. Prove the following

Homotopy addition theorem. *For any map*

$$f: (S^n, K^{n-1}) \to (X, x_0), \quad n \geqslant 2,$$

we alway have

$$[f] = \sum_{i=0}^{n+1} (-1)^i [fe_i^n].$$

For the exceptional case, $\pi_1(X, x_0)$ is not necessarily abelian. However, for any map $f: (S^1, K^0) \to (X, x_0)$, the following relation is obvious:

$$[f] = [fe_2^1] \cdot [fe_0^1] \cdot [fe_1^1]^{-1}.$$

In the remainder of this exercise, we shall give an analog for the maps of cubic boundaries.

Let $n > 0$ and denote by K^{n-1} the $(n-1)$-dimensional skeleton of the boundary n-sphere ∂I^{n+1} of the $(n+1)$-cube I^{n+1}, i.e., K^{n-1} consists of all points $(t_1, \cdots, t_{n+1})$ of I^{n+1} such that $t_i(1-t_i) = 0$ for at least two indices i. Consider any given map

$$f: (\partial I^{n+1}, K^{n-1}) \to (X, x_0).$$

Since f sends the point $(0, \cdots, 0)$ of ∂I^{n+1} into x_0, it represents an element $[f]$ of $\pi_n(X, x_0)$. On the other hand, for each $i = 1, \cdots, n+1$, let η_i and ζ_i denote the homeomorphism of I^n into ∂I^{n+1} defined by

$$\eta_i(t_1, \cdots, t_n) = (t_1, \cdots, t_{i-1}, 0, t_i, \cdots, t_n),$$
$$\zeta_i(t_1, \cdots, t_n) = (t_1, \cdots, t_{i-1}, 1, t_i, \cdots, t_n).$$

Then, the composed maps $f\eta_i$ and $f\zeta_i$ of $(I^n, \partial I^n)$ into (X, x_0) represent elements $[f\eta_i]$ and $[f\zeta_i]$ of $\pi_n(X, x_0)$ respectively. Then, prove that

$$[f] = \sum_{i=1}^{n+1} (-1)^i ([f\zeta_i] - [f\eta_i]), \quad n \geqslant 2.$$

For the exceptional case, $\pi_1(X, x_0)$ is non-abelian in general. However, for an arbitrary map $f: (\partial I^2, K^0) \to (X, x_0)$, the following relation holds:

$$[f] = [f\eta_1] \cdot [f\zeta_2] \cdot [f\zeta_1]^{-1} \cdot [f\eta_2]^{-1}.$$

B. The relative homotopy addition theorem

For any triplet (X, A, x_0) and any integer $n \geqslant 2$, the elements of the group $\pi_n(X, A, x_0)$ can be considered as the homotopy classes of the maps of $(\Delta_n, \partial\Delta_n, v_0)$ into (X, A, x_0), where v_0 denotes the leading vertex of Δ_n.

Let $n \geqslant 2$ and denote by K^{n-1} the $(n-1)$-dimensional skeleton of the boundary n-sphere S^n of Δ_{n+1}. Consider any map

$$f : (S^n, K^{n-1}, v_0) \to (X, A, x_0).$$

If we compose f with $e_i^n : \Delta_n \to \Delta_{n+1}$, we obtain a map fe_i^n of $(\Delta_n, \partial\Delta_n)$ into (X, A) for each $i = 0, 1, \cdots, n + 1$. If $i \neq 0$, fe_i^n maps the leading vertex v_0 of Δ_n into x_0 and hence it represents an element $[fe_i^n]$ of $\pi_n(X, A, x_0)$. Set $x_1 = f(v_1)$. Then, fe_0^n sends v_0 into x_1 and so it represents an element $[fe_0^n]$ of $\pi_n(X, A, x_1)$. Let $\sigma : I \to A$ denote the path joining x_0 to x_1 defined by

$$\sigma(t) = f(1 - t, t, 0, \cdots, 0), \quad t \in I.$$

By (IV; Ex. D), σ induces an isomorphism

$$\sigma_n : \pi_n(X, A, x_1) \approx \pi_n(X, A, x_0).$$

Prove the following

Relative homotopy addition theorem. *For any map*

$$f : (S^n, K^{n-1}, v_0) \to (X, A, x_0), \quad n > 2,$$

we always have

$$j_*[f] = \sigma_n[fe_0^n] + \sum_{i=1}^{n+1} (-1)^i \, [fe_i^n],$$

where $[f]$ is the element of $\pi_n(X, x_0)$ represented by f as a map of (S^n, v_0) into (X, x_0) and j_* is the homomorphism induced by the inclusion map $j : (X, x_0) \subset (X, A, x_0)$.

State and prove the analogous theorem for the maps of $(\partial I^{n+1}, K^{n-1}, v_0)$ into (X, A, x_0).

C. The Hurewicz theorem

Let X be pathwise connected space, A be a pathwise connected subspace of X, and x_0 be a given basic point in A. The pair (X, A) is *m-connected* if $\pi_n(X, A, x_0) = 0$ for every n satisfying $1 \leqslant n \leqslant m$. By means of the relative homotopy addition theorem in the previous exercise, prove the following

Hurewicz theorem. *If $n \geqslant 2$ and (X, A) is $(n-1)$-connected, then the natural homomorphism*

$$\varkappa_n : \pi_n(X, A, x_0) \to H_n(X, A)$$

in (4.1) *is an epimorphism and its kernel is the subgroup generated by the elements of the form $\alpha - w\alpha$, where $\alpha \in \pi_n(X, A, x_0)$ and $w \in \pi_1(A, x_0)$.*

For the special case $n = 2$, prove that the kernel of $\varkappa_2$ contains the commutator subgroup of $\pi_2(X, A, x_0)$.

As corollaries of the theorem given above, we have the following propositions:

1. If (X, A) is $(n-1)$-connected and n-simple for a given $n \geqslant 2$, then

$$\varkappa_n : \pi_n(X, A, x_0) \approx H_n(X, A).$$

2. If $n \geqslant 2$ and X is $(n-1)$-connected, then

$$h_n : \pi_n(X, x_0) \approx H_n(X).$$

The last proposition generalizes (4.4). As an illustrative example, let us consider the space $E = (S^n \times Z) \cup (s_0 \times R)$ of § 7. Then, E is $(n-1)$-connected and hence $\pi_n(E) \approx H_n(E)$. This implies that $\pi_n(E)$ is a free abelian group with a countably infinite set of free generators. Finally, prove

3. If $n \geqslant 2$ and X is $(n-1)$-connected, then h_{n+1} is an epimorphism. [G. W. Whitehead 1].

D. The Whitehead theorem

Let (X, A, x_0) be as assumed at the beginning of the previous exercise. By the aid of the Hurewicz theorem and the following commutative exact ladder

$$\begin{array}{ccccccccccc}
\cdots \xrightarrow{\partial} \pi_n(A,x_0) & \xrightarrow{i_*} & \pi_n(X,x_0) & \xrightarrow{j_*} & \pi_n(X,A,x_0) & \xrightarrow{\partial} & \pi_{n-1}(A,x_0) & \xrightarrow{i_*} \cdots \xrightarrow{j_*} & \pi_1(X,A,x_0) & \xrightarrow{\partial} 0 \\
\downarrow h_n & & \downarrow h_n & & \downarrow \varkappa_n & & \downarrow h_{n-1} & & \downarrow \varkappa_1 & \\
\cdots \xrightarrow{\partial} H_n(A) & \xrightarrow{i_\#} & H_n(X) & \xrightarrow{j_\#} & H_n(X,A) & \xrightarrow{\partial} & H_{n-1}(A) & \xrightarrow{i_\#} \cdots \xrightarrow{j_\#} & H_1(X,A), &
\end{array}$$

prove the following assertions:

1. Let $m > 0$ be a given integer. If i_* is an isomorphism for every $n \leqslant m$, then so is $i_\#$. If i_* is an isomorphism for every $n < m$ and is an epimorphism for $n = m$, then so is $i_\#$.

2. Assume that both X and A are simply connected and $m > 0$ is a given integer. If $i_\#$ is an isomorphism for every $n < m$ and is an epimorphism for $n = m$, then so is i_*. Furthermore, if $i_\#$ is an isomorphism also for $n = m$, then the kernel of i_* in $\pi_n(A, x_0)$ is contained in that of h_n.

Next, let X and Y be pathwise connected spaces. Consider a given map $f : (X, x_0) \to (Y, y_0)$ and the induced homomorphisms

$$f_* : \pi_n(X, x_0) \to \pi_n(Y, y_0), \quad f_\# : H_n(X) \to H_n(Y).$$

Using the mapping cylinder M_f, and (I; 12.1 and 12.2), prove the following assertions:

3. Let $m > 0$ be a given integer. If f_* is an isomorphism for every $n \leqslant m$, then so is $f_\#$. If f_* is an isomorphism for every $n < m$ and is an epimorphism for $n = m$, then so is $f_\#$.

4. Assume that both X and Y are simply connected and $m > 0$ is a given integer. If $f_\#$ is an isomorphism for every $n < m$ and is an epimorphism for

$n = m$, then so is f_*. Furthermore, if $f_\#$ is an isomorphism also for $n = m$, then the kernel of f_* in $\pi_n(X, x_0)$ is contained in that of $h_n : \pi_n(X, x_0) \to H_n(X)$.

Note that the condition that there exists a map $f : (X, x_0) \to (Y, y_0)$ such that f_* is an isomorphism for each $n > 0$ is much stronger than the condition that $\pi_n(X, x_0) \approx \pi_n(Y, y_0)$ for every $n > 0$. In fact, there are pathwise connected spaces which have isomorphic homotopy groups but some of whose homology groups are different [Wang 1].

E. Homotopy groups of adjunction spaces

Let X be a given space and $x_0 \in X$ a given point. Consider an indexed family

$$f_\mu : (S^n, s_0) \to (X, x_0), \quad \mu \in M.$$

Give M the discrete topology and define a map

$$f : S^n \times M \to X$$

by taking $f(s, \mu) = f_\mu(s)$ for every $s \in S^n$ and $\mu \in M$. Let Y denote the adjunction space obtained by adjoining $E^{n+1} \times M$ to X by means of the map f, (I; § 7). The inclusion map i induces the homomorphisms

$$i_* : \pi_m(X, x_0) \to \pi_m(Y, x_0), \quad m > 0.$$

Prove the following assertions:

1. i_* is an isomorphism for every $m < n$.

2. For $m = n$, i_* is an epimorphism and its kernel in $\pi_n(X, x_0)$ is the subgroup generated by the elements wa_μ, where $w \in \pi_1(X, x_0)$ and a_μ is the element of $\pi_n(X, x_0)$ represented by the given map f_μ for all $\mu \in M$. [J. H. C. Whitehead 1, p. 281].

3. If $f_\mu \simeq 0$ for every $\mu \in M$, the i_* is a monomorphism for $m = n + 1$ and its image in $\pi_{n+1}(Y, x_0)$ is a direct summand. The complementary summand is isomorphic to the relative homotopy group $\pi_{n+1}(Y, X, x_0)$ which is a free abelian group with free generators wb_μ, where $w \in \pi_1(X, x_0)$ and $b_\mu \in \pi_{n+1}(Y, X, x_0)$ is the element represented by the map

$$g_\mu : (E^{n+1}, S^n, x_0) \to (Y, X, x_0)$$

defined by $g_\mu(t) = p(t, \mu)$ where $p : E^{n+1} \times M \to Y$ denotes the natural projection. [J. H. C. Whitehead 1, p. 285].

F. Spaces of homotopy type (π, n)

Let π be a given group and $n > 0$ a given integer. If $n > 1$, we assume that π is abelian.

A pathwise connected space X is said to be of *the homotopy type* (π, n) provided

$$\pi_n(X) \approx \pi, \quad \pi_m(X) = 0 \quad \text{if } m \neq n.$$

Thus, by (8.4), if $p : E \to B$ is an n-connective fibering over an $(n - 1)$-

connected space B, then the fiber $F = p^{-1}(b_0)$ over any $b_0 \in B$ is of the homotopy type $(\pi_n(B), n-1)$.

Construct a space X of the homotopy type (π, n) as follows. If $n = 1$, then π can be represented as the quotient group of a free group F over a normal subgroup R of F. If $n > 1$, then π is abelian and can be represented as the quotient group of a free abelian group F over a subgroup R of F. Hence

$$\pi = F/R.$$

Let M denote the set of free generators of F and let b_0 be a space consisting of a single point b_0. Give M the discrete topology and adjoin $E^n \times M$ to b_0 by means of the map $g: S^{n-1} \times M \to b_0$. Let A denote the adjunction space obtained in this way. Then prove that

$$\pi_n(A) = F, \quad \pi_m(A) = 0 \quad \text{if } m < n.$$

For each $r \in R \subset \pi_n(A)$, choose a map $f_r : (S^n, s_0) \to (A, b_0)$ which represents r. Let B denote the space constructed by means of the family $\{ f_r \mid r \in R \}$ as in Ex. E. Then we have

$$\pi_n(B) \approx F/R = \pi, \quad \pi_m(B) = 0 \quad \text{if } m < n.$$

Then, by (8.2), there exists a pathwise connected space X which contains B as a closed subspace and

$$\pi_n(X) \approx \pi, \quad \pi_m(X) = 0 \quad \text{if } m \neq n.$$

This completes the construction.

Finally, prove that if X is a space of the homotopy type (π, n) with $n > 1$, then the space of loops $\Lambda(X)$ at a point $x_0 \in X$ is a space of the homotopy type $(\pi, n-1)$.

G. The realizability theorems

By means of the spaces of homotopy type (π, n), prove the following Realizability Theorem. Let

$$\pi_1, \pi_2, \cdots, \pi_n, \cdots$$

be a sequence of groups. All groups except possibly the first one are abelian and π_1 operates on π_n for every $n \geqslant 2$. Then there exists a pathwise connected space X and a point $x_0 \in X$ such that the following three conditions are satisfied, [J. H. C. Whitehead 5] and [Hu 10]:

(1) There exists, for each $n > 0$, an isomorphism

$$h_n : \pi_n(X, x_0) \approx \pi_n.$$

(2) For any $w \in \pi_1(X, x_0)$ and $\alpha \in \pi_n(X, x_0)$, $n \geqslant 2$, we have

$$h_n(w\alpha) = h_1(w) h_n(\alpha).$$

(3) For every $\alpha \in \pi_m(X, x_0)$ and $\beta \in \pi_n(X, x_0)$, $m \geqslant 2, n \geqslant 2$, the Whitehead product $[\alpha, \beta] = 0$.

By constructing a cone over X, deduce an analogous realizability theorem for relative homotopy groups.

H. Topological realization of semi-simplicial complexes

For a given semi-simplicial complex K as defined in (IV; Ex. H), we can construct a space $|K|$ associated with K as follows. To every integer $m \geqslant 0$ and each m-cell σ of K, let us associate an open m-simplex $|\sigma|$, called the *open m-cell* corresponding to σ, which is defined to be the topological product

$$|\sigma| = \sigma \times Int(\Delta_m), \quad Int(\Delta_0) = \Delta_0,$$

of σ as a single point and the interior of the unit m-simplex Δ_m. Then we define the *closed m-cell* $Cl|\sigma|$ to be the set

$$Cl|\sigma| = \cup_{\tau < \sigma} |\tau|.$$

There is a natural function χ_σ of Δ_m onto $Cl|\sigma|$ defined linearly on each open simplex of the complex Δ_m, [Hu, 10]. Give $Cl|\sigma|$ the identification topology determined by χ_σ, (I; § 6). Next, let us denote by $|K|$ the union of all open cells $|\sigma|$, $\sigma \in K$, and define a topology of $|K|$ as follows. A set W in $|K|$ is called *open* iff $W \cap Cl|\sigma|$ is an open set of $Cl|\sigma|$ for every $\sigma \in K$. This topology of $|K|$ will be called the *Whitehead topology* of $|K|$; it is the largest topology of $|K|$ such that the topology of every closed cell $Cl|\sigma|$ coincides with the relative topology in $|K|$. Following a usual custom in combinatory topology, we identify K with $|K|$ and σ with $|\sigma|$. Thus, we may consider any semi-simplicial complex K as a topological space which is a union of a collection $\{\sigma\}$ of disjoint open cells. For example, the singular complex $S(X)$ of a space X will be considered as a topological space in this way.

For each cell σ of K, it follows immediately that the natural function

$$\chi_0 : \Delta_m \to Cl\,\sigma \subset K,$$

where $m = \dim(\sigma)$, is a continuous map of Δ_m onto $Cl\,\sigma$ and the restriction $\chi_\sigma | Int(\Delta_m)$ is a homeomorphism of $Int(\Delta_m)$ onto the open m-cell σ. We shall call χ_σ the *characteristic map* for the cell $\sigma \in K$.

Prove the following assertions:

1. If X is a compact subspace of a semi-simplicial complex K, then X meets at most a finite number of open cells of K.

2. A function $f : X \to Y$ of a closed or open subspace X of a semi-simplicial complex K into a space Y is continuous iff the restriction $f | X \cap Cl\sigma$ is continuous for every closed cell $Cl\sigma$ of K.

3. A family $f_t : X \to Y$, $(0 \leqslant t \leqslant 1)$, of functions of a closed or open subspace X of a semi-simplicial complex K into a space Y is a homotopy iff the family $f_t | X \cap Cl\sigma$, $(0 \leqslant t \leqslant 1)$, is a homotopy for every closed cell $Cl\sigma$ of K.

4. Every subcomplex L of a semi-simplicial complex K is closed and has the AHEP in K.

5. A necessary and sufficient condition for a semi-simplicial complex K to be a simplicial complex with locally ordered vertices is that, for each m-cell $\sigma \in K$, the characteristic map χ_0 is a homeomorphism of Δ_m onto $Cl\,\sigma$ and, for any two cells σ and τ of K, $Cl\,\sigma \cap Cl\,\tau$ is either empty or a closed cell of K.

6. For each $n \geqslant 2$, the n-th barycentric subdivision, defined in the obvious way, of a semi-simplicial complex K is a simplicial complex with locally ordered vertices. Hence, every semi-simplicial complex is triangulable.

I. The projection $\omega: S(X) \to X$

For any given space X, there is a natural projection ω of the singular complex $S(X)$ onto X as follows. For an arbitrary point p of the space $S(X)$, let σ denote the unique open cell of $S(X)$ which contains p. Then, σ is a singular simplex

$$\sigma: \Delta_m \to X, \quad m = \dim(\sigma)$$

and the characteristic map χ_σ sends $Int\,\Delta_m$ homeomorphically onto the open cell σ of $S(X)$. The natural projection $\omega: S(X) \to X$ is defined by taking

$$\omega(p) = \sigma\chi_\sigma^{-1}(p), \quad p \in \sigma \subset S(X).$$

Prove the following assertions:

1. The projection $\omega: S(X) \to X$ is a map of $S(X)$ onto X. For every subspace A of X, ω carries the subcomplex $S(A)$ onto A.

2. The space X is pathwise connected iff $S(X)$ is connected.

Next, let (X, A, x_0) be any given triplet and let p_0 denote the vertex of $S(X)$ such that $\omega(p_0) = x_0$. Thus we obtain a triplet $(S(X), S(A), p_0)$, and a map

$$\omega: (S(X), S(A), p_0) \to (X, A, x_0)$$

defined by the natural projection ω. Prove the following important proposition, [Giever 1 and J. H. C. Whitehead 7].

3. The induced transformations ω_* of ω are one-to-one correspondences, namely,

$$\omega_*: \pi_m(S(X), S(A), p_0) \approx \pi_m(X, A, x_0),$$

$$\omega_*: \pi_m(S(X), p_0) \approx \pi_m(X, x_0),$$

$$\omega_*: \pi_m(S(A), p_0) \approx \pi_m(A, x_0).$$

The significance of this is that, in computing the homotopy groups of a space X, we may assume without loss of generality that X is triangulable and hence locally contractible. In fact, we may replace X by $S(X)$.

J. Induced cellular maps

A *cellular transformation* $T: K_1 \to K_2$ of a semi-simplicial complex K_1 into another such complex K_2 is a function which assigns to each m-cell σ of K_1 an m-cell $\tau = T(\sigma)$ of K_2 in such a fashion that

$$\tau^{(i)} = T(\sigma^{(i)}), \quad (i = 0, 1, \cdots, m).$$

Prove that a cellular transformation $T : K_1 \to K_2$ induces a unique map $f_T : K_1 \to K_2$ which carries $\sigma \in K_1$ barycentrically on $\tau = T(\sigma) \in K_2$.

A map $f : K_1 \to K_2$ is said to be *cellular* if it carries $K_1{}^m$ into $K_2{}^m$ for every $m = 0, 1, \cdots$. Thus, the map $f_T : K_1 \to K_2$ induced by T is cellular.

Consider a given map $\phi : K \to X$ of a semi-simplicial complex K into a space X. ϕ induces a cellular transformation

$$T_\phi : K \to S(X)$$

as follows: For each m-cell $\sigma \in K$, $\tau = T_\phi(\sigma)$ is defined to be the singular m-simplex

$$\tau = \phi\chi_\sigma : \Delta_m \to X$$

where χ_σ denotes the characteristic map of σ. Verify that $\tau^{(i)} = T_\phi(\sigma^{(i)})$ for every $i = 0, \cdots, m$.

This cellular transformation T_ϕ induces a cellular map $\phi\# : K \to S(X)$ which will be called the *induced cellular map* of ϕ. Prove that $\omega\phi\# = \phi$.

As an application of this, let $X = K$ and take ϕ to be the identity map. Then $\phi\#$ maps K homeomorphically and barycentrically onto a subcomplex of $S(K)$. Hence, we may consider K as a subcomplex of its singular complex $S(K)$. Prove that K is a deformation retract of $S(K)$ and that the cellular transformation T_ϕ defines a chain equivalence. The second assertion implies that the homology and cohomology groups of a semi-simplicial complex K are topological invariants.

Next, let X, Y be spaces and consider a given map $f : X \to Y$. f induces a cellular transformation

$$T_f : S(X) \to S(Y)$$

defined by

$$\tau = T_f(\sigma) = f\sigma : \Delta_m \to Y$$

for every singular m-simplex $\sigma \in S(X)$.

This cellular transformation T_f induces a cellular map $f\# : S(X) \to S(Y)$ which will be called the *induced cellular map* of f. Obviously, $f\#$ is also the induced map of $\phi = f\omega : S(X) \to Y$. Verify that, in the following diagram, we always have $f\omega = \omega f\#$:

$$\begin{array}{ccc} S(X) & \xrightarrow{f\#} & S(Y) \\ \downarrow{\scriptstyle \omega} & & \downarrow{\scriptstyle \omega} \\ X & \xrightarrow{f} & Y \end{array}$$

This implies that, in studying the induced homomorphisms f_* of a given map $f : X \to Y$ on the homotopy groups, we may always assume that X, Y are simplicial complexes and that f is simplicial.

K. Admissible subcomplexes of S(X)

Assume that X is pathwise connected space and that a fixed point x_0 of X has been selected as basic point.

Two singular m-simplexes σ and τ in X are said to be *compatible* if their faces coincide, that is to say, $\sigma^{(i)} = \tau^{(i)}$ for each $i = 0, \cdots, m$. Hence σ and τ are compatible iff

$$\sigma \mid S^{m-1} = \tau \mid S^{m-1}, \quad S^{m-1} = \partial\Delta_m.$$

Further, two compatible singular m-simplex σ and τ are said to be *equivalent* if there exists a homotopy $h_t : \Delta_m \to X$, $(0 \leqslant t \leqslant 1)$, such that $h_0 = \sigma$, $h_1 = \tau$, and $h_t(d) = \sigma(d)$ for every $d \in S^{m-1}$ and every $t \in I$. For $m = 0$, any two singular 0-simplexes are compatible; and, since X is assumed to be pathwise connected, they are also equivalent.

A singular simplex $\sigma : \Delta_m \to X$ is said to be *collapsed* if $\sigma(\Delta_m) = x_0$. The set of all collapsed singular simplexes constitute the subcomplex $S(x_0)$ of $S(X)$ associated with x_0 as a subspace of X.

A subcomplex L of $S(X)$ is called an *admissible subcomplex* of $S(X)$ if the following conditions are satisfied:

(AS1) $S(x_0) \subset L$.

(AS2) If σ is a singular simplex such that $\sigma^{(i)} \in L$ for each $i = 0, \cdots, \dim(\sigma)$, then L contains at least one singular simplex τ which is compatible and equivalent with σ.

If, in the condition (AS2), we require that L contains one and only one singular simplex τ which is compatible and equivalent with σ, then L is called a *minimal subcomplex* of $S(X)$.

Obviously, $S(X)$ is an admissible subcomplex of itself and every minimal subcomplex of $S(X)$ is admissible. Let the admissible subcomplexes of $S(X)$ be partially ordered by means of inclusion. Then a minimal subcomplex M of $S(X)$ is clearly minimal in the sense that, if any admissible subcomplex L is contained in M, then $L = M$.

Prove the following assertions:

1. Every admissible subcomplex L of $S(X)$ contains a minimal subcomplex M of $S(X)$.

2. If L is an admissible subcomplex of $S(X)$, then there is a function D, called a *deformation* of $S(X)$ into L, which assigns to each singular m-simplex σ a singular $(m + 1)$-prism $D(\sigma) = P_\sigma$, [E–S; p. 193], subject to the following conditions:

(i) $D(\sigma^{(i)}) = (P_\sigma)^{(i)}, \quad (i = 0, \cdots, m)$.

(ii) $(P_\sigma)_l = \sigma$.

(iii) $(P_\sigma)_\mu$ is in L.

(iv) If $\sigma \in L$, then $P_\sigma(d, t) = \sigma(d)$ for every $d \in \Delta_m$ and every $t \in I$.

Now let A be a pathwise connected subspace of X which contains the basic point x_0. An admissible subcomplex L of $S(X)$ is said to be *relatively admissible* with respect to $S(A)$ if $L \cap S(A)$ is an admissible subcomplex of $S(A)$. In this case, the deformation D can be so chosen that, for each $\sigma \in S(A)$, P_σ is a singular prism in A.

3. If L is an admissible subcomplex of $S(X)$ which is relatively admissible with respect to $S(A)$, then the inclusion cellular map

$$\lambda : (L, L \cap S(A)) \to (S(X), S(A))$$

is a homotopy equivalence and defines a chain equivalence.

L. The Eilenberg subcomplexes of S(X)

Let X be a pathwise connected space, A a pathwise connected subspace of X, and x_0 a given basic point selected from A.

For each integer $n > 0$, define a subcomplex $S_n(X, A)$ of the singular complex $S(X)$ as follows. A singular simplex $\sigma : \Delta_m \to X$ is in $S_n(X, A)$ iff the following conditions are satisfied:

(ES1) σ sends the vertices of Δ_m into x_0.

(ES2) σ sends all faces of Δ_m of dimension less than n into A.

In particular, if $A = x_0$, then we simply write $S_n(X) = S_n(X, x_0)$. $S_n(X)$ is called *the n-th Eilenberg subcomplex* of $S(X)$; and, in the general case, $S_n(X, A)$ will be called *the n-th relative Eilenberg subcomplex* of $S(X)$ with respect to A. For completeness, we also define $S_0(X, A) = S(X)$. We thus obtain a descending sequence of subcomplexes of $S(X)$, namely,

$$S(X) = S_0(X, A) \supset S_1(X, A) \supset \cdots \supset S_n(X, A) \supset \cdots \supset S_\infty(X, A) = S_1(A).$$

Prove that, if (X, A) is $(n-1)$-connected for a positive integer n, then $S_n(X, A)$ is an admissible subcomplex of $S(X)$ and is also relatively admissible with respect to $S(A)$. In particular, $S_n(X)$ is an admissible subcomplex of $S(X)$ provided that X is $(n-1)$-connected.

CHAPTER VI

OBSTRUCTION THEORY

1. Introduction

Having studied the homotopy groups in the previous chapters, we are now in a better position to investigate our main problems described in the first chapter.

For the sake of simplicity, we shall restrict ourselves to the study of the maps of a finite cell complex K, [S; p. 100], into a pathwise connected space Y. Therefore, we assume throughout the present chapter, that K is a finite cell complex, L a subcomplex of K, and Y a given pathwise connected space with a given point $y_0 \in Y$. We shall denote by K^n *the n-dimensional skeleton* of K which consists of the cells of dimension not exceeding n and use the notation

$$\overline{K}^n = L \cup K^n.$$

For example, let us study the extension problem. Consider an arbitrarily given map

$$f : L \to Y.$$

As described in (I; § 2), the extension problem for f over K in the restricted sense is to determine whether or not f can be extended continuously throughout K. Since L has the AHEP in K according to (I; 9.2), this restricted extension problem is equivalent to the broadened problem described at the end of (I; § 9).

Since K is a cell complex and L is a subcomplex of K, there is a natural approach to attack this problem, known as the *obstruction method.* We first try to extend step by step the map f over the subcomplexes

$$\overline{K}^n, \quad n = 0, 1, 2, \cdots.$$

We shall carry on this stepwise extension until we meet some *obstruction* to further extension. Then we propose to measure this obstruction and try to change the already constructed partial extension of f so that this obstruction might vanish and hence further extension would be possible.

2. The extension index

In order to study the extension problem, let us define the notion of n-extensibility. A map $f : L \to Y$ is said to be *n-extensible over* K for a given integer $n \geqslant 0$ if it has an extension over the subcomplex $\overline{K}^n$ of K.

Since Y is assumed to be pathwise connected, the following proposition is obvious.

Proposition 2.1. *Every map $f : L \to Y$ is 1-extensible over K.*

For a given map $f : L \to Y$, the least upper bound of the set of integers n such that f is n-extensible over K is called the *extension index* of f over K.

Since L has the AHEP in $\bar{K}^n$, the following proposition is obvious.

Proposition 2.2. *Homotopic maps have the same extension index.*

Now let $e : Y \to Y'$ be any map and let $g : (K', L') \to (K, L)$ be any cellular map, [S; p. 161]. Then the following proposition is obvious.

Proposition 2.3. *If a map $f : L \to Y$ is n-extensible over K, then the composed map $f' = efg : L' \to Y'$ is n-extensible over K'.*

In particular, let (K, L) be a subdivision of (K', L') and g be the identity map. If $f : L \to Y$ is n-extensible over K, then $f' = fg : L' \to Y$ is also n-extensible over K'.

By the classical method of simplicial approximation, one can easily deduce from (2.2) and (2.3) the following

Proposition 2.4. *If (K, L) is a simplicial pair, then the extension index of any map $f : L \to Y$ is a topological invariant, i.e., it does not depend on the triangulation of the pair (K, L).*

Because of this it seems that we should prefer simplicial complexes in our treatment. However, to unify the study of the extension problem with that of the homotopy problem and to avoid awkward repetitions, we have to consider cell complexes which are not simplicial.

3. The obstruction $c^{n+1}(g)$

Throughout §§ 3–12, let n be a given positive integer. For the sake of convenience, we assume that the given pathwise connected space Y is n-simple in the sense of (IV; § 16). In this important case, every map of any oriented n-sphere into Y determines an element of the homotopy group $\pi_n(Y)$; the latter group is abelian and will be used as the coefficient group of the cohomology groups in these few sections. If Y fails to be n-simple and $n > 1$, then one can deduce similar results by using local coefficients. See Ex. A at the end of the chapter.

In the present section, let us consider a given map

$$g : \bar{K}^n \to Y.$$

This map g determines an $(n + 1)$-cochain $c^{n+1}(g)$ of K with coefficients in the homotopy group $\pi_n(Y)$ as follows. Let σ be any $(n + 1)$-cell of K. Then the set-theoretic boundary $\partial\sigma$ of σ is an oriented n-sphere. Since $\partial\sigma \subset K^n$, the partial map $g_\sigma = g \mid \partial\sigma$ determines an element $[g_\sigma]$ of $\pi_n(Y)$. Then $c^{n+1}(g)$ is defined by taking

$$[c^{n+1}(g)](\sigma) = [g_\sigma] \in \pi_n(Y)$$

for every $(n+1)$-cell σ of K. This $(n+1)$-cochain $c^{n+1}(g)$ of K is called the *obstruction* of the map g.

Lemma 3.1. *The obstruction $c^{n+1}(g)$ is a relative $(n+1)$-cocycle of K modulo L; in symbols, $c^{n+1}(g) \in Z^{n+1}(K, L; \pi_n(Y))$.*

Proof. Let us first prove that $c^{n+1}(g)$ is in $C^{n+1}(K, L; \pi_n(Y))$. For this purpose, let σ be any $(n+1)$-cell of L. Since g is defined on $Cl\sigma \subset L$, g_σ has an extension over $Cl\sigma$ and hence $[g_\sigma]$ is the zero element of $\pi_n(Y)$. This proves that $c^{n+1}(g)$ is a cochain of K modulo L.

Next, let us prove that $c^{n+1}(g)$ is a cocycle. For this purpose, let σ be any $(n+2)$-cell of K. It suffices to show $[\delta c^{n+1}(g)](\sigma) = 0$.

Let B denote the subcomplex $\partial\sigma$ of K and B^n the n-dimensional skeleton of B. Then we have the homomorphisms

$$C_{n+1}(B) \xrightarrow{\partial} Z_n(B) = Z_n(B^n) = H_n(B^n) \xleftarrow{h} \pi_n(B^n) \xrightarrow{k_*} \pi_n(Y),$$

where h denotes the natural homomorphism and k_* is induced by the partial map $k = g \mid B^n$.

If $n > 1$, then B^n is $(n-1)$-connected and hence h is an isomorphism according to the Hurewicz theorem in (V; 4.4). If $n = 1$, then h is an epimorphism and the kernel of h is contained in that of k_* since $\pi_n(Y)$ is abelian. Hence, in either case, we obtain a well-defined homomorphism

$$\phi = k_* h^{-1} : Z_n(B) \to \pi_n(Y).$$

Since $C_{n-1}(B)$ is a free abelian group, the kernel $Z_n(B)$ of $\partial : C_n(B) \to C_{n-1}(B)$ is a direct summand of $C_n(B)$. Therefore, the homomorphism ϕ has an extension

$$d : C_n(B) \to \pi_n(Y).$$

For every $(n+1)$-cell τ in B, the element $[c^{n+1}(g)](\tau)$ is represented by the partial map $k \mid \partial\tau$. Therefore, it follows that

$$[c^{n+1}(g)](\tau) = k_* h^{-1}(\partial\tau) = d(\partial\tau).$$

This implies that

$$[\delta c^{n+1}(g)](\sigma) = [c^{n+1}(g)](\partial\sigma) = d(\partial\partial\sigma) = 0.$$

Hence $c^{n+1}(g) \in Z^{n+1}(K, L; \pi_n(Y))$. ∎

Because of (3.1), $c^{n+1}(g)$ is usually called the *obstruction cocycle* of g. The following two lemmas are obvious.

Lemma 3.2. *The map $g : \bar{K}^n \to Y$ has an extension over $\bar{K}^{n+1}$ iff $c^{n+1}(g) = 0$.*

Lemma 3.3. *If $g_0, g_1 : \bar{K}^n \to Y$ are homotopic maps, then $c^{n+1}(g_0) = c^{n+1}(g_1)$.*

Now let (K', L') be another cellular pair and $\phi : (K', L') \to (K, L)$ be a proper cellular map, [S; p. 161]. For a given map $g : \bar{K}^n \to Y$, we obtain a composed map $g' = g\phi : \bar{K}'^n \to Y$. Since ϕ is proper and cellular, it induces a unique cochain homomorphism, [S; p. 161],

$$\phi^{\#} : C^{n+1}(K, L; \pi_n(Y)) \to C^{n+1}(K', L'; \pi_n(Y)).$$

Lemma 3.4. $c^{n+1}(g') = \phi^{\#}c^{n+1}(g)$.

To prove this, one must show that the two sides have the same value on an arbitrary $(n + 1)$-cell σ of K'. For this purpose, one should consider the minimal carrier of σ and pick an $(n + 1)$-cell from this minimal carrier; then the lemma is a consequence of a number of trivial commutativity relations. The details of the proof are left to reader, [S; p. 168].

4. The difference cochain

In the present section, we are concerned with two given maps

$$g_0, g_1 : \bar{K}^n \to Y$$

which are homotopic on $\bar{K}^{n-1}$; we shall see that the difference of the obstruction cocycles of g_0 and g_1 is a coboundary. For this purpose, consider a homotopy

$$h_t : \bar{K}^{n-1} \to Y, \quad (0 \leqslant t \leqslant 1),$$

such that $h_0 = g_0 \mid \bar{K}^{n-1}$ and $h_1 = g_1 \mid \bar{K}^{n-1}$.

Regard the closed unit interval I as a cell complex composed of two 0-cells 0 and 1 and the 1-cell I, where $\delta 0 = -I$ and $\delta 1 = I$. Then the topological product

$$J = K \times I$$

is also a cell complex. We shall denote by J^n the n-dimensional skeleton of J and use the notation

$$\bar{J}^n = (L \times I) \cup J^n = (\bar{K}^n \times 0) \cup (\bar{K}^{n-1} \times I) \cup (\bar{K}^n \times 1).$$

Define a map $F : \bar{J}^n \to Y$ by taking

$$F(x, t) = \begin{cases} g_0(x), & (x \in \bar{K}^n, t = 0), \\ h_t(x), & (x \in \bar{K}^{n-1}, t \in I), \\ g_1(x), & (x \in \bar{K}^n, t = 1). \end{cases}$$

Then, according to § 3, this map F determines an obstruction cocycle $c^{n+1}(F)$ of the complex J modulo $L \times I$ with coefficients in $\pi_n(Y)$. It follows from the definition of F that $c^{n+1}(F)$ coincides with $c^{n+1}(g_0) \times 0$ on $K \times 0$ and with $c^{n+1}(g_1) \times 1$ on $K \times 1$.

Let M denote the subcomplex $(K \times 0) \cup (L \times I) \cup (K \times 1)$ of $J = K \times I$. Then it follows that

$$c^{n+1}(F) - c^{n+1}(g_0) \times 0 - c^{n+1}(g_1) \times 1$$

is a cochain of J modulo M with coefficients in $\pi_n(Y)$. Since $\sigma \to \sigma \times I$ is a $1 - 1$ correspondence between the n-cells of $K \setminus L$ and the $(n + 1)$-cells of $J \setminus M$, it defines an isomorphism

$$k : C^n(K, L; \pi_n(Y)) \approx C^{n+1}(J, M; \pi_n(Y)).$$

Hence there is a unique cochain $d^n(g_0, g_1; h_t)$ in $C^n(K, L; \pi_n(Y))$ such that

(i) $$kd^n(g_0, g_1; h_t) = (-1)^{n+1} \{ c^{n+1}(F) - c^{n+1}(g_0) \times 0 - c^{n+1}(g_1) \times 1 \}.$$

This cochain $d^n(g_0, g_1; h_t)$ will be called the *deformation cochain*. In particular, if $g_0 \mid \bar{K}^{n-1} = g_1 \mid \bar{K}^{n-1}$ and $h_t(x) = g_0(x)$ for every $x \in \bar{K}^{n-1}$ and every t, we abbreviate $d^n(g_0, g_1; h_t)$ by $d^n(g_0, g_1)$ and call it the *difference cochain* of g_0 and g_1.

The following lemma is an immediate consequence of (3.2).

Lemma 4.1. *The homotopy $h_t : \bar{K}^{n-1} \to Y$ has an extension $h_t^* : \bar{K}^n \to Y$, $(0 \leqslant t \leqslant 1)$, such that $h_0^* = g_0$ and $h_1^* = g$, iff $d^n(g_0, g_1; h_t) = 0$.*

The importance of $d^n(g_0, g_1; h_t)$ is shown by a coboundary formula given in the following

Lemma 4.2. $\delta d^n(g_0, g_1; h_t) = c^{n+1}(g_0) - c^{n+1}(g_1)$.

Proof. Since $\delta I = 0$, it follows that the isomorphism k commutes with δ. Hence, we have

$$k\delta d^n(g_0, g_1; h_t) = \delta k d^n(g_0, g_1 : h_t).$$

On the other hand, since $c^{n+1}(F)$, $c^{n+1}(g_0)$, $c^{n+1}(g_1)$ are cocycles and since $\delta 0 = - I$ and $\delta 1 = I$, we may apply δ to both sides of (i) and obtain

$$\delta k d^n(g_0, g_1; h_t) = c^{n+1}(g_0) \times I - c^{n+1}(g_1) \times I.$$

Hence we have

$$k\delta d^n(g_0, g_1; h_t) = k[c^{n+1}(g_0) - c^{n+1}(g_1)],$$

Since k is an isomorphism, this implies the lemma. ∎

Next, let us consider any given map $g_0 : \bar{K}^n \to Y$ and any given homotopy $h_t : \bar{K}^{n-1} \to Y$, $(0 \leqslant t \leqslant 1)$, such that $h_0 = g_0 \mid \bar{K}^{n-1}$. We shall establish the following existence lemma.

Lemma 4.3. *For every n-cochain c in $C^n(K, L; \pi_n(Y))$, there exists a map $g_1 : \bar{K}^n \to Y$ such that $g_1 \mid \bar{K}^{n-1} = h_1$ and $d^n(g_0, g_1; h_t) = c$.*

Proof. Let σ be any n-cell of K. Then the boundary

$$S = (\sigma \times 0) \cup (\partial\sigma \times I) \cup (\sigma \times 1)$$

of $\sigma \times I$ is an oriented n-sphere and hence there exists a map $f_\sigma : S \to Y$ which represents the element $c(\sigma)$ of $\pi_n(Y)$. Since the subspace $T = (\sigma \times 0) \cup (\partial\sigma \times I)$ is contractible, any pair of maps defined on T are homotopic. Hence, by the AHEP of T in S, we may assume that

$$f_\sigma(x, t) = \begin{cases} g_0(x), & (\text{if } x \in \sigma, t = 0), \\ h_t(x), & (\text{if } x \in \partial\sigma, t \in I). \end{cases}$$

Thus, we may define a map $g_1 : \bar{K}^n \to Y$ by taking

$$g_1(x) = \begin{cases} h_1(x), & (\text{if } x \in \bar{K}^{n-1}), \\ f_\sigma(x, 1), & (\text{if } x \in \sigma, \sigma \in K). \end{cases}$$

Then it is obvious that $d^n(g_0, g_1; h_t) = c$. ∎

As an immediate consequence of (4.2) and (4.3) we have the following

Corollary 4.4. *For each cocycle* $z \in Z^{n+1}(K, L; \pi_n(Y))$ *such that* $z \sim c^{n+1}(g_0)$ *mod* L, *there exists a map* $g_1 : \bar{K}^n \to Y$ *such that* $g_1 \mid \bar{K}^{n-1} = h_1$ *and* $c^{n+1}(g_1) = z$.

In particular, if the homotopy h_t is given by $h_t = g_0 \mid \bar{K}^{n-1}$ for every $t \in I$, then the corollary gives the existence of a map $g_1 : \bar{K}^n \to Y$ such that $g_0 \mid \bar{K}^{n-1} = g_1 \mid \bar{K}^{n-1}$ and $c^{n+1}(g_1) = z$.

Next, let us consider three given maps $g_0, g_1, g_2 : \bar{K}^n \to Y$ and two homotopies $h_t, j_t : \bar{K}^{n-1} \to Y$, $(0 \leqslant t \leqslant 1)$, such that

$$h_0 = g_0 \mid \bar{K}^{n-1}, \quad h_1 = g_1 \mid \bar{K}^{n-1} = j_0, \quad j_1 = g_2 \mid \bar{K}^{n-1}.$$

Let $k_t : \bar{K}^{n-1} \to Y$, $(0 \leqslant t \leqslant 1)$, denote the homotopy defined by $k_t = h_{2t}$ if $t \leqslant \frac{1}{2}$ and $k_t = j_{2t-1}$ if $t \geqslant \frac{1}{2}$. Then we have the following addition lemma the proof of which is left to the reader, [S; p. 173].

Lemma 4.5. $d^n(g_0, g_2; k_t) = d^n(g_0, g_1; h_t) + d^n(g_1, g_2; j_t)$.

Finally, let us go back to the maps g_0, g_1 and the homotopy h_t described at the beginning of the section. Let (K', L') be another cellular pair and $\phi : (K', L') \to (K, L)$ be a proper cellular map. Let

$$g'_i = g_i\phi : \bar{K}'^n \to Y, \quad h'_t = h_t\phi : \bar{K}'^{n-1} \to Y.$$

Since ϕ is proper and cellular, it induces a unique cochain homomorphism

$$\phi^\# : C^n(K, L; \pi_n(Y)) \to C^n(K', L'; \pi_n(Y)).$$

The following invariance lemma is an easy consequence of (3.4).

Lemma 4.6. $d^n(g'_0, g'_1; h'_t) = \phi^\# d^n(g_0, g_1; h_t)$.

5. Eilenberg's extension theorem

As in § 3, let us consider a given map

$$g : \bar{K}^n \to Y.$$

According to § 3, g determines an obstruction cocycle $c^{n+1}(g)$ in $Z^{n+1}(K, L; \pi_n(Y))$ and hence an *obstruction cohomology class*

$$\gamma^{n+1}(g) \in H^{n+1}(K, L; \pi_n(Y))$$

represented by $c^{n+1}(g)$.

Theorem 5.1. $\gamma^{n+1}(g) = 0$ *iff there exists a map* $h^* : \bar{K}^{n+1} \to Y$ *such that*

$$h^* \mid \bar{K}^{n-1} = g \mid \bar{K}^{n-1}.$$

Proof. *Sufficiency.* Assume the existence of h^* and let $h = h^* \mid \bar{K}^n$. By (3.2), we have $c^{n+1}(h) = 0$. Since $g \mid \bar{K}^{n-1} = h \mid \bar{K}^{n-1}$, the difference cochain $d^n(g, h)$ is defined. By (4.2), $c^{n+1}(h) = 0$ implies that $c^{n+1}(g)$ is the coboundary of $d^n(g, h)$. Hence $\gamma^{n+1}(g) = 0$.

Necessity. Assume that $\gamma^{n+1}(g) = 0$. Then $c^{n+1}(g) \sim 0$ *mod* L. By (4.4), there exists a map $h : \bar{K}^n \to Y$ such that $g \mid \bar{K}^{n-1} = h \mid \bar{K}^{n-1}$ and $c^{n+1}(h) = 0$. Then, by (3.2), h has an extension $h^* : \bar{K}^{n+1} \to Y$. ▮

Now, let us assume that g is an extension of a given map $f: L \to Y$. If the obstruction cocycle $c^{n+1}(g)$ is non-vanishing, then it follows from (3.2) that g cannot be extended over $\bar{K}^{n+1}$ and hence our stepwise extension process faces an obstruction. The significance of Eilenberg's extension theorem (5.1) is that, if $c^{n+1}(g) \sim 0 \; mod \; L$, then this obstruction is removable by modifying the values of g on the open n-cells in $K \setminus L$ only.

6. The obstruction sets for extension

Let $f: L \to Y$ be a given map. We are going to define *the* $(n+1)$*-dimensional obstruction set*

$$O^{n+1}(f) \subset H^{n+1}(K, L; \pi_n(Y))$$

as follows. If f is not n-extensible over K, we define $O^{n+1}(f)$ to be the vacuous set. Now, suppose that f is n-extensible over K. Then there exists an extension $g: \bar{K}^n \to Y$ of f. The cohomology class $\gamma^{n+1}(g)$ in $H^{n+1}(K, L; \pi_n(Y))$ is called an $(n+1)$*-dimensional obstruction element* of f. Then, $O^{n+1}(f)$ is defined as the set of all $(n+1)$-dimensional obstruction elements of f. The following proposition is obvious.

Proposition 6.1. *Homotopic maps have the same* $(n+1)$*-dimensional obstruction set.*

Next, let (K', L') be another cellular pair and $\phi: (K', L') \to (K, L)$ be a proper cellular map. This map ϕ induces a homomorphism

$$\phi^*: H^{n+1}(K, L; \pi_n(Y)) \to H^{n+1}(K', L'; \pi_n(Y)).$$

Then the following proposition is an immediate consequence of (3.4).

Proposition 6.2. *If* $f' = f\phi: L' \to Y$, *then* ϕ^* *sends* $O^{n+1}(f)$ *into* $O^{n+1}(f')$.

In particular, if (K, L) is a subdivision of (K', L') and if ϕ is the identity map, then ϕ^* is an isomorphism and sends $O^{n+1}(f)$onto a subset of $O^{n+1}(f')$ in a 1–1 fashion. Furthermore, if both (K, L) and (K', L') are simplicial, then the identity map ϕ^{-1} is homotopic to a simplicial (and hence proper cellular) map. This implies that ϕ^* sends $O^{n+1}(f)$ onto $O^{n+1}(f')$. Hence, the $(n+1)$-dimensional obstruction set is smaller on a subdivision and is smallest if (K, L) is a simplicial pair. Thus, we have the following

Corollary 6.3. *If* (K, L) *is a simplicial pair and* $f: L \to Y$ *is a map, then* $O^{n+1}(f)$ *is a topological invariant, i.e., it does not depend on the triangulation of* (K, L).

Now, let (K, L) be a triangulable pair without any given triangulation and $f: L \to Y$ be a given map. Because of (6.3), it makes sense to talk about $O^{n+1}(f)$. Further, if $\phi: (K', L') \to (K, L)$ is a map of another triangulable pair (K', L') into (K, L), then one can deduce easily from (6.1) and (6.2) that ϕ^* sends $O^{n+1}(f)$ into $O^{n+1}(f')$, where $f' = f\phi: L' \to Y$.

For the remainder of the section, let us go back to the cellular pair (K, L) and the map $f: L \to Y$ given at the beginning of the section. The following two fundamental lemmas are easy consequences of the definition of $O^{n+1}(f)$ and Eilenberg's extension theorem (5.1).

Lemma 6.4. *The map $f : L \to Y$ is n-extensible over K iff $O^{n+1}(f)$ is non-empty.*

Lemma 6.5. *The map $f : L \to Y$ is $(n + 1)$-extensible over K iff $O^{n+1}(f)$ contains the zero element of $H^{n+1}(K, L; \pi_n(Y))$.*

By recurrent application of these two lemmas, one can easily prove the following

Proposition 6.6. *If Y is r-simple and*

$$H^{r+1}(K, L; \pi_r(Y)) = 0$$

for every r satisfying $n \leqslant r < m$, then the n-extensibility of $f : L \to Y$ over K implies its m-extensibility over K.

In particular, if $K \setminus L$ is of dimension not exceeding m, then the hypothesis of (6.6) implies that a map $f : L \to Y$ has an extension over K iff it is n-extensible over K. Hence, we have the following

Corollary 6.7. *If Y is r-simple and*

$$H^{r+1}(K, L; \pi_r(Y)) = 0$$

for every $r \geqslant 1$, then every map $f : L \to Y$ has an extension over K.

7. The homotopy problem

Let us study two given maps

$$f : K \to Y, \quad g : K \to Y$$

agreeing on L, that is to say, such that $f \mid L = g \mid L$. As described in (I; § 8), the homotopy problem (relative to L) is to determine whether or not f and g are homotopic relative to L, in other words, whether or not there exists a homotopy $h_t : K \to Y$, $(0 \leqslant t \leqslant 1)$, such that

$$h_0 = f, \quad h_1 = g, \quad h_t \mid L = f \mid L \text{ for each } t \in I.$$

The most important special case is that $L = \square$. Then the problem is to determine whether or not two given maps $f, g : K \to Y$ are homotopic.

Since the homotopy problem is a special case of the extension problem, the obstruction method can be applied. Let us begin by defining the notion of n-homotopy as follows. The maps f and g are said to be *n-homotopic* relative to L if $f \mid \bar{K}^n$ and $g \mid \bar{K}^n$ are homotopic relative to L. If f and g are homotopic relative to L, then they are obviously n-homotopic relative to L. Since Y is assumed to be pathwise connected, the following proposition is obvious.

Proposition 7.1. *Every pair of maps $f, g : K \to Y$ with $f \mid L = g \mid L$ are 0-homotopic relative to L.*

The least upper bound of the set of integers n such that f and g are n-homotopic relative to L is called *the homotopy index* of the pair (f, g) relative to L. The following two propositions can be proved as in § 2.

Proposition 7.2. *If $f \simeq f'$ and $g \simeq g'$ relative to L, then the pair (f', g') has the same homotopy index relative to L as the pair (f, g).*

Proposition 7.3. *If (K, L) is a simplicial pair, then the homotopy index of any pair of maps $f, g : K \to Y$ relative to L is a topological invariant.*

8. The obstruction $d^n(f,g;h_t)$

Throughout the present section, we are concerned with two given maps

$$f, g : K \to Y, \quad f \mid L = g \mid L$$

which are $(n - 1)$-homotopic relative to L. Let

$$h_t : \bar{K}^{n-1} \to Y, \quad (0 \leqslant t \leqslant 1),$$

be a given homotopy such that $h_0 = f \mid \bar{K}^{n-1}$, $h_1 = g \mid \bar{K}^{n-1}$, and $h_t \mid L = f \mid L$ for every $t \in I$.

Since f and g are defined on $\bar{K}^n$, the construction in § 4 gives a deformation cochain $d^n(f, g; h_t)$, which will be called the *obstruction* of the homotopy h_t in connection with the pair (f, g).

Lemma 8.1. *The obstruction $d^n(f, g; h_t)$ is a relative n-cocycle of K modulo L; in symbols, $d^n(f, g; h_t) \in Z^n(K, L; \pi_n(Y))$.*

Proof. Since f and g are defined on $\bar{K}^{n+1}$, it follows that $c^{n+1}(f) = 0 = c^{n+1}(g)$. Hence, by (4.2) we have

$$\delta d^n(f, g; h_t) = c^{n+1}(f) - c^{n+1}(g) = 0. \blacksquare$$

Lemma 8.2. *The homotopy h_t has an extension $h_t^* : \bar{K}^n \to Y$, $(0 \leqslant t \leqslant 1)$, such that $h_0^* = f \mid \bar{K}^n$ and $h_1^* = g \mid \bar{K}^n$ iff $d^n(f, g; h_t) = 0$.*

Proof. Consider the pair (J, M), where $J = K \times I$ and $M = (K \times 0) \cup (L \times I) \cup (K \times 1)$. Let

$$\bar{J}^n = M \cup J^n = (K \times 0) \cup (\bar{K}^{n-1} \times I) \cup (K \times 1).$$

Define a map $F : \bar{J}^n \to Y$ by taking

$$F(x, t) = \begin{cases} f(x), & (x \in K, t = 0), \\ h_t(x), & (x \in \bar{K}^{n-1}, t \in I), \\ g(x), & (x \in K, t = 1). \end{cases}$$

Then F determines an obstruction cocycle $c^{n+1}(F)$ of the complex J modulo M. By the formula (i) of § 4, we obtain

$$\text{(i)} \qquad kd^n(f, g; h_t) = (-1)^{n+1} c^{n+1}(F).$$

Since the homotopy h_t has an extension h_t^* iff the map F has an extension

$F^* : \bar{J}^{n+1} \to Y$, the lemma is an immediate consequence of (3.2) and the preceding formula (i) since k is an isomorphism. ∎

The cocycle $d^n(f, g; h_t)$ represents an *obstruction cohomology class*

$$\delta^n(f, g; h_t) \in H^n(K, L; \pi_n(Y)).$$

Analogous to (5.1), we have the following Eilenberg's homotopy theorem.

Theorem 8.3. $\delta^n(f, g; h_t) = 0$ *iff there exists a homotopy* $h_t^* : \bar{K}^n \to Y$, $(0 \leqslant t \leqslant 1)$, *such that*

$$h_0^* = f \mid \bar{K}^n, \quad h_1^* = g \mid \bar{K}^n,$$
$$h_t^* \mid \bar{K}^{n-2} = h_t \mid \bar{K}^{n-2}, \quad (t \in I).$$

Proof. Consider the map F defined in the proof of (8.2). Then h_t^* exists iff there is a map $F^* : \bar{J}^{n+1} \to Y$ such that $F^* \mid \bar{J}^{n-1} = F \mid \bar{J}^{n-1}$ and hence iff $c^{n+1}(F)$ is a coboundary by (5.1). Since the isomorphism k in (i) commutes with the coboundary operator, this implies the theorem. ∎

9. The group $R^n(K,L;f)$

Let $f : K \to Y$ be a given map. We are going to construct a group $R^n(K, L; f)$ as follows. Consider the space Ω of all maps of $\bar{K}^{n-1}$ into Y. Let W denote the subspace of Ω consisting of the maps $g : \bar{K}^{n-1} \to Y$ such that $g \mid L = f \mid L$ and let $w_0 \in W$ denote the restriction $w_0 = f \mid \bar{K}^{n-1}$. Then we define

$$R^n(K, L; f) = \pi_1(W, w_0).$$

An arbitrary element α of $R^n(K, L; f)$ is represented by a homotopy $h_t : \bar{K}^{n-1} \to Y$, $(0 \leqslant t \leqslant 1)$, such that

$$h_0 = f \mid \bar{K}^{n-1} = h_1, \quad h_t \mid L = f \mid L, \quad (0 \leqslant t \leqslant 1).$$

Hence we obtain an obstruction cohomology class $\delta^n(f, f; h_t) \in H^n(K, L; \pi_n(Y))$ which clearly depends only on α. The following lemma is an immediate consequence of (4.5).

Lemma 9.1. *The assignment* $\alpha \to \xi_n(\alpha) = \delta^n(f, f; h_t)$ *defines a homomorphism*

$$\xi_n : R^n(K, L; f) \to H^n(K, L; \pi_n(Y)).$$

Note. If one removes the hypothesis that Y is n-simple, then ξ_n becomes a crossed homomorphism.

Let $J_f^n = J_f^n(K, L; \pi_n(Y))$ denote the image of $R^n(K, L; f)$ under ξ_n. Then J_f^n is a subgroup of $H^n(K, L; \pi_n(Y))$. We shall denote the quotient group by

$$Q_f^n = Q_f^n(K, L; \pi_n(Y)) = H^n(K, L; \pi_n(Y))/J_f^n.$$

Lemma 9.2. *The subgroup* J_f^n *(and hence the quotient group* Q_f^n*) depends only on the* $(n - 1)$*-homotopy class of* f *relative to* L*; that is to say, if* $f, g : K \to Y$ *are* $(n - 1)$*-homotopic relative to* L*, then* $J_f^n = J_g^n$.

Proof. Let $k_t : \bar{K}^{n-1} \to Y$, $(0 \leqslant t \leqslant 1)$, be a homotopy such that $k_0 = f \mid \bar{K}^{n-1}$, $k_1 = g \mid \bar{K}^{n-1}$, and $k_t \mid L = f \mid L$ for each $t \in I$. Because of symmetry, it suffices to prove that $J_f^n \subset J_g^n$. For an arbitrary element α of $R^n(K,L;f)$, the element $\xi_n(\alpha) \in J_f^n$ is represented by the cocycle $d^n(f,f;h_t)$ described as above. Define a homotopy $h_t^* : \bar{K}^{n-1} \to Y$, $(0 \leqslant t \leqslant 1)$, by taking

$$h_t^*(x) = \begin{cases} k_{1-3t}(x), & (x \in \bar{K}^{n-1}, 0 \leqslant t \leqslant \frac{1}{3}), \\ h_{3t-1}(x), & (x \in \bar{K}^{n-1}, \frac{1}{3} \leqslant t \leqslant \frac{2}{3}), \\ k_{3t-2}(x), & (x \in \bar{K}^{n-1}, \frac{2}{3} \leqslant t \leqslant 1). \end{cases}$$

Then $h_0^* = g \mid \bar{K}^{n-1} = h_1^*$ and $h_t^* \mid L = g \mid L$ for every $t \in I$. Hence h_t^* represents an element β of $R^n(K, L; g)$. On the other hand, it follows from (4.5) that

$$d^n(g, g, h_t^*) = - d^n(f, g; k_t) + d^n(f,f; h_t) + d^n(f, g; k_t) = d^n(f, f; h_t).$$

This implies that $\xi_n(\alpha) = \xi_n(\beta) \in J_g^n$. Hence $J_f^n \subset J_g^n$. ∎

10. The obstruction sets for homotopy

Let $f, g : K \to Y$ be two given maps such that $f \mid L = g \mid L$. We are going to define the *n-dimensional obstruction set*

$$O^n(f, g) \subset H^n(K, L; \pi_n(Y))$$

as follows. If f and g are not $(n - 1)$-homotopic relative to L, we define $O^n(f, g)$ to be the vacuous set. Now suppose that f and g are $(n - 1)$-homotopic relative to L. Then there exists a homotopy $h_t : \bar{K}^{n-1} \to Y$, $(0 \leqslant t \leqslant 1)$, such that $h_0 = f \mid \bar{K}^{n-1}$, $h_1 = g \mid \bar{K}^{n-1}$, $h_t \mid L = f \mid L$, $(t \in I)$.

The cohomology class $\delta^n(f, g; h_t)$ in $H^n(K, L; \pi_n(Y))$ is called an *n-dimensional obstruction element* of the pair (f, g). Then, $O^n(f, g)$ is defined to be the set of all n-dimensional obstruction elements of (f, g). The following proposition is obvious.

Proposition 10.1. *If $f \simeq f'$ and $g \simeq g'$ relative to L, then $O^n(f, g) = O^n(f', g')$.*

Next, let (K', L') be another cellular pair and $\phi : (K', L') \to (K, L)$ be a proper cellular map. This map ϕ induces a homomorphism

$$\phi^* : H^n(K, L; \pi_n(Y)) \to H^n(K', L'; \pi_n(Y)).$$

Then the following proposition is an immediate consequence of (4.6).

Proposition 10.2. *If $f' = f\phi$ and $g' = g\phi$, then ϕ^* sends $O^n(f, g)$ into $O^n(f', g')$.*

Corollary 10.3. *If (K, L) is a simplicial pair, then $O^n(f, g)$ is a topological invariant, i.e. it does not depend on the triangulation of (K, L).*

Lemma 10.4. *The maps f and g are $(n - 1)$-homotopic relative to L iff $O^n(f, g)$ is a coset of the subgroup $J_f^n(K, L; \pi_n(Y))$ in the cohomology group $H^n(K, L; \pi_n(Y))$.*

Proof. Since the sufficiency is obvious, it suffices to prove the necessity of the condition. For this purpose, assume that f and g are $(n-1)$-homotopic relative to L. Let $\delta^n(f, g; h_t)$ and $\delta^n(f, g; h'_t)$ be any two n-dimensional obstruction elements of (f, g). Define a homotopy $k_t : \bar{K}^{n-1} \to Y$, $(0 \leqslant t \leqslant 1)$, by taking $k_t = h_{2t}$ if $t \leqslant \frac{1}{2}$ and $k_t = h'_{2-2t}$ if $t \geqslant \frac{1}{2}$. Then k_t represents an element α of $R^n(K, L; f)$. By (4.5), we have

$$\xi_n(\alpha) = \delta^n(f, g; h_t) - \delta^n(f, g; h'_t).$$

Since $\xi_n(\alpha) \in J_f^n$, this proves that $O^n(f, g)$ is contained in the coset $\delta^n(f, g; h_t) + J_f^n$.

On the other hand, let α be any element of $R^n(K, L; f)$. Then α is represented by a homotopy k_t and $\xi_n(\alpha) = \delta^n(f, f; k_t)$. Define a homotopy $h'_t : \bar{K}^{n-1} \to Y$, $(0 \leqslant t \leqslant 1)$, by taking $h'_t = k_{2t}$ if $t \leqslant \frac{1}{2}$ and $h'_t = h_{2t-1}$ if $t \geqslant \frac{1}{2}$. Then we have

$$\delta^n(f, g; h'_t) = \xi_n(\alpha) + \delta^n(f, g; h_t).$$

Hence every element of the coset $\delta^n(f, g; h_t) + J_f^n$ is in $O^n(f, g)$. ▮

Lemma 10.5. (The fundamental homotopy lemma). *The maps f and g are n-homotopic relative to L iff*

$$O^n(f, g) = J_f^n(K, L; \pi_n(Y)).$$

Proof. *Necessity.* Suppose that f and g are n-homotopic relative to L. Then there exists a homotopy $h_f^* : \bar{K}^n \to Y$, $(0 \leqslant t \leqslant 1)$, such that $h_0^* = f \mid \bar{K}^n$, $h_1^* = g \mid \bar{K}^n$, and $h_t^* \mid L = f \mid L$ for every $t \in I$. Let $h_t = h_t^* \mid \bar{K}^{n-1}$. By (8.3), we have $\delta^n(f, g; h_t) = 0$. This implies that $O^n(f, g)$ contains the zero element of $H^n(K, L; \pi_n(Y))$. According to (10.4), it follows that $O^n(f,g) = J_f^n$.

Sufficiency. Suppose $O^n(f, g) = J_f^n$. Then $O^n(f, g)$ contains the zero element of $H^n(K, L; \pi_n(Y))$. Hence there exists a homotopy $h_t : \bar{K}^{n-1} \to Y$, $(0 \leqslant t \leqslant 1)$, such that $h_0 = f \mid \bar{K}^{n-1}$, $h_1 = g \mid \bar{K}^{n-1}$, $h_t \mid L = f \mid L$ for each $t \in I$, and $\delta^n(f, g; h_t) = 0$. By (8.3), there exists a homotopy $h_t^* : \bar{K}^n \to Y$, $(0 \leqslant t \leqslant 1)$, such that $h_0^* = f \mid \bar{K}^n$, $h_1^* = g \mid \bar{K}^n$, and $h_t^* \mid \bar{K}^{n-2} = h_t \mid \bar{K}^{n-2}$ for each $t \in I$. Hence f and g are n-homotopic. ▮

11. The general homotopy theorem

The following general homotopy theorem is an immediate consequence of (10.4) and (10.5).

Theorem 11.1. *Any two maps $f, g : K \to Y$ with $f \mid L = g \mid L$ which are $(n-1)$-homotopic relative to L determine a unique element $\chi^n(f, g)$ of the group $Q_f^n(K, L; \pi_n(Y))$. $\chi^n(f, g) = 0$ iff f and g are n-homotopic relative to L.*

This element $\chi^n(f, g)$ will be called the *n-dimensional characteristic element* of (f, g). By (10.3), $\chi^n(f, g)$ is a topological invariant if (K, L) is a simplicial pair.

By recurrent applications of (11.1), we obtain the following

Proposition 11.2. *Let $f: K \to Y$ be a given map and assume that Y is r-simple and $Q_f^r(K, L; \pi_r(Y)) = 0$ for every r satisfying $n < r \leqslant m$. If two maps $f, g: K \to Y$ with $f \mid L = g \mid L$ are n-homotopic relative to L, then they are also m-homotopic relative to L.*

Corollary 11.3. *If Y is r-simple and $H^r(K, L; \pi_r(Y)) = 0$ for every r satisfying $1 \leqslant r \leqslant \dim(K \setminus L)$, then every two maps $f, g: K \to Y$ with $f \mid L = g \mid L$ are homotopic relative to L.*

As an application of this general result, let us assume that K is connected and L is vacuous. Take $Y = K$. Then we obtain the following

Proposition 11.4. *The following statements are equivalent:*

(*i*) *K is contractible.*

(*ii*) *$\pi_r(K) = 0$ for each $r \geqslant 1$.*

(*iii*) *K is r-simple and $H^r(K; \pi_r(K)) = 0$ for every r satisfying $1 \leqslant r \leqslant$ $\dim K$.*

(*iv*) *K is r-simple and $Q_i^r(K; \pi_r(K)) = 0$ for every r satisfying $1 \leqslant r \leqslant$ $\dim K$, where $i: K \to K$ denotes the identity map.*

12. The classification problem

Let $\mu: L \to Y$ be a given map and let us study the set W of all possible extensions of μ over K. In symbols,

$$W = \{ f: K \to Y \mid \ f \mid L = \mu \}.$$

The maps W are divided into disjoint *homotopy classes relative to L*. As in (I; § 8), the classification problem for these maps W is to enumerate these homotopy classes by means of some convenient invariants.

The relation of n-homotopy relative to L among the maps W divides W into disjoint *n-homotopy classes relative to L*. For each $n \geqslant 1$, every $(n-1)$-homotopy class relative to L contains a certain collection of n-homotopy classes relative to L. For the classification problem, we have to find a reasonable way to count the n-homotopy classes contained in a given $(n-1)$-homotopy class by means of some homology or cohomology invariant.

Throughout the remainder of the section, let θ be a given $(n-1)$-homotopy class relative to L of the maps W. We are going to enumerate the n-homotopy classes relative to L of the maps W which are contained in θ. According to (9.2), θ determines a subgroup $J_\theta^n(K, L; \pi_n(Y))$ of the cohomology group $H^n(K, L; \pi_n(Y))$ and hence the quotient group

$$Q_\theta^n(K, L; \pi_n(Y)) = H^n(K, L; \pi_n(Y))/J_\theta^n(K, L; \pi_n(Y)).$$

Now let us choose a map $f: K \to Y$ from the class θ as our reference map. According to (11.1), every map $g \in \theta$ determines a characteristic element $\chi^n(f, g)$ in the group $Q_\theta^n(K, L; \pi_n(Y))$. An element α of $Q_\theta^n(K, L; \pi_n(Y))$ is said to be *f-admissible* if there is a map $g \in \theta$ with $\chi^n(f, g) = \alpha$. The

f-admissible elements of $Q_\theta^n(K, L; \pi_n(Y))$ form a set A_f^n, which will be called *the f-admissible set* in $Q_\theta^n(K, L; \pi_n(Y))$. The following proposition is obvious.

Proposition 12.1. *For any two maps $f, g : K \to Y$ in the $(n - 1)$-homotopy class θ, A_f^n is the image of A_g^n under the translation determined by their characteristic element $\chi^n(f, g)$, that is to say,*

$$A_f^n = \chi^n(f, g) + A_g^n.$$

Now we are in a position to establish the following general classification theorem.

Theorem 12.2. *Given an $(n - 1)$-homotopy class θ relative to L of the maps W, the n-homotopy classes relative to L of W which are contained in θ are in a $1 - 1$ correspondence with the elements of the f-admissible set A_f^n in $Q_\theta^n(K, L; \pi_n(Y))$, where f is an arbitrarily given map in θ.*

Proof. According to (11.1), every $g \in \theta$ determines an element $\chi^n(f, g)$ in A_f^n. We assert that $\chi^n(f, g)$ depends only on the n-homotopy class relative to L which contains g. In fact, if $g, h \in \theta$ are n-homotopic relative to L, then it follows from (11.1) that

$$\chi^n(f, h) - \chi^n(f, g) = \chi^n(g, h) = 0,$$

and hence $\chi^n(f, g) = \chi^n(f, h)$. Therefore, the correspondence $g \to \chi^n(f, g)$ defines a function τ of the n-homotopy classes relative to L contained in θ into the elements of A_f^n. That τ is onto follows from the definition of A_f^n. It remains to prove that τ is one-to-one. Suppose that $\chi^n(f, g) = \chi^n(f, h)$. Then we have

$$\chi^n(g, h) = \chi^n(f, h) - \chi^n(f, g) = 0.$$

According to (11.1), this implies that g and h are n-homotopic relative to L. ∎

Remark. Since A_f^n is, in general, not effectively computable, the general classification theorem given above does not solve the problem. In each particular case, it remains to compute A_f^n by special methods.

13. The primary obstructions

Throughout §§ 13–17, we shall assume one more condition, namely, that the given space Y is $(n - 1)$-connected for a given positive integer n. According to (V; § 4), this means that $\pi_r(Y) = 0$ for every $r < n$. If $n > 1$, then Y is simply connected and hence n-simple. If $n = 1$, we assume that Y is n-simple and hence the fundamental group $\pi_n(Y)$ is abelian.

Let us consider an arbitrarily given map

$$f : L \to Y.$$

According to (6.6), f is always n-extensible over K. Hence the first non-trivial obstruction is the $(n + 1)$-dimensional $O^{n+1}(f)$. We are going to prove that $O^{n+1}(f)$ consists of a single element of the cohomology group $H^{n+1}(K, L; \pi_n(Y))$.

Let $\theta : L \to Y$ denote the constant map $\theta(L) = y_0$. According to (11.2), f and θ are $(n - 1)$-homotopic and hence f is homotopic to a map which sends the $(n - 1)$-dimensional skeleton L^{n-1} of L into y_0. Since homotopic maps have the same $(n + 1)$-dimensional obstruction set, we may assume that $f(L^{n-1}) = y_0$.

Since $f \mid L^{n-1} = \theta \mid L^{n-1}$, the difference cochain $d^n(f, \theta)$ of § 4 is defined. By (4.2), $d^n(f, \theta)$ is an n-cocycle of L with coefficients in $\pi_n(Y)$. This cocycle $d^n(f, \theta)$ can be equivalently defined as follows. Let σ be any n-cell of L. Since $f(\partial\sigma) = y_0$, the partial map $f_\sigma = f \mid \sigma$ represents an element $[f_\sigma]$ of $\pi_n(Y)$. Then $d^n(f, \theta)$ is given by $[d^n(f, \theta)]\,(\sigma) = [f_\sigma]$.

The cocycle $d^n(f, \theta)$ represents a cohomology class $\varkappa^n(f)$ of L over $\pi_n(Y)$, which will be called the *characteristic element* of f.

Proposition 13.1. *The* $(n + 1)$*-dimensional obstruction set* $O^{n+1}(f)$ *consists of a single element*

$$\omega^{n+1}(f) = \delta^*\varkappa^n(f) \in H^{n+1}(K, L; \pi_n(Y)),$$

which is called the primary obstruction *to extending* f *over* K, *where* δ^* *denotes the coboundary homomorphism*

$$\delta^* : H^n(L; \pi_n(Y)) \to H^{n+1}(K, L; \pi_n(Y)).$$

Proof. First, let us prove that $\delta^*\varkappa^n(f)$ is in $O^{n+1}(f)$. Since $f(L^{n-1}) = y_0$, f has an extension $f^* : \bar{K}^n \to Y$ such that $f^*(\bar{K}^n \setminus L) = y_0$. Let $\theta^* : \bar{K}^n \to Y$ denote the constant map $\theta^*(\bar{K}^n) = y_0$. Then it follows that the difference cochain $d^n(f^*, \theta^*)$ is the trivial extension of $d^n(f, \theta)$, i.e., for any n-cell σ of K, we have

$$[d^n(f^*, \theta^*)](\sigma) = \begin{cases} [d^n(f, \theta)](\sigma), & (\text{if } \sigma \in L), \\ 0, & (\text{if } \sigma \in K \setminus L). \end{cases}$$

Hence $\delta^*\varkappa^n(f)$ is represented by the cocycle

$$\delta d^n(f^*, \theta^*) = c^{n+1}(f^*) - c^{n+1}(\theta^*) = c^{n+1}(f^*).$$

This implies that $\delta^*\varkappa^n(f) = \gamma^{n+1}(f^*) \in O^{n+1}(f)$.

Next, let α be an arbitrary element of $O^{n+1}(f)$. Then there exists an extension $g^* : \bar{K}^n \to Y$ of f such that $\gamma^{n+1}(g^*) = \alpha$. By (11.2), g^* and f^* are $(n - 1)$-homotopic relative to L. Hence we may assume that $g^* \mid \bar{K}^{n-1} = f^* \mid \bar{K}^{n-1}$. By (4.2), we have

$$c^{n+1}(g^*) = c^{n+1}(f^*) + \delta d^n(g^*, f^*).$$

Since $d^n(g^*, f^*)$ is in $C^n(K, L; \pi_n(Y))$ according to § 4, this implies that $\gamma^{n+1}(g^*) = \gamma^{n+1}(f^*)$. Hence $\alpha = \delta^*\varkappa^n(f)$. ∎

Next, let us turn to the homotopy problem and consider a pair of maps

$$f, g : K \to Y, \quad f \mid L = g \mid L.$$

Then their characteristic elements $\varkappa^n(f)$ and $\varkappa^n(g)$ in $H^n(K; \pi_n(Y))$ are uniquely defined and depend only on their homotopy classes.

Proposition 13.2. *The n-dimensional obstruction set $O^n(f, g)$ consists of a single element*

$$\omega^n(f, g) \in H^n(K, L; \pi_n(Y))$$

which is called the primary obstruction *to homotopy relative to L of the pair (f, g). Moreover*

$$\varkappa^n(f) - \varkappa^n(g) = j^*\omega^n(f, g),$$

where $j^* : H^n(K, L; \pi_n(Y)) \to H^n(K; \pi_n(Y))$ denotes the homomorphism induced by the inclusion map $j : K \subset (K, L)$.

Proof. That $O^{n+1}(f, g)$ consists of a single element follows from (13.1) and the definition of the deformation cochain in § 4. For the second part of the proposition, we may assume that

$$f(K^{n-1}) = y_0 = g(K^{n-1}).$$

Let θ denote the constant map. Then we have

$$d^n(f, \theta) - d^n(g, \theta) = d^n(f, g).$$

This implies that $\varkappa^n(f) - \varkappa^n(g) = j^*\omega^n(f, g)$. ∎

Corollary 13.3 *For any given map $f : K \to Y$, we have*

$$J_f{}^n(K, L; \pi_n(Y)) = 0, \quad Q_f{}^n(K, L; \pi_n(Y)) = H^n(K, L, \pi_n(Y)).$$

Finally, if (K, L) is a simplicial pair, then $\omega^{n+1}(f)$, $\varkappa^n(f)$, $\omega^n(f, g)$ are topological invariants according to (6.3) and (10.3).

14. Primary extension theorems

By means of the primary obstructions, we shall be able to strengthen the general results in § 6 and obtain an effective solution of the extension problem for the case where $K \setminus L$ is of dimension not exceeding $n + 1$.

An element of the cohomology group $H^n(L ; G)$ is said to be *extensible over K* if it is contained in the image of the homomorphism

$$i^* : H^n(K; G) \to H^n(L; G).$$

induced by the inclusion map $i : L \subset K$.

Theorem 14.1. *For a given map $f : L \to Y$, the following statements are equivalent:*

(1) *f is $(n + 1)$-extensible over K.*

(2) $\omega^{n+1}(f) = 0$.

(3) *$\varkappa^n(f)$ is extensible over K.*

Proof. The equivalence of (1) and (2) follows from (6.5) and (13.1), and that of (2) and (3) follows from the exactness of the cohomology sequence. ∎

If $K \setminus L$ is of dimension $n + 1$, then we have the following corollary which includes Hopf extension theorem (II; § 8) as a special case. See § 17.

Corollary 14.2. *A map $f : L \to Y$ is extensible over K iff its characteristic element $\varkappa^n(f)$ is extensible over K.*

The following generalization of (14.2) is an immediate consequence of (6.6).

Theorem 14.3. *If Y is r-simple and $H^{r+1}(K, L; \pi_r(Y)) = 0$ for every r satisfying $n < r < \dim (K \setminus L)$, then a necessary and sufficient condition for a given map $f : L \to Y$ to have an extension over K is that the characteristic element $\varkappa^n(f)$ is extensible over K.*

In particular, the hypothesis of (14.3) holds if $\pi_r(Y) = 0$ whenever $n < r < \dim (K \setminus L)$.

15. Primary homotopy theorems

The following theorem is an immediate consequence of the fundamental homotopy lemma (10.5) and the corollary (13.3).

Theorem 15.1. *Two maps $f, g : K \to Y$ with $f \mid L = g \mid L$ are n-homotopic relative to L iff $\omega^n(f, g) = 0$.*

If L is empty, then this can be given in the following form:

Corollary 15.2. *For any two given maps $f, g : K \to Y$, the following statements are equivalent:*

(1) *f and g are n-homotopic.*
(2) $\omega^n(f, g) = 0$.
(3) $\varkappa^n(f) = \varkappa^n(g)$.

If K is of dimension n, then we have the following corollary which includes the Hopf homotopy theorem (II; § 8) as a special case. See § 17.

Corollary 15.3. *Two maps $f, g : K \to Y$ are homotopic iff $\varkappa^n(f) = \varkappa^n(g)$.*

The following generalization of (15.3) is an immediate consequence of (11.2).

Theorem 15.4. *If Y is r-simple and $H^r(K; \pi_r(Y)) = 0$ for each r satisfying $n < r \leqslant \dim K$, then a necessary and sufficient condition for a given pair of maps $f, g : K \to Y$ to be homotopic is that $\varkappa^n(f) = \varkappa^n(g)$.*

In particular, the hypothesis of (15.4) is satisfied if $\pi_r(Y) = 0$ whenever $n < r \leqslant \dim K$.

16. Primary classification theorems

Lemma 16.1. *If Y is r-simple and $H^{r+1}(K, L; \pi_r(Y)) = 0$ for every r satisfying $n < r < \dim (K \setminus L)$, then, for each map $f : K \to Y$ and each element $\alpha \in H^n(K, L; \pi_n(Y))$, there exists a map $g : K \to Y$ such that $f \mid L = g \mid L$ and $\omega^n(f, g) = \alpha$.*

Proof. Choose a cocycle $z \in Z^n(K, L; \pi_n(Y))$ which represents α. According to (4.3), there is a map $h : \bar{K}^n \to Y$ such that $f \mid \bar{K}^{n-1} = h \mid \bar{K}^{n-1}$ and $d^n(f, h) = z$. By (4.2), we have

$$c^{n+1}(f) - c^{n+1}(h) = \delta d^n(f, h) = \delta z = 0.$$

Since f is defined over $K \supset \bar{K}^{n+1}$, we obtain $c^{n+1}(f) = 0$. This implies that $c^{n+1}(h) = 0$ and hence h has an extension $h^* : \bar{K}^{n+1} \to Y$. By (6.6), one deduces that h has an extension $g : K \to Y$. Then it is clear that $f \mid L = g \mid L$ and $\omega^n(f, g) = \alpha$. ∎

If L is empty, then the preceding lemma can be formulated as follows.

Lemma 16.2. *If Y is r-simple and $H^{r+1}(K; \pi_r(Y)) = 0$ for every r satisfying $n < r < \dim K$, then, for each element α in $H^n(K; \pi_n(Y))$, there exists a map $f : K \to Y$ such that $\varkappa^n(f) = \alpha$.*

Now let $f : K \to Y$ be a given map and let us consider the totality W of the maps $g : K \to Y$ such that $f \mid L = g \mid L$. Since $\pi_m(Y) = 0$ for every $m < n$, there is only one $(n - 1)$-homotopy class of the maps W relative to L. We are going to enumerate the n-homotopy classes of the maps W relative to L.

Theorem 16.3. *If Y is r-simple and $H^{r+1}(K, L; \pi_r(Y)) = 0$ for each r satisfying $n < r < \dim (K \setminus L)$, then the n-homotopy classes relative to L of the maps W are in a one-to-one correspondence with the elements of the group $H^n(K, L; \pi_n(Y))$. The correspondence is given by the assignment $g \to \omega^n(f, g)$.*

Proof. By (13.3), we have

$$Q_f^n(K, L; \pi_n(Y)) = H^n(K, L, \pi_n(Y)).$$

On the other hand, (16.1) implies that every element of $Q_f^n(K, L; \pi_n(Y))$ is f-admissible. Hence, we have

$$A_f^n = H^n(K, L; \pi_n(Y)).$$

Then, (16.3) follows from (12.2). The second assertion is obvious. ∎

If L is empty, then (16.3) can be conveniently restated as follows.

Corollary 16.4. *If Y is r-simple and $H^{r+1}(K; \pi_r(Y)) = 0$ for each r satisfying $n < r < \dim K$, then the n-homotopy classes of the maps $f : K \to Y$ are in a one-to-one correspondence with the elements of the group $H^n(K; \pi_n(Y))$. The correspondence is determined by the assignment $f \to \varkappa^n(f)$.*

If K is of dimension n, then we have the following corollary which includes the Hopf classification theorem (II; § 8) as a special case. See § 17.

Corollary 16.5. *The homotopy classes of the maps $f : K \to Y$ are in a one-to-one correspondence with the elements of the group $H^n(K; \pi_n(Y))$. The correspondence is determined by the assignment $f \to \varkappa^n(f)$.*

The following generalization of (16.5) is an immediate consequence of (11.2) and (16.3).

Theorem 16.6 *If Y is r-simple and*

$$H^{r+1}(K, L; \pi_r(Y)) = 0 = H^r(K, L; \pi_r(Y))$$

for every r satisfying $n < r \leqslant \dim (K \setminus L)$, then the homotopy classes of the maps W relative to L are in a one-to-one correspondence with the elements of *the group* $H^n(K, L; \pi_n(Y))$. *The correspondence is determined by the assignment* $g \to \omega^n(f, g)$.

In particular, the hypothesis of (16.6) is satisfied if $\pi_r(Y) = 0$ whenever $n < r \leqslant \dim (K \setminus L)$.

17. The characteristic element of Y

Throughout this last section of the chapter, we assume that Y is a finite cell complex in addition to the conditions assumed at the beginning of § 13. Then, according to § 13, the identity map $\iota : Y \to Y$ determined a unique characteristic element

$$\varkappa^n(\iota) = \omega^n(\iota, \theta) \in H^n(Y; \pi_n(Y)),$$

where θ denotes the constant map $\theta(Y) = y_0$. This will be called *the characteristic element of* Y and will be denoted by $\varkappa^n(Y)$.

For example, if Y is the n-sphere, then $\pi_n(Y) \approx Z$. It can be easily verified that the characteristic element $\varkappa^n(Y)$ is a generator of the free cyclic group $H^n(Y)$ which depends on the given orientation of Y.

Now, consider an arbitrary map $f : L \to Y$. This map f induces a homomorphism

$$f^* : H^n(Y; \pi_n(Y)) \to H^n(L; \pi_n(Y)).$$

The following lemma can be proved by means of (4.6) and simplicial approximation.

Lemma 17.1. $\varkappa^n(f) = f^*[\varkappa^n(Y)]$.

This lemma justifies the assertions of §§ 14–16 that (14.2), (15.3) and (16.5) include the Hopf theorems given in (II; § 8).

EXERCISES

A. Minor generalizations of the theory.

1. Make use of the characteristic maps χ_σ in (V; Ex. H) and establish the obstruction theory for the case where K is any topologically realized semi-simplicial complex instead of a finite cell complex and L is a subcomplex of K.

2. Study the notion of a *CW-complex*, [J. H. C. Whitehead 4], and generalize the obstruction theory to the case where K is any CW-complex and L is a subcomplex of K.

3. By the use of local coefficients, construct the obstruction theory for the case where Y is not necessarily n-simple.

B. The fundamental group of a semi-simplicial complex.

Let K be a given connected semi-simplicial complex and v_0 a given vertex of K. Then, as in (II; Ex. A) for finite simplicial complexes, the fundamental group $\pi_1(K, v_0)$ may be given in terms of generators and relations as follows.

A *broken line* joining a vertex of K to another is a path which consists of a finite number of edges of K. Since K is connected, v_0 can be joined to any vertex v of K by a broken line $\beta(v)$. Suppose that these broken lines $\beta(v)$ have been chosen and that $\beta(v_0)$ consists of only a single point v_0. To each edge e of K, let us consider the loop

$$\lambda_e = \beta(e_{(0)})\bullet e\cdot[\beta(e_{(1)})]^{-1}$$

which represents a element g_e of $\pi_1(K, v_0)$. Prove that the set $\{ g_e \mid e \in K \}$ generates $\pi_1(K, v_0)$.

It remains to find the relations among the generators $\{ g_e \}$. For each edge $e \in K$, the loop λ_e consists of a finite number of edges of K; more precisely,

$$\lambda_e = e_1^{\alpha_1} e_2^{\alpha_2} \cdots e_n^{\alpha_n}$$

where $\alpha_i = \pm 1$ for each $i = 1, 2, \cdots, n$. Then, deduce the relation

$$(R_e) \qquad g_e = g_{e_1}^{\alpha_1} g_{e_2}^{\alpha_2} \cdots g_{en}^{\alpha_n}.$$

On the other hand, for each 2-cell σ of K, verify the following relation

$$(R_\sigma) \qquad g_{\sigma(0,1)}\, g_{\sigma(1,2)} = g_{\sigma(0,2)}.$$

Now prove that $\pi_1(K, v_0)$ can be defined as the group with generators $\{ g_e \}$ and relations $\{ R_e \}$ and $\{ R_\sigma \}$.

By means of this expression of $\pi_1(K, v_0)$, establish the following results:

1. *Condition for 2-extensibility*. Let L be a connected subcomplex of K containing v_0 and $f : (L, v_0) \to (Y, y_0)$ be a map into a pathwise connected space Y. Then f and the inclusion map $i : L \subset K$ induce the homomorphisms

$$f_* : \pi_1(L, v_0) \to \pi_1(Y, y_0), \quad i_* : \pi_1(L, v_0) \to \pi_1(K, v_0).$$

Prove that the map f is 2-extensible over K iff there exists a homomorphism $h : \pi_1(K, v_0) \to \pi_1(Y, y_0)$ such that $f_* = hi_*$ and that, for any such homomorphism h, there exists an extension $g : \bar{K}^2 \to Y$ of f such that $g_* = h$.

2. *Condition for 1-homotopy*. Let us consider any two maps $f, g : (K, v_0) \to (Y, y_0)$. They induce the homomorphisms

$$f_*, g_* : \pi_1(K, v_0) \to \pi_1(Y, y_0).$$

Prove that f and g are 1-homotopic iff there exists an element $\xi \in \pi_1(Y, y_0)$ such that

$$g_*(\alpha) = \xi^{-1}\cdot f_*(\alpha)\cdot\xi, \quad \alpha \in \pi_1(K, v_0).$$

For an indication of the proof, see [Hu, 7].

C. Generalizing the obstruction theory by using Čech cohomology theory

The obstruction theory studied in this chapter is formulated for finite cell complexes so as to avoid the complications involved in limiting processes. One can extend the theory to more general spaces by using Čech cohomology theory as follows.

Assume that X is a compact Hausdorff space, A a closed subspace of X, and Y a given ANR. For every finite open covering $\alpha = \{ a_1, \cdots, a_r \}$ of X, the nerve K_a of α is a finite simplicial complex. The non-void sets $A \cap a_i$,

$(i = 1, \cdots, r)$, form an open covering of A whose nerve L_a is a subcomplex of K_a. With the aid of the bridge theorems in (II; Ex. B), we can formulate an obstruction theory as follows.

A map $f : A \to Y$ is said to be *n-extensible over* X if it has a bridge α and a bridge map $\psi_a : L_a \to Y$ which is n-extensible over K_a in the sense of § 2.

Prove the following assertion which gives set-theoretic meaning to the n-extensibility defined above.

1. If X is metrizable with $\dim (X \setminus A) \leqslant q$ and if $f : A \to Y$ is n-extensible over X, then there exists a closed subspace B of X contained in $X \setminus A$ with $\dim B < q - n$ and such that f has an extension $g : X \setminus B \to Y$. [Hu 5; p. 344].

Next, assume for sake of convenience that Y is n-simple and let $\pi_n = \pi_n(Y)$. Define the $(n + 1)$-*dimensional obstruction set* $O^{n+1}(f)$ of a given map $f : A \to Y$ in the Čech cohomology group $H^{n+1}(X, A; \pi_n)$ as follows.

If f is not n-extensible over X, define $O^{n+1}(f)$ to be the vacuous set. Assume f to be n-extensible. There exists a bridge α with a bridge map $\psi_a : L_a \to Y$ which has an extension $\psi'_a : \bar{K}_a^n \to Y$. The obstruction cocycle $c^{n+1}(\psi'_a)$ of K_a modulo L_a represents an element $\gamma^{n+1}(\psi'_a)$ of $H^{n+1}(X, A; \pi_n)$, called an $(n + 1)$-*dimensional obstruction element* of f. Then, $O^{n+1}(f)$ is defined to be the set of all $(n + 1)$-dimensional obstruction elements of f. Homotopic maps have the same $(n + 1)$-dimensional obstruction set.

Prove the following assertion from which one can deduce further results as in §§ 6, 13 and 14.

2. A map $f : A \to Y$ is $(n + 1)$-extensible over X iff $O^{n+1}(f)$ contains the zero element of $H^{n+1}(X, A; \pi_n)$. [Hu 5; p. 346].

Now let us turn to the homotopy problem. Two maps $f, g : X \to Y$ are said to be n-homotopic if there exists a bridge α for both f and g with bridge maps $\phi_a : K \to Y$ and $\psi_a : K \to Y$ respectively such that $\phi_a \mid K^n = \psi_a \mid K^n$.

Prove the following assertion which gives set-theoretic meaning to n-homotopy.

3. If X is metrizable with $\dim X \leqslant q$ and $f, g : X \to Y$ are n-homotopic, then there exists a closed subspace B of X such that $\dim B < q - n$ and $f \mid X \setminus B \simeq g \mid X \setminus B$. [Hu 5; p. 349].

Next, assume that Y is n-simple and let $\pi_n = \pi_n(Y)$. Define the *n-dimensional obstruction set* $O^n(f, g)$, of a pair of given maps $f, g : X \to Y$ in the Čech cohomology group $H^n(X; \pi_n)$ and prove the following assertion from which one can deduce further results on homotopy and classification as in §§ 10, 11, 12 and 15.

4. The maps $f, g : X \to Y$ are n-homotopic iff $O^{n+1}(f, g)$ contains the zero element of $H^n(X; \pi_n)$. [Hu 5; p. 350].

D. Generalizing the obstruction theory by using the singular complex

The generalizations of the obstruction theory by using Čech cohomology as illustrated in the preceding exercise are very satisfactory except that one must assume either the domain X or the range Y to be reasonably smooth. To avoid these conditions, one can also generalize the obstruction theory by using the singular complex $S(X)$ of the domain X as follows. [Olum 1 and Hu 7].

Consider two spaces X, Y and a subspace A of X which might not be closed. According to (V; Ex. I), we have a projection

$$\omega : (S(X), S(A)) \to (X, A).$$

A map $f: A \to Y$ is said to be *n-extensible over* X if the map $f\omega: S(A) \to Y$ is n-extensible over $S(X)$ in the sense of Ex. A1. Prove that a map $f: A \to Y$ is n-extensible over X iff, for every map $\phi : (K, L) \to (X, A)$ of a semi-simplicial pair (K, L) into (X, A), the map $f\phi : L \to Y$ is n-extensible over K. Then, prove that, if (X, A) is a semi-simplicial pair, the definition of n-extensibility defined here is equivalent to that of Ex. A1.

Pick $x_0 \in A$ and $y_0 = f(x_0) \in Y$. Then the map $f: A \to Y$ and the inclusion $i: A \subset X$ induce the following homomorphisms

$$f_* : \pi_1(A, x_0) \to \pi_1(Y, y_0), \quad i_* : \pi_1(A, x_0) \to \pi_1(X, x_0).$$

Assume X, A, Y to be pathwise connected and prove that f is 2-extensible over X iff there exists a homomorphism $h : \pi_1(X, x_0) \to \pi_1(Y, y_0)$ such that $f_* = hi_*$. [Hu 7; p. 177].

Assume Y to be pathwise connected and n-simple with $\pi_n = \pi_n(Y, y_0)$. Define for any given map $f: A \to Y$ the $(n+1)$-*dimensional obstruction set* $O^{n+1}(f)$ in the singular cohomology group $H^{n+1}(X, A; \pi_n)$ in obvious way, and prove that f is $(n+1)$-extensible over X iff $O^{n+1}(f)$ contains the zero element of the group $H^{n+1}(X, A; \pi_n)$. Deduce further results on the extension problem.

Now, let us turn to the homotopy problem. Two maps $f, g: X \to Y$ are said to be *n-homotopic* if the maps $f\omega, g\omega : S(X) \to Y$ are n-homotopic in the sense of Ex. A1. Prove that f, g are n-homotopic iff, for every map $\phi: K \to X$ of a semi-simplicial complex K into X, the maps $f\phi, g\phi$ are n-homotopic. Then, prove that, if X is a semi-simplicial complex, the definition of n-homotopy given here is equivalent to that of Ex. A1.

Pick $x_0 \in X$ and $y_0 \in Y$ and assume $f(x_0) = y_0 = g(x_0)$. Then, the maps $f, g: X \to Y$ induce the following homomorphisms

$$f_*, g_* : \pi_1(X, x_0) \to \pi_1(Y, y_0).$$

Assume X, Y to be pathwise connected and prove that $f, g: X \to Y$ are 1-homotopic iff there exists an element $\xi \in \pi_1(Y, y_0)$ such that

$$g_*(\alpha) = \xi^{-1} \cdot f_*(\alpha) \cdot \xi, \quad \alpha \in \pi_1(X, x_0).$$

Assume Y to be pathwise connected and n-simple with $\pi_n = \pi_n(Y, y_0)$. Define for a given pair of maps $f, g: X \to Y$ the *n-dimensional obstruction set* $O^n(f, g)$ in the singular cohomology group $H^n(X; \pi_n)$ in the obvious way, and prove that $f, g: X \to Y$ are n-homotopic iff $O^n(f, g)$ contains the zero element of $H^n(X;\pi_n)$. Deduce further results on the homotopy problem and the classification problem.

E. The obstruction theory of deformation

Throughout this exercise, let (Y, B) be a given pair and (K, L) a given finite cellular pair. A map $f:(K, L) \to (Y, B)$ is said to be *deformable into* B if there exists a homotopy $f_t:(K, L) \to (Y, B)$, $(0 \leqslant t \leqslant 1)$, such that $f_0 = f$ and $f_1(K) \subset B$. By the aid of the AHEP, one can easily prove that , in this definition, we may require that $f_t(x) = f(x)$ for every $x \in L$ and $t \in I$.

In order to apply the obstruction method, let us define the notion of n-deformability. Let $\bar{K}^n = K^n \cup L$. A map $f:(K, L) \to (Y, B)$ is said to be *n-deformable into* B if the partial map $f \mid (\bar{K}^n, L)$ is deformable into B. A map f is said to *n-normal* if $f(\bar{K}^{n-1}) \subset B$. Hence f is $(n - 1)$-deformable into B iff it is homotopic (relative to L) to an n-normal map.

Hereafter, let us assume that both Y and B are pathwise connected and pick $y_0 \in B$. Then it is obvious that every map $f:(K, L) \to (Y, B)$ has to be 0-deformable into B. Prove that every map $f:(K, L) \to (Y, B)$ is 1-deformable into B provided that $\pi_1(Y, B, y_0) = 0$. The latter is equivalent to the condition that the induced homomorphism

$$i_* : \pi_1(B, y_0) \to \pi_1(Y, y_0)$$

is an epimorphism.

Next, let $n > 1$ and assume that (Y, B) is n-simple in the sense of (IV; Ex. E). Then the n-th relative homotopy group $\pi_n = \pi_n(Y, B)$ is abelian and may be used as coefficient group.

For each n-normal map $f:(K, L) \to (Y, B)$, define an n-cochain $c^n(f)$ of K with coefficients in π_n in the obvious way. Prove the following assertions, [Hu 9; p. 194]:

1. $c^n(f)$ is a cocycle of K modulo L.

2. $c^n(f) = 0$ iff there exists a homotopy $f_t : K \to Y$, $(0 \leqslant t \leqslant 1)$, such that $f_0 = f$, $f_1(\bar{K}^n) \subset B$, and $f_t(x) = f(x)$ for every $x \in \bar{K}^{n-1}$ and $t \in I$.

3. $c^n(f)$ is a coboundary of K modulo L iff there exists a homotopy $f_t : K \to Y$, $(0 \leqslant t \leqslant 1)$, such that $f_0 = f$, $f_1(\bar{K}^n) \subset B$, and $f_t(x) = f(x)$ for every $x \in \bar{K}^{n-2}$ and $t \in I$.

Next, define the *n-dimensional obstruction set* $O^n(f)$ in the n-dimensional cohomology group $H^n(K, L;\pi_n)$ in the obvious way and prove the following assertions:

4. The map $f:(K, L) \to (Y, B)$ is $(n - 1)$-deformable into B iff $O^n(f)$ is non-empty.

5. The map $f:(K, L) \to (Y, B)$ is n-deformable into B iff $O^n(f)$ contains the zero element of $H^n(K, L;\pi_n)$.

6. Every map $f:(K, L) \to (Y, B)$ is deformable into B if $\pi_1(Y, B, y_0) = 0$ and if, for each $n > 1$, (Y, B) is n-simple and $H^n(K, L;\pi_n) = 0$, where $\pi_n = \pi_n(Y, B)$.

7. Assume that K and L are pathwise connected and $x_0 \in L$. Then, L is a deformation retract of K iff $\pi_n(K, L; x_0) = 0$ for every $n \geqslant 1$.

Develop this obstruction theory of deformation as far as possible and then generalize this theory as in Exs. A, C and D. In particular, deduce a proof of the following assertion, [Hu 9; p. 216].

8. If Y and B are ANR's and if B is closed, then the following statements are equivalent:

(*i*) B is a deformation retract of Y.

(*ii*) There exists a homotopy $h_t : (Y, B) \to (Y, B)$, $(0 \leqslant t \leqslant 1)$, such that h_0 is the identity map and $h_1(Y) \subset B$.

(*iii*) $\pi_n(Y, B, y_0) = 0$ for every $n \geqslant 1$.

F. Maps into a space of homotopy type (π, n)

Let Y be a space of homotopy type (π, n) as defined in (V; Ex. F) with a given point $y_0 \in Y$, and let K be a finite cellular complex with a given subcomplex L and a given vertex $v_0 \in L$. If $n = 1$, we assume that both K and L are connected. The extension, homotopy and classification problems are completely solved for this special case as follows.

1. *The case* $n = 1$. Let $f : (L, v_0) \to (Y, y_0)$ be a given map. Then f and the inclusion map $i : L \subset K$ induce the homomorphisms

$$f_* : \pi_1(L, v_0) \to \pi_1(Y, y_0), \quad i_* : \pi_1(L, v_0) \to \pi_1(K, v_0).$$

Prove that f is extensible over K iff there exists a homomorphism $h : \pi_1(K, v_0) \to \pi_1(Y, y_0)$ such that $f_* = hi_*$ and that, for any such homomorphism h, there exists an extension $g : K \to Y$ of f such that $g_* = h$.

For the homotopy problem, let $f, g : K \to Y$ be two given maps. Without loss of generality, we may assume that $f(v_0) = y_0 = g(v_0)$. Then, we have the induced homomorphisms

$$f_*, g_* : \pi_1(K, v_0) \to \pi_1(Y, y_0).$$

Prove that f and g are homotopic iff f_* and g_* are *equivalent*, that is to say, there exists an element ξ in $\pi_1(Y, y_0)$ such that

$$g_*(\alpha) = \xi^{-1} \cdot f_*(\alpha) \cdot \xi, \quad \alpha \in \pi_1(K, v_0).$$

For the classification problem, let us consider the set H of all homomorphisms of $\pi_1(K, v_1)$ into $\pi_1(Y, y_1)$. The equivalence relation defined above divides H into disjoint equivalence classes. Prove that the homotopy classes of the maps $K \to Y$ are in a natural one-to-one correspondence with the equivalence classes of the homomorphisms H.

2. *The case* $n > 1$. Then Y is n-simple and we may write $\pi = \pi_n(Y)$. The following assertions are special cases of (14.3), (15.4) and (16.4).

A map $f : L \to Y$ is extensible over K iff its characteristic element $\varkappa^n(f)$ is extensible over K.

Two maps $f, g : K \to Y$ are homotopic iff $\varkappa^n(f) = \varkappa^n(g)$.

The homotopy classes of the maps $K \to Y$ are in a one-to-one correspondence with the elements of the cohomology group $H^n(K; \pi)$ determined by the assignment $f \to \varkappa^n(f)$.

3. Generalize the preceding results as in Ex. A and Ex. C.

G. Homology groups of (π, n)

Let K and L be two semi-simplicial complexes of the homotopy type (π, n) and (τ, n) respectively. Then, according to the previous exercise, the classification problem of the maps $K \to L$ is solved as follows: the homotopy classes of the maps $K \to L$ are in a natural one-to-one correspondence with the equivalence classes of the homomorphisms $\pi \to \tau$. If $n > 1$, then π and τ are abelian and therefore these homotopy classes are in a natural one-to-one correspondence with the homomorphisms $\pi \to \tau$ themselves. An important direct consequence of this result is that K and L are homotopically equivalent iff π and τ are isomorphic.

It follows that, for any space X of the homotopy type (π, n) and any coefficient group G, the singular homology group $H_m(X; G)$ and the singular cohomology group $H^m(X; G)$ depend only on the integers m, n and the groups π, G. They will be called the *m-th homology group* and the *m-th cohomology group* of the pair (π, n) with coefficients in G, denoted by $H_m(\pi, n; G)$ and $H^m(\pi, n; G)$ respectively. The integer n will be simply deleted from the notation if $n = 1$. Similarly, we shall omit the group G if it is the group Z of all integers. In particular, if $n = 1$ and $G = Z$, these are denoted by $H_m(\pi)$ and $H^m(\pi)$ and called the *m-th homology group* and the *m-th cohomology group* of the group π.

Finally, let $n = 1$ and let G be an abelian group on which π acts as left operators. Choose a space X of the homotopy $(\pi, 1)$. Since the fundamental group of X is π which acts on G, the singular groups $H_m(X; G)$, $H^m(X; G)$ with local coefficients in G are defined. Prove that these groups do not depend on the choice of the space X and hence may be denoted by $H_m(\pi; G)$ and $H^m(\pi; G)$.

H. The complex $K(\pi)$ of a group π

Let π be a given abstract group written multiplicatively. Define a semi-simplicial complex $K(\pi)$ as follows. [Eilenberg and MacLane 1].

1. *The homogeneous definition.* For each $n \geqslant 0$, consider the set Φ^n of all ordered sets $(x_0, \cdots, x_n)$ of $n + 1$ elements of π called the *n-sets* of π. Two n-sets $(x_0, \cdots, x_n)$ and $(y_0, \cdots, y_n)$ of π are said to be *equivalent* if there exists an element x of π such that $y_i = xx_i$ for each $i = 0, \cdots, n$. The n-sets Φ^n of π are thus divided into disjoint equivalence classes, called the *n-cells* of π. If $n > 0$, the i-th face of the n-cell $\sigma = [x_0, \cdots, x_n]$ of π, $0 \leqslant i \leqslant n$, is defined to be the following $(n - 1)$-cell

$$\sigma^{(i)} = [x_0, \cdots, x_{i-1}, x_{i+1}, \cdots, x_n].$$

Verify the condition (SSC) of (IV; Ex. H). Hence the n-cells of π, $(n = 0, 1, \cdots)$, constitute a semi-simplicial complex $K(\pi)$, called *the complex* of π.

2. *The non-homogeneous definition.* For each $n \geqslant 0$, consider the set of all n-tuples $(x_1, \cdots, x_n)$ of elements $x_i \in \pi$. If $n > 0$, the i-th face of the n-tuple $\sigma = (x_1, \cdots, x_n)$, $0 \leqslant i \leqslant n$, is defined to be the $(n - 1)$-tuple $\sigma^{(i)}$ as follows:

$$\sigma^{(0)} = (x_2, \cdots, x_n), \quad \sigma^{(n)} = (x_1, \cdots, x_{n-1}),$$

$$\sigma^{(i)} = (x_1, \cdots, x_{i-1}, x_i x_{i+1}, x_{i+2}, \cdots, x_n), \quad (0 < i < n).$$

Verify the condition (SSC) and hence obtain a semi-simplicial complex. Prove that this semi-simplicial complex may be identified with $K(\pi)$ by means of the correspondence

$$[x_0, \cdots, x_n] \longleftrightarrow (x_0^{-1} x_1, x_1^{-1} x_2, \cdots, x_{n-1}^{-1} x_n).$$

3. *The matric definition.* For each $n \geqslant 0$, consider the set of all $(n + 1) \times (n + 1)$ matrices $\sigma = \| d_{ij} \|$, $(i, j = 0, \cdots, n)$, with $d_{ij} \in \pi$ satisfying the condition

$$d_{ij} d_{jk} = d_{ik}, \quad (i, j, k = 0, \cdots, n).$$

If $n > 0$, the i-th face of σ, $0 \leqslant i \leqslant n$, is defined to be the matrix $\sigma^{(i)}$ obtained from σ by erasing the i-th row and the i-th column. Verify (SSC). Hence, these matrices σ constitute a semi-simplicial complex. Prove that this semi-simplicial complex may be identified with $K(\pi)$ by means of the correspondence

$$[x_0, \cdots, x_n] \longleftrightarrow \| d_{ij} \|, \quad d_{ij} = z_i^{-1} x_j.$$

4. Prove that $K(\pi)$ is a space of the homotopy type $(\pi, 1)$. See (V; Ex. H).

5. Let π and τ be two groups and $h : \pi \to \tau$ a homomorphism. Prove that the assignment $[x_0, \cdots, x_n] \to [h(x_0), \cdots, h(x_n)]$ defines a cellular map $f_h : K(\pi) \to K(\tau)$. If two homomorphisms $h, k : \pi \to \tau$ are equivalent, prove that the cellular maps f_h, f_k are homotopic.

I. Contractible complexes on which π acts freely

Let K be a contractible (finite or infinite) simplicial complex with the weak topology, [E–S; p. 75]. Assume that a group π acts freely on K as a group of simplicial homeomorphisms. Let X denote the orbit space K/π with projection $p : K \to X$. See (I; §6). Then prove that K is the universal covering space over X relative to p. Hence X is a space of the homotopy type $(\pi, 1)$.

Next, prove the existence of such complexes by the following construction. Let π be a given group. Construct a simplicial complex K as follows: its vertices are the pairs (ξ, n), where ξ is an element of π and n is a positive integer; and $(\xi_0, n_0), \cdots, (\xi_q, n_q)$ are vertices of a simplex of K iff $n_0 < \cdots < n_q$. Give K the weak topology. If F is any finite subcomplex of K and (ξ, n) is a vertex of K such that $n > m$ for every vertex (η, m) of F, then the join of F with (ξ, n) is in K. By using this and the weak topology, prove that $\pi_q(K) = 0$ for every $q \geqslant 0$. Hence K is contractible. For each $\xi \in \pi$, define a simplicial map $\xi : K \to K$ by means of the vertex assignment $(\eta, n) \to$

$(\xi\eta, n)$. In this way, π acts on K as a group of left operators. If $\xi \neq 1$, prove that the homeomorphism ξ has no fixed point. Hence π acts freely on K.

J. The influence of the fundamental group

Let X be a pathwise connected space and $x_0 \in X$ a given point as the common basic point for all homotopy groups. Study the influence of the fundamental group $\pi = \pi_1(X)$ upon the structure of homology and co-homology groups of the space X as follows.

According to (V; 8.2), X can be imbedded in a space Y of the homotopy type $(\pi, 1)$ such that the inclusion map $i : X \subset Y$ induces an isomorphism $i_* : \pi_1(X) \approx \pi_1(Y)$.

Let $n > 1$ be a given integer such that $\pi_m(X) = 0$ whenever $1 < m < n$. Verify the following assertions:

1. By the exactness of the homotopy sequence, $\pi_m(Y, X) = 0$ for $m \leqslant n$ and $\partial_* : \pi_m(Y, X) \approx \pi_{m-1}(X)$ for $m > n$.

2. By the relative Hurewicz theorem (V; Ex. C), $H_m(Y, X) = 0$ for $m \leqslant n$ and the natural homomorphism $g : \pi_{n+1}(Y, X) \to H_{n+1}(Y, X)$ is an epimorphism.

3. By the exactness of the homology sequence, the induced homomorphism

$$i_\# : H_m(X) \to H_m(Y) \approx H_m(\pi)$$

is an isomorphism for $m < n$ and is an epimorphism for $m = n$.

4. In the following commutative ladder

$$\begin{array}{ccccccccc}
\cdots \longrightarrow & \pi_{n+1}(Y, X) & \xrightarrow{\partial_*} & \pi_n(X) & \xrightarrow{i_*} & \pi_n(Y) & \longrightarrow \cdots \\
& \downarrow g & & \downarrow h & & \downarrow k & \\
\cdots \longrightarrow & H_{n+1}(Y, X) & \xrightarrow{\partial_\#} & H_n(X) & \xrightarrow{i_\#} & H_n(Y) & \longrightarrow \cdots
\end{array}$$

where g, h, k are natural homomorphisms, the *spherical subgroup* $\Sigma_n(X) = Im(h)$ of $H_n(X)$ coincides with $Im(\partial_\#) = Ker(i_\#)$.

5. The induced homomorphism

$$H^m(X) \xleftarrow{i^\#} H^m(Y) \approx H^m(\pi)$$

is an isomorphism for $m < n$ and is a monomorphism for $m = n$.

6. The subgroup $\Lambda^n(X) = Im(i^\#)$ of $H^n(X)$ consists of all elements which annihilate $\Sigma_n(X)$.

Summarizing these assertions, we obtain the classical theorem: *If* $\pi_1(X) \approx \pi$ *and* $\pi_m(X) = 0$ *whenever* $1 < m < n$, *then*

$$H_m(X) \approx H_m(\pi), \quad H^m(X) \approx H^m(\pi), \quad m < n,$$
$$H_n(X)/\Sigma_n(X) \approx H_n(\pi), \quad \Lambda^n(X) \approx H^n(\pi).$$

K. Computation of $H_m(\pi; G)$ and $H^m(\pi; G)$

Let G be a given *π-group*; that is to say, G is an abelian group on which π acts as left operators. A π-group G is said to be *π-free* with $\{ g_a \}$ as a

π-basis if the elements ξg_a for all $\xi \in \pi$ and $g_a \in \{ g_a \}$ form a basis of G.

A homomorphism $f : A \to B$ of π-groups is called a *π-homomorphism* it it commutes with the operators. A π-group G is said to be *π-injective* if, for every π-group B and every π-subgroup A of B, every π-homomorphism $f : A \to G$ has a π-extension $g : B \to G$.

For any π-group G, we shall denote by $I_\pi(G)$ the subgroup of G which consists of the elements $g \in G$ such that $\xi g = g$ for all $\xi \in \pi$, and denote by $J_\pi(G)$ the quotient group of G over its subgroup $L_\pi(G)$ generated by the elements $\xi g - g$ for all $\xi \in \pi$ and $g \in G$.

Prove the following assertions:

1. $H_0(\pi; G) \approx J_\pi(G), \quad H^0(\pi; G) \approx I_\pi(G)$.

2. If G is π-free, then $H_m(\pi; G) = 0$ for each $m > 0$; if G is π-injective, then $H^m(\pi; G) = 0$ for each $m > 0$.

3. If $\pi = 1$, then $H_m(\pi; G) = 0 = H^m(\pi; G)$ for each $m \geqslant 0$; if π is a free (non-abelian) group, then $H_m(\pi; G) = 0 = H^m(\pi; G)$ for each $m \geqslant 2$.

4. If π is a free abelian group with r generators and if π operates simply on G, then

$$H_m(\pi; G) = 0 = H^m(\pi; G), \quad \text{for } m > r,$$

$$H_m(\pi; G) \approx G^{\binom{r}{m}} \approx H^m(\pi; G), \quad \text{for } m \leqslant r,$$

where G^j denotes the direct sum $G + \cdots + G$ of j terms of G.

Now let π be a cyclic group of finite order r and G a π-group. Let ξ be a generator of π and define a π-homomorphism $\Delta : G \to G$ by taking $\Delta(g) = g + \xi g + \cdots + \xi^{r-1} g$ for each $g \in G$. Denote the image of Δ by $M_\pi(G)$ and the kernel of Δ by $N_\pi(G)$. Then $M_\pi(G) \subset I_\pi(G)$ and $L_\pi(G) \subset N_\pi(G)$.

5. If π is a cyclic group of finite order r, then for each $p \geqslant 0$ we have

$$H_{2p+1}(\pi; G) \approx I_\pi(G)/M_\pi(G) \approx H^{2p+2}(\pi; G),$$

$$H_{2p+2}(\pi; G) \approx N_\pi(G)/L_\pi(G) \approx H^{2p+1}(\pi; G).$$

In particular,

$$H_{2p+1}(\pi) \approx \pi \approx H^{2p+2}(\pi), \quad H_{2p+2}(\pi) = 0 = H^{2p+1}(\pi).$$

6. Universal coefficient theorem. If π operates simply on G, then

$$H_m(\pi; G) \approx H_m(\pi) \otimes G + Tor\,(H_{m-1}(\pi), G),$$

$$H^m(\pi; G) \approx Hom\,(H_m(\pi), G) + Ext\,(H_{m-1}(\pi), G).$$

7. Künneth relations. For the direct product $\pi \times \tau$ of two groups π and τ, we have

$$H_m(\pi \times \tau) = \sum_{p+q=m} H_p(\pi) \otimes H_q(\tau) + \sum_{p+q=m-1} Tor\,(H_p(\pi), H_q(\tau)).$$

The assertions 4–7 allow for a complete computation of the groups $H_m(\pi; G)$ and $H^m(\pi; G)$ for a finitely generated abelian group π which operates simply on G.

L. The Eilenberg-MacLane complex $K(\pi, n)$

Let π be a given abelian group written additively and n a positive integer. Define a semi-simplicial complex $K(\pi, n)$ as follows.

An m-cell σ of $K(\pi, n)$ is an n-cocycle of the unit m-simplex Δ_m with coefficients in π. To define the i-th face $\sigma^{(i)}$ of σ, $0 \leqslant i \leqslant m$, consider the cellular map $e^i_m : \Delta_{m-1} \to \Delta_m$ as defined in [E–S; p. 185]. It induces a homomorphism

$$(e^i_m)^{\#} : Z^n(\Delta_m; \pi) \to Z^n(\Delta_{m-1}; \pi).$$

Define $\sigma^{(i)} = (e^i_m)^{\#}\sigma$. If $m > 1$ and $i < j$, verify the condition (SSC). Thus we obtain a semi-simplicial complex which is called the *Eilenberg-MacLane complex* $K(\pi, n)$, [Eilenberg-MacLane 1].

Prove that $K(\pi, 1)$ is essentially $K(\pi)$ and that $K(\pi, n)$ is a space of the homotopy type (π, n).

M. The influence of $\pi_n(X)$

Let $n \geqslant 2$ be a given integer and X an $(n - 1)$-connected space. Study the influence of $\pi = \pi_n(X)$ upon the structure of homology and cohomology groups of X as follows.

According to (V; 8.2), X can be imbedded in a space Y of the homotopy type (π, n) such that the inclusion map $i : X \subset Y$ induces an isomorphism $i_* : \pi_n(X) \approx \pi_n(Y)$.

Let $q > n$ be a given integer such that $\pi_m(X) = 0$ whenever $n < m < q$. Verify the following assertions as in Ex. J:

1. $\pi_m(Y, X) = 0$ for $m \leqslant q$ and $\partial_* : \pi_m(Y, X) \approx \pi_{m-1}(X)$ for $m > q$.

2. $H_m(Y, X) = 0$ for $m \leqslant q$ and the natural homomorphism $g : \pi_{q+1}(Y,X) \to H_{q+1}(Y, X)$ is an isomorphism.

3. The induced homomorphism

$$i_{\#} : H_m(X) \to H_m(Y) \approx H_m(\pi, n)$$

is an isomorphism for $m < q$ and is an epimorphism for $m = q$.

4. In the following commutative ladder

$$\begin{array}{ccccccccc}
\cdots \longrightarrow \pi_{q+1}(Y) & \xrightarrow{j_*} & \pi_{q+1}(Y, X) & \xrightarrow{\partial_*} & \pi_q(X) & \xrightarrow{i_*} & \pi_q(Y) \longrightarrow \cdots \\
\downarrow f & & \downarrow g & & \downarrow h & & \downarrow k \\
\cdots \longrightarrow H_{q+1}(Y) & \xrightarrow{j_{\#}} & H_{q+1}(Y, X) & \xrightarrow{\partial_{\#}} & H_q(X) & \xrightarrow{i_{\#}} & H_q(Y) \longrightarrow \cdots
\end{array}$$

where f, g, h, k are natural homomorphisms, the spherical subgroup $\Sigma_q(X) = Im(h)$ of $H_q(X)$ coincides with $Im(\partial_{\#}) = Ker(i_{\#})$.

5. The induced homomorphism $i^{\#} : H^m(Y) \to H^m(X)$ is an isomorphism for $m < q$ and is a monomorphism for $m = q$ with $\Lambda^q(X)$ as image.

Summarizing the assertions (1)–(5), we obtain

$$H_m(X) \approx H_m(\pi, n), H^m(X) \approx H^m(\pi, n), \quad m < q,$$

$$H_q(X)/\Sigma_q(X) \approx H_q(\pi, n), \quad \Lambda^q(X) \approx H^q(\pi, n).$$

6. The composed homomorphism

$$\partial_* g^{-1} j_\# : H_{q+1}(\pi, n) \to \pi_q(X),$$

considered as an element of $H^{q+1}(\pi, n; \pi_q(X))$, does not depend on the choice of Y and is called *the invariant* $k_n^{q+1}(X)$ of the space X. The following conditions are equivalent:

(*i*) $k_n^{q+1}(X) = 0$.

(*ii*) $h : \pi_q(X) \to H_q(X)$ is a monomorphism.

(*iii*) $i_\# : H_{q+1}(X) \to H_{q+1}(\pi, n)$ is an epimorphism.

7. If $q = n + 1$, then $h : \pi_q(X) \to H_q(X)$ is an epimorphism by (V; Ex. C). This implies that $H_{n+1}(\pi, n) = 0$. Furthermore, $k_n^{n+2}(X) = 0$ iff h is an isomorphism.

CHAPTER VII

COHOMOTOPY GROUPS

1. Introduction

Let (X, A) and (Y, B) be any two pairs each consisting of a space and a subspace. As in (I; § 8), we denote by the symbol

$$\pi(X, A; Y, B)$$

the set of all homotopy classes of the maps of (X, A) into (Y, B) relative to $\{A, B\}$. As described in (I; § 8), $\pi(X, A; Y, B)$ is a functor contravariant in (X, A) and covariant in (Y, B).

If X is the n-sphere S^n, A consists of a single point s_0 of S^n and B consists of a single point y_0 of Y, then $\pi(X, A; Y, B)$ is the underlying set of the n-th homotopy group $\pi_n(Y, y_0)$ of Chapter IV. As shown in the preceding two chapters, the group operation in $\pi_n(Y, y_0)$ is very helpful in studying the extension and classification problems especially for the maps of S^n into Y.

On the other hand, if Y is the m-sphere S^m and B consists of a single point s_0 of S^m, then $\pi(X, A; Y, B)$ will be simply denoted by $\pi^m(X, A)$ and called the *m-th cohomotopy set* of (X, A). Under suitable conditions on (X, A), we shall see in § 5 that $\pi^m(X, A)$ forms an abelian group which will be called the *m-th cohomotopy group* of (X, A). This group operation was first sketched by Borsuk [3] and later studied in detail by Spanier [1]. It is useful in studying the extension and classification problems for the maps of X into S^m.

Between the homotopy groups and the cohomotopy groups, there is an informal duality given in § 12 by means of the usual composition operation. Precisely, for every pair (m, n) such that $\pi_n(X, A, x_0)$ and $\pi^m(X, A)$ are abelian groups, the composition operation yields a homomorphism

$$\pi_n(X, A, x_0) \otimes \pi^m(X, A) \to \pi_n(S^m, s_0).$$

In particular, if $m = n$, then the right side can be replaced by the additive group Z of integers.

2. The cohomotopy set $\pi^m(X, A)$

Let S^m denote the unit m-sphere in the $(m + 1)$-dimensional euclidean space and let s_0 denote the point $(1, 0, \cdots, 0)$. Then, as described in the introduction, the *m-th cohomotopy set* $\pi^m(X, A)$ of a pair (X, A) is defined to be the set of all homotopy classes of the maps of (X, A) into (S^m, s_0) relative to A. It is a contravariant functor in (X, A). If A is empty, then it will be simply denoted by $\pi^m(X)$ and called the *m-th cohomotopy set* of the space X.

The set $\pi^m(X, A)$ has an exceptional element, namely, the homotopy class of the constant map $0 : X \to s_0$. This exceptional element will be denoted by the symbol 0 and called the *zero element* of $\pi^m(X, A)$.

Since $\{ s_0 \}$ is a closed set of S^m, one can easily see that

$$\pi^m(X, \bar{A}) = \pi^m(X, A),$$

where $\bar{A}$ denotes the closure of A in X. Thus, we may assume that A is a closed subspace of X; however, we will not so assume unless explicitly mentioned.

Finally, $\pi^0(X, A)$ can be considered as the set of all open and closed subspaces of X not meeting A. If so considered, then the zero element 0 of $\pi^0(X, A)$ is the empty subspace of X.

3. The induced transformations

Since $\pi^m(X, A)$ is a contravariant functor in (X, A), every map $f : (X, A) \to (Y, B)$ induces a transformation

$$f^* : \pi^m(Y, B) \to \pi^m(X, A),$$

called the *induced transformation* of f on the m-dimensional cohomotopy sets. For any element α in $\pi^m(Y, B)$, choose a representative map $\phi : (Y, B) \to (S^m, s_0)$, then $f^*(\alpha)$ is represented by the composed map $\phi f : (X, A) \to (S^m, s_0)$. The zero element of $\pi^m(Y, B)$ is evidently mapped into that of $\pi^m(X, A)$ by f^*; in other words, f^* preserves the zero element 0. Furthermore, f^* depends only on the homotopy class of f relative to $\{ A, B \}$.

In the remainder of this section, we shall prove some important properties of the induced transformations.

Let us consider a pair (X, A), where the subspace A of X is non-empty. Identifying A to a point q_A, we obtain a space X_A. Let

$$f : (X, A) \to (X_A, q_A)$$

denote the natural projection. Then f is a map which sends $X \setminus A$ homeomorphically onto $X_A \setminus q_A$.

Lemma 3.1. *The induced transformation f^* carries $\pi^m(X_A, q_A)$ onto $\pi^m(X, A)$ in a one-to-one fashion.*

Proof. Let $\alpha \in \pi^m(X, A)$ be an arbitrary element represented by a map $\phi : (X, A) \to (S^m, s_0)$. We may define a map $\psi : (X_A, q_A) \to (S^m, s_0)$ by

$$\psi(y) = \begin{cases} \phi(f^{-1}(y)), & \text{if } y \in X_A \setminus q_A, \\ s_0, & \text{if } y = q_A. \end{cases}$$

The continuity of ψ follows from that of ϕ by (I; § 6). The map ψ represents an element β of $\pi^m(X_A, q_A)$. Since $\psi f = \phi$, it follows that $f^*(\beta) = \alpha$. Hence f^* is onto.

Next, let $\xi, \eta : (X_A, q_A) \to (S^m, s_0)$ be two maps such that ξf and ηf are homotopic relative to A. Then there exists a map

$$F : (X \times I, A \times I) \to (S^m, s_0)$$

such that $F(x, 0) = \xi f(x)$ and $F(x, 1) = \eta f(x)$ for every $x \in X$. Define $G : (X_A \times I, q_A \times I) \to (S^m, s_0)$ by taking

$$G(y, t) = \begin{cases} F(f^{-1}(y), t), & \text{if } y \in X_A \setminus q_A, \\ s_0, & \text{if } y = q_A. \end{cases}$$

Then G is continuous and gives a homotopy between ξ and η relative to q_A. This proves that f^* is one-to-one. ▮

The significance of (3.1) is that, in the computation of the cohomotopy sets of a given pair (X, A), we may assume that A consists of a single point a if A is non-empty. For the case $m > 1$, since S^m is simply connected, one can prove without difficulty that the inclusion map $i : X \subset (X, a)$ induces a one-to-one transformation i^* of $\pi^m(X, a)$ onto $\pi^m(X)$ provided that a certain homotopy extension properties are satisfied. In particular, these homotopy extension properties are satisfied if X is a paracompact Hausdorff space and a is any given point of X.

Theorem 3.2. (The map excision theorem). *If $f : (X, A) \to (Y, B)$ is a relative homeomorphism, that is to say, f maps $X \setminus A$ homeomorphically onto $Y \setminus B$, and if X is a compact space, A is non-empty, Y is a regular Hausdorff space and B is closed, then the induced transformation f^* sends $\pi^m(Y, B)$ onto $\pi^m(X, A)$ in a one-to-one fashion.*

Proof. Since A is non-empty, so is B. It follows that we may identify A and B to single points q_A and q_B respectively. Let

$$\xi : (X, A) \to (X_A, q_A), \quad \eta : (Y, B) \to (Y_B, q_B)$$

denote the natural projections of identification. One can easily verify that f induces a one-to-one map

$$g : (X_A, q_A) \to (Y_B, q_B)$$

of (X_A, q_A) onto (Y_B, q_B) such that $\eta f = g\xi$. Since X is compact, so is X_A. Since Y is a regular Hausdorff space and B is closed, it follows that Y_B is a Hausdorff space. Hence, g is a homeomorphism.

By a property of contravariant functors, we have $f^*\eta^* = \xi^*g^*$. Since g is a homeomorphism, it is obvious that g^* is a one-to-one correspondence. On the other hand, ξ^* and η^* are one-to-one correspondences according to (3.1). Therefore, $f^* = \xi^*g^*\eta^{*-1}$ sends $\pi^m(Y, B)$ onto $\pi^m(X, A)$ in a one-to-one fashion. ▮

The auxiliary conditions in (3.2) are essential. In fact, we have the following counter examples.

First, let Y denote the unit m-cell of the euclidean m-space and B its boundary $(m - 1)$-sphere. Let $y_0, y_1 \in B$ where $y_0 \neq y_1$. Let $X = Y \setminus y_0$

and $A = B \setminus y_0$. Then clearly there is a deformation retraction of (X, A) into the pair (y_1, y_1). Therefore $\pi^m(X, A) = 0$ although $\pi^m(Y, B) \approx Z$. This shows that the induced transformation $i^* : \pi^m(Y, B) \to \pi^m(X, A)$ of the inclusion map $i : (X, A) \subset (Y, B)$, which is a relative homeomorphism, cannot be one-to-one. In this example, X fails to be compact.

Next, let Y denote the unit $(m + 1)$-cell of the euclidean $(m + 1)$-space and X the boundary m-sphere. Let A consist of a single point of X and let $B = (Y \setminus X) \cup A$. Then $\pi^m(X, A) \approx Z$ while $\pi^m(Y, B) = 0$. This shows that the induced transformation $i^* : \pi^m(Y, B) \to \pi^m(X, A)$ of the inclusion map $i : (X, A) \subset (Y, B)$, which is a relative homeomorphism, cannot be onto. In this example, B fails to be closed.

Corollary 3.3. (The excision theorem). *Let (X, A) be a pair where X is a compact Hausdorff space and A is a closed subspace of X. If V is any open set of X contained in A, then the induced transformation e^* of the inclusion map*

$$e : (X \setminus V, A \setminus V) \subset (X, A)$$

carries $\pi^m(X, A)$ onto $\pi^m(X \setminus V, A \setminus V)$ in a one-to-one fashion.

Proof. Since e is a relative homeomorphism, (3.3) follows from (3.2) if $A \setminus V$ is non-empty. If $A \setminus V$ is empty, then $A = V$ and hence A is both open and closed in X. Therefore, every map $\phi : X \setminus A \to S^m$ has a unique extension $\psi : (X, A) \to (S^m, s_0)$. This proves that e^* is a one-to-one correspondence. ▮

Remark. The condition in (3.3) that X is a compact Hausdorff space can be removed if we assume that the closure of V is contained in the interior of A. The verification is left to the reader.

4. The coboundary operator

In the present section, we are going to construct a coboundary operator δ for cohomotopy sets which will be analogous to the coboundary operator for cohomology groups. Unfortunately, this coboundary operator δ is not defined for a completely arbitrary pair (X, A). Since, in the construction of δ, one has to use some form of Tietze's extension theorem, it is necessary to assume a normality condition on the pair (X, A).

By a *binormal pair* (X, A), we mean a binormal space X, (I; § 9), and a closed subspace A of X. Then, by definition, both X and $X \times I$ are normal spaces. In particular, (X, A) is a binormal pair if X is a semi-simplicial complex and A is a subcomplex of X, or more generally, if X is a paracompact Hausdorff space and A is a closed subspace of X. Throughout the present section, we assume that (X, A) is an arbitrarily given binormal pair.

Let S^{m+1} denote the $(m + 1)$-sphere obtained by joining S^m to two points, the north pole and the south pole. Denote by E_+^{m+1} and E_-^{m+1} the north and the south hemispheres respectively. Then we have

$$S^{m+1} = E_+^{m+1} \cup E_-^{m+1}, \quad s_0 \in S^m = E_+^{m+1} \cap E_-^{m+1}.$$

Consider the following diagram

$$\begin{array}{ccc} \pi(A: S^m) & & \pi(X, A; S^{m+1}, s_0) \\ \Big\uparrow \alpha & & \Big\downarrow \gamma \\ \pi(X, A; E_+^{m+1}, S^m) & \xrightarrow{\beta} & \pi(X, A; S^{m+1}, E_-^{m+1}) \end{array}$$

where the transformations are natural, namely, α is defined by restriction while β and γ are induced by inclusions.

Since E_+^{m+1}, E_-^{m+1} are contractible to the point s_0 and (X, A) is a binormal pair, it is straightforward to verify that both α and γ are one-to-one correspondences. Thus, we obtain a transformation

$$\delta = \gamma^{-1}\beta\alpha^{-1} : \pi^m(A) \to \pi^{m+1}(X, A),$$

which is called the *coboundary operator*.

Geometrically, the coboundary operator δ can be determined as follows. Let $e \in \pi^m(A)$. Choose a map $\phi : A \to S^m$ which represents e. Since E_+^{m+1} is contractible, ϕ has an extension

$$\phi\# : (X, A) \to (E_+^{m+1}, S^m).$$

Then, $\phi\#$ represents $\alpha^{-1}(e)$. Let

$$\xi_t : (E_+^{m+1}, S^m) \to (S^{m+1}, E_-^{m+1}), \quad 0 \leqslant t \leqslant 1,$$

be a homotopy such that ξ_0 is the inclusion map and ξ_1 maps S^m into s_0 and $E_+^{m+1} \setminus S^m$ onto $S^{m+1} \setminus s_0$ homeomorphically. It is obvious that such a homotopy ξ_t exists and that ξ_1 is essentially unique, i.e., unique up to a homotopy relative to S^m. Then the map $\xi_1\phi\# : (X, A) \to (S^{m+1}, s_0)$ represents the element $\delta(e) = \gamma^{-1}\beta\alpha^{-1}(e)$.

Obviously, δ sends the zero element of $\pi^m(A)$ into that of $\pi^{m+1}(X, A)$; in other words, δ preserves the zero element.

Now let (X, A) and (Y, B) be two binormal pairs and $f : (X, A) \to (Y, B)$ a given map. Let $g : A \to B$ denote the map defined by f. Since $\delta = \gamma^{-1}\beta\alpha^{-1}$ and since α, β, γ obviously commute with the induced transformations, we obtain the following

Proposition 4.1. *The commutativity relation $f^*\delta = \delta g^*$ holds in the rectangle:*

$$\begin{array}{ccc} \pi^m(A) & \xrightarrow{\delta} & \pi^{m+1}(X, A) \\ \Big\uparrow g^* & & \Big\uparrow f^* \\ \pi^m(B) & \xrightarrow{\delta} & \pi^{m+1}(Y, B). \end{array}$$

5. The group operation in $\pi^m(X, A)$

For convenience in using the obstruction method, we shall assume throughout this section that (X, A) is a *cellular pair*, that is to say, X is a finite cell complex and A is a subcomplex of X. Generalizations to more general pairs will be indicated in the exercises at the end of the chapter.

A cellular pair (X, A) is said to be *n-coconnected* if the cohomology group $H^q(X, A; G) = 0$ for every integer $q \geqslant n$ and every coefficient group G. In particular, (X, A) is n-coconnected if $X \setminus A$ is of dimension less than n. However, the latter condition is undesirable because dimension is not a homotopy invariant. By the universal coefficient theorems, it follows that (X, A) is n-coconnected iff the integral cohomology group $H^q(X, A) = 0$ for every $q \geqslant n$.

For any given pair (Y, y_0) consisting of a space Y and a point $y_0 \in Y$, consider the set $\pi(X, A; Y, y_0)$ of all homotopy classes of the maps $(X, A) \to (Y, y_0)$. Then every map $f: (Y, y_0) \to (Y', y'_0)$ induces a transformation

$$f_*: \pi(X, A; Y, y_0) \to \pi(X, A; Y', y'_0)$$

by means of composition. In particular, let us consider a triplet (Y, B, y_0) where Y and B are pathwise connected. Then the inclusion map $i: (B, y_0) \subset (Y, y_0)$ induces a transformation

$$i_*: \pi(X, A; B, y_0) \to \pi(X, A; Y, y_0).$$

We recall that (Y, B) is said to be *n-connected* if $\pi_q(Y, B) = 0$ for every $q \leqslant n$. Then the following lemma can be easily established by the obstruction method described in (VI; Ex. E).

Lemma 5.1. *If (X, A) is n-coconnected and (Y, B) is n-connected, then i_* sends $\pi(X, A; B, y_0)$ onto $\pi(X, A; Y, y_0)$ in a one-to-one fashion.*

Throughout the remainder of this section, let m be any given positive integer and assume that (X, A) is $(2m - 1)$-coconnected. Under this assumption, we shall define a group operation in $\pi^m(X, A)$. For this purpose, let us consider the triplet

$$Y = S^m \times S^m, \quad B = S^m \vee S^m, \quad y_0 = (s_0, s_0).$$

Then, by (IV; 3.4), (Y, B) is $(2m - 1)$-connected and hence i_* is a one-to-one correspondence according to (5.1).

Next, consider the map $j: (B, y_0) \to (S^m, s_0)$ defined by $j(s, s_0) = s = j(s_0, s)$ for every s in S^m. Then j induces a transformation

$$j_*: \pi(X, A; B, y_0) \to \pi^m(X, A).$$

Finally, we shall define a transformation

$$h_*: \pi^m(X, A) \times \pi^m(X, A) \to \pi(X, A; Y, y_0)$$

as follows. Let (α, β) be any pair of elements in $\pi^m(X, A)$. Choose representative maps $\phi, \psi: (X, A) \to (S^m, s_0)$ for α, β respectively. Then the map $\phi \times \psi: (X, A) \to (Y, y_0)$ represents an element of $\pi(X, A; Y, y_0)$ which obviously depends only on the ordered pair (α, β). We define $h_*(\alpha, \beta)$ to be this element.

Composing these three transformations, we obtain a transformation

$$k = j_* i_*^{-1} h_*: \pi^m(X, A) \times \pi^m(X, A) \to \pi^m(X, A).$$

For any pair (α, β) of elements in $\pi^m(X, A)$, the element $k(\alpha, \beta)$ of $\pi^m(X, A)$ will be called the *sum* of α, β and denoted by $\alpha + \beta$.

Theorem 5.2. *If (X, A) is a $(2m - 1)$-coconnected cellular pair, then $\pi^m(X, A)$ forms an abelian group with the addition $\alpha + \beta$ as its group operation and 0 as its group-theoretic neutral element. If $\alpha \in \pi^m(X, A)$ is represented by a map $\phi : (X, A) \to (S^m, s_0)$, then its negative $-\alpha$ is represented by the composed map $r\phi$, where $r : (S^m, s_0) \to (S^m, s_0)$ is an arbitrary map of degree -1.*

Proof. (a) *Commutativity.* Let us consider the homeomorphism $\xi : (Y, y_0) \to (Y, y_0)$ defined by $\xi(s, t) = (t, s)$ for every pair of points (s, t) in S^m. Since ξ maps B onto itself, it defines a homeomorphism $\eta : (B, y_0) \to (B, y_0)$. Obviously we have $i\eta = \xi i$ and $j\eta = j$. Therefore, the following diagram is commutative:

$$\begin{array}{ccccc}
\pi(X, A; Y, y_0) & \xleftarrow{i_*} & \pi(X, A; B, y_0) & \searrow^{j_*} & \\
\downarrow \xi_* & & \downarrow \eta_* & & \pi^m(X, A) \\
\pi(X, A; Y, y_0) & \xleftarrow{i_*} & \pi(X, A; B, y_0) & \nearrow_{j_*} &
\end{array}$$

Let α, β be any two elements of $\pi^m(X, A)$. It follows from the definition of h_* that

$$\xi_* h_*(\alpha, \beta) = h_*(\beta, \alpha).$$

Hence
$$\begin{aligned}\beta + \alpha &= j_* i_*^{-1} h_*(\beta, \alpha) = j_* i_*^{-1} \xi_* h_*(\alpha, \beta) \\ &= j_* \eta_* i_*^{-1} h_*(\alpha, \beta) = j_* i_*^{-1} h_*(\alpha, \beta) = \alpha + \beta.\end{aligned}$$

(b) *Associativity.* Consider the triplet (Z, C, z_0), where

$$Z = S^m \times S^m \times S^m, \quad C = S^m \vee S^m \vee S^m, \quad z_0 = (s_0, s_0, s_0).$$

Then, by (I; Ex. S), it is easy to verify that (Z, C) is $(2m - 1)$-connected and hence, by (5.1), the inclusion map $\varkappa : (C, z_0) \subset (Z, z_0)$ induces a one-to-one transformation

$$\varkappa_* : \pi(X, A; C, z_0) \to \pi(X, A; Z, z_0).$$

Let $\lambda : (C, z_0) \to (S^m, s_0)$ denote the map defined by

$$\lambda(s_1, s_2, s_3) = \begin{cases} s_1, & (\text{if } s_2 = s_0 = s_3), \\ s_2, & (\text{if } s_1 = s_0 = s_3), \\ s_3, & (\text{if } s_1 = s_0 = s_2). \end{cases}$$

Then λ induces a transformation

$$\lambda_* : \pi(X, A; C, z_0) \to \pi^m(X, A).$$

Now, let α, β, γ be any three elements of $\pi^m(X, A)$ and choose representative maps

$$\phi, \psi, \chi : (X, A) \to (S^m, s_0)$$

for α, β, γ respectively. Then the map $\phi \times \psi \times \chi$ represents an element

$\mu_*(\alpha, \beta, \gamma)$ of $\pi(X, A; Z, z_0)$ which depends only on the triple (α, β, γ). Thus we obtain an element $\lambda_* \varkappa_*^{-1} \mu_*(\alpha, \beta, \gamma)$ of $\pi^m(X, A)$. We are going to prove that

$$(\alpha + \beta) + \gamma = \lambda_* \varkappa_*^{-1} \mu_*(\alpha, \beta, \gamma) = \alpha + (\beta + \gamma).$$

For this purpose, let $\rho : (Z, z_0) \to (Y, y_0)$ denote the map defined by $\rho(s_1, s_2, s_3) = (s_1, s_2)$. Since ρ sends C onto B, it defines a map $\theta : (C, z_0) \to (B, y_0)$. Then the rectangle

$$\begin{array}{ccc} \pi(X, A; Z, z_0) & \xleftarrow{\varkappa_*} & \pi(X, A; C, z_0) \\ \downarrow \rho_* & & \downarrow \theta_* \\ \pi(X, A; Y, y_0) & \xleftarrow{i_*} & \pi(X, A; B, y_0) \end{array}$$

is commutative. Furthermore, by the definition of ρ, it is clear that $\rho_* \mu_*(\alpha, \beta, \gamma) = h_*(\alpha, \beta)$.

Choose a map $f : (X, A) \to (C, z_0)$ representing the element $\varkappa_*^{-1} \mu_*(\alpha, \beta, \gamma)$. Then λf represents the element $\lambda_* \varkappa_*^{-1} \mu_*(\alpha, \beta, \gamma)$ and $j\theta f$ represents the element $\alpha + \beta$. Let $\omega : (Z, z_0) \to (S^m, s_0)$ denote the map defined by $\omega(s_1, s_2, s_3) = s_3$. Then the map ωf clearly represents γ. Since f sends X into C, it is easy to see that $j\theta f \times \omega f$ carries (X, A) into (B, y_0) and that $j(j\theta f \times \omega f) = \lambda f$. This implies that $(\alpha + \beta) + \gamma = \lambda_* \varkappa_*^{-1} \mu_*(\alpha, \beta, \gamma)$.

Similarly, we can prove $\alpha + (\beta + \gamma) = \lambda_* \varkappa_*^{-1} \mu_*(\alpha, \beta, \gamma)$. This completes the proof of associativity.

(c) *Existence of neutral element.* Let α be any element of $\pi^m(X, A)$ and choose a representive map $\phi : (X, A) \to (S^m, s_0)$ of α. Let $\zeta : (X, A) \to (S^m, s_0)$ denote the constant map $\zeta(X) = s_0$. Then $\phi \times \zeta$ sends (X, A) into (B, y_0) and clearly $j(\phi \times \zeta) = \phi$. This implies that $\alpha + 0 = \alpha$ for every $\alpha \in \pi^m(X, A)$.

(d) *Existence of negatives.* Let $\alpha \in \pi^m(X, A)$ be represented by $\phi : (X, A) \to (S^m, s_0)$ and let $r : (S^m, s_0) \to (S^m, s_0)$ be any map of degree -1. Then $r\phi$ represents an element $\beta \in \pi^m(X, A)$ which depends only on α. We are going to prove that $\alpha + \beta = 0$.

Without loss of generality, we may assume that r is defined by

$$r(x_0, \cdots, x_{m-1}, x_m) = (x_0, \cdots, x_{m-1}, -x_m).$$

Let E_+^m and E_-^m denote the hemi-spheres of S^m defined by $x_m \geqslant 0$ and $x_m \leqslant 0$ respectively. Then r maps E_+^m onto E_-^m and E_-^m onto E_+^m.

Let $h_t : (S^m, s_0) \to (S^m, s_0)$, $0 \leqslant t \leqslant 1$, be a homotopy such that h_0 is the identity map and $h_1(E_-^m) = s_0$. Then $h_1\phi$ represents α and $h_1 r\phi$ represents β. Let $M_1 = \phi^{-1}(E_+^m)$ and $M_2 = \phi^{-1}(E_-^m)$, then we have

$$X = M_1 \cup M_2, \qquad A \subset M_1 \cap M_2, \qquad h_1\phi(M_2) = s_0 = h_1 r\phi(M_1).$$

Hence the map $h_1\phi \times h_1 r\phi$ sends X into B and the composed map $\psi = j(h_1\phi \times h_1 r\phi)$ represents the element $\alpha + \beta$. Furthermore, it is obvious that $\psi(x) = h_1\phi(x)$ if $x \in M_1$ and $\psi(x) = h_1 r\phi(x)$ if $x \in M_2$. Then, define a homotopy $\psi_t : (X, A) \to (S^m, s_0)$, $0 \leqslant t \leqslant 1$, by taking

$$\psi_t(x) = \begin{cases} h_t\phi(x), & (\text{if } x \in M_1), \\ h_t r\phi(x), & (\text{if } x \in M_2). \end{cases}$$

Then ψ_0 also represents $\alpha + \beta$. Since h_0 is the identity map, it follows that ψ_0 sends X into E_+^m and hence ψ_0 is homotopic (relative to A) with the constant map $e(X) = s_0$. This proves that $\alpha + \beta = 0$. ∎

For the special cases $m = 1$ or 3, S^m is a topological group with s_0 as neutral element. Then there is a natural multiplication in the cohomotopy set $\pi^m(X, A)$ of any pair (X, A) defined as follows. Let α, β be any two elements of $\pi^m(X, A)$ represented by $\phi, \psi : (X, A) \to (S^m, s_0)$ respectively. Let $\chi : (X, A) \to (S^m, s_0)$ denote the map defined by $\chi(x) = \phi(x)\psi(x)$. Then χ represents an element $[\chi]$ of $\pi^m(X, A)$ which obviously depends only on α and β. This element $[\chi]$ will be called the *product* of α and β, and will be denoted by $\alpha\beta$. Obviously, $\pi^m(X, A)$ forms a group with this multiplication as group operation.

Proposition 5.3. *If $m = 1$ or 3 and if (X, A) is a $(2m - 1)$-coconnected cellular pair, then $\alpha\beta = \alpha + \beta$ for any pair (α, β) of elements of $\pi^m(X, A)$.*

Proof. Let $\mu : (Y, y_0) \to (S^m, s_0)$ denote the map defined by $\mu(s, t) = st$ for any pair (s, t) of elements in the topological group S^m. By (5.1), there exists a homotopy $f_t : (X, A) \to (Y, y_0)$, $0 \leqslant t \leqslant 1$, such that $f_0 = \phi \times \psi$ and $f_1(X) \subset B$, where ϕ, ψ are representative maps of α, β respectively. Then μf_0 represents $\alpha\beta$ and jf_1 represents $\alpha + \beta$. Since $f_1(X) \subset B$, it follows that $\mu f_1 = jf_1$. Hence the homotopy μf_t implies that $\alpha\beta = \alpha + \beta$. ∎

In the remainder of the section, we shall establish that the induced transformations and coboundary operators are homomorphisms.

Proposition 5.4. *If (X, A) and (X', A') are two $(2m - 1)$-coconnected cellular pairs and if $f : (X, A) \to (X', A')$ is a map, then the induced transformation $f^* : \pi^m(X', A') \to \pi^m(X, A)$ is a homomorphism.*

Proof. Let α, β be any two elements of $\pi^m(X', A')$ and choose representative maps $\phi, \psi : (X', A') \to (S^m, s_0)$ for α, β respectively. By (5.1) there exists a homotopy $g_t : (X', A') \to (Y, y_0)$, $0 \leqslant t \leqslant 1$, such that $g_0 = \phi \times \psi$ and $g_1(X) \subset B$. Then jg_1 represents $\alpha + \beta$ and hence $jg_1 f$ represents $f^*(\alpha + \beta)$. On the other hand, since $\phi f, \psi f$ represent $f^*(\alpha), f^*(\beta)$ respectively and since $g_0 f = \phi f \times \psi f$, it follows that $jg_1 f$ also represents $f^*(\alpha) + f^*(\beta)$. Hence $f^*(\alpha + \beta) = f^*(\alpha) + f^*(\beta)$. ∎

Proposition 5.5. *If (X, A) is a $(2m + 1)$-coconnected cellular pair such that A is $(2m - 1)$-coconnected, then the coboundary operator $\delta : \pi^m(A) \to \pi^{m+1}(X, A)$ is a homomorphism.*

Proof. Let α, β be any two elements of $\pi^m(A)$ and choose representative maps $\phi, \psi : A \to S^m$ respectively for α, β. By (5.1), there exists a homotopy

$g_t : A \to S^m \times S^m$, $0 \leqslant t \leqslant 1$, such that $g_0 = \phi \times \psi$ and $g_1(A) \subset S^m \vee S^m$. Then $jg_1 : A \to S^m$ represents the element $\alpha + \beta$.

Since $E_+^{m+1} \vee E_+^{m+1}$ is contractible over itself to (s_0, s_0), g_1 has an extension

$$g_1\# : (X, A) \to (E_+^{m+1} \vee E_+^{m+1}, S^m \vee S^m).$$

Let $j\# : E_+^{m+1} \vee E_+^{m+1} \to E_+^{m+1}$ denote the map analogous to $j : S^m \vee S^m \to S^m$; then $j\#g_1\#$ is an extension of jg_1. Hence, $\xi_1 j\#g_1\#$ represents $\delta(\alpha+\beta)$.

Let $p_i : E_+^{m+1} \times E_+^{m+1} \to E_+^{m+1}$, $(i = 1,2)$, denote the projection defined by $p_i(y_1, y_2) = y_i$. Then the map $p_1 g_1\# : (X, A) \to (E_+^{m+1}, S^m)$ is an extension $p_1 g_1$ which obviously represents α. Similarly, $p_2 g_1\#$ is an extension of $p_2 g_1$ which represents β. Therefore, $\xi_1 p_1 g_1\#$ represents $\delta(\alpha)$ and $\xi_1 p_2 g_1\#$ represents $\delta(\beta)$. Let

$$\Psi : (E_+^{m+1} \vee E_+^{m+1}, S^m \vee S^m) \to (S^{m+1} \vee S^{m+1}, (s_0, s_0))$$

denote the map defined by $\Psi(y_1, y_2) = (\xi_1(y_1), \xi_1(y_2))$. Then we have

$$\xi_1 p_1 g_1\# \times \xi_1 p_2 g_1\# = \Psi g_1\#.$$

Let $J : S^{m+1} \vee S^{m+1} \to S^{m+1}$ denote the map analogous to $j : S^m \vee S^m \to S^m$. Then the map $J\Psi g_1\#$ represents $\delta(\alpha) + \delta(\beta)$.

Since clearly $J\Psi g_1\# = \xi_1 j\#g_1\#$, it follows that

$$\delta(\alpha + \beta) = \delta(\alpha) + \delta(\beta). \blacksquare$$

6. The cohomotopy sequence of a triple

By a *binormal triple* $(X; A, B)$, we mean a binormal space X and two closed subspaces A and B such that $A \supset B$. Throughout the present section, let (X, A, B) be a given binormal triple. Then (X, A), (X, B) and (A, B) are binormal pairs.

By § 3, the inclusion maps $i : (A, B) \subset (X, B)$ and $j : (X, B) \subset (X, A)$ induce transformations

$$i^* : \pi^m(X, B) \to \pi^m(A, B), \quad j^* : \pi^m(X, A) \to \pi^m(X, B)$$

defined for every $m \geqslant 0$. On the other hand, the inclusion map $k : A \subset (A, B)$ induces a transformation

$$k^* : \pi^m(A, B) \to \pi^m(A).$$

If we compose k^* with the coboundary operator

$$\delta : \pi^m(A) \to \pi^{m+1}(X, A),$$

we obtain a transformation

$$\delta^* : \pi^m(A, B) \to \pi^{m+1}(X, A)$$

which will be called *the coboundary operator for the triple* (X, A, B). Thus, $\delta^* = \delta k^*$.

Therefore, we obtain the following sequence of sets and transformations:

$$\pi^0(X,A) \xrightarrow{j^*} \pi^0(X,B) \xrightarrow{i^*} \pi^0(A,B) \xrightarrow{\delta^*} \pi^1(X,A) \xrightarrow{j^*} \pi^1(X,B) \xrightarrow{i^*} \cdots$$
$$\cdots \xrightarrow{\delta^*} \pi^m(X,A) \xrightarrow{j^*} \pi^m(X, B) \xrightarrow{i^*} \pi^m(A.B) \xrightarrow{\delta^*} \pi^{m+1}(X, A) \xrightarrow{j^*} \cdots$$

which will be called *the cohomotopy sequence of the triple* (X, A, B). In particular, if $B = \square$, it is called *the cohomotopy sequence of the pair* (X, A).

As usual, we define the *image* of the transformation $\delta^* : \pi^m(A, B) \to \pi^{m+1}(X, A)$ to be the set $Im(\delta^*) = \delta^*(\pi^m(A, B))$ and the *kernel* $Ker(\delta^*)$ of δ^* to be the inverse image $\delta^{*-1}(0)$ of the zero element 0 of $\pi^{m+1}(X, A)$. Analogously, one may define the images and the kernels of i^* and j^*. This being done, it makes sense to ask whether or not the cohomotopy sequence is *exact* in the usual sense, namely the image of each transformation should coincide with the kernel of the subsequent transformation. As usual, to prove its exactness, one has to prove the following six statements:

$$Im\,(j^*) \subset Ker\,(i^*), \tag{1}$$

$$Im\,(i^*) \subset Ker\,(\delta^*), \tag{2}$$

$$Im\,(\delta^*) \subset Ker\,(j^*), \tag{3}$$

$$Im\,(j^*) \supset Ker\,(i^*), \tag{4}$$

$$Im\,(i^*) \supset Ker\,(\delta^*), \tag{5}$$

$$Im\,(\delta^*) \supset Ker\,(j^*). \tag{6}$$

Theorem 6.1. *The statements* (1)–(4) *hold for every binormal triple* (X,A,B).

Proof. To prove (1), let $\alpha \in \pi^m(X, A)$ be represented by $\phi : (X, A) \to (S^m, s_0)$. Then, by definition, $i^*j^*(\alpha)$ is represented by the partial map $\phi \mid A$. Since $\phi(A) = s_0$, we get $i^*j^*(\alpha) = 0$. This proves (1).

To prove (2), let $\beta \in \pi^m(X, B)$ be represented by $\psi : (X, B) \to (S^m, s_0)$. Then the partial map $\phi = \psi \mid A$ represents $i^*(\beta)$. Define an extension $\phi^\# : (X, A) \to (E_+^{m+1}, S^m)$ of ϕ by taking $\phi^\#(x) = \psi(x)$ for each $x \in X$. Then, $\phi^\#(X) \subset S^m$ and hence $\xi_1\phi^\#(X) = s_0$. Since $\xi_1\phi^\#$ represents $\delta^*i^*(\beta)$, we deduce $\delta^*i^*(\beta) = 0$. This proves (2).

To prove (3), let $\gamma \in \pi^m(A, B)$ be represented by $\chi : (A, B) \to (S^m, s_0)$. Take an extension $\chi^\# : (X, A) \to (E_+^{m+1}, S^m)$ of χ. Then $\xi_1\chi^\#$ represents the element $\delta^*(\gamma)$ of $\pi^{m+1}(X, A)$. Since $\chi^\#(B) = \chi(B) = s_0$, we may define a map $F : (X \times 0) \cup (B \times I) \cup (X \times 1) \to E_+^{m+1}$ by

$$F(x, t) = \begin{cases} \chi^\#(x), & (x \in X, t = 0), \\ s_0, & (x \in B, t \in I), \\ s_0, & (x \in X, t = 1). \end{cases}$$

Since (X, B) is binormal and E_+^{m+1} is solid, F has an extension $F^\# : X \times I \to E_+^{m+1}$. Define a homotopy $f_t : (X, B) \to (S^{m+1}, s_0)$, $(0 \leqslant t \leqslant 1)$, by taking

$$f_t(x) = \xi_1 F^\#(x, t), \quad (x \in X, t \in I).$$

Then $f_0 = \xi_1\chi^\# j$ and $f_1(X) = s_0$. Hence $j^*\delta^*(\gamma) = 0$. This proves (3).

To prove (4), let α be an element of $\pi^m(X, B)$ such that $i^*(\alpha) = 0$. Pick a map $\phi : (X, B) \to (S^m, s_0)$ which represents α; then $\phi i = \phi \mid A$ represents $i^*(\alpha) = 0$. Hence there exists a partial homotopy $\psi_t : (A, B) \to (S^m, s_0)$ $(0 \leqslant t \leqslant 1)$, of ϕ such that $\psi_0 = \phi \mid A$ and $\psi_1(A) = s_0$. Since (X, A) is a binormal pair and S^m is a compact ANR, it follows from (I; Ex. N) that ψ_t has an extension $\phi_t : (X, B) \to (S^m, s_0)$, $(0 \leqslant t \leqslant 1)$, such that $\phi_0 = \phi$. Hence α is also represented by ϕ_1. Since $\phi_1(A) = \psi_1(A) = s_0$, ϕ_1 represents an element $\beta \in \pi^m(X, A)$ and $j^*(\beta) = \alpha$. This proves (4). ∎

There are examples which show that (5) and (6) may fail (see Ex. B at the end of the chapter); under additional hypotheses, (6) and (5) will be proved in § 8 and § 9 respectively.

7. An important lemma

Lemma 7.1. *Assume that A is a (finitely) triangulable space which is $2m$-coconnected and that X denotes the cone over A with vertex v. Then the coboundary operator δ carries $\pi^m(A)$ onto $\pi^{m+1}(X, A)$.*

Proof. First, let us prove the lemma under a further condition that either $\dim A \leqslant 2m - 1$ or $m = 1$.

For this purpose, let α be any element of $\pi^{m+1}(X, A)$, then α is represented by a map

$$f : (X, A) \to (S^{m+1}, s_0).$$

Give S^{m+1} a finite triangulation such that s_0 is a vertex, S^m is a subcomplex and the north pole u of S^{m+1} is an interior point of some $(m + 1)$-simplex which is not in the star of s_0. By the usual process of simplicial approximation, we may assume that f is a simplicial map on some finite triangulation of the pair (X, A) such that the cone K over the $(2m - 1)$-dimensional skeleton $B = A^{2m-1}$ of A is a subcomplex of X. Let

$$\phi : (K, B) \to (S^{m+1}, s_0)$$

denote the restriction of f on the pair (K, B). Since ϕ is a simplicial map and u is an interior point of an $(m + 1)$-simplex, the inverse image $C = \phi^{-1}(u)$ is of dimension less than m. It is obvious that B and C are disjoint.

Let $\theta : B \times I \to K$ denote the identification map which defines the cone K over B. Then θ carries $B \times 1$ on the vertex v and carries the remainder of $B \times I$ homeomorphically onto $K \setminus v$. Since C is compact, disjoint from B, and of dimension less than m, there exist a positive real number $r < 1$ and a closed subset D of B with $\dim D < m$ such that every point of $C \setminus v$ can be expressed in the form $\theta(d, t)$ with $d \in D$ and $r \leqslant t < 1$. Let E denote the set of all points $\theta(d, t)$ with $d \in D$ and $r \leqslant t \leqslant 1$. Then we have

$$C \subset E, \quad B \cap E = \square, \quad \dim E \leqslant m.$$

The proof depends on the construction of a few homotopies of ϕ sketched as follows. The details are left to the reader, [Spanier 1, p. 229].

Since $\dim E \leqslant m$, ϕ is homotopic to a map $\psi : (K, B) \to (S^{m+1}, s_0)$ relative to B such that $\psi^{-1}(u) = C$ and $s_0 \notin \psi(E)$.

Since $s_0 \notin \psi(E)$, there is an open set W of K such that $E \subset W$ and $s_0 \notin \psi(\overline{W})$. Pushing $\psi(E)$ to u along line-segments in the euclidean space $S^{m+1} \setminus s_0$ by the aid of a continuous real function $\rho : K \to I$ such that $\rho^{-1}(0) = E$ and $\rho^{-1}(1) = K \setminus W$, one can construct a map $\chi : (K, B) \to (S^{m+1}, s_0)$ which is homotopic with ψ relative to B and such that $\chi^{-1}(u) = E$.

By means of a continuous real function $\eta : K \to I$ such that $\eta^{-1}(0) = D$, one can construct a map $M : (K, B) \to (K, B)$ which sends E into the vertex v, carries $K \setminus E$ homeomorphically onto $K \setminus v$, and is homotopic with the identity map relative to B. Then $\varkappa = \chi M^{-1} : (K, B) \to (S^{m+1}, s_0)$ is a map homotopic with χ relative to B and such that $\varkappa^{-1}(u) = v$.

Since $\varkappa(v) = u$, there exists a positive real number $s < 1$ such that $\varkappa\theta(x, t)$ is in the north hemisphere E^{m+1}_+ for every $x \in B$ and $s \leqslant t \leqslant 1$. Let

$$P = \{ \theta(x, t) \mid x \in B \text{ and } s \leqslant t \leqslant 1 \},$$

$$Q = \{ \theta(x, t) \mid x \in B \text{ and } 0 \leqslant t \leqslant s \}.$$

Then $\varkappa(P) \subset E^{m+1}_+$ and $\varkappa(Q) \subset S^{m+1} \setminus s_0$. Pushing $\varkappa(Q)$ along geodesic arcs from u until Q is mapped into the south hemisphere E^{m+1}_-, one can easily see that $\varkappa$ is homotopic with a map

$$\lambda : (K, B) \to (S^{m+1}, s_0)$$

relative to B such that $\lambda(P) \subset E^{m+1}_+$ and $\lambda(Q) \subset E^{m+1}_-$.

Since $\phi \simeq \psi \simeq \chi \simeq \varkappa \simeq \lambda$ relative to B, it follows from the homotopy extension property that the representative map f of α can be so chosen that $f \mid K = \lambda$ and hence we may assume that

$$f(P) \subset E^{m+1}_+, \quad f(Q) \subset E^{m+1}_-, \quad f(P \cap Q) \subset S^m.$$

Now we are going to construct a map $g : A \to S^m$ as follows. First, define a map $\mu : B \to S^m$ by taking $\mu(x) = f\theta(x, s)$ for every $x \in B$. If $\dim A \leqslant 2m\text{-}1$, then $B = A$ and we set $g = \mu$. On the other hand, if $m = 1$, then μ determines an obstruction cocycle $c^2(\mu)$ of A. Since f is defined throughout X, it follows from (II; Ex. D) that $c^2(\mu) = 0$ and hence μ is 2-extensible over A. Then, by (II; 7.4), μ has an extension $g : A \to S^m$. This map g represents an element β of $\pi^m(A)$. We are going to prove that $\delta(\beta) = \alpha$.

Define a map $h : (K, B) \to (E^{m+1}_+, S^m)$ by setting $h\theta(x, t) = f\theta(x, s + t - st)$ for every $x \in B$ and $t \in I$. Since $h \mid B = g \mid B$, we may define an extension $k : K \cup A \to E^{m+1}_+$ by taking $k(x) = g(x)$ if $x \in A$ and $k(x) = h(x)$ if $x \in K$. By Tietze's extension theorem, k has an extension $g^\# : (X, A) \to (E^{m+1}_+, S^m)$. Composing with the relative homeomorphism ξ_1 of § 4, we obtain a map

$$f^\# = \xi_1 g^\# : (X, A) \to (S^{m+1}, s_0).$$

Since $g^\# \mid A = g$, $f^\#$ represents the element $\delta(\beta)$ according to § 4.

We have to prove that f and $f^\#$ are homotopic relative to A. Since $g^\# \mid K = h$, we can easily see that $f \mid K \simeq \xi_1 h = f^\# \mid K$ relative to B. Since $f(A) = s_0 = f^\#(A)$, this implies that $f \mid K \cup A \simeq f^\# \mid K \cup A$ relative to A. The cones over the simplexes of A constitute a triangulation of X. In this special triangulation of X, we have $K \cup A = X^{2m} \cup A$. Hence f and $f^\#$ are $2m$-homotopic relative to A. Since A is $2m$-coconnected and X is contractible to a point, it follows that (X, A) is $(2m+1)$-coconnected. Hence, by (VI; 11.2), f and $f^{\#}$ are homotopic relative to A.

Thus we have proved the lemma for the special case that either dim $A \leqslant 2m - 1$ or $m = 1$.

Next, let us prove the lemma with the condition that $m > 1$ and A is simply connected.

In this case, we shall first prove that A is dominated by its $(2m-1)$-dimensional skeleton B. Since $2m - 1 \geqslant 3$ and A is simply connected, so is B. Hence the pair (A, B) is n-simple for every $n \geqslant 2m$. Consider the inclusion map

$$e : (A, A^{2m-2}) \subset (A, B),$$

and apply the obstruction method described in (VI; Ex. E) for deforming e into B. Since

$$H^n(A, A^{2m-2}) = H^n(A) = 0$$

for every $n \geqslant 2m$, it follows that e is deformable into B with A^{2m-2} held fixed. In particular, there is a map $j : A \to B$ such that the composition ij with the inclusion map $i : B \subset A$ is homotopic to identity map on A. This proves that B dominates A.

Since X and K are the cones over A and B respectively, the maps i, j extend to maps

$$\hat{i} : (K, B) \to (X, A), \quad \hat{j} : (X, A) \to (K, B)$$

in the obvious way so that $\hat{i}\,\hat{j}$ is homotopic to the identity map on (X, A). This gives a commutative diagram

$$\begin{array}{ccccc}
\pi^m(A) & \xrightarrow{i^*} & \pi^m(B) & \xrightarrow{j^*} & \pi^m(A) \\
\downarrow \delta & & \downarrow \delta & & \downarrow \delta \\
\pi^{m+1}(X, A) & \xrightarrow{\hat{i}^*} & \pi^{m+1}(K, B) & \xrightarrow{\hat{j}^*} & \pi^{m+1}(X, A).
\end{array}$$

Given $\alpha \in \pi^{m+1}(X, A)$, then $\hat{i}^*(\alpha) \in \pi^{m+1}(K, B)$. Since $\dim B \leqslant 2m - 1$, we have prove that δ sends $\pi^m(B)$ onto $\pi^{m+1}(K, B)$. Hence there exists an element $\gamma \in \pi^m(B)$ such that $\delta(\gamma) = \hat{i}^*(\alpha)$. Let $\beta = j^*(\gamma) \in \pi^m(A)$. Then we have

$$\delta(\beta) = \delta j^*(\gamma) = \hat{j}^*\delta(\gamma) = \hat{j}^*\hat{i}^*(\alpha) = \alpha.$$

Thus we have proved the lemma also for the case that $m > 1$ and A is simply connected.

Finally, let us prove the lemma for the general case $m > 1$. Let $A' \subset X$ denote the union of A and the cone over its 1-skeleton. Then A' is simply connected and $2m$-coconnected. Consider the inclusion maps $k : A \subset A'$ and $\hat{k} : (X, A) \subset (X, A')$. We obtain the commutative rectangle

$$\begin{array}{ccc}
\pi^m(A') & \xrightarrow{k^*} & \pi^m(A) \\
\downarrow \delta & & \downarrow \delta \\
\pi^{m+1}(X, A') & \xrightarrow{\hat{k}^*} & \pi^{m+1}(X, A).
\end{array}$$

Since $\dim(A' \setminus A) \leq 2$ and $m + 1 > 2$, it follows that $\hat{k}^*$ is an epimorphism. Let X' denote the cone over A'. Then (X, A') is homotopically equivalent to (X', A'). Hence δ sends $\pi^m(A')$ onto $\pi^{m+1}(X, A')$ since A' is simply connected and $2m$-coconnected. Then it follows from the commutativity of the rectangle that δ carries $\pi^m(A)$ onto $\pi^{m+1}(X, A)$. ▮

8. The statement (6)

A triple (X, A, B) is said to be (finitely) *triangulable* if there exists a finite triangulation of X such that A and B are both subcomplexes.

Lemma 8.1. *If (X, A, B) is a triangulable triple such that (A, B) is $2m$-coconnected, then the statement* (6) *of* § 6 *holds. More explicitly in the sequence*

$$\pi^m(A, B) \xrightarrow{\delta^*} \pi^{m+1}(X, A) \xrightarrow{j^*} \pi^{m+1}(X, B)$$

the image of δ^* contains the kernel of j^*.

Proof. We shall be dealing with subsets of $X \times I$ and will use the following abridged notation. If Y is a subset of X, we define

$$Y_I = Y \times I, \quad Y_0 = Y \times 0, \quad Y_1 = Y \times 1.$$

Let $f: (X, A, B) \to (X_I, X_1 \cup A_0 \cup B_I, X_1 \cup B_I)$ denote the map defined by $f(x) = (x, 0)$ for every $x \in X$. Let

$$f_1 = f \mid (A, B), \quad f_2 = f \mid (X, A).$$

According to § 3 and § 4, commutativity holds in the following diagram:

$$\begin{array}{ccccc}
\pi^m(A, B) & \xrightarrow{\delta^*} & \pi^{m+1}(X, A) & \xrightarrow{j^*} & \pi^{m+1}(X, B) \\
\uparrow {\scriptstyle f_1^*} & & \uparrow {\scriptstyle f_2^*} & & \\
\pi^m(X_1 \cup A_0 \cup B_I, X_1 \cup B_I) & \xrightarrow{\delta_1^*} & \pi^{m+1}(X_I, X_1 \cup A_0 \cup B_I) & &
\end{array}$$

Since f_1 is essentially the excision map obtained by excising $(B_I \setminus B_0) \cup X_1$ from $X_1 \cup A_0 \cup B_I$, it follows from (3.3) that f_1^* is one-to-one and onto. By the definition of homotopy, one can easily see that

$$Im\,(f_2^*) = Ker\,(j^*).$$

By the commutativity relation $\delta^* f_1^* = f_2^* \delta_1^*$, it suffices to prove that the coboundary operator δ_1^* is onto.

Let $g: (X_1 \cup A_I, X_1 \cup A_0 \cup B_I) \subset (X_I, X_1 \cup A_0 \cup B_I)$ denote the inclusion map. Since $X_1 \cup A_I$ is a deformation retract of X_I according to (I; 10.1), it is easy to see that the induced transformation g^* is one-to-one and onto. By the commutativity of the triangle

$$\begin{array}{ccc}
\pi^m(X_1 \cup A_0 \cup B_I, X_1 \cup B_I) & \xrightarrow{\delta_1^*} & \pi^{m+1}(X_I, X_1 \cup A_0 \cup B_I) \\
{\scriptstyle \delta_2^*} \searrow & & \swarrow {\scriptstyle g^*} \\
& \pi^{m+1}(X_1 \cup A_I, X_1 \cup A_0 \cup B_I), &
\end{array}$$

it remains to prove that the coboundary operator δ_2^* is onto.

Identify B to a point q and X_1 to a point r, and let

$$h : (X_1 \cup A_I, X_1 \cup A_0 \cup B_I, X_1 \cup B_I) \to (\hat{A}_B, A_B \cup \hat{q}, \hat{q})$$

denote the identification map, where if Y is a subspace of $A_0 = A$, $\hat{Y}$ denotes the join of Y with the point r, and A_B denotes A_0 with B_0 identified to a point q. Let

$$h_1 = h \mid (X_1 \cup A_0 \cup B_I, X_1 \cup B_I), h_2 = h \mid (X_1 \cup A_I, X_1 \cup A_0 \cup B_I).$$

According to § 3 and § 4, the following rectangle is commutative:

$$\begin{array}{ccc} \pi^m(X_1 \cup A_0 \cup B_I, X_1 \cup B_I) & \xrightarrow{\delta_2^*} & \pi^{m+1}(X_1 \cup A_I, X_1 \cup A_0 \cup B_I) \\ \uparrow h_1^* & & \uparrow h_2^* \\ \pi^m(A_B \cup \hat{q}, \hat{q}) & \xrightarrow{\delta_3^*} & \pi^{m+1}(\hat{A}_B, A_B \cup \hat{q}). \end{array}$$

Since h_1 and h_2 are obviously relative homeomorphisms, it follows from (3.2) that both h_1^* and h_2^* are one-to-one and onto. Hence, it remains to prove that δ_3^* is onto.

Consider the inclusion map

$$k : (\hat{A}_B, A_B, \square) \subset (\hat{A}_B, A_B \cup \hat{q}, \hat{q})$$

and let $k_1 = k \mid (A_B, \square)$ and $k_2 = k \mid (\hat{A}_B, A_B)$. Then, we obtain a commutative rectangle:

$$\begin{array}{ccc} \pi^m(A_B \cup \hat{q}, \hat{q}) & \xrightarrow{\delta_3^*} & \pi^{m+1}(\hat{A}_B, A_B \cup \hat{q}) \\ \downarrow k_1^* & & \downarrow k_2^* \\ \pi^m(A_B) & \xrightarrow{\delta_4^*} & \pi^{m+1}(\hat{A}_B, A_B). \end{array}$$

Let $\alpha \in \pi^m(A_B)$ be an arbitrary element represented by $\phi : A_B \to S^m$. Since S^m is pathwise connected, we may assume that $\phi(q) = s_0$. Then ϕ has a unique extension $\hat{\phi} : (A_B \cup \hat{q}, \hat{q}) \to (S^m, s_0)$. Hence k_1^* is onto. Since A_B is a strong deformation retract of $A_B \cup \hat{q}$, it follows that k_2^* is one-to-one and onto. Since (A, B) is $2m$-coconnected, so is A_B. Hence, by (7.1), δ_4^* is onto. This implies that $\delta_3^* = k_2^{*-1}\delta_4^* k_1^*$ is also onto. ∎

9. The statement (5)

Lemma 9.1. *If (X, A, B) is a triangulable triple such that (X, A) is $2m$-coconnected, then the statement* (5) *of* § 6 *holds. More explicitly, in the sequence*

$$\pi^m(X, B) \xrightarrow{i^*} \pi^m(A, B) \xrightarrow{\delta^*} \pi^{m+1}(X, A),$$

the image of i^* contains the kernel of δ^*.

Proof. Let $\alpha \in \pi^m(A, B)$ be an arbitrary element with $\delta^*(\alpha) = 0$. Let $\phi : (A, B) \to (S^m, s_0)$ be a map which represents α. We are going to extend ϕ throughout X by means of stepwise inductive construction over the n-dimensional skeletons $\bar{X}^n = X^n \cup A$ of X.

It is obvious that ϕ can be extended over $\bar{X}^m$. Assume that n is an integer with $m \leqslant n < 2m - 1$ and that ϕ has an extension $\phi_n : (\bar{X}^n, B) \to (S^m, s_0)$. Consider the inclusion maps

$$(A, B) \xrightarrow{i_n} (\bar{X}^n, B) \xrightarrow{j_n} (\bar{X}^n, A) \xrightarrow{h_n} (X, A) \xrightarrow{k_n} (X, \bar{X}^n)$$

and the coboundary operators δ^*, $\delta_1{}^*$, $\delta_2{}^*$ given in the following diagram:

$$\begin{array}{ccccc} \pi^m(\bar{X}^n, A) & \xrightarrow{j_n{}^*} & \pi^m(\bar{X}^n, B) & \xrightarrow{i_n{}^*} & \pi^m(A, B) \\ & {\scriptstyle\delta_1{}^*}\searrow & \swarrow{\scriptstyle\delta_2{}^*} & & \swarrow{\scriptstyle\delta^*} \\ & & \pi^{m+1}(X, \bar{X}^n) & \xrightarrow{k_n{}^*} & \pi^{m+1}(X, A). \end{array}$$

This diagram has the following properties:

(a) $\delta^* i_n{}^* = k_n{}^* \delta_2{}^*$ by § 3 and § 4.

(b) $\delta_2{}^* j_n{}^* = \delta_1{}^*$ by the definition of $\delta_1{}^*$ and $\delta_2{}^*$.

(c) $Im\ (\delta_1{}^*) = Ker\ (k_n{}^*)$ by § 6 and § 8.

(d) $Im\ (j_n{}^*) = Ker\ (i_n{}^*)$ by § 6.

The map ϕ_n represents an element β of $\pi^m(\bar{X}^n, B)$. Since ϕ_n is an extension of ϕ, we have $i_n{}^*(\beta) = \alpha$. Hence

$$k_n{}^* \delta_2{}^*(\beta) = \delta^* i_n{}^*(\beta) = \delta^*(\alpha) = 0.$$

By (c), there exists an element γ in $\pi^m(\bar{X}^n, A)$ such that $\delta_1{}^*(\gamma) = \delta_2{}^*(\beta)$. Since $\dim (\bar{X}^n \setminus A) = n < 2m - 1$ and $(X, \bar{X}^n)$ is $2m$-coconnected, it follows that $\pi^m(\bar{X}^n, A)$ and $\pi^{m+1}(X, \bar{X}^n)$ are abelian groups according to (5.2). By the proof of (5.5), $\delta_1{}^*$ is a homomorphism. Hence

$$\delta_2{}^*(\beta) + \delta_2{}^* j_n{}^*(-\gamma) = \delta_2{}^*(\beta) + \delta_1{}^*(-\gamma) = \delta_2{}^*(\beta) - \delta_1{}^*(\gamma) = 0.$$

Let $\psi : (\bar{X}^n, A) \to (S^m, s_0)$ represent $-\gamma$. Consider $f = \phi_n \times \psi$. Since $f(A) \subset S^m \times s_0 \subset S^m \vee S^m$, it follows from (5.1) that f is homotopic to a map g relative to A such that $g(\bar{X}^n) \subset S^m \vee S^m$. Consider the composed map

$$\chi = jg : (\bar{X}^n, B) \to (S^m, s_0),$$

where $j : S^m \vee S^m \to S^m$ denotes the map in § 5. Since $g \mid A = f \mid A$ and $\psi(A) = s_0$, we have $\chi \mid A = \phi_n \mid A = \phi$. Hence χ is also an extension of ϕ over $\bar{X}^n$. χ represents an element β' of $\pi^m(\bar{X}^n, B)$. As in the proof of (5.5), we have

$$\delta_2{}^*(\beta') = \delta_2{}^*(\beta) + \delta_2{}^* j_n{}^*(-\gamma) = 0.$$

Let $l_n{}^* : \pi^m(X, \bar{X}^n) \to \pi^m(\bar{X}^{n+1}, \bar{X}^n)$ be induced by inclusion and $\delta_3{}^* : \pi^m(\bar{X}^n, B) \to \pi^{m+1}(\bar{X}^{n+1}, \bar{X}^n)$ the coboundary operator. Then $\delta_3{}^* = l_n{}^* \delta_2{}^*$ and hence $\delta_3{}^*(\beta') = 0$. According to Ex. D at the end of the chapter, χ has an extension ϕ_{n+1} over $\bar{X}^{n+1}$. This completes the inductive construction of an extension ϕ_{2m-1} of ϕ over $\bar{X}^{2m-1}$. Since (X, A) is $2m$-coconnected, it follows that ϕ has an extension $\phi\#$ throughout X. Hence α is contained in the image of i^*. ∎

Summarizing the results of (6.1), (8.1) and (9.1) we state the following

Theorem 9.2. *If (X, A, B) is a triangulable triple such that*

$$H^n(A, B) = 0 = H^n(X, A)$$

for every $n \geqslant 2m$, then the following part of the cohomotopy sequence is exact:

$$\pi^m(X,A) \xrightarrow{j^*} \pi^m(X,B) \xrightarrow{i^*} \pi^m(A,B) \xrightarrow{\delta^*} \pi^{m+1}(X,A) \xrightarrow{j^*} \pi^{m+1}(X,B) \xrightarrow{i^*} \cdots$$

Corollary 9.3. *If (X, A) is a triangulable pair such that $H^n(A) = 0 = H^n(X, A)$ for every $n \geqslant 2m$, then the following part of the cohomotopy sequence is exact:*

$$\pi^m(X, A) \xrightarrow{j^*} \pi^m(X) \xrightarrow{i^*} \pi^m(A) \xrightarrow{\delta} \pi^{m+1}(X, A) \xrightarrow{j^*} \pi^{m+1}(X) \xrightarrow{i^*} \cdots.$$

10. Higher cohomotopy groups

Theorem 10.1. *If (X, A) is an m-coconnected cellular pair, then $\pi^m(X,A) = 0$.*

Proof. Since S^m is $(m-1)$-connected, every two maps $\phi, \psi : (X, A) \to (S^m, s_0)$ are $(m-1)$-homotopic relative to A. Then, by (VI; 11.3), ϕ and ψ are homotopic relative to A. ▮

The significance of this theorem is that, for a (finite) cellular triple (X, A, B), the cohomotopy sequence of (X, A, B) ends with a term 0.

Corollary 10.2. *If X is an m-coconnected triangulable space, then $\pi^m(X) = 0$.*

In particular, the corollary includes the obvious special case: If X consists of a single point, then $\pi^m(X) = 0$ for every $m > 0$; on the other hand, $\pi^0(X)$ clearly consists of two elements.

11. Relations with cohomology groups

In (V; § 4), we constructed a natural homomorphism h_m from the homotopy group $\pi_m(X, A)$ into the homology group $H_m(X, A)$ over integral coefficients. In the present section, we shall define a similar operation

$$h^m : \pi^m(X, A) \to H^m(X, A; \pi_m(S^m, s_0)), \quad m > 0,$$

which will be a homomorphism if (X, A) is an $(2m-1)$-coconnected cellular pair.

Since $\pi_m(S^m, s_0)$ is free cyclic and the identity map on S^m represents a generator, it can be identified with the group Z of integers in a natural way. Hence we may denote the cohomology group $H^m(X, A; \pi_m(S^m, s_0))$ simply by $H^m(X, A)$.

In order to construct the natural homomorphism h^m, let us denote by χ_m the characteristic element of the cohomology group $H^m(S^m, s_0)$ as defined in (VI; § 17). This characteristic element χ_m can also be defined by the natural homomorphism $h_m : \pi_m(S^m, s_0) \to H_m(S^m, s_0)$ as follows. By Hurewicz's theorem, (V; §4), h_m is an isomorphism and hence the inverse h_m^{-1} is a well-defined homomorphism of $H_m(S^m, s_0)$ into $\pi_m(S^m, s_0) = Z$. Therefore, h_m^{-1} determines a generator of $H^m(S^m, s_0)$ which can be easily proved to be χ_m.

Now let us define the natural operation h^m as follows. Let $\alpha \in \pi^m(X, A)$ be an element represented by a map $\phi : (X, A) \to (S^m, s_0)$. ϕ induces a homomorphism

$$\phi^* : H^m(S^m, s_0) \to H^m(X, A).$$

If we use singular cohomology, than ϕ^* depends only on α according to the homotopy axiom. In this case, we define

$$h^m(\alpha) = \phi^*(\chi_m).$$

This completes the definition of the natural operation h^m.

Proposition 11.1. *If (X, A) is a $(2m - 1)$-coconnected cellular pair, then h^m is a homomorphism.*

Proof. Let $p_i : S^m \times S^m \to S^m$, $(i = 1, 2)$, denote the projections defined by $p_i(y_1, y_2) = y_i$; let $j : S^m \vee S^m \to S^m$ denote the map defined in § 5; and consider the inclusion map $k : S^m \vee S^m \subset S^m \times S^m$. These maps induce the homomorphisms

$$p_i^* : H^m(S^m, s_0) \to H^m(S^m \times S^m, (s_0, s_0)),$$

$$j^* : H^m(S^m, s_0) \to H^m(S^m \vee S^m, (s_0, s_0)),$$

$$k^* : H^m(S^m \times S^m, (s_0, s_0)) \to H^m(S^m \vee S^m, (s_0, s_0)).$$

It is easy to see that $j^*(\chi_m) = k^*p_1^*(\chi_m) + k^*p_2^*(\chi_m)$.

Now let $\alpha, \beta \in \pi^m(X, A)$ be represented by $\phi, \psi : (X, A) \to (S^m, s_0)$ respectively. By (5.1), $\phi \times \psi$ is homotopic to a map $g : (X, A) \to (S^m \vee S^m, (s_0, s_0))$ relative to A. Then we have $\phi \simeq p_1kg$ and $\psi \simeq p_2kg$ relative to A. Hence

$$\begin{aligned} h^m(\alpha) + h^m(\beta) &= \phi^*(\chi_m) + \psi^*(\chi_m) \\ &= g^*k^*p_1^*(\chi_m) + g^*k^*p_2^*(\chi_m) \\ &= g^*j^*(\chi_m). \end{aligned}$$

On the other hand, since $jg : (X, A) \to (S^m, s_0)$ represents $\alpha + \beta$ by definition, we have

$$h^m(\alpha + \beta) = (jg)^*(\chi_m) = g^*j^*(\chi_m).$$

This implies that $h^m(\alpha + \beta) = h^m(\alpha) + h^m(\beta)$. ∎

Proposition 11.2. *For any map $f : (X, A) \to (Y, B)$, the following rectangle is commutative:*

$$\begin{array}{ccc} \pi^m(X, A) & \xleftarrow{f^*} & \pi^m(Y, B) \\ \downarrow h^m & & \downarrow h^m \\ H^m(X, A) & \xleftarrow{f^*} & H^m(Y, B). \end{array}$$

Proof. Let $\alpha \in \pi^m(Y, B)$ be represented by $\phi : (Y, B) \to (S^m, s_0)$. Then $f^*(\alpha)$ is represented by ϕf and hence

$$h^m f^*(\alpha) = (\phi f)^*(\chi_m) = f^*\phi^*(\chi_m) = f^*h^m(\alpha). \ ∎$$

Proposition 11.3. *For any binormal pair (X, A), the following rectangle is commutative:*

$$\begin{array}{ccc} \pi^m(A) & \xrightarrow{\delta} & \pi^{m+1}(X, A) \\ \downarrow h^m & & \downarrow h^m \\ H^m(A) & \xrightarrow{\delta} & H^{m+1}(X, A). \end{array}$$

Proof. Consider the map ξ_1 of § 4. Then we have

$$H^m(S^m, s_0) \xrightarrow{\delta^*} H^{m+1}(E_+^{m+1}, S^m) \xleftarrow{\xi_1^*} H^{m+1}(S^{m+1}, s_0).$$

By the definition of the characteristic elements χ_m, it can be seen that $\delta^*(\chi_m) = \xi_1^*(\chi_{m+1})$.

Now let α be any element of $\pi^m(A)$. Choose a representative map $\phi : A \to S^m$. Then, by § 4, $\delta(\alpha)$ is represented by $\xi_1\psi : (X, A) \to (S^{m+1}, s_0)$, where $\psi : (X, A) \to (E_+^{m+1}, S^m)$ is any extension of ϕ. Hence we have

$$\begin{aligned} h^{m+1}\delta(\alpha) &= (\xi_1\psi)^*(\chi_{m+1}) = \psi^*\xi_1^*(\chi_{m+1}) \\ &= \psi^*\delta^*(\chi_m) = \delta\phi^*(\chi_m) = \delta h^m(\alpha). \ \blacksquare \end{aligned}$$

As a consequence of the last two propositions, we have the following

Proposition 11.4. *For every binormal triple (X, A, B), the natural operators h^m, $m = 1, 2, \cdots$, define a transformation of the cohomotopy sequence of (X, A, B) into its integral cohomology sequence, that is to say, each rectangle of the following ladder is commutative:*

$$\begin{array}{ccccccccccc} \pi^1(X,A) & \to \cdots \to & \pi^m(X,A) & \to & \pi^m(X,B) & \to & \pi^m(A,B) & \to & \pi^{m+1}(X,A) & \to & \cdots \\ \downarrow h^1 & & \downarrow h^m & & \downarrow h^m & & \downarrow h^m & & \downarrow h^{m+1} & & \\ H^1(X,A) & \to \cdots \to & H^m(X,A) & \to & H^m(X,B) & \to & H^m(A,B) & \to & H^{m+1}(X,A) & \to & \cdots \end{array}$$

Theorem 11.5. (Hopf theorem). *If m is a positive integer and (X, A) is an $(m + 1)$-coconnected cellular pair, then h^m sends $\pi^m(X, A)$ onto $H^m(X, A)$ in a one-to-one fashion.*

This is merely a special case of (VI; 16.6). The significance of this theorem is that, for an $(m + 1)$-coconnected cellular pair (X, A), we have

$$\pi^m(X, A) \approx H^m(X, A), \quad \pi^n(X, A) = 0 \text{ for } n > m.$$

12. Relations with homotopy groups

If we neglect the group operations, $\pi_n(S^m, s_0)$ and $\pi^m(S^n, s_0)$ are identical. If $n \leqslant 2m - 2$, $\pi^m(S^n, s_0)$ has a group operation as defined in § 5; on the other hand, $\pi_n(S^m, s_0)$ is always a group. We are going to see that these two group operations are the same.

Let α, β be any two elements of $\pi^m(S^n, s_0)$. Let E_+^n and E_-^n denote the north and the south hemispheres of S^n. As in (IV; § 2), we can choose representatives $\phi, \psi : (S^n, s_0) \to (S^m, s_0)$ of α, β respectively such that

$$\phi(E_-^n) = s_0 = \psi(E_+^n).$$

Then the map $g = \phi \times \psi$ sends X into $S^m \vee S^m$ and jg represents the sum of α and β in $\pi^m(S^n, s_0)$, where j denotes the map defined in § 5. On the other hand, since

$$jg(x) = \begin{cases} \phi(x), & \text{if } x \in E_+^n, \\ \psi(x), & \text{if } x \in E_-^n, \end{cases}$$

jg also represents the sum of α and β in $\pi_n(S^m, s_0)$. Hence we obtain the following

Proposition 12.1. *The cohomotopy group $\pi^m(S^n, s_0)$ is exactly the homotopy group $\pi_n(S^m, s_0)$, that is*

$$\pi^m(S^n, s_0) = \pi_n(S^m, s_0).$$

Now let us consider a pair (X, A) and a given point $x_0 \in A$. Let $\alpha \in \pi_m(X, A, x_0)$ and $\beta \in \pi^n(X, A)$ be represented by the maps

$$\phi : (I^m, I^{m-1}, J^{m-1}) \to (X, A, x_0),$$
$$\psi : (X, A) \to (S^n, s_0).$$

The composition $\psi\phi$ is a map of $(I^m, \partial I^m)$ into (S^n, s_0) and therefore represents an element $[\psi\phi]$ of $\pi_m(S^n, s_0)$ which obviously depends only on α and β. We denote

$$\beta \bigcirc \alpha = [\psi\phi]$$

and call $\beta \bigcirc \alpha$ *the composition of* α *and* β. If $m = n$, then $\beta \bigcirc \alpha \in \pi_m(S^m, s_0)$ and hence is an integer.

The following proposition is an immediate consequence of the definition of addition in $\pi_m(X, A, s_0)$ and in $\pi_m(S^n, s_0)$.

Proposition 12.2. *If $\alpha_1, \alpha_2 \in \pi_m(X, A, x_0)$ and $\beta \in \pi^n(X, A)$, then we have*

$$\beta \bigcirc (\alpha_1 + \alpha_2) = (\beta \bigcirc \alpha_1) + (\beta \bigcirc \alpha_2).$$

Now, if we assume that $\pi^n(X, A)$ forms a group according to § 5, then we have the following

Proposition 12.3. *If $\alpha \in \pi_m(X, A, x_0)$ and $\beta_1, \beta_2 \in \pi^n(X, A)$, then we have*

$$(\beta_1 + \beta_2) \bigcirc \alpha = (\beta_1 \bigcirc \alpha) + (\beta_2 \bigcirc \alpha).$$

Proof. Let $\phi : (I^m, I^{m-1}, J^{m-1}) \to (X, A, x_0)$ represent α and $\psi_1, \psi_2 : (X, A) \to (S^n, s_0)$ represent β_1, β_2. Let $g : (X, A) \to (S^n \vee S^n, (s_0, s_0))$ be a normalization of $\psi_1 \times \psi_2$; then $g\phi$ is a normalization of

$$(\psi_1 \times \psi_2)\phi = \psi_1\phi \times \psi_2\phi : (I^m, \partial I^m) \to (S^n, s_0).$$

This and (12.1) imply the proposition. ∎

Combining (12.2) and (12.3), we obtain that the composition operation defines a homomorphism

$$\pi_m(X, A, x_0) \otimes \pi^n(X, A) \to \pi_m(S^n, s_0).$$

The following properties of the composition operation can be easily verified.

Proposition 12.4. *Let* $f:(X,A,x_0) \to (Y,B,y_0)$ *be a map and* $\alpha \in \pi_m(X,A,x_0)$, $\beta \in \pi^n(Y,B)$. *Then we have*

$$(f^*\beta) \bigcirc \alpha = \beta \bigcirc (f_*\alpha),$$

that is, the induced transformations f^* *and* f_* *are "dual" to each other.*

Proposition 12.5. *Let* (X,A,B) *be a binormal triple with* $x_0 \in B$. *If* $\alpha \in \pi_m(X,A,x_0)$ *and* $\beta \in \pi^{n-1}(A,B)$, *then we have*

$$(\delta^*\beta) \bigcirc \alpha = \Sigma\,[\beta \bigcirc (\partial_*\alpha)\,],$$

where $\Sigma : \pi_{m-1}(S^{n-1}) \to \pi_m(S^n)$ *denotes the suspension of Freudenthal,* (V; § 11).

Thus, in case Σ is an isomorphism, ∂_* and δ^* can be looked on as dual operations.

EXERCISES

A. Generalizations to compact pairs

For the sake of simplicity, most of the operations and results in this chapter are formulated and proved for triangulable pairs. However, by using Čech cohomology theory, these can be generalized to the finite dimensional compact pairs (X,A), that is to say, X is a finite dimensional compact Hausdorff space and A is a closed subspace of X. The pair (X,A) is said to be *n-coconnected* if the integral Čech cohomology group $H^q(X,A) = 0$ for every $q \geqslant n$.

Prove that, if (X,A) is a $(2m - 1)$-coconnected finite-dimensional compact pair, then it is possible to define a commutative group operation in $\pi^m(X,A)$ by the method of § 5. See [Massey 1, p. 283]. Also prove that the induced transformations f^* and the coboundary operator δ are homomorphisms for finite-dimensional compact pairs satisfying similar conditions concerning coconnectivity.

Now let (X,A) be a compact pair with $\dim X < 2m - 1$. Consider any finite open covering $\alpha = \{\, U \,\}$ of X. Let K_α denote the geometric nerve of α and L_α the geometric nerve of the open covering $\alpha \cap A = \{\, U \cap A \,\}$ of A. Then (K_α, L_α) is a finite simplicial pair. Since $\dim X < 2m - 1$, the finite open coverings α of X such that $\dim K_\alpha < 2m - 1$ form a cofinal subset M of the directed set of all finite open coverings of X. For each $\alpha \in M$, $\pi^m(K_\alpha, L_\alpha)$ is an abelian group; and hence we obtain a direct system of abelian groups $\{\, \pi_m(K_\alpha, L_\alpha) \mid \alpha \in M \,\}$. Prove that $\pi^m(X,A)$ is isomorphic to the limit group of this direct system, [Spanier 1; p. 227]. Consequently, the cohomotopy groups satisfy the continuity axiom. This can be used to extend results from finitely triangulable pairs to compact pairs. For example, generalize (9.3) to compact pairs.

B. Connection with Freudenthal's suspension

Consider the inclusion maps $p:(S^{m+1}, s_0) \subset (S^{m+1}, E_-^{m+1})$ and $q:(E_+^{m+1}, S^m) \subset (S^{m+1}, E_-^{m+1})$. By (3.3), q^* is one-to-one and onto and hence q^{*-1} is well-defined. Prove the commutativity of the following diagram:

$$\begin{array}{ccccccc} \pi^n(S^m, s_0) & \xrightarrow{\delta^*} & \pi^{n+1}(E_+^{m+1}, S^m) & \xrightarrow{q^{*-1}} & \pi^{n+1}(S^{m+1}, E_-^{m+1}) & \xrightarrow{p^*} & \pi^{n+1}(S^{m+1}, s_0) \\ \updownarrow (=) & & & & & & \updownarrow (=) \\ \pi_m(S^n, s_0) & & & \xrightarrow{\Sigma} & & & \pi_{m+1}(S^{n+1}, s_0) \end{array}$$

where Σ denotes the suspension. Since p^* is also one-to-one and onto, this shows that the coboundary operator δ^* is essentially the suspension Σ.

Now consider the following part of the cohomotopy sequence of the triple (E_+^{m+1}, S^m, s_0):

$$\pi^n(E_+^{m+1}, s_0) \xrightarrow{i^*} \pi^n(S^m, s_0) \xrightarrow{\delta^*} \pi^{n+1}(E_+^{m+1}, S^m) \xrightarrow{j^*} \pi^{n+1}(E_+^{m+1}, s_0).$$

Since E_+^{m+1} is contractible, we have $Im(i^*) = 0$ and $Ker(j^*) = \pi^{n+1}(E_+^{m+1}, S^m)$. If $m = 2$ and $n = 1$, then (6) of § 6 is false since $\pi_2(S^1) = 0$ while $\pi_3(S^2) \approx Z$. On the other hand, if $m = 3$ and $n = 2$, then (5) of § 6 is false since $\pi_4(S^3) \approx Z_2$ while $\pi_3(S^2) \approx Z$.

C. Connection with cochain groups

Let (K, L) be a finite cellular pair. Denote $\bar{K}^m = K^m \cup L$. Let α be any element of $\pi^n(\bar{K}^m, \bar{K}^{m-1})$ and pick a representative map $\phi : (\bar{K}^m, \bar{K}^{m-1}) \to (S^n, s_0)$ for α. For each simplex σ_i^m of K, the partial map $\phi_i = \phi \mid (\sigma_i^m, \partial\sigma_i^m)$ represents an element $[\phi_i]$ of $\pi_m(S^n, s_0)$ which depends only on α and σ_i^m. Hence α determines an m-dimensional cochain $\psi(\alpha)$ of K modulo L with coefficients in $\pi_m(S^n, s_0)$. Prove that $\pi^n(\bar{K}^m, \bar{K}^{m-1})$ forms an abelian group with addition defined as in § 5 and that the correspondence $\alpha \to \psi(\alpha)$ defines an isomorphism

$$\psi : \pi^n(\bar{K}^m, \bar{K}^{m-1}) \approx C^m(K, L; \pi_m(S^n, s_0)).$$

Furthermore, verify that ψ commutes with the induced homomorphisms. Prove also that the following rectangle is commutative:

$$\begin{array}{ccc} \pi^n(\bar{K}^m, \bar{K}^{m-1}) & \xrightarrow{\delta^*} & \pi^{n+1}(\bar{K}^{m+1}, \bar{K}^m) \\ \downarrow \psi & & \downarrow \psi \\ C^m(K, L; \pi_m(S^n, s_0)) & \xrightarrow{\Sigma\delta} & C^{m+1}(K, L; \pi_{m+1}(S^{n+1}, s_0)). \end{array}$$

D. Connection with the obstruction

Let (K, L) be a pair as in Ex. C. Let α be any element of $\pi^n(\bar{K}^m)$ and $\phi : \bar{K}^m \to S^n$ be a representative map of α. By (VI; § 4), ϕ determines an obstruction cocycle

$$c^{m+1}(\phi) \in C^{m+1}(K, L; \pi_m(S^n, s_0)),$$

which depends only on α. Prove that

$$\psi\delta(\alpha) = \Sigma\,(c^{m+1}(\phi)),$$

where $\delta : \pi^n(\bar{K}^m) \to \pi^{n+1}(\bar{K}^{m+1}, \bar{K}^m)$ is the coboundary operator, ψ denotes the isomorphism in Ex. C, and Σ is the homomorphism determined by the suspension in the coefficient group. Hence, if Σ is an isomorphism, $\psi\delta(\alpha)$ is essentially the obstruction cocycle $c^{m+1}(\phi)$. Therefore, ϕ has an extension over $\bar{K}^{m+1}$ iff $\delta(\alpha)$ is the zero element of $\pi^{n+1}(\bar{K}^{m+1}, \bar{K}^m)$.

E. The structure of $\pi^n(X,A)$

Let (X, A) be a given $(2n - 1)$ coconnected cellular pair. Define a sequence of subgroups

$$\pi^n(X, A) = D^{n-1} \supset D^n \supset \cdots \supset D^{2n-2} = 0$$

as follows: an element of $\pi^n(X, A)$ is in D^m iff it can be represented by a map $f : (X, A) \to (S^n, s_0)$ such that $f(\bar{X}^m) = s_0$. As in [Hu 6] and [Chen 1], define the *presentable subgroup* and the *regular subgroup* of $H^m(X, A; \pi_m(S^n, s_0))$ and consider their quotient group

$$J_*{}^m(X, A; \pi_m(S^n, s_0)) = P_*{}^m(X, A; \pi_m(S^n, s_0))/R_*{}^m(X, A; \pi_m(S^n, s_0)).$$

Then prove that

$$D^{m-1}/D^m \approx J_*{}^m(X, A; \pi_m(S^n, s_0))$$

for every $m = n, \cdots, 2n - 2$.

F. Relations between induced homomorphisms

Let (X, A) be a triangulable pair such that $H^n(X) = 0 = H^n(A)$ for every $n \geqslant 2m - 1$, where m is a given positive integer. Prove that the following four statements are equivalent:

1. The induced homomorphism $i^* : H^n(X) \to H^n(A)$ of the inclusion map $i : A \subset X$ is an isomorphism for every $n > m$ and is an epimorphism for $n = m$.
2. $H^n(X, A) = 0$ for every $n > m$.
3. $\pi^n(X, A) = 0$ for every $n > m$.
4. The induced homomorphism $i^* : \pi^n(X) \to \pi^n(A)$ of the inclusion map $i : A \subset X$ is an isomorphism for every $n > m$ and is an epimorphism for $n = m$.

Next, let X and Y be any two $(2m - 1)$-coconnected triangulable spaces and $f : X \to Y$ a map; prove that the following two statements are equivalent.

5. The induced homomorphism $f^* : H^n(Y) \to H^n(X)$ is an isomorphism for every $n > m$ and is an epimorphism for $n = m$.
6. The induced homomorphism $f^* : \pi^n(Y) \to \pi^n(X)$ is an isomorphism for every $n > m$ and is an epimorphism for $n = m$.

CHAPTER VIII

EXACT COUPLES AND SPECTRAL SEQUENCES

1. Introduction

Spectral sequences were originally devised by Leray to exhibit relations among the Čech groups of the various spaces of a fibering; but they have proved useful, indeed crucial, in numerous other investigations. Serre, for example, established the same results for the singular groups, and, using relations between homotopy groups and singular groups, he was able to obtain important information about the homotopy groups of spheres; and Eilenberg, Massey, and Spanier found spectral sequences to be useful in the study of the homotopy groups of a complex.

Leray's original device involved imbedding the homology or cohomology groups in question in a much larger system of groups and homomorphisms, namely the *spectral sequence*; this sequence, after the first term, is itself an invariant of the fibering. Being quite complicated, it is not amenable to successful computation except in special cases; but it always provides much information, and it furnishes a pattern for deeper investigation.

Massey improved Leray's device by imbedding in a still larger but much more versatile system which he called an *exact couple*. The present chapter will be devoted to this formulation of the machinery of spectral sequences; an outline of Leray's direct construction appears in Ex. A. There follows in Chapter IX Serre's version of homology and cohomology theory of a fibering, together with several immediate applications; results from Chapter IX will then be used in the computation of certain of the groups $\pi_m(S^n)$.

2. Differential groups

Let A be an abelian group. An endomorphism

$$d : A \to A$$

is said to be a *differential operator* on A if

$$dd = 0,$$

i.e., $d[d(a)] = 0$ for each $a \in A$. An abelian group A furnished with a given differential operator d is called a *differential group*, or simply a *d-group*. With the d-groups as objects and the homomorphisms which commute with d as mappings, we obtain a category $\mathcal{G}_d$ called *the category of differential groups*. For the definition of a category, see [E–S; p. 109].

If $f : A \to B$ is a mapping in $\mathcal{G}_d$ such that $A \subset B$ and $f(a) = a$ for each $a \in A$, then A is called a *subgroup* of the d-group B. Let $C = B/A$; then we can define a differential operator on C so that the projection $g : B \to C$ becomes a mapping in $\mathcal{G}_d$. When furnished with this differential operator, C is called the *quotient d-group*. We obtain an exact sequence

$$0 \longrightarrow A \xrightarrow{f} B \xrightarrow{g} C \longrightarrow 0. \tag{2.1}$$

Conversely, if an exact sequence (2.1) of d-groups and mappings is given, then A can be identified with a subgroup of the d-group B by means of the monomorphism f and the quotient d-group B/A can be identified with C by means of the epimorphism g.

Given a d-group A, we shall denote by $\mathcal{Z}(A)$ the kernel of d (called the group of *cycles* of A), and by $\mathcal{B}(A)$ the image of d (called the group of *boundaries* of A). The condition $dd = 0$ implies that $\mathcal{B}(A) \subset \mathcal{Z}(A)$ and that they are subgroups of the d-group A with $d = 0$ on each of them. Hence we may define the quotient d-group

$$\mathcal{H}(A) = \mathcal{Z}(A)/\mathcal{B}(A)$$

with $d = 0$ which is called the *derived group* of the d-group A.

Let $f : A \to B$ be a mapping in $\mathcal{G}_d$. Then the commutativity $df = fd$ implies that f maps $\mathcal{Z}(A)$ into $\mathcal{Z}(B)$ and $\mathcal{B}(A)$ into $\mathcal{B}(B)$ and hence f induces a mapping

$$\mathcal{H}(f) : \mathcal{H}(A) \to \mathcal{H}(B).$$

One can easily verify that the operation $\mathcal{H}$ is a covariant functor from $\mathcal{G}_d$ to itself, [E–S; p. 111].

Two mappings $f, g : A \to B$ in $\mathcal{G}_d$ are said to be *homotopic* (notation: $f \simeq g$) if there is a homomorphism $\xi : A \to B$ such that

$$f - g = d\xi + \xi d.$$

The homomorphism ξ is called a *homotopy* (notation: $\xi : f \simeq g$). One can easily prove that $f \simeq g$ implies $\mathcal{H}(f) = \mathcal{H}(g)$.

Let an exact sequence (2.1) in $\mathcal{G}_d$ be given. We shall define a mapping

$$\partial : \mathcal{H}(C) \to \mathcal{H}(A)$$

as follows. Let $\alpha \in \mathcal{H}(C)$ and choose $x \in \mathcal{Z}(C)$ which represents α. There is some $y \in B$ with $g(y) = x$. Since $gd(y) = d(gy) = 0$, there is a $z \in A$ with $f(z) = d(y)$. Since $fd(z) = d(fz) = 0$, $d(z) = 0$ and z represents a $\beta \in \mathcal{H}(A)$ which depends only on α. Then ∂ is defined by $\partial(\alpha) = \beta$. Since the differential operators on $\mathcal{H}(C)$ and $\mathcal{H}(A)$ are trivial, ∂ is a mapping in $\mathcal{G}_d$. Thus we obtain a triangle in $\mathcal{G}_d$:

$$\begin{array}{ccc} \mathcal{H}(A) & \xrightarrow{\mathcal{H}(f)} & \mathcal{H}(B) \\ & {\scriptstyle\partial}\nwarrow \quad \swarrow{\scriptstyle\mathcal{H}(g)} & \\ & \mathcal{H}(C) & \end{array}$$

One can verify that this triangle is *exact* in the sense that the kernel of each homomorphism is precisely the image of the preceding.

3. Graded and bigraded groups

An abelian group A is said to be *graded*, or to have a *graded structure*, if there is prescribed to each integer n, (positive, zero, or negative), a subgroup A_n of A such that A can be written as a (weak) direct sum

$$A = \sum_n A_n.$$

The elements of the subgroup A_n are said to be *homogeneous of degree n*.

Similarly, an abelian group A is said to be *bigraded*, or to have a *bigraded structure*, if there is prescribed to each (ordered) pair (m, n) of integers a subgroup $A_{m,n}$ of A such that A can be written as a direct sum

$$A = \sum_{m,n} A_{m,n}.$$

The elements of the subgroup $A_{m,n}$ are said to be *homogeneous of degree* (m, n).

When dealing with graded or bigraded groups, only a certain limited class of homomorphisms are of interest, namely, the homogeneous homomorphisms. If

$$A = \sum_{m,n} A_{m,n}, \quad B = \sum_{m,n} B_{m,n}$$

are bigraded groups, then a homomorphism $f : A \to B$ is said to be *homogeneous of degree* (p, q) if

$$f(A_{m,n}) \subset B_{m+p,\, n+q}$$

for every pair (m, n). With bigraded groups as objects and homogeneous homomorphisms as mappings, we obtain a category $\mathscr{G}_b$ called the *category of bigraded groups*. Similarly, one can define the *category* $\mathscr{G}_g$ *of graded groups*.

If $f : A \to B$ is a mapping in $\mathscr{G}_b$ of degree $(0, 0)$ such that $A \subset B$ and $f(a) = a$ for each $a \in A$, then A is called a *subgroup* of the bigraded group B. Let $C = B/A$. Then one can verify that C is isomorphic to the direct sum of the groups $B_{m,n}/A_{m,n}$. We agree to identify these naturally isomorphic groups. Then C is bigraded and the projection $g : B \to C$ is a mapping in $\mathscr{G}_b$ of degree $(0, 0)$. With this bigraded structure, C is called the *quotient bigraded group*. Note that both the kernel and the image of a mapping in $\mathscr{G}_b$ are subgroups of the corresponding bigraded groups. One can easily formulate the analogous concepts for graded groups.

In algebraic topology, we have to deal with graded (or bigraded) groups with a differential operator which is homogeneous. In this case, the derived group is also graded (or bigraded). For example, let us consider the group $C(X)$ of all singular chains in a space X with integral coefficients as defined in [E-S; p. 187]. $C(X)$ has a natural graded structure

$$C(X) = \sum_n C_n(X).$$

where $C_n(X)$ is the group of n-chains if $n \geqslant 0$ and $C_n(X) = 0$ if $n < 0$. The boundary homomorphism $\partial : C_n(X) \to C_{n-1}(X)$ extends to a differential operator

$$d = \partial : C(X) \to C(X)$$

which is homogeneous of degree -1. Hence the kernel $Z(X)$ of d and the image $B(X)$ of d are subgroups of the graded d-group $C(X)$ with graded structures

$$Z(X) = \sum_n Z_n(X), \quad B(X) = \sum_n B_n(X).$$

Furthermore, the derived group $H(X)$ of $C(X)$ has a graded structure

$$H(X) = \sum_n H_n(X),$$

where $H_n(X)$ is the n-dimensional singular homology group of X if $n \geqslant 0$ and $H_n(X) = 0$ for $n < 0$.

4. Exact couples

By an *exact couple*, we mean a system

$$\mathscr{C} = < D, E; i, j, k >$$

which consists of two abelian groups D and E, and three homomorphisms

$$i : D \to D, \quad j : D \to E, \quad k : E \to D$$

such that the following triangle is exact:

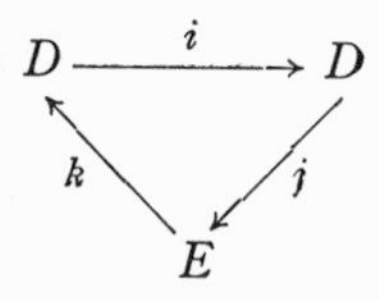

There is an operation which assigns to an exact couple $\mathscr{C}$ another exact couple

$$\mathscr{C}' = < D', E'; i', j', k' >$$

called the *derived couple* of $\mathscr{C}$ constructed as follows. Define an endomorphism

$$d : E \to E$$

by $d = jk$. Since $kj = 0$ by exactness, it follows that $dd = jkjk = j(kj)k = 0$. This implies that d is a differential operator on E. Let

$$D' = i(D), \quad E' = \mathscr{H}(E);$$

then D' is a subgroup of D and E' is the derived group of the d-group E. Since $D' \subset D$, $i(D') \subset D'$. We may define an endomorphism

$$i' : D' \to D'$$

by $i' = i \mid D'$. Since one can easily verify that

$$k[\mathscr{Z}(E)] \subset D', \quad k[\mathscr{B}(E)] = 0,$$

k induces a homomorphism $\quad k' : E' \to D'$.

Let $x \in D'$ and choose a $y \in D$ with $i(y) = x$. Then $j(y)$ is in $\mathscr{Z}(E)$ and the coset of $j(y)$ mod $\mathscr{B}(E)$ does not depend on the choice of y. Denote this coset by $j'(x)$. The assignment $x \to j'(x)$ defines a homomorphism

$$j' : D' \to E'.$$

This completes the construction of $\mathscr{C}'$. The verification that i', j' k' form an exact triangle is straightforward and is left to the reader.

This process of derivation can be applied to $\mathscr{C}'$ to obtain a *second derived couple* $\mathscr{C}''$, and so on. In this way, we obtain a sequence of exact couples

$$\mathscr{C}^n = < D^n, E^n; i^n, j^n, k^n >, \quad n = 1, 2, \cdots,$$

defined inductively by

$$\mathscr{C}^1 = \mathscr{C}, \quad \mathscr{C}^n = (\mathscr{C}^{n-1})', \quad n > 1.$$

The sequence $\{ \mathscr{C}^n \}$ has two important properties. Firstly, the groups D^n form a decreasing sequence

$$D = D^1 \supset D^2 \supset \cdots \supset D^n \supset D^{n+1} \supset \cdots$$

with $i^n : D^n \to D^n$ defined by the restriction of i to D^n. The intersection of the groups D^n is denoted by D^∞.

Secondly, it has been shown above that the endomorphism

$$d^n = j^n k^n : E^n \to E^n$$

is a differential operator on E^n and E^{n+1} is the derived group of E^n with respect to d^n. Hence we obtain a sequence of differential groups

$$E = E^1, E^2, \cdots, E^n, \cdots$$

such that $E^{n+1} = \mathscr{H}(E^n)$ for each $n > 0$. This will be called the *spectral sequence* associated with the exact couple $\mathscr{C}$.

In the spectral sequence $\{ E^n \}$ there exist natural homomorphisms

$$h_n : \mathscr{Z}(E^n) \to E^{n+1}$$

defined by assigning to each element of $\mathscr{Z}(E^n)$ its coset modulo $\mathscr{B}(E^n)$. Thus h_n is an epimorphism of a subgroup of E^n onto E^{n+1}. We can define an epimorphism h_n^p of a subgroup of E^n onto E^{n+p} by the formula

$$h_n^p = h_{n+p-1} h_{n+p-2} \cdots h_n.$$

Then $h_n^1 = h_n$. The precise domain $\mathscr{Z}_n^p$ of definition of h_n^p can be defined inductively as follows:

$$\mathscr{Z}_n^1 = \mathscr{Z}(E^n),$$

$$\mathscr{Z}_n^p = \{ a \in \mathscr{Z}_n^{p-1} \mid h_n^{p-1}(a) \in \mathscr{Z}(E^{n+p-1}) \}, \quad p > 1.$$

Let $\bar{E}^n$ denote the intersection of all subgroups $\mathscr{Z}_n^p$, $p = 1, 2, \cdots$, then, for each $a \in \bar{E}^n$, $h_n^p(a)$ is defined for all values of p. Define $\bar{h}_n : \bar{E}^n \to \bar{E}^{n+1}$ to be the restriction of h_n to $\bar{E}^n$. Then the sequence of groups $\{ \bar{E}^n \}$ and homomorphisms $\{ \bar{h}_n \}$ constitutes a direct sequence of groups in the usual sense, [H–W; p. 132]. The limit group E^∞ of this direct sequence of groups will be called the *limit group* of the spectral sequence $\{ E^n \}$.

In particular, if there is an integer $r \geqslant 1$ such that $d^n = 0$ for each $n \geqslant r$, then

$$\mathscr{Z}(E^n) = E^n, \quad h_n : E^n \approx E^{n+1}$$

and therefore $\bar{E}^n = E^n$ and $\bar{h}_n = h_n$ for each $n \geqslant r$. This implies that $E^\infty \approx E^r$.

Returning to the general case again, we can express the limit group E^∞ in terms of the original exact couple $\mathscr{C}$ as follows. Consider the epimorphisms

$$i^{(n)} = i^n i^{n-1} \cdots i^1 : D \to D^{n+1} \subset D.$$

Since $i^n = i \mid D^n$ for each $n \geqslant 1$, $i^{(n)}$ is actually the n-fold iteration of the homomorphism $i : D \to D$. Let

$$D^\infty = \bigcap_{n=1}^{\infty} D^n = \bigcap_{n=1}^{\infty} Im\,[i^{(n)}], \quad D^0 = \bigcup_{n=1}^{\infty} Ker\,[i^{(n)}],$$

where $Im\,[i^{(n)}]$ and $Ker\,[i^{(n)}]$ denote the image and the kernel of $i^{(n)}$ respectively. Then it is easy to verify that

$$E^\infty \approx k^{-1}D^\infty / jD^0.$$

In the following chapters as well as in the exercises at the end of this chapter, we shall give detailed accounts of some of the important exact couples. For the moment, let us be content with a simple illustrative example.

Consider a (finitely) *filtered space*, i.e., a space X furnished with a finite increasing sequence of subspaces

$$(\Phi) \qquad \square = X_{-1} \subset X_0 \subset X_1 \subset \cdots \subset X_r = X.$$

Using total singular homology groups over a given coefficient group G, we define

$$D = \sum_p H(X_p), \quad E = \sum_p H(X_p, X_{p-1}).$$

Then the homology sequence of the pairs (X_p, X_{p-1}) give rise to an exact couple

$$\mathscr{C}_H(\Phi) = \langle D, E; i, j, k \rangle$$

called the *homology exact couple* of the filtered space X.

5. Bigraded exact couples

In the exact couples $\mathscr{C} = \langle D, E; i, j, k \rangle$ which we shall deal with in the sequel, the groups D, E are usually bigraded and the homomorphisms i, j, k are homogeneous. In this case, in the successive derived couples $\mathscr{C}^n$ of $\mathscr{C}$, the groups D^n, E^n are also bigraded, i.e.,

$$D^n = \sum_{p,q} D^n_{p,q} \quad E^n = \sum_{p,q} E^n_{p,q}$$

and the homomorphisms i^n, j^n, k^n, and $d^n = j^n k^n$ are homogeneous. The elements of $D^n_{p,q}$ and $E^n_{p,q}$ are said to be *homogeneous of degree* (p, q). We shall call p the *primary degree*, q the *complementary degree*, and $p + q$ the *total degree*.

It is easy to verify the following relations about the degree of homogeneity:

$$deg\,(i^n) = deg\,(i), \quad deg\,(k^n) = deg\,(k),$$

$$deg\,(j^n) = deg\,(j) - (n - 1)\,[deg\,(i)].$$

Hence the degree of the differential operator d^n is given by the formula:

$$deg\,(d^n) = deg\,(j) + deg\,(k) - (n - 1)\,[deg\,(i)].$$

The following two special classes of bigraded exact couples are important. A bigraded exact couple $\mathscr{C} = \langle D, E; i, j, k \rangle$ will be called a *∂-couple* if i is of degree $(1, -1)$, j is of degree $(0, 0)$ and k is of degree $(-1, 0)$. In this case, the degrees of i^n, j^n, k^n, and d^n are listed as follows:

($\partial 1$) i^n is of degree $(1, -1)$;
($\partial 2$) j^n is of degree $(-n + 1, n - 1)$;
($\partial 3$) k^n is of degree $(-1, 0)$;
($\partial 4$) d^n is of degree $(-n, n - 1)$.

Similarly, $\mathscr{C}$ will be called a *δ-couple* if i is of degree $(-1, 1)$, j is of degree $(0, 0)$, and k is of degree $(1, 0)$. Then the degrees of i^n, j^n, k^n, and d^n are listed as follows:

($\delta 1$) i^n is of degree $(-1, 1)$;
($\delta 2$) j^n is of degree $(n - 1, -n + 1)$;
($\delta 3$) k^n is of degree $(1, 0)$;
($\delta 4$) d^n is of degree $(n, -n + 1)$.

A ∂-couple $\mathscr{C} = \langle D, E; i, j, k \rangle$ is a quite elaborate structure; it can be developed into a "lattice-like" diagram as follows:

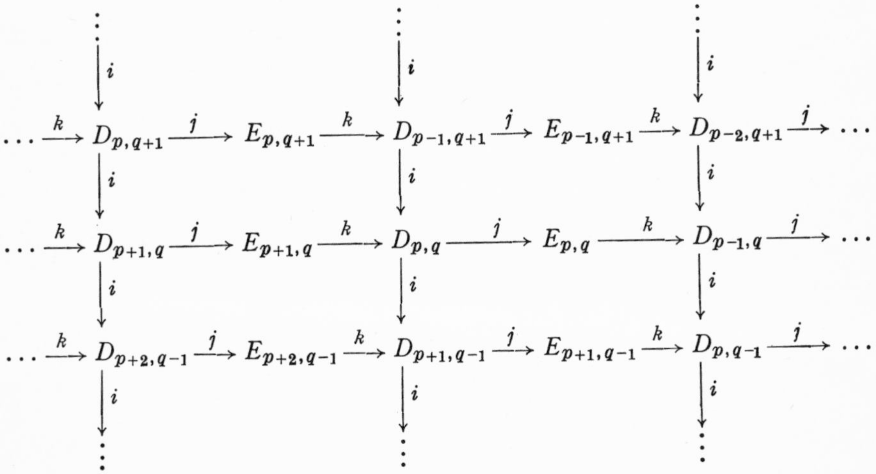

$$\begin{array}{ccccccccccccc}
 & & \vdots & & & & \vdots & & & & \vdots & & \\
 & & \downarrow{\scriptstyle i} & & & & \downarrow{\scriptstyle i} & & & & \downarrow{\scriptstyle i} & & \\
\cdots & \xrightarrow{k} & D_{p,q+1} & \xrightarrow{j} & E_{p,q+1} & \xrightarrow{k} & D_{p-1,q+1} & \xrightarrow{j} & E_{p-1,q+1} & \xrightarrow{k} & D_{p-2,q+1} & \xrightarrow{j} & \cdots \\
 & & \downarrow{\scriptstyle i} & & & & \downarrow{\scriptstyle i} & & & & \downarrow{\scriptstyle i} & & \\
\cdots & \xrightarrow{k} & D_{p+1,q} & \xrightarrow{j} & E_{p+1,q} & \xrightarrow{k} & D_{p,q} & \xrightarrow{j} & E_{p,q} & \xrightarrow{k} & D_{p-1,q} & \xrightarrow{j} & \cdots \\
 & & \downarrow{\scriptstyle i} & & & & \downarrow{\scriptstyle i} & & & & \downarrow{\scriptstyle i} & & \\
\cdots & \xrightarrow{k} & D_{p+2,q-1} & \xrightarrow{j} & E_{p+2,q-1} & \xrightarrow{k} & D_{p+1,q-1} & \xrightarrow{j} & E_{p+1,q-1} & \xrightarrow{k} & D_{p,q-1} & \xrightarrow{j} & \cdots \\
 & & \downarrow{\scriptstyle i} & & & & \downarrow{\scriptstyle i} & & & & \downarrow{\scriptstyle i} & & \\
 & & \vdots & & & & \vdots & & & & \vdots & &
\end{array}$$

The steps from upper left to lower right are exact sequences, for example,

$$\cdots \xrightarrow{k} D_{p-1,q+1} \xrightarrow{i} D_{p,q} \xrightarrow{j} E_{p,q} \xrightarrow{k} D_{p-1,q} \xrightarrow{i} D_{p,q-1} \xrightarrow{j} \cdots$$

is an exact sequence. Similarly, one can develop the derived couples $\mathscr{C}^n$, $n = 2, 3, \cdots$, into analogous diagram.

It is more or less evident that the notions of ∂-couple and δ-couple are dual to one another. Thus, let $\mathscr{C} = \langle D, E, i, j, k \rangle$ be any given bigraded exact couple. We shall construct another bigraded exact couple $\mathscr{C}^* = \langle D^*, E^*; i^*, j^*, k^* \rangle$, called the *dual of* $\mathscr{C}$ by merely reindexing the groups $D_{p,q}$ and $E_{p,q}$ as follows:

$$D^*_{p,q} = D_{-p,-q}, \quad E^*_{p,q} = E_{-p,-q}.$$

Then we obtain $D^* = D$ and $E^* = E$; therefore, we may take $i^* = i$, $j^* = j$, and $k^* = k$. Their degrees are given by

$$deg\,(i^*) = -\,deg\,(i), \quad deg\,(j^*) = -\,deg\,(j), \quad deg\,(k^*) = -\,deg\,(k).$$

Hence the dual of a ∂-couple is a δ-couple and vice versa.

For example, the homology exact couple $\mathscr{C}_H(\Phi)$ of § 4 is a ∂-couple Indeed, the groups D and E are bigraded, namely,

$$D_{p,q} = H_{p+q}(X_p), \quad E_{p,q} = H_{p+q}(X_p, X_{p-1});$$

and the homomorphisms i, j, k are obviously homogeneous of degree $(1, -1)$, $(0, 0)$, $(-1, 0)$ respectively.

6. Regular couples

A ∂-couple $\mathscr{C} = \langle D, E; i, j, k \rangle$ is said to be *regular* provided:

$(R\partial 1)$ $D_{p,q} = 0$ if $p < 0$;

$(R\partial 2)$ $E_{p,q} = 0$ if $q < 0$.

(R∂1) and the exactness of

$$D_{p,q} \xrightarrow{j} E_{p,q} \xrightarrow{k} D_{p-1,q}$$

imply that $E_{p,q} = 0$ if $p < 0$. Hence, for any positive integer n, we have

$(R\partial 3)$ $$E^n_{p,q} = 0 = E^\infty_{p,q}, \quad \text{if } p < 0 \text{ or } q < 0.$$

If $n > p$, each element of $E^n_{p,q}$ is a cycle since d^n is of degree $(-n, n-1)$. If $n > q + 1$, no non-zero element of $E^n_{p,q}$ can be a boundary under d^n. Hence

$(R\partial 4)$ $$E^n_{p,q} = E^{n+1}_{p,q} = \cdots = E^\infty_{p,q} \quad \text{if } n > \max\,(p, q+1),$$

Since each i^r is of degree $(1, -1)$, we have

(R∂5) $$D^n_{p,q} = i^{n-1}\, i^{n-2} \cdots i^1\, (D_{p-n+1,q+n-1}).$$

This and (R∂1) imply

(R∂6) $$D^n_{p,q} = 0 = D^\infty_{p,q} \quad \text{if } n > p + 1.$$

Consider the homomorphism $i : D_{p,q} \to D_{p+1,q-1}$. (R∂2) and the exactness of

$$E_{p+1,q} \xrightarrow{k} D_{p,q} \xrightarrow{i} D_{p+1,q-1} \xrightarrow{j} E_{p+1,q-1}$$

imply that i is an epimorphism if $q \leqslant 0$ and an isomorphism if $q < 0$. Thus we obtain a sequence

$$0 \to D_{0,m} \xrightarrow{i} \cdots \xrightarrow{i} D_{m,0} \xrightarrow{i} D_{m+1,-1} \overset{i}{\approx} D_{m+2,-2} \overset{i}{\approx} \cdots \overset{i}{\approx} D_{m+r,-r} \overset{i}{\approx} \cdots$$

where $i : D_{m,0} \to D_{m+1,-1}$ is an epimorphism. This suggests defining a graded group

$$\mathscr{H}(\mathscr{C}) = \sum_m \mathscr{H}_m(\mathscr{C}), \quad \mathscr{H}_m(\mathscr{C}) = D_{m+1,-1} = D^2_{m+1,-1}.$$

Then $D^n_{m+1,-1}$ is a subgroup of $\mathscr{H}_m(\mathscr{C})$ and (R∂5) gives a homomorphism

$$\lambda_{p,q} : D_{p,q} \to \mathscr{H}_{p+q}(\mathscr{C}), \quad (q \geqslant 0),$$

where $\lambda_{p,q} = i^{q+1} i^q \cdots i^1$ is actually the $(q + 1)$-fold iteration of the homomorphism $i : D \to D$. Denote the image of $\lambda_{p,q}$ by

$$\mathscr{H}_{p,q}(\mathscr{C}) = \lambda_{p,q}(D_{p,q}) = D^{q+2}_{p+q+1,-1}.$$

If $p < 0$, then $\mathscr{H}_{p,q}(\mathscr{C}) = 0$ since $D_{p,q} = 0$. If $q = 0$, then $\mathscr{H}_{p,q}(\mathscr{C}) = \mathscr{H}_{p+q}(\mathscr{C})$ since $\lambda_{p,q}$ reduces to the epimorphism i. Hence, for each $m \geqslant 0$, we obtain a finite decreasing sequence:

(R∂7) $\mathscr{H}_m(\mathscr{C}) = \mathscr{H}_{m,0}(\mathscr{C}) \supset \mathscr{H}_{m-1,1}(\mathscr{C}) \supset \cdots \supset \mathscr{H}_{0,m}(\mathscr{C}) \supset \mathscr{H}_{-1,m+1}(\mathscr{C}) = 0.$

Let $n > \max(p, q + 2)$ and consider the diagram:

$$\begin{array}{ccccccccc} E^n_{p+n-1,q-n+2} & \xrightarrow{k^n} & D^n_{p+n-2,q-n+2} & \xrightarrow{i^n} & D^n_{p+n-1,q-n+1} & \xrightarrow{j^n} & E^n_{p,q} & \xrightarrow{k^n} & D^n_{p-1,q} \\ & & \uparrow \alpha & & \uparrow \beta & & & & \\ & & \mathscr{H}_{p-1,q+1}(\mathscr{C}) & \xrightarrow{\gamma} & \mathscr{H}_{p,q}(\mathscr{C}) & & & & \end{array}$$

The first row is exact since it is a part of $\mathscr{C}^n$. Since i maps $D_{p+q+r,-r}$ isomorphically onto $D_{p+q+r+1,-r-1}$ for each $r \geqslant 1$, we obtain two natural isomorphisms α and β. γ denotes the inclusion and the rectangle is commutative. Since n is greater than *max* $(p, q + 2)$, we get

$$E^n_{p+n-1,q-n+2} = 0, \; E^n_{p,q} = E^\infty_{p,q}, \; D^n_{p-1,q} = 0$$

according to (R∂3), (R∂4), and (R∂6). Hence we have

(R∂8) $$\mathscr{H}_{p,q}(\mathscr{C}) / \mathscr{H}_{p-1,q+1}(\mathscr{C}) \approx E^\infty_{p,q}.$$

In other words, *$\mathscr{H}_{p,q}(\mathscr{C})$ is an extension of $\mathscr{H}_{p-1,q+1}(\mathscr{C})$ by $E^\infty_{p,q}$.*

Similarly, a δ-couple $\mathscr{C} = \langle D, E; i, j, k \rangle$ is said to be *regular* provided:

(Rδ1) $D_{p,q} = 0$ if $q < 0$;

(Rδ2) $E_{p,q} = 0$ if $p < 0$.

By methods dual to those used above, one can prove the following assertions.

(Rδ3) $$E^n_{p,q} = 0 = E^\infty_{p,q} \quad \text{if } p < 0 \text{ or } q < 0.$$

(Rδ4) $$E^n_{p,q} = E^{n+1}_{p,q} = \cdots = E^\infty_{p,q} \quad \text{if } n > \max(p, q+1).$$

(Rδ5) $$D^n_{p,q} = i^{n-1} i^{n-2} \cdots i^1 (D_{p+n-1,q-n+1}).$$

(Rδ6) $$D^n_{p,q} = 0 = D^\infty_{p,q} \text{ if } n > q+1.$$

The homomorphism $i: D_{p,q} \to D_{p-1,q+1}$ is an isomorphism if $p \leqslant 0$. Thus we obtain a sequence

$$0 \to D_{m,0} \xrightarrow{i} D_{m-1,1} \xrightarrow{i} \cdots \xrightarrow{i} D_{0,m} \overset{i}{\approx} D_{-1,m+1} \overset{i}{\approx} \cdots \overset{i}{\approx} D_{-r,m+r} \overset{i}{\approx} \cdots$$

This suggests defining a graded group

$$\mathscr{H}(\mathscr{C}) = \sum_m \mathscr{H}_m(\mathscr{C}), \quad \mathscr{H}_m(\mathscr{C}) = D_{0,m} \approx D^2_{-1,m+1}.$$

Let $\mathscr{H}_{p,q}(\mathscr{C})$ denote the subgroup $D^{p+1}_{0,p+q}$ of $\mathscr{H}_{p+q}(\mathscr{C})$. Then, for each $m \geqslant 0$, we obtain a finite decreasing sequence:

(Rδ7) $$\mathscr{H}_m(\mathscr{C}) = \mathscr{H}_{0,m}(\mathscr{C}) \supset \mathscr{H}_{1,m-1}(\mathscr{C}) \supset \cdots \supset \mathscr{H}_{m,0}(\mathscr{C}) \supset \mathscr{H}_{m+1,-1}(\mathscr{C}) = 0$$

which satisfies the relations:

(Rδ8) $$\mathscr{H}_{p,q}(\mathscr{C}) / \mathscr{H}_{p+1,q-1}(\mathscr{C}) \approx E^\infty_{p,q}.$$

Let us consider again the ∂-couple $\mathscr{C}_H(\Phi)$ of § 5. ($R\partial1$) is obviously true while (R∂2) is false in general. However, in case X is a finite simplicial complex and X_p is a subcomplex of X containing the p-dimensional skeleton of X, then (R∂2) is also satisfied. Hence, in this case, $\mathscr{C}_H(\Phi)$ is a regular ∂-couple. If, in particular, X_p is the p-dimensional skeleton of X, then $E_{p,q} = 0$ for each $q \neq 0$ and $E_{p,0} = H_p(X_p, X_{p-1})$ is the group of p-chains of X over G. One can verify that the differential operator $d: E_{p,0} \to E_{p-1,0}$ is precisely the boundary operator on p-chains. Hence we obtain

$$E^2_{p,0} = H_p(X), \quad E^2_{p,q} = 0 \text{ if } q \neq 0.$$

If $n \geqslant 2$, $d^n = 0$ since it increases the complementary degree q. Hence we obtain

$$H(X) = E^2 = E^3 = \cdots = E^\infty.$$

7. The graded groups R($\mathscr{C}$) and S($\mathscr{C}$)

Let $\mathscr{C} = \langle D, E; i, j, k \rangle$ be a regular ∂-couple. Define a graded group

$$R(\mathscr{C}) = \sum_q R_q(\mathscr{C}), \quad R_q(\mathscr{C}) = E_{0,q}.$$

Since d^n is of degree $(-n, n-1)$, each element of $E^n_{0,q}$ is a cycle. Thus we obtain epimorphisms

$$R_q(\mathscr{C}) = E^1_{0,q} \xrightarrow{\varkappa_1} E^2_{0,q} \xrightarrow{\varkappa_2} \cdots \xrightarrow{\varkappa_{q+1}} E^{q+2}_{0,q} = E^\infty_{0,q}.$$

Let $\varkappa$ denote the composition. By (R∂8) of § 6, we have

$$E^\infty_{0,q} \approx \mathscr{H}_{0,q}(\mathscr{C}) \subset \mathscr{H}_q(\mathscr{C}).$$

Denote the composed monomorphism by ι. Then we have

(R∂9) $$R_q(\mathscr{C}) \xrightarrow{\varkappa} E^\infty_{0,q} \xrightarrow{\iota} \mathscr{H}_q(\mathscr{C}).$$

On the other hand, define a graded group

$$S(\mathscr{C}) = \sum_p S_p(\mathscr{C}), \quad S_p(\mathscr{C}) = E^2_{p,0}.$$

If $n \geqslant 2$, no non-zero element of $E^n_{p,0}$ can be a boundary under d^n. Thus we obtain the monomorphisms

$$E^\infty_{p,0} = E^{p+1}_{p,0} \xrightarrow{\iota_p} E^p_{p,0} \xrightarrow{\iota_{p-1}} \cdots \xrightarrow{\iota_3} E^3_{p,0} \xrightarrow{\iota_2} E^2_{p,0} = S_p(\mathscr{C}).$$

Let ι denote the composition. By (R∂8) of § 6, we have

$$E^\infty_{p,0} \approx \mathscr{H}_{p,0}(\mathscr{C})/\mathscr{H}_{p-1,1}(\mathscr{C}) = \mathscr{H}_p(\mathscr{C})/\mathscr{H}_{p-1,1}(\mathscr{C}).$$

Let $\varkappa$ denote the natural epimorphism. Since

$$\mathscr{H}_p(\mathscr{C}) = D^2_{p+1,-1}, \quad S_p(\mathscr{C}) = E^2_{p,0}.$$

we have $j^2 : \mathscr{H}_p(\mathscr{C}) \to S_p(\mathscr{C})$. One can verify commutativity in the triangle

(R∂10)
$$\begin{array}{ccc} \mathscr{H}_p(\mathscr{C}) & \xrightarrow{j^2} & S_p(\mathscr{C}) \\ & {\scriptstyle\varkappa}\searrow \quad \nearrow{\scriptstyle\iota} & \\ & E^\infty_{p,0} & \end{array}$$

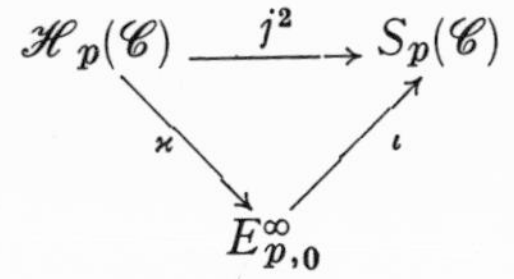

Thus j^2 is factored into the composition of an epimorphism and a monomorphism.

If $\mathscr{C}$ is a regular δ-couple, then we define the graded groups $R(\mathscr{C})$ and $S(\mathscr{C})$ exactly as above. Since d^n is of degree $(n, -n+1)$, no non-zero element of $E^n_{0,q}$ can be a boundary under d^n and every element of $E^n_{p,0}$ with $n \geqslant 2$ is a cycle. Hence, one may establish the epimorphisms $\varkappa$ and the monomorphisms ι in an analogous way as above with the roles of $R(\mathscr{C})$ and $S(\mathscr{C})$ interchanged. Furthermore, since $\mathscr{H}_q(\mathscr{C}) = D_{0,q}$ and $R_q(\mathscr{C}) = E_{0,q}$, we have $j : \mathscr{H}_q(\mathscr{C}) \to R_q(\mathscr{C})$. Thus we get

(Rδ9)
$$S_p(\mathscr{C}) \xrightarrow{\varkappa} E^\infty_{p,0} \xrightarrow{\iota} \mathscr{H}_p(\mathscr{C})$$

and a commutative triangle

(Rδ10)
$$\begin{array}{ccc} \mathscr{H}_q(\mathscr{C}) & \xrightarrow{j} & R_q(\mathscr{C}) \\ & {\scriptstyle\varkappa}\searrow \quad \nearrow{\scriptstyle\iota} & \\ & E^\infty_{0,q} & \end{array}$$

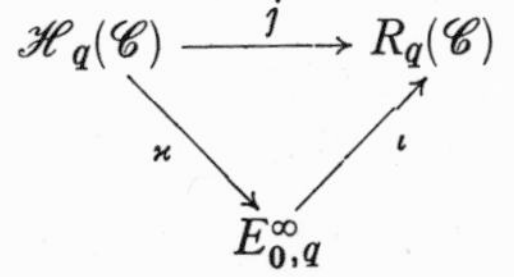

Note. One might define $R_q(\mathscr{C})$ to be $E^2_{0,q}$ instead of $E_{0,q}$. Then all the results in this section stand as they are except that the j in (Rδ10) should be replaced by j^2. Our choice of $E_{0,q}$ is based on the fact that it gives a natural expression for $R_q(\mathscr{C})$ if $\mathscr{C}$ is defined by a filtered graded differential group. See § 11 below.

8. The fundamental exact sequence

Let $\mathscr{C} = \langle D, E; i, j, k \rangle$ be a regular ∂-couple and $\nu > 0$ an integer. Under certain conditions on the bigraded group E^ν, we shall obtain a useful exact sequence which will be referred to as *the fundamental exact sequence.*

Theorem 8.1. *If E^2 has only two rows which might be non-trivial, more precisely, if there are two integers $a < b$ such that*

$$E^2_{p,q} = 0 \quad \text{if } p \neq a \text{ and } p \neq b,$$

then we obtain a fundamental exact sequence

$$\cdots \to E^2_{a,m-a} \xrightarrow{\phi_m} \mathscr{H}_m(\mathscr{C}) \xrightarrow{\psi_m} E^2_{b,m-b} \xrightarrow{\chi_m} E^2_{a,m-1-a} \xrightarrow{\phi_{m-1}} \mathscr{H}_{m-1}(\mathscr{C}) \to \cdots$$

Proof. The hypothesis implies that $E^n_{p,q} = 0$ for each $n \geqslant 2$ and hence $E^\infty_{p,q} = 0$ if $p \neq a$ and $p \neq b$. Then (R∂7) and (R∂8) of § 6 give rise to an exact sequence

$$0 \to E^\infty_{a,m-a} \to \mathscr{H}_m(\mathscr{C}) \to E^\infty_{b,m-b} \to 0.$$

Since d^n is of degree $(-n, n-1)$ and $a < b$, every element $E^n_{a,m-a}$ is a cycle if $n \geqslant 2$. Hence we obtain epimorphisms

$$E^2_{a,m-a} \xrightarrow{\varkappa^2} E^3_{a,m-a} \xrightarrow{\varkappa^3} \cdots \xrightarrow{\varkappa^n} E^{n+1}_{a,m-a} = E^\infty_{a,m-a},$$

where $n = \max(a, m - a + 1)$. Similarly, if $n \geqslant 2$, no non-zero element of $E^n_{b,m-b}$ can be a boundary under d^n and we obtain monomorphisms

$$E^\infty_{b,m-b} = E^{n+1}_{b,m-b} \xrightarrow{\iota^n} E^n_{b,m-b} \xrightarrow{\iota^{n-1}} \cdots \xrightarrow{\iota^2} E^2_{b,m-b},$$

where $n = \max(b, m - b+1)$. Thus we obtain an exact sequence

$$E^2_{a,m-a} \xrightarrow{\phi_m} \mathscr{H}_m(\mathscr{C}) \xrightarrow{\psi_m} E^2_{b,m-b}.$$

To determine the kernel of ϕ_m, we have to study that of $\varkappa^r$, $r = 2, \cdots, n$. The kernel of $\varkappa^r$ consists of those elements of $E^r_{a,m-a}$ which are boundaries under d^r. There are only two terms of E^r of total degree $m + 1$ which might be non-trivial, namely, $E^r_{a,m+1-a}$ and $E^r_{b,m+1-b}$. The elements of the first are cycles; and d^r maps the second into $E^r_{a,m-a}$ iff $r = b - a$. Hence ϕ_m is a monomorphism if $b - a = 1$. Now let $r = b - a \geqslant 2$; we have an exact sequence

$$E^r_{b,m+1-b} \xrightarrow{d^r} E^r_{a,m-a} \xrightarrow{\varkappa^r} E^{r+1}_{a,m-a}.$$

Furthermore, $E^r_{a,m-a} = E^2_{a,m-a}$ and $E^{r+1}_{a,m-a} = E^\infty_{a,m-a}$. A similar argument shows that $E^r_{b,m+1-b} = E^2_{b,m+1-b}$. Thus we obtain an exact sequence

$$E^2_{b,m+1-b} \xrightarrow{\chi_{m+1}} E^2_{a,m-a} \xrightarrow{\phi_m} \mathscr{H}_m(\mathscr{C}),$$

where $\chi_{m+1} = 0$ if $b - a = 1$ and $\chi_{m+1} = d^r$ if $b - a = r \geqslant 2$.

By similar methods, one can prove the exactness of the sequence

$$\mathscr{H}_m(\mathscr{C}) \xrightarrow{\psi_m} E^2_{b,m-b} \xrightarrow{\chi_m} E^2_{a,m-1-a}.$$

The fundamental exact sequence is obtained by putting the various parts together. ∎

It was proved above that $\chi_{m+1} = 0$ if $b - a = 1$. Hence we have the following

Corollary 8.2. *If in the hypothesis of* (8.1), $b - a = 1$, *then, for each* m, *we have an exact sequence*

$$0 \to E^2_{a,m-a} \xrightarrow{\phi_m} \mathscr{H}_m(\mathscr{C}) \xrightarrow{\psi_m} E^2_{b,m-b} \to 0.$$

Hence $\mathscr{H}_m(\mathscr{C})$ *is an extension of* $E^2_{a,m-a}$ *by* $E^2_{b,m-b}$.

Analogously, one can prove the following

Theorem 8.3. *If* E^2 *has only two columns which might be non-trivial, more precisely, if there are two integers* $a < b$ *such that*

$$E^2_{p,q} = 0, \quad \text{if } q \neq a \text{ and } q \neq b,$$

then we obtain a fundamental exact sequence

$$\cdots \to E^2_{m-b,b} \xrightarrow{\phi_m} \mathscr{H}_m(\mathscr{C}) \xrightarrow{\psi_m} E^2_{m-a,a} \xrightarrow{\chi_m} E^2_{m-1-b,b} \xrightarrow{\phi_{m-1}} \mathscr{H}_{m-1}(\mathscr{C}) \to \cdots$$

Corollary 8.4. *If, in the hypothesis of* (8.3), $b - a = 1$, *then, for each* m, *we have an exact sequence*

$$0 \to E^2_{m-b,b} \xrightarrow{\phi_m} \mathscr{H}_m(\mathscr{C}) \xrightarrow{\psi_m} E^2_{m-a,a} \to 0.$$

Hence $\mathscr{H}_m(\mathscr{C})$ *is an extension of* $E^2_{m-b,b}$ *by* $E^2_{m-a,a}$.

Although the preceding two theorems cover most of the applications, it is sometimes necessary to deal with fundamental exact sequences obtained under weaker conditions.

The two-term condition. Let λ, μ, ν be integers such that $\lambda < \mu$ and $\nu \geqslant 1$. We shall say that $\mathscr{C}$ satisfies the two-term condition $\{\lambda, \mu; \nu\}$ if the bigraded group E^ν has the following properties. For each integer m such that $\lambda \leqslant m \leqslant \mu$, $E^\nu_{p,q} = 0$ if $p + q = m$ and (p, q) is different from two given pairs (a_m, b_m) and (c_m, d_m), where

$$a_m + b_m = m = c_m + d_m, \quad a_m < c_m.$$

Moreover, we require that the following two conditions should also be fulfilled:

(1) $\quad E^\nu_{p,q} = 0$ if $p + q = m - 1$, $p \leqslant a_m - \nu$, and $\lambda \leqslant m \leqslant \mu$;

(2) $\quad E^\nu_{p,q} = 0$ if $p + q = m + 1$, $p \geqslant c_m + \nu$, and $\lambda \leqslant m \leqslant \mu$.

With some obvious modifications of the proof of (8.1) one can prove the following theorem which includes (8.1) and (8.3) as special cases.

Theorem 8.5. *If* $\mathscr{C}$ *satisfies the two-term condition* $\{\lambda, \mu; \nu\}$, *then we have a fundamental exact sequence*

$$E^\nu_{a_\mu,b_\mu} \to \cdots \to E^\nu_{a_m,b_m} \to \mathscr{H}_m(\mathscr{C}) \to E^\nu_{c_m,d_m} \to E^\nu_{a_{m-1},b_{m-1}} \to \cdots \to E^\nu_{c_\lambda,d_\lambda}$$

Corollary 8.6. *Let* $\nu \geqslant 1$ *and* $\{a_m\}$ *be a sequence of integers such that*

$$a_m < a_{m-1} + \nu$$

for each $m \geqslant 0$. *If* $E^{\nu}_{p,q} = 0$ *for every pair* (p, q) *such that* $p + q = m$ *and* $p \neq a_m$, *then*

$$\mathscr{H}_m(\mathscr{C}) \approx E^{\nu}_{a_m,b_m}, \quad b_m = m - a_m.$$

Proof. Put $c_m = a_m + 1$ and $d_m = b_m - 1$. Then $\mathscr{C}$ satisfies the two-term condition $\{0, \mu; \nu\}$ for every $\mu > 0$. Since $E^{\nu}_{c_m,d_m} = 0$ for each $m \geqslant 0$, the fundamental exact sequence implies the conclusion of this corollary. ∎

If $\mathscr{C}$ is a regular δ-couple, then the two-term condition is stated by precisely the same words. The fundamental exact sequence in (8.5) becomes

$$E^{\nu}_{a_\mu,b_\mu} \leftarrow \cdots \leftarrow E^{\nu}_{a_m,b_m} \leftarrow \mathscr{H}_m(\mathscr{C}) \leftarrow E^{\nu}_{c_m,d_m} \leftarrow E^{\nu}_{a_{m-1},b_{m-1}} \leftarrow \cdots \leftarrow E^{\nu}_{c_\lambda,d_\lambda}$$

and similarly in (8.1) and (8.3).

Regular ∂- and δ-couples are considered further in Ex. B at the end of the chapter.

9. Mappings of exact couples

Let $\mathscr{C}_r = \langle D_r, E_r; i_r, j_r, k_r \rangle, r = 1, 2,$ be two exact couples. By a *mapping*

$$(\phi, \psi) : \mathscr{C}_1 \to \mathscr{C}_2$$

we mean a pair of homomorphisms $\phi : D_1 \to D_2$ and $\psi : E_1 \to E_2$ such that

$$\phi i_1 = i_2\phi, \quad \psi j_1 = j_2\phi, \quad \phi k_1 = k_2\psi.$$

If $d_r = j_r k_r$, then $\psi d_1 = d_2\psi$ and ψ is a mapping in the sense of § 2. ψ induces a homomorphism

$$\psi' : E_1' \to E_2'.$$

Since

$$\phi(D_1') = \phi i_1(D_1) = i_2\phi(D_1) \subset i_2(D_2) = D_2',$$

we may define a homomorphism $\phi' : D_1' \to D_2'$ by $\phi' = \phi \mid D_1'$. It can be readily verified that the pair (ϕ', ψ') of homomorphisms constitutes a mapping of the derived couples. This mapping $(\phi', \psi') : \mathscr{C}_1' \to \mathscr{C}_2'$ is called the *derived mapping* of (ϕ, ψ). Thus, the set of all exact couples and their mappings forms a category, and the operation of derivation is a covariant functor.

If we iterate this process, we obtain a sequence of successive derived mappings

$$(\phi^n, \psi^n) : \mathscr{C}_1{}^n \to \mathscr{C}_2{}^n, \quad (n = 1, 2, \cdots).$$

In particular, the homomorphisms $\psi^n : E_1{}^n \to E_2{}^n$, $(n = 1, 2, \cdots)$, of the spectral sequences commute with the differential operators d^n, and ψ^{n+1} is the induced homomorphism of ψ^n.

If the exact couples $\mathscr{C}_1$ and $\mathscr{C}_2$ are bigraded, then only a limited class of mappings are of interest, namely, the homogeneous mappings. The mapping (ϕ, ψ) is said to be *homogeneous of degree* (p, q) if both ϕ and ψ are homogeneous of the same degree (p, q). By reindexing one of the given exact couples, we may always assume that (ϕ, ψ) is of degree $(0, 0)$; in this case, we say that (ϕ, ψ) *preserves degree*. Then each of the derived mappings (ϕ^n, ψ^n) also preserves degree.

Now let $\mathscr{C}_1$, $\mathscr{C}_2$ be regular ∂-couples, and $(\phi, \psi) : \mathscr{C}_1 \to \mathscr{C}_2$ a mapping which preserves degree. Then the homomorphism ϕ^2 defines a homomorphism

$$(\phi, \psi)^* : \mathscr{H}(\mathscr{C}_1) \to \mathscr{H}(\mathscr{C}_2)$$

which carries $\mathscr{H}_m(\mathscr{C}_1)$ into $\mathscr{H}_m(\mathscr{C}_2)$ and $\mathscr{H}_{p,q}(\mathscr{C}_1)$ into $\mathscr{H}_{p,q}(\mathscr{C}_2)$. Furthermore, ψ and ψ^2 define homomorphisms

$$R(\mathscr{C}_1) \to R(\mathscr{C}_2), \quad S(\mathscr{C}_1) \to S(\mathscr{C}_2).$$

These obviously commute with the epimorphisms $\varkappa$ and the monomorphisms ι in § 7.

Proposition 9.1. $(\phi, \psi)^*$ *is an isomorphism if either of the following two equivalent conditions is satisfied*:

$$(i)\ \phi^2 : D_1{}^2 \approx D_2{}^2; \quad (ii)\ \psi^2 : E_1{}^2 \approx E_2{}^2.$$

Proof. It is obvious that (i) implies that $(\phi, \psi)^*$ is an isomorphism. That $(i) \Rightarrow (ii)$ is an immediate consequence of the "five" lemma, [E–S; p. 16]. Finally, that $(ii) \Rightarrow (i)$ follows from an easy induction on the primary degree by using (R∂1) and the "five" lemma. ▮

Note. (9.1) remains true if we replace the superscripts 2 in (i) and (ii) by any positive integer n. In fact, it can be easily seen from § 6, that (i) implies that $(\phi, \psi)^*$ is an isomorphism while the equivalence of (i) and (ii) is proved exactly as above.

Obviously, similar results can be obtained for regular δ-couples.

Now, let us go back to the general case where $\mathscr{C}_1$, $\mathscr{C}_2$ are any two exact couples and consider two mappings

$$(\phi, \psi), (\sigma, \tau) : \mathscr{C}_1 \to \mathscr{C}_2.$$

They are said to be *homotopic* (notation: $(\phi, \psi) \simeq (\sigma, \tau)$) if there is a homomorphism $\xi : E_1 \to E_2$ such that

$$\sigma(x) - \phi(x) = k_2 \xi j_1(x),$$

$$\tau(y) - \psi(y) = \xi d_1(y) + d_2 \xi(y)$$

for every $x \in D_1$ and $y \in E_1$. Then one can easily prove the following obvious but important

Proposition 9.2. *If the mappings* (ϕ, ψ) *and* (σ, τ) *are homotopic, then the derived mappings* (ϕ', ψ') *and* (σ', τ') *are equal.*

Hence, $(\phi^n, \psi^n) = (\sigma^n, \tau^n)$ for every $n \geqslant 2$. In particular, if $\mathscr{C}_1, \mathscr{C}_2$ are regular ∂-couples or regular δ-couples and if (ϕ, ψ), (σ, τ) preserve the degree, then $(\phi, \psi) \simeq (\sigma, \tau)$ implies that $(\phi, \psi)^* = (\sigma, \tau)^*$.

10. Filtered differential groups

Let A be a d-group. By an *increasing filtration* in A, we mean an increasing sequence of subgroups $\{A^p\}$ of the d-group A, p ranging over all integers, with their union equal to A; in symbols, we have

$$\cup_p A^p = A, \quad A^p \subset A^{p+1}, \quad d(A^p) \subset A^p.$$

If an increasing filtration $\{A^p\}$ is given in A, A will be called an *increasing filtered d-group.*

Let A be an increasing filtered d-group. For each $a \in A$, the greatest lower bound of the integers p such that $a \in A^p$ is called the *weight* of a and is denoted by $w(a)$. The following properties are obvious:

$$w(a - b) \leqslant \max [w(a), w(b)], \quad w(da) \leqslant w(a).$$

Conversely, if there is given a function w defined on a d-group A with integral values (including $-\infty$) which has the properties given above, we can define an increasing filtration $\{A^p\}$ by taking

$$A^p = \{ a \in A \mid w(a) \leqslant p \}.$$

Let A and B be increasing filtered d-groups. A homomorphism $f: A \to B$ is called a *mapping* if it commutes with d and preserves the filtration, that is,

$$fd = df, \quad f(A^p) \subset B^p.$$

The set of all increasing filtered d-groups and their mappings constitutes a category $\mathscr{F}_i$.

Subgroups and quotient groups of groups in $\mathscr{F}_i$ are defined in a manner analogous to that of § 2. They are also groups in $\mathscr{F}_i$. In particular, for each A in $\mathscr{F}_i$, the derived group $\mathscr{H}(A)$ is also in $\mathscr{F}_i$. One can verify that the operation $\mathscr{H}$ of § 2 defines a covariant functor from $\mathscr{F}_i$ to itself.

To each A in $\mathscr{F}_i$, there is an *associated graded group* $\mathscr{G}(A)$ of A defined by

$$\mathscr{G}(A) = \hat{A} = \sum_p \hat{A}_p, \quad \hat{A}_p = A^p/A^{p-1}$$

and each mapping $f: A \to B$ induces in an obvious way a homogeneous homomorphism of degree 0

$$\mathscr{G}(f) = \hat{f}: \mathscr{G}(A) \to \mathscr{G}(B).$$

One can verify that the operation $\mathscr{G}$ is a covariant functor from the category $\mathscr{F}_i$ to the category $\mathscr{G}_g$ of § 3. The associated graded group of the derived group $\mathscr{H}(A)$ is important and will be denoted by

$$\mathscr{G}\mathscr{H}(A) = \hat{\mathscr{H}}(A) = \sum_p \hat{\mathscr{H}}_p(A).$$

Since A^{p-1}, A^p, $\hat{A}_p$ are d-groups, the exact sequence

$$0 \to A^{p-1} \to A^p \to \hat{A}_p \to 0$$

gives rise to an exact triangle

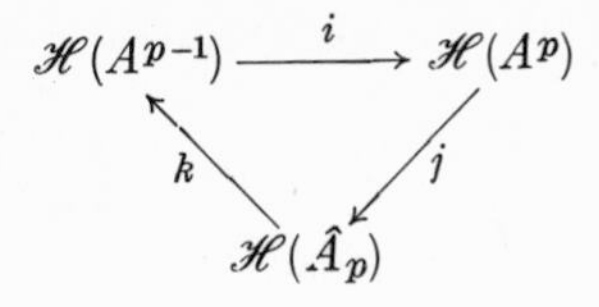

according to § 2. Define two graded groups

$$D = \sum_p D_p, \quad D_p = \mathscr{H}(A^p),$$

$$E = \sum_p E_p, \quad E_p = \mathscr{H}(\hat{A}_p).$$

Then the homomorphisms i, j, k of the preceding exact triangle define homomorphisms

$$i : D \to D, \quad j : D \to E, \quad k : E \to D$$

and we obtain an exact couple

$$\mathscr{C}(A) = \langle D, E; i, j, k \rangle$$

which will be referred to as the *exact couple associated with* A. The groups D and E are graded and the homomorphisms i, j, k are homogeneous of degree $1, 0, -1$ respectively.

Now let $f : A \to B$ be a mapping in $\mathscr{F}_i$. Then f defines two mappings $f_p : A^p \to B^p$ and $\hat{f}_p : \hat{A}_p \to \hat{B}_p$ in $\mathscr{G}_d$. One can easily verify that, by passing to direct sums, the derived homomorphisms

$$\mathscr{H}(f_p) : \mathscr{H}(A^p) \to \mathscr{H}(B^p), \quad \mathscr{H}(\hat{f}_p) : \mathscr{H}(\hat{A}_p) \to \mathscr{H}(\hat{B}_p)$$

define a mapping

$$\mathscr{C}(f) = (\phi_f, \psi_f) : \mathscr{C}(A) \to \mathscr{C}(B)$$

in the sense of § 9. One verifies that the operation $\mathscr{C}$ is a covariant functor from $\mathscr{F}_i$ to the category of § 9.

Similarly, a *decreasing filtration* in a d-group A consists of a decreasing sequence of subgroups $\{ A^p \}$ of A such that

$$\cap_p A^p = 0, \quad A^p \supset A^{p+1}, \quad d(A^p) \subset A^p$$

The *weight* $w(a)$ of $a \in A$ is defined to be the least upper bound of the integers p such that $a \in A^p$.

By an obvious analogy, one can formulate all of the preceding concepts for the category $\mathscr{F}_d$ of *decreasing filtered d-groups* and their mappings.

By a *filtered d-group* A, we mean either an increasing or a decreasing filtered d-group. The associated exact couple $\mathscr{C}(A)$ constructed above and its associated spectral sequence $\{ E^n, n = 1, 2, \cdots \}$ will be referred to as those of the filtered d-group A.

11. Filtered graded differential groups

By an *increasing* (or *decreasing*) *filtration in a graded d-group* A, we mean an increasing (or decreasing) filtration $\{ A^p \}$ of the d-group A such that each A^p is a subgroup of the graded group A. Such a filtration $\{A^p\}$ is said to be *regular* if

(RF1) $$0 \leqslant w(a) \leqslant \deg(a)$$

for every non-zero homogeneous element a of A. In other words, the weight and the degree are both non-negative, and the weight does not exceed the degree.

The filtered d-groups with which we have to deal in the sequel usually belong to two limited classes, namely, the *filtered ∂-complexes* and the *filtered δ-complexes*. A filtered ∂-complex is a graded d-group, where d is of degree -1, with a regular increasing filtration; a filtered δ-complex is a graded d-group, where d is of degree 1, with a regular decreasing filtration.

Theorem 11.1. *The associated exact couple*

$$\mathscr{C}(A) = \langle D, E; i, j, k \rangle$$

of a regular ∂-complex A is a regular ∂-couple.

Proof. Since A^p and $\hat{A}_p = A^p/A^{p-1}$ are graded d-groups, the derived groups $\mathscr{H}(A^p)$ and $\mathscr{H}(\hat{A}_p)$ are also graded. So we may put

$$D_{p,q} = \mathscr{H}_{p+q}(A^p), \quad E_{p,q} = \mathscr{H}_{p+q}(\hat{A}_p).$$

Thus the groups D and E of $\mathscr{C}(A)$ are bigraded. One can easily verify that i is of degree $(1, -1)$, j is of degree $(0, 0)$, and k is of degree $(-1, 0)$. This proves that $\mathscr{C}(A)$ is a ∂-couple.

For an increasing filtration $\{A^p\}$ of A, the regularity condition (RF1) is equivalent to the following two conditions:

(RF∂2) $$A^p = 0 \quad \text{if } p < 0;$$

(RF∂3) $$A_p \subset A^p \quad \text{for all } p.$$

(RF∂2) implies $D_{p,q} = 0$ if $p < 0$; (RF∂3) implies $E_{p,q} = 0$ if $q < 0$. Hence $\mathscr{C}(A)$ is regular. ▮

Next, let us investigate the graded groups $\mathscr{H}(\mathscr{C})$, $R(\mathscr{C})$, and $S(\mathscr{C})$ of the regular ∂-couple $\mathscr{C} = \mathscr{C}(A)$ associated with a given regular ∂-complex A.

The inclusion $A^p \subset A$ induces homomorphisms

$$h_{p,q} : D_{p,q} = \mathscr{H}_{p+q}(A^p) \to \mathscr{H}_{p+q}(A).$$

By (RF∂3), one can verify that $h_{p,q}$ is an epimorphism if $q \leqslant 0$ and an isomorphism if $q < 0$. Hence

$$h_{m+1,-1} : \mathscr{H}_m(\mathscr{C}) \approx \mathscr{H}_m(A).$$

Passing to direct sums, we obtain a homogeneous isomorphism of degree 0

$$h : \mathscr{H}(\mathscr{C}) \approx \mathscr{H}(A).$$

According to § 10, $\mathscr{H}(A)$ is filtered; hence $\mathscr{H}_{p+q}(A)$ is also filtered. It is clear that the image $h_{p,q}(D_{p,q})$ is the subgroup of $\mathscr{H}_{p+q}(A)$ with filtration p. This will be denoted by $\mathscr{H}_{p,q}(A)$. Then it is easy to show that $h_{m+1,-1}$ maps

$\mathscr{H}_{p,q}(\mathscr{C})$ isomorphically onto $\mathscr{H}_{p,q}(A)$ if $m = p + q$. Thus, the following theorem is a consequence of (R∂7) and (R∂8) in § 6.

Theorem 11.2. *The filtration of $\mathscr{H}_m(A)$ is given by the finite sequence*

(RF∂4) $\mathscr{H}_m(A) = \mathscr{H}_{m,0}(A) \supset \mathscr{H}_{m-1,1}(A) \supset \cdots \supset \mathscr{H}_{0,m}(A) \supset \mathscr{H}_{-1,m+1}(A) = 0$

and its associated graded group $\mathscr{G}\mathscr{H}_m(A)$ is isomorphic with $\Sigma_{p+q=m} E^{\infty}_{p,q}$ as given by the relation

(RF∂5) $$\mathscr{H}_{p,q}(A)/\mathscr{H}_{p-1,q+1}(A) \approx E^{\infty}_{p,q}.$$

Since $A^{-1} = 0$ by (RF∂2), we have

$$R_q(\mathscr{C}) = E_{0,q} = \mathscr{H}_q(A^0)$$

and hence $R(\mathscr{C}) = \mathscr{H}(A^0)$. Then one can verify that the composition $\iota\varkappa$ of (R∂9) in § 7 gives the derived homomorphism

$$\mathscr{H}(f) : \mathscr{H}(A^0) \to \mathscr{H}(A)$$

of the inclusion $f : A^0 \subset A$. Thus $\mathscr{H}(f)$ is factored into the composition of an epimorphism and a monomorphism.

Define a graded group $\tilde{A} = \Sigma \tilde{A}_p$ by $\tilde{A}_p = E_{p,0}$. Since d is of degree $(-1, 0)$ on E, $\tilde{A}$ is a subgroup of the d-group E and

$$S_p(\mathscr{C}) = E^2_{p,0} = \mathscr{H}_p(\tilde{A}).$$

Hence we have $S(\mathscr{C}) = \mathscr{H}(\tilde{A})$.

There is a natural epimorphism $g : A \to \tilde{A}$ defined as follows. Let $x \in A_p$. Then $x \in A^p$ by (RF∂3), and $dx \in A_{p-1} \subset A^{p-1}$. This implies that x is a cycle of $\hat{A}_p$ and represents an element $g(x)$ of $\tilde{A}_p = E_{p,0}$. The assignment $x \to g(x)$ defines an epimorphism $g : A \to \tilde{A}$ which preserves the degree and commutes with d. One can verify that the derived homomorphism

$$\mathscr{H}(g) : \mathscr{H}(A) \to \mathscr{H}(\tilde{A})$$

reduces to the homomorphism $j^2 : \mathscr{H}(\mathscr{C}) \to S(\mathscr{C})$ in (R∂10) of § 7 after identifying $\mathscr{H}(A)$ with $\mathscr{H}(\mathscr{C})$ and $\mathscr{H}(\tilde{A})$ with $S(\mathscr{C})$. Thus $\mathscr{H}(g)$ is factored into the composition of an epimorphism and a monomorphism. As a consequence of this, the kernel of $\mathscr{H}(g)$ in $\mathscr{H}_p(A)$ is $\mathscr{H}_{p-1,1}(A)$ and the image in $\mathscr{H}_p(\tilde{A}) = E^2_{p,0}$ is the subgroup $E^{\infty}_{p,0}$.

One can establish the foregoing results for regular δ-complexes analogously. The theorems (11.1) and (11.2) can be stated and proved with some obvious modifications. The graded groups $\mathscr{H}(\mathscr{C})$, $R(\mathscr{C})$, and $S(\mathscr{C})$ are identified to be

$$\mathscr{H}(\mathscr{C}) = \mathscr{H}(A), \quad R(\mathscr{C}) = \mathscr{H}(\hat{A}_0), \quad S(\mathscr{C}) = \mathscr{H}(\tilde{A}).$$

where $\hat{A}_0 = A^0/A^1 = A/A^1$ and $\tilde{A} = \Sigma E_{p,0}$. After these identifications, the homomorphism $j : \mathscr{H}(\mathscr{C}) \to R(\mathscr{C})$ in (R∂10) of § 7 reduces to the derived homomorphism

$$\mathscr{H}(f) : \mathscr{H}(A) \to \mathscr{H}(\hat{A}_0)$$

of the natural projection $f : A \to \hat{A}_0$. There is a natural monomorphism

$g : \tilde{A} \to A$ which preserves degree and commutes with d. The derived homomorphism

$$\mathscr{H}(g) : \mathscr{H}(\tilde{A}) \to \mathscr{H}(A)$$

is given by the composition $\iota\varkappa$ in (Rδ9) of § 7.

As an example of regular ∂-complexes and δ-complexes, let us consider a filtered space

$$(\Phi) \qquad \square = X_{-1} \subset X_0 \subset X_1 \subset \cdots \subset X_r = X$$

where X is a finite simplicial complex and X_p is a subcomplex of X. Assume that X_p contains all p-simplexes of X. Then, for any given coefficient group G, the group $A = C(X)$ of chains form a regular ∂-complex with

$$A^p = C(X_p), \quad A_p = C_p(X).$$

One can easily see that $\mathscr{C}(A)$ is precisely the exact couple $\mathscr{C}_H(\Phi)$ of § 4. Hence $\mathscr{C}_H(\Phi)$ is also a regular ∂-couple in this case. Similarly, the group of cochains of X over G forms a regular δ-complex.

12. Mappings of filtered graded d-groups

Let A and B be two regular ∂-complexes. A homomorphism $f : A \to B$ is said to be a *mapping* if it commutes with d and preserves both degree and filtration, i.e.

$$fd = df, \quad f(A_m) \subset B_m, \quad f(A^p) \subset B^p.$$

The set of all regular ∂-complexes and their mappings constitutes a category $\mathscr{K}_\partial$ which is a subcategory of $\mathscr{F}_i$.

Now let $f : A \to B$ be a mapping in $\mathscr{K}_\partial$. According to § 10, f induces a mapping

$$\mathscr{C}(f) = (\phi_f, \psi_f) : \mathscr{C}(A) \to \mathscr{C}(B)$$

in the sense of § 9. On the other hand, we have a derived homomorphism

$$\mathscr{H}(f) : \mathscr{H}(A) \to \mathscr{H}(B).$$

The following proposition is a corollary of (9.1) and its note.

Proposition 12.1. *For any mapping $f : A \to B$ in $\mathscr{K}_\partial$, $\mathscr{H}(f)$ is an isomorphism if there exists an integer $n \geqslant 1$ such that either of the following two equivalent conditions is satisfied*:

$$(i)\ \phi_f^n : D^n(A) \approx D^n(B); \quad (ii)\ \psi_f^n : E^n(A) \approx E^n(B).$$

Similarly, one can define the mappings of regular δ-complexes and the category $\mathscr{K}_\delta$ of regular δ-complexes and their mappings. The preceding proposition is also true for any mapping in $\mathscr{K}_\delta$.

EXERCISES

A. Direct construction of the spectral sequence of a filtered differential group

Let A be a filtered d-group with an increasing filtration $\{ A^p \}$. Introduce the following notations:

$$Z_p = \{ x \in A^p \mid d(x) = 0 \},$$

$$Z_p^n = \{ x \in A^p \mid d(x) \in A^{p-n} \}, \quad (n \geqslant 0),$$

$$B_p^n = d(Z_{p+n}^n), \quad (n \geqslant 0).$$

Verify the following inclusions and equalities:

$$Z_p \subset Z_{p+1},$$

$$d(A^p) = B_p^0 \subset B_p^1 \subset \cdots \subset B_p^n \subset B_p^{n+1} \subset \cdots \subset Z_p,$$

$$Z_p \subset \cdots \subset Z_p^n \subset Z_p^{n-1} \subset \cdots \subset Z_p^1 \subset Z_p^0 = A^p,$$

$$Z_{p-1}^n = Z_p^{n+1} \cap A^{p-1} \subset Z_p^{n+1},$$

$$B_{p-1}^n = B_p^{n-1} \cap A^{p-1} \subset B_p^{n-1}.$$

Then define the following quotient groups:

$$D_p^n = Z_{p-n+1}/B_{p-n+1}^{n-1}, \quad (n \geqslant 1),$$

$$E_p^n = Z_p^n/(Z_{p-1}^{n-1} + B_p^{n-1}), \ (n \geqslant 1),$$

and the graded groups

$$D^n = \sum_p D_p^n, \qquad E^n = \sum_p E_p^n.$$

The inclusions $Z_{p-n} \subset Z_{p-n+1}$ and $B_{p-n}^{n-1} \subset B_{p-n+1}^{n-1}$ define a homomorphism $i^n : D^n \to D^n$ which is homogeneous of degree 1. The inclusions $Z_{p-n+1} \subset Z_{p-n+1}^n$ and $B_{p-n+1}^{n-1} \subset (Z_{p-n}^{n-1} + B_{p-n+1}^{n-1})$ define a homomorphism $j^n : D^n \to E^n$ which is homogeneous of degree $-(n-1)$. Finally, since $d(Z_p^n) \subset Z_{p-n}$, $d(Z_{p-1}^{n-1}) = B_{p-n}^{n-1}$, and $d(B_p^{n-1}) = 0$, the differential operator d induces a homomorphism $k^n : E^n \to D^n$ which is homogeneous of degree -1. Prove that

$$< D^n, E^n ; i^n, j^n k^n, >$$

is an exact couple for each $n = 1, 2, \cdots$, and in fact is precisely the associated exact couple $\mathscr{C}^n(A)$ defined in § 10. The differential operator d also induces the homomorphism

$$d^n = j^n k^n : E^n \to E^n$$

of degree $-n$. Thus one obtains Leray's direct construction of the spectral sequence $\{ E^n \}$ of the filtered d-group A. See [Leray 1].

Formulate a similar construction for the case that the filtration $\{ A^p \}$ is decreasing.

B. Regular couples satisfying two-term conditions

Let $\mathscr{C} = \langle D, E; i, j, k \rangle$ be either a regular ∂-couple or a regular δ-couple. Prove the following assertions, [Moore 1; p. 327]:

1. Let $\mu > 0$ and $\nu \geqslant 2$. If $E^{\nu}_{p,q} = 0$ whenever $p \neq 0, q \neq 0$, and $p + q \leqslant \mu$, then $\mathscr{C}$ satisfies the two-term condition $\{0, \mu; \nu\}$ with

$$a_0 = -1, \quad b_0 = 1, \quad c_0 = 0, \quad d_0 = 0,$$

$$a_m = 0, \quad b_m = m, \quad c_m = m, \quad d_m = 0, \quad (0 < m \leqslant \mu).$$

2. Let ν, p_0, p_1, p_2 and q_0 be integers such that $\nu \geqslant 1, p_0 > p_1 > p_2 > 0$, and $q_0 > 0$. Then $\mathscr{C}$ satisfies the two-term condition $\{0, p_0 + q_0 - 2; \nu\}$ if the following two conditions are satisfied:

(*i*) $E^{\nu}_{p,q} = 0$ if $p < p_0, p \neq p_1$, and $p \neq p_2$;

(*ii*) $E^{\nu}_{p,q} = 0$ if $q < q_0$.

3. Let ν, p_0, p_1, q_0 and q_1 be integers such that $\nu \geqslant 1, p_0 > p_1 > 0$ and $q_0 > q_1 > 0$. Then $\mathscr{C}$ satisfies the two-term condition $\{0, p_0 + q_0 - 1; \nu\}$ if the following two conditions are satisfied:

(*i*) $E^{\nu}_{p,q} = 0$ if $p < p_0$ and $p \neq p_1$;

(*ii*) $E^{\nu}_{p,q} = 0$ if $q < q_0$ and $q \neq q_1$.

C. Multiplicative structures

Let A be a regular δ-complex. A multiplication in A which makes A a ring is said to be *allowable* if it satisfies the following three conditions:

(M1) If $x \in A^p$ and $y \in A^q$, then $xy \in A^{p+q}$.

(M2) If $x \in A_p$ and $y \in A_q$, then $xy \in A_{p+q}$.

(M3) d is an *anti-derivation*, i.e., for $x \in A_p$ and $y \in A$, we have

$$d(xy) = (dx)y + (-1)^p x(dy).$$

Assume that an allowable multiplication has been given in A. By using (M1)–(M3), define multiplicative structures in the bigraded groups D^n and E^n of the exact couples

$$\mathscr{C}^n(A) = \langle D^n, E^n; i^n, j^n, k^n \rangle$$

for each $n = 1, 2, \cdots$. Prove the following assertions, [Massey, 2]:

1. If x is homogeneous of degree (p, q) and y is homogeneous of degree (r, s), then xy is homogeneous of degree $(p + r, q + s)$.

2. i^n is a *transducer* of D^n, i.e.,

$$i^n(xy) = x[i^n(y)] = [i^n(x)]y.$$

3. j^n is *multiplicative*, i.e. $j^n(xy) = j^n(x) \cdot j^n(y)$.

4. $d^n = j^n k^n$ is an *anti-derivation* of E^n, i.e., for $x \in E_{p,q}$ and $y \in E^n$, we have

$$d^n(xy) = [d^n(x)]y + (-1)^{p+q} x\, [d^n(y)].$$

5. For $x \in D^n_{p,q}$ and $y \in E^n$, we have

$$k^n[j^n(x)\cdot y] = (-1)^{p+q} x\cdot k^n(y),$$
$$k^n[y\cdot j^n(x)] = k^n(y)\cdot x.$$

D. The exact couples of a bundle space over a finite simplicial complex

Let X be a bundle space over a base space B with projection $\omega : X \to B$ and fiber F as defined in (III, § 4). Assume that B is a finite simplicial complex with B^p denoting its p-dimensional skeleton and that G is an abelian group. Define the bigraded groups D and E by taking the singular homology groups

$$D_{p,q} = H_{p+q}(X_p; G), \quad E_{p,q} = H_{p+q}(X_p, X_{p-1}; G),$$

where $X_p = \omega^{-1}(B^p)$, and the homomorphisms $i : D \to D$, $j : D \to E$, and $k : E \to D$ as in the example of § 5. Thus one obtains an exact couple

$$\mathscr{C} = \langle D, E; i, j, k \rangle$$

which is called the *homology exact couple* of the bundle space X over B. Consider the derived couples $\mathscr{C}^n = \langle D^n, E^n; i^n, j^n, k^n \rangle$ of C and prove the following assertions:

1. $\mathscr{C}^n$ does not depend on the triangulation of B if $n \geqslant 2$.
2. $E^2_{p,q}$ is naturally isomorphic to the homology group $H_p(B; H_q(F; G))$ of the complex B with local coefficients in $H_q(F; G)$.
3. $\mathscr{C}$ is a regular ∂-couple with $D_{p,q} \approx H_{p+q}(X; G)$ if $q < 0$.

Similarly, one can construct the *cohomology exact couple* and prove the analogous assertions.

E. The homotopy exact couple

Let K be a connected cellular complex and v be a vertex of K. Denote by K^n the n-dimensional skeleton of K. Then, for each pair (K^p, K^{p-1}), there is an exact homotopy sequence

$$\cdots \to \pi_m(K^{p-1},v) \xrightarrow{i} \pi_m(K^p,v) \xrightarrow{j} \pi_m(K^p,K^{p-1},v) \xrightarrow{k} \pi_{m-1}(K^{p-1},v) \to \cdots.$$

Define two bigraded groups D and E as follows:

$$D_{p,q} = \begin{cases} \pi_{p+q}(K^p, v), & \text{if } p \geqslant 0 \text{ and } p+q \geqslant 2, \\ 0, & \text{for other values of } p \text{ and } q; \end{cases}$$

$$E_{p,q} = \begin{cases} \pi_{p+q}(K^p, K^{p-1}, v), & \text{if } p \geqslant 1 \text{ and } p+q \geqslant 3, \\ j[\pi_{p+q}(K^p, v)], & \text{if } p \geqslant 1 \text{ and } p+q = 2, \\ 0, & \text{for other values of } p \text{ and } q. \end{cases}$$

The homomorphisms i, j, k in the homotopy sequences of the pairs (K^p, K^{p-1}) may be extended in a unique way to define homomorphisms

$$i : D \to D, \quad j : D \to E, \quad k : E \to D.$$

Prove that the following triangle is exact:

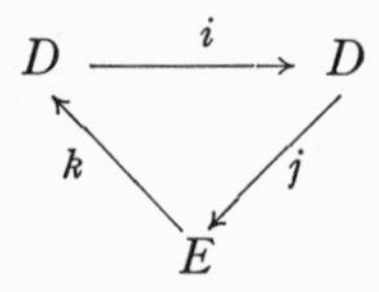

Thus we obtain an exact couple $\mathscr{C}(K, v) = < D, E; i, j, k >$ which is called *the homotopy exact couple* of K at v. Prove the following assertions [Massey 1]:

1. The homomorphisms i, j, k in $\mathscr{C}(K, v)$ are homogeneous of degree $(1, -1)$, $(0, 0)$, $(-1, 0)$ respectively. Hence $\mathscr{C}(K, v)$ is a ∂-couple in the sense of § 5.

2. $D_{p,q} = 0$ if $p < 2$ or $p + q < 2$; $D_{p,q} \approx \pi_{p+q}(K, v)$ if $q < 0$ and $p + q > 1$.

3. $E_{p,q} = 0$ if $p < 2$ or $q < 0$ or $p + q < 2$.

4. $\mathscr{C}(K, v)$ is a regular ∂-couple in the sense of § 6.

5. If v_0 and v_1 are any two vertices of K, and if $\sigma : I \to K^1$ is a path joining v_0 to v_1, then ξ induces a natural isomorphic mapping

$$\mathscr{C}(\xi) = (\phi_\xi, \psi_\xi) : \mathscr{C}(K, v_1) \approx \mathscr{C}(K, v_0)$$

in the sense of § 9. Furthermore, if two paths $\xi, \eta : I \to K^1$ from v_0 to v_1 are homotopic in K with endpoints fixed, then $\mathscr{C}(\xi) = \mathscr{C}(\eta)$. These facts can be concisely expressed by saying that the set of homotopy exact couples $\mathscr{C}(K, v)$ for various vertices v of K constitutes a *local system* of exact couples in K. The same remarks hold for the successive derived couples $\mathscr{C}^n(K, v)$.

6. Let K, L be connected cellular complexes, v, w be vertices of K, L respectively, and $f : (K, v) \to (L, w)$ be a cellular map. Then f induces a mapping

$$\mathscr{C}(f) = (\phi_f, \psi_f) : \mathscr{C}(K, v) \to \mathscr{C}(L, w).$$

Let $\mathscr{K}$ denote the category in which the objects are all pairs (K, v) of a cellular complex K with a vertex $v \in K$ and the mappings are all cellular maps. Let $\mathscr{C}_\partial$ denote the category of all regular ∂-couples and their mappings. Then the operation $(K, v) \to \mathscr{C}(K, v)$ and $f \to \mathscr{C}(f)$ is a covariant functor from $\mathscr{K}$ to $\mathscr{C}_\partial$.

7. Let $\tilde{K}$ be a connected covering complex over K with projection $\omega : \tilde{K} \to K$ and $\tilde{v}$ be a vertex of $\tilde{K}$ with $\omega(\tilde{v}) = v$. Then

$$\mathscr{C}(\omega) : \mathscr{C}(\tilde{K}, \tilde{v}) \approx \mathscr{C}(K, v).$$

The significance of this result is that there is no essential lack of generality if we assume that K is simply connected when discussing the properties of $\mathscr{C}(K, v)$ and its derived couples.

8. The derived couple $\mathscr{C}^2(K, v) = < D^2, E^2; i^2, j^2, k^2 >$ is an invariant of the homotopy type of K.

9. $D^2_{p,q} = 0$ if $p < 3$ or $p + q < 2$.

10. $D^2_{p,q} \approx \pi_{p+q}(K, v)$ if $q < 0$ and $p + q > 1$.

11. $E^2_{p,q} = 0$ if $p < 2$ or $q < 0$.

12. If K is simply connected, then $E^2_{p,0}$ is isomorphic to the homology group $H_p(K)$ of the complex K for each $p \geqslant 2$.

13. If K is r-connected, then $D^2_{p,q} = 0$ for $p \leqslant r + 1$ and $E^2_{p,q} = 0$ for $p \leqslant r$.

14. The exact couple $\mathscr{C}^2(K, v)$ contains the following exact sequence of J. H. C. Whitehead:

$$\cdots \xrightarrow{k^2} D^2_{p,0} \xrightarrow{i^2} D^2_{p+1,-1} \xrightarrow{j^2} E^2_{p,0} \xrightarrow{k^2} D^2_{p-1,0} \xrightarrow{i^2} \cdots.$$

If K is simply connected, then, by 10 and 12, there are natural isomorphisms

$$D^2_{p+1,-1} \approx \pi_p(K, v), \quad E^2_{p,0} \approx H_p(K)$$

for each $p \geqslant 2$. Hence we obtain the following exact sequence for a simply connected K:

$$\cdots \xrightarrow{k^2} D^2_{p,0} \xrightarrow{i^2} \pi_p(K, v) \xrightarrow{j^2} H_p(K) \xrightarrow{k^2} D^2_{p-1,0} \xrightarrow{i^2} \cdots.$$

As a consequence of this and 12, we obtain the result that, if $n \geqslant 2$ and K is $(n - 1)$-connected, then $j^2 : \pi_p(K, v) \to H_p(K)$ is an isomorphism for $p \leqslant n$ and an epimorphism for $p = n + 1$. This includes essentially the Hurewicz theorem.

15. Since $\mathscr{C} = \mathscr{C}(K, v)$ is a regular ∂-couple we may apply the results of § 6 and § 7. Thus we obtain that

$$\mathscr{H}_p(\mathscr{C}) \approx \pi_p(K, v), \quad R_p(\mathscr{C}) = 0, \quad S_p(\mathscr{C}) \approx H_p(K)$$

for every $p \geqslant 2$ and that these groups are all trivial whenever $p < 2$. Then, (R∂10) of § 7 gives a commutative triangle for each $p \geqslant 2$:

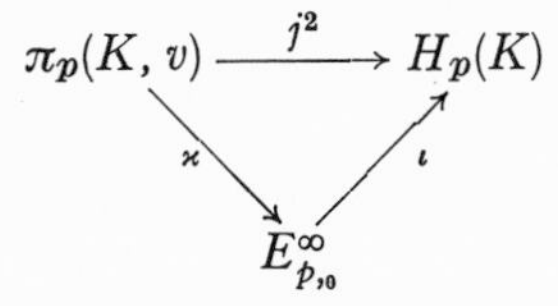

This implies that $E^\infty_{p,0}$ is isomorphic to the image of $\pi_p(K, v)$ in $H_p(K)$ under j^2.

16. The inclusion $K^p \subset K$ induces homomorphisms

$$\lambda_{p,q} : \pi_{p+q}(K^p, v) \to \pi_{p+q}(K, v).$$

Let us denote the image of $\lambda_{p,q}$ by $\pi_{p,q}(K, v)$. Then

$$\mathscr{H}_{p,q}(\mathscr{C}) \approx \pi_{p,q}(K, v) \quad \text{if } p \geqslant 2 \text{ and } p + q \geqslant 2;$$

$$\mathscr{H}_{p,q}(\mathscr{C}) = 0 \quad \text{if } p < 2 \text{ or } p + q < 2.$$

Hence (R∂7) of § 6 gives a finite sequence

$$\pi_m(K, v) = \pi_{m,0}(K, v) \supset \pi_{m-1,1}(K, v) \supset \cdots \supset \pi_{1,m-1}(K. v) = 0$$

for every $m \geqslant 2$. Furthermore (R∂8) of § 6 implies that

$$\pi_{p,q}(K, v)/\pi_{p-1,q+1}(K, v) \approx E^\infty_{p,q}.$$

If $q = 0$, the triangle in 15 shows that $\pi_{p-1,1}(K, v)$ is the kernel of the homomorphism j^2.

17. Generalize the foregoing results to semi-simplicial complexes or, more generally, to CW-complexes.

18. Finally, let X be a pathwise connected space and x_0 a given point in X. Consider the singular complex $K = S(X)$. Then K is a connected semi-simplicial complex and x_0 determines a vertex v of K. $\mathscr{C}(K, v)$ is a homotopy invariant of the pair (X, x_0) and will be called the *homotopy exact couple* of X at x_0, denoted by $\mathscr{C}(X, x_0)$. By means of the relations

$$\pi_m(K, v) \approx \pi_m(X, x), \quad H_m(K) = H_m(X),$$

deduce results concerning $\pi_m(X, x_0)$ and $H_m(X)$ analogous to those given above for the pair (K, v).

The usefulness of the homotopy exact couple and its associated spectral sequence is limited by the fact that we know very little about the groups $E_{p,q}$.

F. The cohomotopy exact couple

Let K be a finite cellular complex of dimension n. Let r denote the least integer greater than $\frac{1}{2}(n + 1)$. Then we have the following exact cohomotopy sequence of the triple (K, K^{p+1}, K^p):

$$\pi^r(K,K^{p+1}) \xrightarrow{i} \cdots \to \pi^m(K,K^{p+1}) \xrightarrow{i} \pi^m(K,K^p) \xrightarrow{j} \pi^m(K^{p+1},K^p)$$
$$\xrightarrow{k} \pi^{m+1}(K,K^{p+1}) \to \cdots$$

where i and j are homomorphisms induced by the inclusions and k denotes the coboundary operator. Define two bigraded groups D and E as follows:

$$D_{p,q} = \begin{cases} \pi^{p+q}(K, K^p), & \text{if } p + q \geqslant r, \\ 0, & \text{if } p + q < r; \end{cases}$$

$$E_{p,q} = \begin{cases} \pi^{p+q}(K^{p+1}, K^p), & \text{if } p + q \geqslant r, \\ i^{-1}(0) \subset \pi^r(K, K^{p+1}), & \text{if } p + q = r - 1, \\ 0, & \text{if } p + q < r - 1. \end{cases}$$

The homomorphisms i, j, k in the cohomotopy sequence may be extended in an obvious way to define homomorphisms i, j, k of the triangle

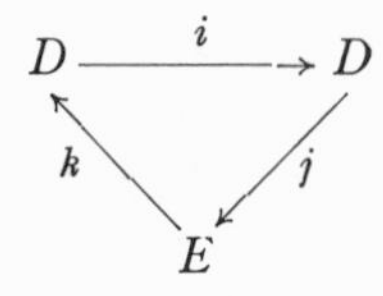

which are homogeneous of degrees $(-1, 1)$, $(0, 0)$ and $(1, 0)$ respectively. Prove that the triangle is exact. Thus we obtain a δ-couple $\mathscr{C}^*(K) = \langle D, E; i, j, k \rangle$ which is not necessarily regular in the sense of § 6; it is called the *cohomotopy exact couple* of K, [Massey 1] and [Peterson 1].

Study $\mathscr{C}^*(K)$ and its derived couples as in the previous exercise. In particular, the exact couple $\mathscr{C}^{*2}(K)$ contains the following exact sequence

$$\cdots \xrightarrow{k^2} D^2_{p-1,1} \xrightarrow{i^2} D^2_{p-2,2} \xrightarrow{j^2} E^2_{p-1,1} \xrightarrow{k^2} D^2_{p,1} \xrightarrow{i^2} \cdots$$

Prove that, for each $p \geqslant r$, we have

$$D^2_{p-1,1} = \Gamma^p(K), \quad D^2_{p-2,2} = \pi^p(K), \quad E^2_{p-1,1} \approx H^p(K),$$

where $\Gamma^p(K)$ denotes the image of the homomorphism $i : \pi^p(K, K^p) \to \pi^p(K, K^{p-1})$. Hence the preceding exact sequence becomes

$$\Gamma^r \longrightarrow \cdots \longrightarrow \Gamma^p \longrightarrow \pi^p(K) \xrightarrow{h} H^p(K) \xrightarrow{\Delta} \Gamma^{p+1} \longrightarrow \cdots$$

where h denotes the natural homomorphism defined in (VI; § 11). Prove that $\Gamma^p = 0$ for all $p \geqslant n$. This implies that h is an isomorphism for $p = n$ and is an epimorphism for $p = n - 1$. Thus we obtain the Hopf theorems as corollaries.

G. The Gamma functor of abelian groups

A function $f : A \to B$ of an abelian group A into another B is said to be *quadratic* if

$$f(a+b+c) - f(b+c) - f(c+a) - f(a+b) + f(a) + f(b) + f(c) = 0$$

for all elements a, b, c in A. Let $\phi_f : A \times A \to B$ denote the function defined by

$$\phi_f(a, b) = f(a + b) - f(a) - f(b)$$

for every pair of elements in A; in other words, ϕ_f is the *deviation* of f from a homomorphism. Prove that f is quadratic iff ϕ_f is bilinear. Hence every homomorphism is quadratic. Also prove that compositions of quadratic functions with homomorphisms are quadratic.

A quadratic function $f : A \to B$ is said to be *homogeneous* if

$$f(-a) = f(a)$$

for every $a \in A$. Prove that every quadratic function is the sum of a homomorphism and a homogeneous quadratic function.

Define an abelian group $\Gamma(A)$ as follows. Let $F(A)$ denote the free abelian group generated by a set of generators in a $1 - 1$ correspondence with the elements $a \in A$. Let $R(A)$ denote the subgroup of $F(A)$ generated by the elements of the forms

$$w(a) - w(-a),$$

$$w(a+b+c) - w(b+c) - w(c+a) - w(a+b) + w(a) + w(b) + w(c)$$

for all elements a, b, c in A, where $w(a)$ denotes the generator of $F(A)$ which corresponds to the element $a \in A$. Then the group $\Gamma(A)$ is defined by

$$\Gamma(A) = F(A)/R(A).$$

For each $a \in A$, let $[a]$ denote the element $w(a) + R(A)$ in $\Gamma(A)$. Prove that the assignment $a \to [a]$ gives a homogeneous quadratic function $\gamma : A \to \Gamma(A)$.

If $f : A \to B$ is any given homogeneous quadratic function, prove that there exists a homomorphism $h : \Gamma(A) \to B$ such that $h\gamma = f$.

Next, let $f : A \to B$ be a homomorphism. Prove that the assignment $[a] \to [fa]$ gives a homomorphism

$$\Gamma(f) : \Gamma(A) \to \Gamma(B)$$

called the homomorphism induced by f. Verify that the assignments $A \to \Gamma(A)$ and $f \to \Gamma(f)$ define a covariant functor Γ from the category of abelian groups and homomorphisms into itself and hence $f : A \approx B$ implies $\Gamma(f) : \Gamma(A) \approx \Gamma(B)$.

Prove the following elementary properties, [J. H. C. Whitehead 7]:

1. If A is additive group of rationals, then $\Gamma(A) \approx A$.

2. If every element in A is of finite order and also divisible by its order, then $\Gamma(A) = 0$. In particular, if A is the group of rationals mod 1, then $\Gamma(A) = 0$.

3. If A is a free abelian group with a free basis $\{ a_i \}$ indexed by an ordered set $\{ i \}$, then $\Gamma(A)$ is the free abelian group freely generated by the elements $\gamma(a_i), \phi_\gamma(a_j, a_k)$ with $j < k$.

4. If A is a finite cyclic group of order h and if a is a generator of A, then $\Gamma(A)$ is a cyclic group of order h or $2h$, according as h is odd or even, and is generated by $\gamma(a)$.

5. If A is the direct sum ΣA_p of its subgroups $\{ A_p \}$ indexed by an ordered set $\{ p \}$, then

$$\Gamma(A) = \sum_p \Gamma(A_p) + \sum_{q<r} A_q \otimes A_r.$$

Properties (3)–(5) provide a complete computation of the group $\Gamma(A)$ for A an abelian group with a finite number of generators.

6. If f is an epimorphism, so is $\Gamma(f)$.

7. If A is finitely generated, then $\Gamma(f) : \Gamma(A) \approx \Gamma(B)$ implies that $f : A \approx B$.

8. If π is a group of operators for A, then π also operates on $\Gamma(A)$ according to the rule $\xi\gamma(a) = \gamma(\xi a)$ for $\xi \in \pi$ and $a \in A$.

9. If π operates on A and B and if f commutes with the operators, then so does $\Gamma(f)$.

10. Γ commutes with the direct limit functor.

Finally, prove that the group $D^2_{3,0}$ in (14) of Ex. E is isomorphic to $\Gamma(\pi_2(K, v))$.

H. Transgression and suspension

Let A be a regular ∂-complex. Consider the natural epimorphism $g : A \to \tilde{A}$ defined in § 11. Let $A^0_p = A^0 \cap A_p$. If $p < 0$, then g maps A^0_p into 0 and hence defines a homomorphism

$$h : A_p/A^0_p \to \tilde{A}_p.$$

If $p > 1$, then h commutes with d and hence induces a derived homomorphism

$$h_* : \mathscr{H}_p(A/A^0) \to \mathscr{H}_p(\tilde{A}), \quad (p \geqslant 2).$$

Prove that the images of $g_* = \mathscr{H}(g)$ and h_* in $\mathscr{H}_p(\tilde{A})$ are

$$g_*[\mathscr{H}_p(A)] = E^{p+1}_{p,0}, \quad h_*[\mathscr{H}_p(A/A^0)] = E^p_{p,0}.$$

Next, consider h_* together with the boundary homomorphism ∂:

$$\mathscr{H}_p(\tilde{A}) \xleftarrow{h_*} \mathscr{H}_p(A/A^0) \xrightarrow{\partial} \mathscr{H}_{p-1}(A^0), \quad (p \geqslant 2).$$

Denote by J the image of h_*, K the kernel of h_*, L the image of ∂, and M the kernel of ∂.

Let $x \in J$. Choose $y \in \mathscr{H}_p(A/A^0)$ with $h_*(y) = x$, and consider $\partial(y)$. It is an element of $\mathscr{H}_{p-1}(A^0)$ which, when y varies, describes a coset mod $\partial(K)$. Hence we obtain a homomorphism

$$T : J \to \mathscr{H}_{p-1}(A^0)/\partial(K)$$

called the *transgression*.

As the image of h_*, we have $J = E^p_{p,0}$. By § 7, $E^p_{0,p-1}$ is a quotient group of $\mathscr{H}_{p-1}(A^0)$. Denote the projection by $\varkappa$. Prove that the rectangle

$$\begin{array}{ccc} \mathscr{H}_p(A/A^0) & \xrightarrow{\partial} & \mathscr{H}_{p-1}(A^0) \\ \downarrow{\scriptstyle h_*} & & \downarrow{\scriptstyle \varkappa} \\ E^p_{p,0} & \xrightarrow{d^p} & E^p_{0,p-1} \end{array}$$

is commutative and $\partial(K)$ is the kernel of $\varkappa$. Hence $\varkappa$ induces an isomorphism

$$\varkappa_* : \mathscr{H}_{p-1}(A)/\partial(K) \approx E^p_{0,p-1}, \quad (p \geqslant 2).$$

After this identification, the transgression T reduces to the differential operator

$$d^p : E^p_{p,0} \to E^p_{0,p-1}.$$

For an arbitrary ∂-couple, we may define the *transgression* to be this differential operator d^p.

Analogous to the transgression, define a homomorphism

$$\Sigma : L \to \mathscr{H}_p(\tilde{A})/h_*(M)$$

called the *suspension*. If $\mathscr{H}_p(A) = 0 = \mathscr{H}_{p-1}(A)$, then ∂ becomes an isomorphism and the suspension reduces to

$$\Sigma = h_* \partial^{-1} : \mathscr{H}_{p-1}(A^0) \to \mathscr{H}_p(\tilde{A}).$$

In this case, the transgression $d^p = T$ is an isomorphism and in the diagram

$$\begin{array}{ccc} \mathscr{H}_{p-1}(A^0) & \xrightarrow{\Sigma} & \mathscr{H}_p(\tilde{A}) \\ \downarrow{\scriptstyle \varkappa} & & \uparrow{\scriptstyle \iota} \\ E^p_{0,p-1} & \xleftarrow{d_p} & E^p_{p,0} \end{array}$$

the image of Σ coincides with the image of the monomorphism ι and the kernel of Σ is precisely the kernel of the epimorphism $\varkappa$. Furthermore, prove that the following relation holds:

$$\Sigma = \iota (d^p)^{-1} \varkappa.$$

Establish analogous results for a regular δ-complex A.

I. Properties of Steenrod squares

Let (X, A) denote any pair consisting of a space X and a subspace A of X, and Z_2 denote the cyclic group of order 2. Reproduce the definition of the Steenrod square operations

$$Sq^i : H^n(X, A; Z_2) \to H^{n+i}(X, A; Z_2), \quad i \geqslant 0,$$

and prove the following properties of these operations, [Steenrod 2 and Cartan 1]:

1. $Sq^i \circ f^* = f^* \circ Sq^i$ for any map $f: (X, A) \to (Y, B)$.

2. $Sq^i \circ \delta = \delta \circ Sq^i$, where δ denotes the coboundary operator in the cohomology sequence.

3. $Sq^i(\alpha \cup \beta) = \sum_{j+k=i} Sq^j(\alpha) \cup Sq^k(\beta)$.

4. $Sq^i(\alpha) = 0$ if $\dim(\alpha) < i$.

5. $Sq^i(\alpha) = \alpha \cup \alpha$ if $\dim(\alpha) = i$.

6. $Sq^0(\alpha) = \alpha$.

7. Sq^1 coincides with the coboundary operator induced by the exact sequence

$$0 \to Z_2 \to Z_4 \to Z_2 \to 0$$

and hence we have an exact sequence

$$\cdots \to H^n(X,A; Z_4) \to H^n(X,A; Z_2) \xrightarrow{Sq^1} H^{n+1}(X,A; Z_2) \to H^{n+1}(X,A; Z_4) \to \cdots$$

CHAPTER IX

THE SPECTRAL SEQUENCE OF A FIBER SPACE

1. Introduction

We turn now to the study of the relations among the homology and cohomology groups of the various spaces of a fibering. As already indicated, the principal tool is the machinery of spectral sequences developed in the preceding chapter.

Since we shall follow Serre, we use the singular groups; our principal hypotheses will be that the fibering have the covering homotopy property and that the fiber be pathwise connected. It is convenient to use cubes rather than simplexes in defining the singular groups, and hence we begin with an outline of this construction of the singular complex of a space. The associated exact couple of a fibering is then introduced, and the term E^2 of this exact couple is computed (§§ 3–10; the main result appears in §§ 5–6).

A variety of applications follow. Relations among the Poincaré polynomials of the fiber space, base space, and fiber are deduced in § 11; and several exact sequences, including those of Gysin and Wang, are constructed in the next three sections.

In the final sections of this chapter (§§ 15–18), regular covering spaces are treated: there is a difficulty, namely that the fiber fails to be connected, but this is avoided by the introduction of certain auxillary fiberings. A spectral sequence due to H. Cartan is constructed and used to deduce certain facts about the action of finite groups on a sphere and the celebrated results on the determination of certain homology and cohomology groups in terms of the fundamental group.

2. Cubical singular homology theory

In the traditional simplicial singular homology theory, [E–S; pp. 185–211], the unit n-simplex Δ_n is used as the anti-image in defining the n-chains of a space. However, to study the spectral sequence of a fiber space, it will be more convenient to use the cubical definition in which the n-cube I^n plays the role of Δ_n. In the present section, we shall give a sketch of the cubical theory. As to the equivalence of the cubical theory and the simplicial theory, a proof is given in [Eilenberg and MacLane 2].

By a *singular n-cube* in a space X, we mean a map $u : I^n \to X$. If $n = 0$, then u is interpreted as a single point in X. If $n > 0$, we define the *i-th lower* and *upper faces* $\lambda_i^0 u$ and $\lambda_i^1 u$ of u to be the singular $(n-1)$-cubes given by

$$(\lambda_i^{\varepsilon} u)(t_1, \cdots, t_{n-1}) = u(t_1, \cdots, t_{i-1}, \varepsilon, t_i, \cdots, t_{n-1})$$

for every $i = 1, 2, \cdots, n$, $\varepsilon = 0, 1$, and $(t_1, \cdots, t_{n-1}) \in I^{n-1}$. Then, for $i < j$, we have

$$\lambda_i^{\varepsilon} \lambda_j^{\eta} = \lambda_{j-1}^{\eta} \lambda_i^{\varepsilon}$$

where ε and η may be either 0 or 1.

Define $Q_n(X)$ to be the free abelian group generated by all singular n-cubes in X if $n \geqslant 0$ and $Q_n(X) = 0$ if $n < 0$. Then the operation

$$\partial u = \sum_{i=1}^{n} (-1)^i (\lambda_i^1 u - \lambda_i^0 u)$$

determines a homomorphism

$$\partial : Q_n(X) \to Q_{n-1}(X)$$

for every n. It is straightforward to verify that $\partial\partial = 0$. This yields a chain complex $\{ Q_n(X), \partial \}$, [E–S; p. 124]. Unlike the simplicial theory, this chain complex $\{ Q_n(X), \partial \}$ does not give the "correct" homology groups of X; for example, if X consists of a single point, a simple computation reveals that the n-th integral homology group of $\{ Q_n(X), \partial \}$ is infinite cyclic for every $n \geqslant 0$. Hence, we have to "normalize" $\{ Q_n(X), \partial \}$.

For each singular $(n - 1)$-cube u in X, $n > 0$, we define a singular n-cube Du in X by taking

$$(Du)(t_1, \cdots, t_{n-1}, t_n) = u(t_1, \cdots, t_{n-1}).$$

A singular n-cube v in X is said to be *degenerate* if $v = Du$ for some u. In other words, v is degenerate iff it does not depend on the last coordinate t_n of the point $(t_1, \cdots, t_n)$ in I^n. The degenerate singular n-cubes in X, $n > 0$, generate a subgroup $D_n(X)$ of $Q_n(X)$. Since

$$\lambda_i^{\varepsilon} D = D \lambda_i^{\varepsilon}, \quad i < n, \quad \lambda_n^{\varepsilon} D = 1,$$

it follows that

$$\partial Du = \sum_{i=1}^{n} (-1)^i (\lambda_i^1 Du - \lambda_i^0 Du) = \sum_{i=1}^{n-1} (-1)^i (D \lambda_i^1 u - D \lambda_i^0 u)$$

and hence ∂ carries $D_n(X)$ into $D_{n-1}(X)$. So, the degenerate singular n-cubes for all n form a subcomplex $\{ D_n(X), \partial \}$ of the chain complex $\{ Q_n(X), \partial \}$.

For each integer n, the quotient group

$$C_n(X) = Q_n(X)/D_n(X)$$

is obviously a free abelian group and will be called *the group of normalized cubical singular n-chains in* X. Since ∂ carries $D_n(X)$ into $D_{n-1}(X)$, it induces a homomorphism

$$\partial : C_n(X) \to C_{n-1}(X)$$

for every n, which will be called the *boundary homomorphism*. Since $\partial\partial = 0$, we obtain a chain complex $\{ C_n(X), \partial \}$. For an arbitrarily given abelian group G, the homology group $H_n(X; G)$ and the cohomology group $H^n(X; G)$

of this chain complex over G are called respectively *the n-dimensional cubical singular homology and cohomology group of the space X over the coefficient group G.*

Let $C(X)$ denote the direct sum of the groups $C_n(X)$ for all n. Then, $C(X)$ is a graded differential group with a homogeneous differential operator

$$\partial : C(X) \to C(X)$$

of degree -1. We shall call $C(X)$ *the group of all normalized cubical singular chains in X.*

For any subspace X_0 of X, the group $C_n(X_0)$ can be considered as a subgroup of the group $C_n(X)$. The quotient group

$$C_n(X, X_0) = C_n(X)/C_n(X_0)$$

is obviously isomorphic to the free abelian group generated by the non-degenerate singular n-cubes in X not contained in X_0. Since ∂ carries $C_n(X)$ into $C_{n-1}(X)$ and $C_n(X_0)$ into $C_{n-1}(X_0)$, it induces a *boundary homomorphism*

$$\partial : C_n(X, X_0) \to C_{n-1}(X, X_0).$$

Since $\partial\partial = 0$, we obtain a chain complex $\{ C_n(X, X_0), \partial \}$ and hence the relative homology groups $H_n(X, X_0; G)$ and the relative cohomology groups $H^n(X, X_0; G)$ over an arbitrarily given abelian group G. Furthermore, if $G = \{ G_x \mid x \in X \}$ denotes a local system of abelian groups in X, then one can define the groups $H_n(X, X_0; G)$ and $H^n(X, X_0; G)$ with local coefficients in G in a way similar to that in the traditional singular homology theory. See [Steenrod 1] and [Eilenberg 3].

The direct sum $C(X, X_0)$ of the groups $C_n(X, X_0)$ is a graded differential group, called *the group of all normalized cubical singular chains in X modulo* X_0. The direct sum $H(X, X_0; G)$ of the groups $H_n(X, X_0; G)$ will be called the *total singular homology group of X modulo X_0 over G*, and the direct sum $H^*(X, X_0; G)$ of the groups $H^n(X, X_0; G)$ will be called the *total singular cohomology group of X modulo X_0 over G.*

For the important special case that both X and X_0 are pathwise connected, pick a point x_0 of X as basic point and assume $x_0 \in X_0$ if X_0 is non-empty. Just as in the traditional singular theory, it may be proved that we may consider only the singular cubes with all vertices at x_0. Let $G = \{ G_x \mid x \in X \}$ denote a local system of abelian groups in X. Since all vertices of the singular cubes are at x_0, the coefficients of the chains and the cochains with local coefficients in G are all in the group G_{x_0} on which the fundamental group $\pi_1(X, x_0)$ operates. For this particular case, the homology groups with local coefficients can be simply defined as follows.

Let G denote an abelian group on which $\pi_1(X, x_0)$ acts as left operators. For each n, consider the group

$$C_n(X, X_0; G) = C_n(X, X_0) \otimes G,$$

where $C_n(X, X_0)$ denotes the group of normalized cubical singular n-chains in X modulo X_0 (defined by the n-cubes with all vertices at x_0). Define a boundary homomorphism

$$\partial : C_n(X, X_0; G) \to C_{n-1}(X, X_0; G)$$

as follows. Consider a generator $u \otimes g$ of the group $C_n(X, X_0) \otimes G$, where $u : I^n \to X$ is a nondegenerate singular cube not in X_0 with all vertices at x_0 and g is an element of G. For each $i = 1, \cdots, n$, we define a loop $\sigma_i u : I \to X$ by taking

$$\sigma_i u(t) = u(t_1, \cdots, t_n)$$

where $t_i = t$ and $= t_j 0$ for every $j \neq i$. The loop $\sigma_i u$ represents an element $[\sigma_i u]$ of $\pi_1(X, x_0)$. Then ∂ is defined by

$$\partial(u \otimes g) = \sum_{i=1}^{n} (-1)^i \{ \lambda_i^1 u \otimes [\sigma_i u] g - \lambda_i^0 u \otimes g \}.$$

It can be easily verified that $\partial\partial = 0$. Hence we obtain a chain complex $\{ C_n(X, X_0; G), \partial \}$. The n-th homology group of this chain complex is defined to be *the n-dimensional homology group $H_n(X, X_0; G)$ of X modulo X_0 with local coefficients in G*. Analogously, one may define *the n-dimensional cohomology group $H^n(X, X_0; G)$ of X modulo X_0 with local coefficients in G.*

In the sequel, we shall need a slight refinement of the notion of the degeneracy of a singular cube. A singular cube $u : I^n \to X$ is said to be of *degeneracy* q iff there exists a non-degenerate singular cube $v : I^{n-q} \to X$ such that $u = D^q v$, where D^q denotes the q-fold iteration of the operation D. Hence the degeneracy of a non-degenerate cube is 0, and degeneracy of $u : I^n \to X$ is not less than q if

$$u(t_1, \cdots, t_n) = u(t_1, \cdots, t_{n-q}, 0, \cdots, 0)$$

for every point $(t_1, \cdots, t_n)$ of I^n.

3. A filtration in the group of singular chains in a fiber space

Throughout §§ 3–14, let $\omega : X \to B$ be a given fibering as defined in (III; § 3). In other words, X is a fiber space over a base space B with ω as projection. Pick a point $x_0 \in X$ and let $b_0 = \omega(x_0) \in B$. The subspace

$$F = \omega^{-1}(b_0)$$

of X will be called simply *the fiber*. We assume that both B and F are pathwise connected. As an immediate consequence of the existence of covering paths, X is also pathwise connected. Hence, according to a remark of § 2, we may consider only the singular cubes with all vertices at x_0 or b_0 when dealing with the cubical singular theory. We assume once for all that every singular cube considered in the sequal is of this limited kind.

A singular cube $u : I^n \to X$ is said to be of *weight* p (in notation : $w(u) = p$) if the singular cube $\omega u : I^n \to B$ is of degeneracy $n - p$; thus u is of weight p iff $u(t_1, \cdots, t_n)$ remains within a fixed fiber as $t_{p+1}, \cdots, t_n$ vary, but moves from fiber to fiber as t_p varies. For each singular cube u in X, the weight $w(u)$ satisfies the condition

$$0 \leqslant w(u) \leqslant \dim (u). \tag{3.1}$$

It is also straightforward to verify the following two relations:

$$w(\lambda_i^\varepsilon u) < w(u) \text{ if } 1 \leqslant i \leqslant w(u) \text{ and } \varepsilon = 0, 1; \tag{3.2}$$

$$w(\lambda_i^\varepsilon u) = w(u) \text{ if } w(u) < i \leqslant \dim (u) \text{ and } \varepsilon = 0, 1. \tag{3.3}$$

Let B_0 be a pathwise connected subspace of the base space B which contains b_0 if B_0 is not empty. Let $X_0 = \omega^{-1}(B_0)$; then X_0 is also pathwise connected.

Let $C = C(X, X_0)$ denote the group of all normalized cubical singular chains in X modulo X_0. Then there is a natural isomorphism between C and the free abelian group generated by the non-degenerate singular cubes in X not contained in X_0. Let us identify these two groups by means of this natural isomorphism. Then C is a graded differential group with $C_m = C_m(X, X_0)$ and with a homogeneous differential operator ∂ of degree -1. Furthermore, we may define an increasing filtration $\{C^p\}$ by taking C^p to be the subgroup of C generated by the non-degenerate singular cubes of weight not exceeding p. It is obvious that

$$\cup_p C^p = C, \quad C^p \subset C^{p+1}, \quad \partial(C^p) \subset C^p$$

and that C^p is a subgroup of the graded group C. In fact $C^p_m = C^p \cap C_m$. If we extend the weight function w over C by defining $w(c)$, $c \in C$, to be the greatest lower bound of the integers p such that $c \in C^p$, then (3.1) implies

$$0 \leqslant w(c) \leqslant \dim (c) \tag{3.4}$$

for every non-zero homogeneous element $c \in C$. Furnished with this filtration $\{C^p\}$, C becomes a regular ∂-complex in the sense of (VIII; § 11).

4. The associated exact couple

Consider the regular ∂-complex $C = C(X, X_0)$ of the previous section. According to (VIII; 11.1), the associated exact couple

$$\mathscr{C}(C) = \langle D, E; i, j, k \rangle$$

is a regular ∂-couple. We shall determine the structure of the bigraded differential group E. Since

$$E_{p,q} = \mathscr{H}_{p+q}(\hat{C}^p), \quad \hat{C}^p = C^p/C^{p-1},$$

we have to analyze the graded differential group $\hat{C}^p$.

By definition, $\hat{C}^p_m = C^p_m/C^{p-1}_m$. Let u be a non-degenerate singular cube in C^p_m, then u represents an element $[u]$ in $\hat{C}^p_m$. By (3.2) and (3.3), it is evident that the differential operator ∂ on $\hat{C}^p$ sends $[u]$ into the following element of $\hat{C}^p_{m-1}$:

$$(i) \qquad \partial[u] = \sum_{i=p+1}^{m} (-1)^i \{ [\lambda_i^1 u] - [\lambda_i^0 u] \}.$$

Now let $u : I^m \to X$ be a singular cube with $w(u) \leqslant p$. Set $q = m - p$ and define two singular cubes

$$B_p u : I^p \to B, \quad F_p u : I^q \to F$$

as follows:

$$B_p u(t_1, \cdots, t_p) = \omega u(t_1, \cdots, t_p, 0, \cdots, 0);$$

$$F_p u(t_1, \cdots, t_q) = u(0, \cdots, 0, t_1, \cdots, t_q).$$

For the second, since

$$\omega u(0, \cdots, 0, t_1, \cdots, t_q) = \omega u(0, \cdots, 0) = b_0,$$

we have $F_p u(t_1, \cdots, t_q) \in F$. Thus, for a given u, we can define $B_p u$ and $F_p u$ for all p such that $w(u) \leqslant p \leqslant \dim(u)$. The following properties of $B_p u$ and $F_p u$ can be easily verified:

(1) If u is in X_0, then $B_p u$ is in B_0.

(2) If $w(u) < p$, then $B_p u$ is degenerate.

(3) If u is degenerate and $q = 0$, then $B_p u$ is degenerate; if u is degenerate and $q > 0$, then $F_p u$ is degenerate.

(4) If $i > p$, then $B_p \lambda_i^\varepsilon u = B_p u$ and $F_p \lambda_i^\varepsilon u = \lambda_{i-p}^\varepsilon F_p u$ for $\varepsilon = 0, 1$.

Let $K^p = C_p(B, B_0) \otimes C(F)$; then K^p is a graded group with $K^p_m = C_p(B, B_0) \otimes C_{m-p}(F)$. Define a differential operator d_F on K^p by

$$(ii) \qquad d_F(b \otimes f) = (-1)^p \, b \otimes \partial f$$

for each generator $b \otimes f$ of K^p. Then d_F is of degree -1. Define a homomorphism $\phi : C^p \to K^p$ by taking

$$\phi(u) = B_p u \otimes F_p u$$

for every generator u of C^p. This definition is justified by the properties (1) and (3) listed above. According to (2), ϕ carries C^{p-1} into the zero element of K^p and hence induces a homomorphism

$$\psi : \hat{C}^p \to K^p$$

by passing to the quotient group $\hat{C}^p = C^p/C^{p-1}$. By using (1) through (4), one can verify that ψ commutes with the differential operators ∂ in $\hat{C}^p$ and d_F in K^p. Hence ψ is a mapping in the sense of (VIII; § 12).

Since $C_p(B, B_0)$ is free, we have

$$\mathscr{H}_{p+q}(K^p) = C_p(B, B_0) \otimes H_q(F)$$

where $H_q(F)$ denotes the q-dimensional singular homology group of F. Hence ψ induces a homomorphism

$$\chi_{p,q} : E_{p,q} \to C_p(B, B_0) \otimes H_q(F)$$

for each pair of integers (p, q).

Theorem 4.1. *$\chi_{p,q}$ is an isomorphism of $E_{p,q}$ onto $C_p(B, B_0) \otimes H_q(F)$.*

To prove this theorem, it suffices to construct a mapping $\mu : K^p \to \hat{C}^p$ such that $\psi\mu$ is the identity endomorphism of K^p and $\mu\psi$ is homotopic to the identity endomorphism of $\hat{C}^p$.

Lemma A. *For every pair of singular cubes $u : I^p \to B$ and $v : I^q \to F$, we can construct a singular cube*

$$z = M(u, v) : I^{p+q} \to X$$

satisfying the conditions:

(A1) $w(z) \leqslant p$.

(A2) $B_p z = u$ *and* $F_p z = v$.

(A3) $\lambda^{\varepsilon}_{p+i} z = M(u, \lambda_i^{\varepsilon} v)$ *for each* $i \leqslant q$ *and* $\varepsilon = 0, 1$.

(A4) *If v is degenerate, so is z.*

(A5) *If u is in B_0, then z is in X_0.*

The proof of this lemma will be given in § 8.

The singular cube $z = M(u, v)$ represents an element $M[u, v]$ of $\hat{C}^p$ because of (A1). If u is degenerate, then (A2) implies that $w(z) < p$ and hence $M[u, v] = 0$. If v is degenerate, then so is z and hence $M[u, v] = 0$. If u is in B_0, then z is in X_0 and hence $M[u, v] = 0$. Therefore, we may define the homomorphism μ by taking

$$\mu(u \otimes v) = M[u, v]$$

for each generator $u \otimes v$ of $C_p(B, B_0) \otimes C(F)$.

The condition (A3) implies that μ commutes with the differential operators. Hence μ is a mapping. The condition (A2) shows that $\psi\mu$ is the identity on K^p. It remains to prove that $\mu\psi$ is homotopic to the identity of $\hat{C}^p$.

Lemma B. *For every singular cube $u : I^{p+q} \to X$ with $w(u) \leqslant p$, we can construct a singular cube*

$$v = D_p u : I^{p+q+1} \to X$$

satisfying the conditions:

(B1) $w(v) \leqslant p$.

(B2) $B_p v = B_p u$.

(B3) $v(0, \cdots, 0\ t, t_1, \cdots, t_q) = u(0, \cdots, 0, t_1, \cdots, t_q)$.

(B4) $\lambda^0_{p+1} v = u$ *and* $\lambda^1_{p+1} v = M(B_p u, F_p u)$.

(B5) $\lambda^{\varepsilon}_{i+1} v = D_p \lambda_i^{\varepsilon} u$ *for each* $i > p$ *and* $\varepsilon = 0, 1$.

(B6) *If $q > 0$ and u is degenerate, so is v.*

(B7) *If u is in X_0, so is v.*

The proof of this lemma will be given in § 9.

Now the proof of the theorem can be completed. The correspondence

$u \to (-1)^p D_p u$ defines a homogeneous endomorphism $\xi : \hat{C}^p \to \hat{C}^p$ of degree 1. To justify this, it suffices to note the facts: if u is degenerate and $q > 0$, then $D_p u$ is degenerate by (B6); if u is degenerate and $q = 0$, then (B2) implies that $w(D_p u) < p$; if u is in X_0, then $D_p u$ is also in X_0 by (B7); and if $w(u) < p$, then $w(D_p u) < p$ according to (B2). By means of the formula (i) and the conditions (B4) and (B5), one can easily verify that

$$\partial(\xi c) + \xi(\partial c) = c - \mu\psi(c)$$

for every $c \in \hat{C}^p$. Hence $\mu\psi$ is homotopic with the identity endomorphism of $\hat{C}^p$. ▮

The isomorphism $\chi_{p,q} : E_{p,q} \approx C_p(B, B_0) \otimes H_q(F)$ indicates the significance of the various degrees. We shall call the primary degree p the *base degree*, and the secondary degree q the *fiber degree*. The total degree $p + q$ is usually called the *dimension*.

5. The derived couple

In the preceding section, we studied the associated exact couple $\mathscr{C}(C) = \langle D, E; i, j, k \rangle$ of the regular ∂-complex C and established an important isomorphism

$$\chi : E \approx C(B, B_0) \otimes H(F)$$

which is homogeneous of degree (0, 0). According to (VIII; § 4), $\mathscr{C}(C)$ has a derived couple

$$\mathscr{C}^2(C) = \langle D^2, E^2; i^2, j^2, k^2 \rangle.$$

The key to all applications is the bigraded group E^2, and hence we shall determine its structure. To this end, we must analyze the differential operator $d = jk$ on E. If we identify E and $C(B, B_0) \otimes H(F)$ by means of the isomorphism χ, then our problem is to determine how d operates on $C(B, B_0) \otimes H(F)$.

For this purpose, let us first show that the fundamental group $\pi_1(B, b_0)$ acts on $H(F)$ as a group of left operators.

Lemma C. *For each loop $\sigma : I \to B$ with $\sigma(0) = b_0 = \sigma(1)$, there exists a homogeneous homomorphism $D_\sigma : C(F) \to C(X)$ of degree* 1 *which assigns to each singular n-cube v in F a singular $(n + 1)$-cube $D_\sigma v$ in X satisfying the conditions*:

(C1) $\lambda_1^0 D_\sigma v = v$.

(C2) $(\omega D_\sigma v)(t, t_1, \cdots, t_n) = \sigma(t)$.

(C3) $D_\sigma \lambda_i^\varepsilon v = \lambda_{i+1}^\varepsilon D_\sigma v$, $(\varepsilon = 0, 1)$.

(C4) *If v is degenerate, so is $D_\sigma v$.*

The proof of this lemma will be given in § 10.

Such a homomorphism D_σ will be called a *deformation of $C(F)$ covering σ*. By (C2), D_σ determines a homogeneous endomorphism $J_\sigma : C(F) \to C(F)$ as follows. For each singular n-cube v in F, $J_\sigma v$ is the singular n-cube in F defined by

$$(J_\sigma v)(t_1, \cdots, t_n) = (D_\sigma v)(1, t_1, \cdots, t_n).$$

By (C4), it follows that $J_\sigma v$ is degenerate if v is degenerate. By (C3), we have $J_\sigma \partial = \partial J_\sigma$. Hence J_σ induces an endomorphism of $H(F)$ which, by the next lemma, depends only on the element $[\sigma]$ of $\pi_1(B, b_0)$.

Lemma D. *If two loops σ, τ represent the same element of $\pi_1(B, b_0)$, then J_σ and J_τ are chain homotopic.*

The proof of this lemma will be given in § 10.

Thus, it is clear that $\pi_1(B, b_0)$ acts on $H(F)$ as a group of left operators. So, we may define a differential operator d_B on $C(B, B_0) \otimes H(F)$ by taking d_B to be the boundary operator on chains with local coefficients in $H(F)$ as defined in § 2.

Now we can settle our problem by saying that, if we identify the groups E and $C(B, B_0) \otimes H(F)$ by means of χ, then the differential operator d coincides with d_B. Precisely, this can be stated in the form of the following

Lemma 5.1. *The isomorphism χ is a mapping, that is to say, $d_B \chi = \chi d$.*

Proof. We shall prove that $d_B = \chi d \chi^{-1}$. Consider the generator $u \otimes h$ of $C(B, B_0) \otimes H(F)$ described above. The homology class $h \in H_q(F)$ can be represented by a singular q-cycle

$$f = \sum_{j=1}^{r} a_j v_j,$$

where the v_j's are q-dimensional singular cubes in F and the a_j's are integers. Consider the singular $(p + q)$-chain

$$z = \sum_{j=1}^{r} a_j M(u, v_j)$$

in X. By the definition of μ and χ, z represents the elements

$$\mu(u \otimes f) \in \hat{C}^p, \quad \chi^{-1}(u \otimes h) \in E_{p,q}.$$

The boundary of the chain z is given by

$$\partial z = \sum_{j=1}^{r} \sum_{i=1}^{p+q} (-1)^i a_j \{ \lambda_i^1 M(u, v_j) - \lambda_i^0 M(u, v_j) \}.$$

If $i > p$, it follows from (A3) of § 4 that

$$\lambda_i^\varepsilon M(u, v_j) = M(u, \lambda_{i-p}^\varepsilon v_j), \quad (\varepsilon = 0, 1).$$

Since f is a cycle, the expression

$$\sum_{j=1}^{r} \sum_{i=1}^{q} (-1)^i a_j (\lambda_i^1 v_j - \lambda_i^0 v_j)$$

is equal to a linear combination of degenerate singular cubes in F. Hence, if we neglect the degenerate terms, we can write ∂z in the following form:

$$\partial z = \sum_{j=1}^{r} \sum_{i=1}^{p} (-1)^i a_j \{ \lambda_i^1 M(u, v_j) - \lambda_i^0 M(u, v_j) \}.$$

According to (3.2), ∂z is a cycle in C^{p-1}. Projecting to the quotient group $\hat{C}^{p-1}$, ∂z represents a cycle of $\hat{C}^{p-1}$ and hence an element y of $E_{p-1,q}$. By the definition of d, we have

$$y = d\chi^{-1}(u \otimes h).$$

Hence the element $\chi(y) = \chi d\chi^{-1}(u \otimes h)$ is represented by the cycle

$$x = \sum_{j=1}^{r} \sum_{i=1}^{p} (-1)^i a_j \{ B_{p-1}\lambda_i^1 M(u, v_j) \otimes F_{p-1}\lambda_i^1 M(u, v_j) - B_{p-1}\lambda_i^0 M(u, v_j) \otimes F_{p-1}\lambda_i^0 M(u, v_j) \}$$

of $C_{p-1}(B, B_0) \otimes C(F)$ under d_F. We have to consider the singular cubes $B_{p-1}\lambda_i^\varepsilon M(u, v_j)$ and $F_{p-1}\lambda_i^\varepsilon M(u, v_j)$ for $i \leqslant p$. The first is evidently $\lambda_i^\varepsilon u$; the second is defined by

$$(F_{p-1}\lambda_i^\varepsilon M(u,v_j))\ (t_1,\cdots,t_q) = M(u,v_j)\ (0,\cdots, 0, \varepsilon, 0,\cdots, 0, t_1,\cdots, t_q),$$

where ε is at the i-th place. Obviously we have $F_{p-1}\lambda_i^0 M(u, v_j) = v_j$; hence

$$x = \sum_{i=1}^{r} (-1)^i \{ \lambda_i^1 u \otimes g_i - \lambda_i^0 u \otimes f \},$$

where g_i denotes the cycle

$$g_i = \sum_{i=1}^{r} a_j F_{p-1}\lambda_i^1 M(u, v_j)$$

of $C(F)$.

To determine the homology class of g_i, let us construct for each q-dimensional singular cube v in F a $(q + 1)$-dimensional singular cube Dv in X defined by

$$(Dv)\ (t, t_1,\cdots, t_q) = M(u, v)\ (0,\cdots, 0, t, 0,\cdots, 0, t_1,\cdots, t_q),$$

where t is at the i-th place. Since

$$(\omega Dv)\ (t, t_1,\cdots, t_q) = (\sigma_i u)\ (t),$$

where $\sigma_i u : I \to B$ denotes the path defined as in § 2, one can easily verify that the assignment $v \to Dv$ determines a deformation

$$D : C(F) \to C(X)$$

covering the path $\sigma_i u$. Since

$$(Dv_j)\ (0, t_1,\cdots, t_q) = v_j(t_1,\cdots, t_q),$$

$$(Dv_j)\ (1, t_1,\cdots, t_q) = F_{p-1}\lambda_i^1 M(u, v_j)\ (t_1,\cdots, t_q),$$

it follows that g_i represents the element $[\sigma_i u]h$ of $H_q(F)$. Therefore, we obtain

$$\chi d\chi^{-1}(u \otimes h) = \sum_{i=1}^{r} (-1)^i \{ \lambda_i^1 u \otimes [\sigma_i u]h - \lambda_i^0 u \otimes h \} = d_B(u \otimes h).$$

Hence we have $\chi d\chi^{-1} = d_B$. ∎

As an immediate consequence of the lemma, we have the following

Theorem 5.2. *The isomorphism χ induces for each pair (p, q) of integers an isomorphism*

$$\varkappa_{p,q} : E^2_{p,q} \approx H_p(B, B_0; H_q(F))$$

of $E^2_{p,q}$ onto the p-dimensional singular homology group of B modulo B_0 with local coefficients in $H_q(F)$.

The isomorphisms $\varkappa_{p,q}$ reveal an important fact, namely that the bigraded group E^2 is the same for all fiber spaces having the same base space B, the same subspace B_0 of B, the same fiber F, and the same operations of $\pi_1(B)$ on $H(F)$. In particular, if $\pi_1(B)$ operates simply on $H(F)$, the bigraded group E^2 is the same as that of the product space $B \times F$. The deviation from the product structure is bound up in the deviation of the differential operator d^2 from 0.

6. Homology with arbitrary coefficients

In the preceding two sections, we restricted ourselves to integral coefficients for the sake of simplicity. However, the results can be easily generalized to arbitrary coefficients as follows.

Let G be an abelian group and C denote the regular ∂-complex $C(X, X_0)$ of the previous sections. Consider the group

$$A = C \otimes G$$

of normalized singular chains of X mod X_0 with coefficients in G. Then A is a graded group with

$$A_m = C_m \otimes G$$

and a differential operator ∂ of degree -1. Define an increasing filtration $\{ A^p \}$ by taking

$$A^p = C^p \otimes G.$$

Thus, A becomes a regular ∂-complex. We shall study the associated exact couple

$$\mathscr{C}(A) = < D, E; i, j, k >$$

and its derived couple

$$\mathscr{C}^2(A) = < D^2, E^2; i^2, j^2, k^2 >.$$

Since C^p is a direct summand of C, it follows that

$$\hat{A}^p = A^p / A^{p-1} = \hat{C}^p \otimes G.$$

Next, let us consider the group

$$K^p = C_p(B, B_0) \otimes C(F) \otimes G.$$

Define a differential operator d_F on K^p by

$$d_F(b \otimes f \otimes g) = (-1)^p b \otimes \partial f \otimes g$$

for every generator $b \otimes f \otimes g$ of K^p. Then

$$\mathscr{H}_{p+q}(K^p) = C_p(B, B_0) \otimes H_q(F; G),$$

where $H_q(F; G)$ denotes the q-dimensional singular homology group of F with coefficients in G.

The homomorphism ψ of § 4 defines in this case a mapping $\psi: \hat{A}^p \to K^p$ and hence induces the homomorphisms

$$\chi_{p,q} : E_{p,q} \to C_p(B, B_0) \otimes H_q(F; G).$$

The lemmas of § 4 imply the following

Theorem 6.1. *$\chi_{p,q}$ is an isomorphism of $E_{p,q}$ onto $C_p(B, B_0) \otimes H_q(F; G)$.*

By the lemmas C and D in § 5, $\pi_1(B, b_0)$ acts on $H(F; G)$ as a group of left operators. Let d_B denote the boundary operator in $C(B, B_0) \otimes H(F; G)$ as chains with local coefficients in $H(F; G)$. Then, one can easily verify that (5.1) is also true in this general case. Hence we have the following

Theorem 6.2. *The isomorphism χ induces for each pair (p, q) of integers an isomorphism*

$$\varkappa_{p,q} : E^2_{p,q} \approx H_p(B, B_0; H_q(F; G))$$

of $E^2_{p,q}$ onto the p-dimensional singular homology group of B modulo B_0 with local coefficients in $H_q(F; G)$.

An important special case of (6.2) is that $A = C(X) \otimes G$. Then B_0 is empty and $\varkappa_{p,q}$ is an isomorphism of $E^2_{p,q}$ onto the homology group $H_p(B; H_q(F; G))$ with local coefficients in $H_q(F; G)$.

Analogous results hold for cohomology with arbitrary coefficients. See Ex. A at the end of the chapter.

In most of the applications, we shall deal with the case where $\pi_1(B, b_0)$ operates simply on the homology and cohomology groups of the fiber F. In particular, this is the case if B is simply connected or if X is a fiber bundle over B with a pathwise connected structural group, [Serre 1; p. 445].

Now assume that the coefficient group G is either the additive group of integers or a field. Then the isomorphism $\varkappa_{p,q}$ in (6.2) is actually an isomorphism of G-modules. If $\pi_1(B, b_0)$ operates simply on $H_q(F; G)$, then the group $H_p(B, B_0; H_q(F; G))$ in (6.2) reduces to the singular homology group with coefficients in $H_q(F; G)$ in the usual sense. Hence, by the Universal Coefficient Theorem, [E–S; p. 161], we obtain the following

Theorem 6.3. *If $\pi_1(B, b_0)$ operates simply on $H_q(F; G)$, then*

$$E^2_{p,q} \approx H_p(B, B_0; G) \otimes_G H_q(F; G) + Tor_G(H_{p-1}(B, B_0; G), H_q(F; G)).$$

The torsion product Tor_G has the important property that $Tor_G(L, M) = 0$ if L or M is a free G-module, [E–S; p. 134]. Hence we have the following

Corollary 6.4. *If $H_{p-1}(B, B_0; G)$ or $H_q(F; G)$ is a free G-module, then*

$$E^2_{p,q} \approx H_p(B, B_0; G) \otimes_G H_q(F; G).$$

Note that the hypothesis of (6.4) is always true if G is a field.

7. The spectral homology sequence

Let us consider the regular ∂-complex

$$A = C \otimes G, \quad C = C(X, X_0)$$

of the preceding section. As in (VIII; § 4), we denote by

$$\mathscr{C}^n(A) = < D^n, E^n; i^n, j^n, k^n >$$

the successive derived couples of the associated exact couple $\mathscr{C}(A) = \mathscr{C}^1(A)$ of A. Then, E^n is a bigraded differential group with $d^n = j^n k^n$ as differential operator and is the n-th term of the associated spectral sequence

$$\{ E^n \mid n = 1, 2, \cdots \}$$

of $\mathscr{C} = \mathscr{C}(A)$, which will be called the *spectral homology sequence* of the fiber space X modulo X_0 over the coefficient group G.

Since A is a regular ∂-complex, we may apply the results of (VIII; §§10–11). In particular,

$$\mathscr{H}(\mathscr{C}) = \mathscr{H}(A) = H(X, X_0; G).$$

Then $H(X, X_0; G)$ is filtered with E^∞ as its associated graded group; more precisely,

$$H_m(X, X_0; G) = \mathscr{H}_{m,0}(\mathscr{C}) \supset \mathscr{H}_{m-1,1}(\mathscr{C}) \supset \cdots \supset \mathscr{H}_{0,m}(\mathscr{C}) \supset \mathscr{H}_{-1,m+1}(\mathscr{C}) = 0,$$

$$\mathscr{H}_{p,q}(\mathscr{C}) / \mathscr{H}_{p-1,q+1}(\mathscr{C}) \approx E^\infty_{p,q}.$$

Hereafter, we shall use the notation $H_{p,q}(X, X_0; G) = \mathscr{H}_{p,q}(\mathscr{C})$.

For further studies in this section, we have to specify whether or not X_0 is empty. Let us first consider the case that X_0 is empty.

According to (VIII; § 11), $R_q(\mathscr{C}) = \mathscr{H}_q(A^0)$. A singular cube u in X is of weight $w(u) = 0$ if and only if the image of ωu is a single point. Since all vertices of u are assumed at x_0, this single point must be b_0 and hence u is in F. Therefore, $A^0 = C(F) \otimes G$. Then it follows that

$$R_q(\mathscr{C}) = H_q(F; G).$$

Next, $S_p(\mathscr{C}) = \mathscr{H}_p(\tilde{A})$ with $\tilde{A}_p = E_{p,0}$. However, by means of $\chi_{p,0}$, we may identify $E_{p,0}$ with $C_p(B) \otimes H_0(F; G)$. Since F is pathwise connected, $H_0(F; G)$ may be identified with G on which $\pi_1(B)$ operates trivially. Hence $\tilde{A} = C(B) \otimes G$ and

$$S_p(\mathscr{C}) = H_p(B; G).$$

Then, we may apply the results of (VIII; § 7) and obtain the following two commutative triangles

where θ_*, ω_* are induced by the inclusion $\theta : F \subset X$ and the projection $\omega : X \to B$, the $\varkappa$'s are epimorphisms, and the ι's are monomorphisms. It is an immediate consequence of these triangles that *$H_{0,q}(X; G)$ is the image of θ_* and $H_{p-1,1}(X; G)$ is the kernel of ω_*.*

It remains to study the case that X_0 is not empty. Then $F \subset X_0$. One can easily see that $A^0 = 0$ and hence $E^n_{0,q} = 0 = E^\infty_{0,q}$; in particular, $R_q(\mathscr{C}) = 0$. On the other hand, we have

$$S_p(\mathscr{C}) = H_p(B, B_0; G).$$

Thus, we still have a non-trivial commutative triangle

$$\begin{array}{ccc} H_p(X, X_0; G) & \xrightarrow{\omega_*} & H_p(B, B_0; G) \\ & {\scriptstyle\varkappa}\searrow \quad \swarrow{\scriptstyle\iota} & \\ & E^\infty_{p,0} & \end{array}$$

The kernel of ω_* is $H_{p-1,1}(X, X_0; G)$ and the image of ω_* is isomorphic to $E^\infty_{p,0}$.

Similarly, one can study the spectral cohomology sequence. See Ex. B.

8. Proof of Lemma A

We shall prove the lemma by induction on the integer q.

First, assume that $q = 0$. Then the singular cube v reduces to the point x_0; and the problem reduces to constructing, for each singular cube $u: I^p \to B$, a singular cube $z = M(u, v) : I^p \to X$ such that $\omega z = u$.

Let V_0 denote the leading vertex $(0, \cdots, 0)$ of I^p. Since V_0 is a strong deformation retract of I^p, we may apply (III; 3.1) and obtain a map $y: I^p \to X$ such that $\omega y = u$ and $y(V_0) = x_0$. Let V_a denote the various vertices of I^p. Since $u(V_a) = b_0$ and $\omega y = u$, it follows that $y(V_a)$ is a point of F. Since F is pathwise connected, there exists a path $\sigma_a : I \to F$ such that $\sigma_a(0) = y(V_a)$ and $\sigma_a(1) = x_0$.

Let Q denote the subspace of I^p which consists of all vertices of I^p. Define a homotopy $f_t : I^p \to B$, $(0 \leqslant t \leqslant 1)$, and a homotopy $g_t : Q \to X$, $(0 \leqslant t \leqslant 1)$, by taking

$$f_t = u, \quad g_t(V_a) = \sigma_a(t)$$

for every $t \in I$ and every vertex V_a of I^p. Then we have

$$g_0 = y \mid Q, \quad \omega g_t = f_t \mid Q$$

for every $t \in I$. According to (III; 3.1), there exists a homotopy $g_t^*: I^p \to X$, $(0 \leqslant t \leqslant 1)$, such that

$$g_0^* = y, \quad g_t = g_t^* \mid Q, \quad \omega g_t^* = f_t$$

for every $t \in I$.

Let $z = g_1^*$. Then z is a singular cube in X with all vertices at x_0 and such that $\omega z = u$. This proves the lemma for $q = 0$.

Next let $q > 0$ and assume that we have constructed $M(u, v)$ for all u and all v with $\dim(v) < q$ satisfying the conditions of the lemma. We are going to construct a singular cube $z = M(u, v)$ for a given pair of singular cubes $u : I^p \to B$ and $v : I^q \to F$ as follows.

To construct the singular cube $z = M(u, v)$, let us first dispose of the special case where v is degenerate. Then v does not depend on the last variable and $\lambda_q{}^0(v) = \lambda_q{}^1(v)$. Define $z = M(u, v)$ by means of the formula:

$$z(t_1, \cdots, t_{p+q}) = M(u, \lambda_q{}^0 v)\,(t_1, \cdots, t_{p+q-1}).$$

We have to verify the conditions (A1)–(A5) of the lemma. The conditions (A1), (A4) and (A5) are obviously satisfied. The condition (A2) is verified as follows:

$$\begin{aligned} B_p z(t_1, \cdots, t_p) &= \omega z(t_1, \cdots, t_p, 0, \cdots, 0) \\ &= \omega M(u, \lambda_q{}^0 v)\,(t_1, \cdots .\, t_p, 0, \cdots, 0) \\ &= u(t_1, \cdots, t_1); \\ F_p z(t_1, \cdots, t_q) &= z(0, \cdots, 0, t_1, \cdots, t_q) \\ &= M(u, \lambda_q{}^0 v)\,(0, \cdots, 0, t_1, \cdots, t_{q-1}) \\ &= \lambda_q{}^0 v(t_1, \cdots, t_{q-1}) = v(t_1, \cdots, t_q). \end{aligned}$$

To verify the condition (A3), we consider two cases. If $i = q$, then we have

$$\begin{aligned} \lambda^\varepsilon_{p+q} z\,(t_1, \cdots, t_{p+q-1}) &= z(t_1, \cdots, t_{p-q-1}, \varepsilon) \\ &= M(u, \lambda_q{}^\varepsilon v)\,(t_1, \cdots, t_{p+q-1}). \end{aligned}$$

If $i < q$, we have

$$\begin{aligned} \lambda^\varepsilon_{p+i} z\,(t_1, \cdots, t_{p+q-1}) &= z(t_1, \cdots, t_{p+i-1}, \varepsilon, t_{p+i}, \cdots, t_{p+q-1}) \\ &= M(u, \lambda_q{}^0 v)\,(t_1, \cdots, t_{p+i-1}, \varepsilon, t_{p+i} \cdot \cdots, t_{p+q-1}) \\ &= \lambda^\varepsilon_{p+i} M(u, \lambda_q{}^0 v)\,(t_1, \cdots, t_{p+q-2}) \\ &= M(u, \lambda_i{}^\varepsilon \lambda_q{}^0 v)\,(t_1, \cdots, t_{p+q-2}). \end{aligned}$$

On the other hand, since $i < q$ and v is degenerate, it follows that $\lambda_i{}^\varepsilon v$ is also degenerate and so is $M(u, \lambda_i{}^\varepsilon v)$. Therefore, we have

$$\begin{aligned} M(u, \lambda_i{}^\varepsilon v)\,(t_1, \cdots, t_{p+q-1}) &= M(u, \lambda_i{}^\varepsilon v)\,(t_1, \cdots, t_{p+q-2}, 0) \\ &= \lambda^0_{p+q-1} M(u, \lambda_i{}^\varepsilon v)\,(t_1, \cdots, t_{p+q-2}) \\ &= M(u, \lambda^0_{q-1} \lambda_i{}^\varepsilon v)\,(t_1, \cdots, t_{p+q-2}). \end{aligned}$$

Since $\lambda_i{}^\varepsilon \lambda_q{}^0 v = \lambda^0_{q-1} \lambda_i{}^\varepsilon v$, this completes the verification of the condition (A3).

It remains to construct $z = M(u, v)$ for a non-degenerate v.

Let $P = I^{p+q} = I^p \times I^q$ and $Q = V_0 \times I^q \cup I^p \times \partial I^q$, where V_0 denotes the vertex $(0, \cdots, 0)$ of I^p and ∂I^q the set-theoretic boundary of I^q. Then Q is contractible, because it is clearly deformable into $V_0 \times I^q \cup V_0 \times \partial I^q = V_0 \times I^q$ which is contractible. Hence, Q is a strong deformation retract of P and we may apply (III; 3.1) once again.

Define maps $f : P \to B$ and $g : Q \to X$ as follows:

$$f(t_1, \cdots, t_p, s_1, \cdots, s_q) = u(t_1, \cdots, t_p);$$

$$g(0, \cdots, 0, s_1, \cdots, s_q) = v(s_1, \cdots, s_q), \quad \text{on } V_0 \times I^q;$$

$$g(t_1, \cdots, t_p, s_1, \cdots, s_{i-1}, \varepsilon, s_i, \cdots, s_{q-1})$$
$$= M(u, \lambda_i^\varepsilon v)\,(t_1, \cdots, t_p, s_1, \cdots, s_{q-1}), \text{on } I^p \times \partial I^q.$$

It is straighforward to verify that g is one-valued and hence, by (I; 5.1), g is continuous. Besides, it is obvious that $\omega g = f \mid Q$. Therefore, according to (III; 3.1), there exists a map $z : I^{p+q} \to X$ such that $\omega z = f$ and $z \mid Q = g$.

Since $q > 0$, Q contains all vertices of I^{p+q}. This implies that $z = M(u,v)$ is a singular cube in X with all vertices at x_0. The conditions (A1) through (A5) are all obviously satisfied. This completes the inductive construction of $M(u, v)$ and proves the lemma.

9. Proof of Lemma B

The proof of this lemma is analogous to that of Lemma A and is based on a construction $u \to D_p u$ by induction on the integer q. It will be sketched as follows.

First assume that $q = 0$. Let

$$P = I^{p+1} = I^p \times I, \quad Q = (I^p \times 0) \cup (V_0 \times I) \cup (I^p \times 1) \subset P.$$

where V_0 denotes the leading vertex $(0, \cdots, 0)$ of I^p. Then Q is a strong deformation retract of P. Define two maps $f : P \to B$ and $g : Q \to X$ by taking

$$f(t_1, \cdots, t_p, t) = \omega u(t_1, \cdots, t_p), \qquad \text{on } I^p \times I;$$

$$g(t_1, \cdots, t_p, 0) = u(t_1, \cdots, t_p), \qquad \text{on } I^p \times 0;$$

$$g(0, \cdots, 0, t) = x_0, \qquad \text{on } V_0 \times I;$$

$$g(t_1, \cdots, t_p, 1) = M(B_p u, F_p u)\,(t_1, \cdots, t_p), \quad \text{on } I^p \times 1.$$

Applying (III; 3.1), we obtain a map $v : P \to X$ such that $\omega v = f$ and $v \mid Q = g$. Since Q contains all vertices of P, v is a singular cube with all vertices x_0. Let $D_p u = v$ and the conditions (B1)–(B7) are obvious.

Next let $q > 0$ and assume that we have constructed $D_p u$ for every u with $w(u) \leqslant p$ and $\dim(u) < p + q$. Now let u be a singular cube in X with $w(u) \leqslant p$ and $\dim(u) = p + q$. To construct $v = D_p u$, let us dispose of the special case where u is degenerate. Then u does not depend on the last variable and $\lambda_{p+q}^0 u = \lambda_{p+q}^1 u$. Define $v = D_p u$ by taking

$$v(t_1, \cdots, t_{p+q+1}) = D_p \lambda_{p+q}^0 u\,(t_1, \cdots, t_{p+q}).$$

By means of the hypotheses of induction, one can easily verify the conditions (B1)–(B7).

It remains to construct $v = D_p u$ for a non-degenerate u with $w(u) \leqslant p$ and $\dim(u) = p + q$. For this purpose, let

$$P = I^p \times I \times I^q = I^{p+q+1},$$

$$Q = (I^p \times 0 \times I^q) \cup (I^p \times 1 \times I^q) \cup (V_0 \times I \times I^q) \cup (I^p \times I \times \partial I^q),$$

where V_0 denotes the leading vertex of I^p and ∂I^q the set-theoretic boundary of I^q. Define two maps $f : P \to B$ and $g : Q \to X$ by taking

$$f(t_1, \cdots, t_p, t, s_1, \cdots, s_q) = B_p u(t_1, \cdots, t_p).$$

$$g(t_1, \cdots, t_p, 0, s_1, \cdots, s_q) = u(t_1, \cdots, t_p, s_1, \cdots, s_q),$$

$$g(t_1, \cdots, t_p, 1, s_1, \cdots, s_q) = M(B_p u, F_p u)\,(t_1, \cdots, t_p, s_1, \cdots, s_q),$$

$$g(0, \cdots, 0, t, s_1, \cdots, s_q) = u(0, \cdots, 0, s_1, \cdots, s_q) = F_p u(s_1, \cdots, s_q),$$

$$g(t_1, \cdots, t_p, t, s_1, \cdots, s_{i-1}, \varepsilon, s_i, \cdots, s_{q-1}) =$$
$$B_p \lambda^{\varepsilon}_{p+i} u\,(t_1, \cdots, t_p, t, s_1, \cdots, s_{q-1}).$$

Since Q is a strong deformation retract of P and $\omega g = f \mid Q$, there exists a map $v : P \to X$ such that $\omega v = f$ and $v \mid Q = g$. Since Q contains all vertices of P, v is a singular cube with all vertices at x_0. Let $v = D_p u$ and the conditions (B1) through (B7) can be easily verified. This completes the proof.

10. Proof of Lemmas C and D

To prove Lemma C, we consider σ as a singular 1-cube in B. For each singular n-cube v in F, we take

$$D_\sigma v = M(\sigma, v)$$

given by Lemma A. Then $D_\sigma v$ is a singular $(n + 1)$-cube in X. The assignment $v \to D_\sigma v$ defines a homogeneous homomorphism

$$D_\sigma : C(F) \to C(X)$$

of degree 1. The conditions (A1)–(A4) imply the conditions (C1)–(C4). This proves Lemma C.

To prove Lemma D, let σ, τ be two loops in B representing the same element of $\pi_1(B, b_0)$. Then there exists a singular 2-cube $u : I^2 \to B$ such that

$$u(0, t) = b_0 = u(1, t), \quad u(t, 0) = \sigma(t), \quad u(t, 1) = \tau(t)$$

for each $t \in I$. Let D_σ and D_τ be deformations of $C(F)$ covering the paths σ and τ respectively. For each singular n-cube v in F, we shall construct a singular $(n + 2)$-cube Qv in X satisfying the conditions:

(D1) $w(Qv) \leqslant 2$.

(D2) $B_2 Qv = u$.

(D3) $\lambda_1^0 Qv(t, t_1, \cdots, t_n) = v(t_1, \cdots, t_n)$.

(D4) $\lambda_2^0 Qv = D_\sigma v$; $\lambda_2^1 Qv = D_\tau v$.

(D5) $Q\lambda_i^\varepsilon v = \lambda^\varepsilon_{i+2} Qv$, $(\varepsilon = 0, 1; i = 1, \cdots, n)$.

(D6) If v is degenerate, so is Qv.

The cube Qv will be constructed by induction on the dimension n of v.

First assume $n = 0$. Then v reduces to the point x_0. Consider the subspace

$$A = (I \times 0) \cup (I \times 1) \cup (0 \times I)$$

of I^2 and define a map $f : A \to X$ by taking

$$f(t, 0) = D_\sigma v(t), \quad f(t, 1) = D_\tau v(t), \quad f(0, t) = x_0$$

for every $t \in I$. Since $\omega f = u \mid A$ and since A is a strong deformation retract of I^2, f has an extension $g : I^2 \to X$ covering u. We define Qv to be the singular 2-cube g. The conditions (D1)–(D6) are obviously satisfied.

Next assume that $n > 0$ and that Qv has been constructed for every singular cube v in F of dimension less than n. Let v be a singular cube in F of dimension n; we are going to construct Qv. If v is degenerate, we set

$$Qv(s, t, t_1, \cdots, t_n) = Q\lambda_n{}^0 v(s, t, t_1, \cdots, t_{n-1}).$$

If v is non-degenerate, let us consider the subspace

$$A = (I \times 0 \times I^n) \cup (I \times 1 \times I^n) \cup (0 \times I \times I^n) \cup (I \times I \times \partial I^n)$$

of $I^{n+2} = I^2 \times I^n$ and define two maps $f : A \to X$ and $\phi : I^{n+2} \to B$ by setting

$$\phi(s, t, t_1, \cdots, t_n) = u(s, t),$$
$$f(s, 0, t_1, \cdots, t_n) = D_\sigma v(s, t_1, \cdots, t_n),$$
$$f(s, 1, t_1, \cdots, t_n) = D_\tau v(s, t_1, \cdots, t_n),$$
$$f(0, t, t_1, \cdots, t_n) = v(t_1, \cdots, t_n),$$
$$f(s, t, t_1, \cdots, t_{i-1}, \varepsilon, t_i, \cdots, t_{n-1}) = Q\lambda_i{}^\varepsilon v(s, t, t_1, \cdots, t_{n-1}).$$

Since $\omega f = \phi \mid A$ and A is a strong deformation retract of I^{n+2}, f has an extension $g : I^{n+2} \to X$ covering ϕ. We define Qv to be the $(n + 2)$-cube g. The verification of the conditions (D1)–(D6) is left to the reader. The inductive construction of Qv is complete.

Now consider the endomorphisms

$$J_\sigma, J_\tau : C(F) \to C(F)$$

defined by the deformations D_σ, D_τ respectively. Let K denote the homogeneous endomorphism of $C(F)$ of degree 1 defined by

$$(Kv)(t, t_1, \cdots, t_n) = (Qv)(1, t, t_1, \cdots, t_n)$$

for each singular cube $v : I^n \to F$. Then, one can easily verify that

$$\partial Kv + K\partial v = J_\tau v - J_\sigma v.$$

Hence, J_σ and J_τ are chain homotopic. This completes the proof of Lemma D.

Remark. If X is a bundle space over B with respect to the projection $\omega : X \to B$, then the proofs of the lemmas A–D can be simplified as follows.

For a given singular cube $u: I^p \to B$, one considers the bundle space U over I^p induced by u. By a theorem of Feldbau [S; p. 53], U is equivalent to the product space $I^p \times F$ and hence the constructions can be easily carried out in U instead of X.

11. The Poincaré polynomials

Throughout this section, let G be a field. We shall consider vector spaces M over G graded by the spaces M_p; the dimension of M_p is called the p-th *Betti number* of M and is denoted by $R_p(M)$. Then M is of finite dimension iff the Betti numbers $\{ R_p(M) \}$ are all finite and only a finite number of them are different from zero. We assume once for all that, unless there is a statement to the contrary, a given graded vector space M is of finite dimension over G and $M_p = 0$ if $p < 0$. For such an M, we define the *Poincaré polynomial* $\psi(M)$ and the *Euler characteristic* $\chi(M)$ by

$$\psi(M) = \sum_p R_p(M)t^p, \quad \chi(M) = \sum_p (-1)^p R_p(M).$$

A few elementary properties of the Poincaré polynomials are listed as follows. Proofs are left to the reader.

(1) If L is a subspace of M and $N = M/L$, then $\psi(N) = \psi(M) - \psi(L)$.

(2) For any two graded vector spaces M and N, we have $\psi(M \otimes_G N) = \psi(M)\psi(N)$.

(3) If M is a graded vector space with a linear differential operator $d: M \to M$ of degree -1, then we have $\psi(\mathscr{H}(M)) = \psi(M) - (1+t)\psi(d(M))$.

For any two polynomials f and g, the symbol $f \leqslant g$ will mean that $g - f$ is a polynomial with non-negative coefficients.

(4) Let M be the same as in (3). If L is a subspace of the graded vector space M such that $L \cap d(M) = 0$, then $N = L \cap Z(M)$ is isomorphic to the subspace of $\mathscr{H}(M)$ represented by the cycles N and $\psi(L) - \psi(N) \leqslant t\psi(d(M))$.

(5) Let M be the same as in (3). If L is a subspace of $Z(M)$ and N denotes the subspace of $\mathscr{H}(M)$ represented by the cycles L, then we have $\psi(L) - \psi(N) \leqslant \psi(d(M))$.

Let (Y, Y_0) be a pair of a space Y and a subspace Y_0 of Y. If $H(Y, Y_0; G)$ is of finite dimension, then its Betti numbers, its Poincaré polynomial, and its Euler characteristic will be defined to be those of the pair (Y, Y_0) over G denoted by $R_p(Y, Y_0; G)$, $\psi(Y, Y_0; G)$, and $\chi(Y, Y_0; G)$ respectively.

Now let us consider the spectral homology sequence of § 7. Assume that $\pi_1(B, b_0)$ oprates simply on $H(F; G)$ and that $H(B, B_0; G)$ and $H(F; G)$ are of finite dimension. Then we have the following two theorems:

Theorem 11.1. $R_m(X, X_0; G) \leqslant \sum_{p+q=m} R_p(B, B_0; G) R_q(F; G)$.

Theorem 11.2. $\chi(X, X_0; G) = \chi(B, B_0; G)\chi(F; G)$.

In words, (11.1) states that the Betti numbers of the fiber space X cannot be greater than those of the product space $B \times F$ and (11.2) states that the

Euler characteristic of the fiber space X is the same as that of the product space $B \times F$.

These two theorems are immediate corollaries of the general theorem below on Poincaré polynomials. Since $H(B, B_0; G)$ and $H(F; G)$ are assumed to be of finite dimension, their Poincaré polynomials are of the form:

$$\psi(B, B_0; G) = b_\alpha t^\alpha + b_{\alpha+1} t^{\alpha+1} + \cdots + b_\beta t^\beta, \quad (\beta \geqslant \alpha \geqslant 0),$$

$$\psi(F; G) = 1 + c_1 t + c_2 t^2 + \cdots + c_\gamma t^\gamma, \quad (\gamma \geqslant 0).$$

We shall also consider the following four polynomials:

$$P = b_{\alpha+2} t^{\alpha+2} + \cdots + b_\beta t^\beta, \quad Q = b_\alpha t^\alpha + \cdots + b_{\beta-2} t^{\beta-2},$$

$$U = c_1 t + c_2 t^2 + \cdots + c_\gamma t^\gamma, \quad V = 1 + c_1 t + \cdots + c_{\gamma-1} t^{\gamma-1}.$$

If M is a vector space over G bigraded by the subspaces $M_{p,q}$, then M can be graded by means of the total degree $p + q$ and its Poincaré polynomial $\psi(M)$ is defined in terms of this grading. Now let

$$\Delta = \sum_{n \geqslant 2} d^n(E^n).$$

Then Δ is a bigraded vector space over G of finite dimension. The general theorem mentioned above can be stated as follows.

Theorem 11.3. *If $H(B, B_0; G)$ and $H(F; G)$ are of finite dimension, then so is $H(X, X_0; G)$ and its Poincaré polynomial is given by*

$$\psi(X, X_0; G) = \psi(B, B_0; G)\psi(F; G) - (1 + t)\psi(\Delta).$$

Furthermore, the polynomial $\psi(\Delta)$ satisfies the following inequalities:

$$\psi(\Delta) \leqslant t^{-1}PV, \quad \psi(\Delta) \leqslant QU.$$

Proof. Consider the spectral homology sequence of § 7. By (6.4), we have

$$E^2 \approx H(B, B_0; G) \otimes_G H(F; G).$$

This implies that E^∞ is of finite dimension and so is $H(X, X_0: G)$. According to (2), we have

$$\psi(E^2) = \psi(B, B_0; G)\psi(F; G).$$

By (3), we obtain

$$\psi(E^{n+1}) = \psi(E^n) - (1 + t)\psi(d^n(E^n)).$$

Hence it follows that

$$\psi(X, X_0; G) = \psi(E^\infty) = \psi(E^2) - (1 + t)\psi(\Delta).$$

This proves the first part of the theorem and it remains to establish the two inequalities.

For each $n \geqslant 2$, consider the subspace

$$M^n = \sum_q E^n_{\alpha,q} + \sum_q E^n_{\alpha+1,q} + \sum_p E^n_{p,\gamma}$$

of E^n. Since d^n is of degree $(-n, n-1)$, every element of M^n is a cycle and hence we have the epimorphisms

$$M^2 \to M^3 \to \cdots \to M^n \to M^{n+1} \to \cdots \to M^\infty.$$

By (5), we have
$$\psi(M^n) - \psi(M^{n+1}) \leqslant \psi(d^n(E^n))$$

for each $n \geqslant 2$ and hence

$$\begin{aligned}\psi(M^2) &\leqslant \psi(M^\infty) + \psi(\Delta) \leqslant \psi(E^\infty) + \psi(\Delta) \\ &= \psi(E^2) - (1 + t)\psi(\Delta) + \psi(\Delta) \\ &= \psi(E^2) - t\psi(\Delta).\end{aligned}$$

Since clearly we have

$$M^2 \approx [H_\alpha(B,B_0;G) + H_{\alpha+1}(B,B_0;G)] \otimes_G H(F;G) + H(B,B_0;G) \otimes_G H_\gamma(F;G),$$

it follows that $\psi(M^2) \geqslant \psi(E^2) - PV$. Hence we deduce

$$\psi(\Delta) \leqslant t^{-1}PV.$$

Next, let us consider, for each $n \geqslant 2$, the subspace

$$N^n = \sum_q E^n_{\beta-1,q} + \sum_q E^n_{\beta,q} + \sum_p E^n_{p,0}$$

of E^n. No non-zero element of N^n can be a boundary under d^n and hence we have the monomorphisms

$$N^\infty \to \cdots \to N^{n+1} \to N^n \to \cdots \to N^3 \to N^2.$$

By (4), we have
$$\psi(N^n) - \psi(N^{n+1}) \leqslant t\psi(d^n(E^n))$$

for each $n \geqslant 2$ and hence we obtain

$$\psi(N^2) \leqslant \psi(E^2) - \psi(\Delta)$$

as before. On the other hand, we have

$$\psi(N^2) \geqslant \psi(E^2) - QU.$$

Hence we obtain $\psi(\Delta) \leqslant QU$. ∎

Corollary 11.4. $R_{\beta+\gamma}(X, X_0; G) = R_\beta(B, B_0; G)\, R_\gamma(F; G)$.

$R_{\beta+\gamma-1}(X, X_0; G) = R_\beta(B, B_0; G)\, R_{\lambda-1}(F; G) + R_{\beta-1}(B, B_0; G) R_\lambda(F; G)$.

As an application of (11.3), let us study the fiber spaces where the base space and the fiber are homology spheres over G. For this purpose, let us assume that
$$\psi(B; G) = 1 + t^p, \quad \psi(F; G) = 1 + t^q$$

with $p > 1$ and $q > 0$. Leaving aside the critical case $q = p - 1$, the Poincaré polynomial of X over G is completely determined by B and F:

Proposition 11.5. $\psi(X; G) = (1 + t^p)(1 + t^q)$ if $q \neq p - 1$.

Proof. $P = t^p$, $Q = 1$, $U = t^q$, $V = 1$. According (11.3), we have

$$\psi(X; G) = (1 + t^p)(1 + t^q) - (1 + t)\psi(\Delta)$$

with $\psi(\Delta) \leqslant t^{p-1}$ and $\psi(\Delta) \leqslant t^q$. Since $q \neq p - 1$, the latter conditions on $\psi(\Delta)$ imply that $\psi(\Delta) = 0$. ▮

For the critical case $q = p - 1$, the conditions on $\psi(\Delta)$ imply that either $\psi(\Delta) = 0$ or $\psi(\Delta) = t^{p-1}$. Hence we obtain the following

Proposition 11.6. *If $q = p - 1$, then $\psi(X; G)$ is either $(1 + t^p)(1 + t^{p-1})$ or $1 + t^{2p-1}$.*

A consequence of (11.5) and (11.6) is that the only possibility for a homology n-sphere X to be a fiber space over a simply connected homology p-sphere B with a homology q-sphere F as fiber is the critical case $q = p - 1$ and $n = 2p - 1$, [S; p. 145]. Examples for $p = 2, 4, 8$ are the Hopf fiberings, (III; § 5).

12. Gysin's exact sequences

Let G denote either the additive group of integers or a field. Assume that the fiber F is a homology r-sphere for some $r \geqslant 1$ and that $\pi_1(B, b_0)$ operates simply on the total homology group $H(F; G)$ and the total cohomology group $H^*(F; G)$. Then we have the following

Theorem 12.1. *There is an exact sequence*

$$\cdots \xrightarrow{\omega_*} H_{m+1}(B, B_0; G) \xrightarrow{\phi_*} H_{m-r}(B, B_0; G) \xrightarrow{\psi_*} H_m(X, X_0; G) \xrightarrow{\omega_*} H_m(B, B_0; G) \xrightarrow{\phi_*} \cdots$$

called Gysin's homology sequence, where ω_* is induced by the projection $\omega : (X, X_0) \to (B, B_0)$.

Proof. Since F is a homology r-sphere, it follows that E^2 contains only two columns which might be non-trivial, namely

$$E^2_{p,0} \approx H_p(B, B_0 : G), \quad E^2_{p,r} \approx H_p(B, B_0; G) \otimes_G H_r(F; G) \approx H_p(B, B_0; G).$$

Applying (VIII; 8.3) and the commutative triangle at the end of § 7, we obtain the theorem. ▮

Similarly, one can establish the following

Theorem 12.2. *There is an exact sequence*

$$\cdots \xrightarrow{\phi^*} H^m(B, B_0; G) \xrightarrow{\omega^*} H^m(X, X_0; G) \xrightarrow{\psi^*} H^{m-r}(B, B_0; G) \xrightarrow{\phi^*} H^{m+1}(B, B_0; G) \xrightarrow{\omega^*} \cdots$$

called Gysin's cohomology sequence, where ω^* is induced by the projection $\omega : (X, X_0) \to (B, B_0)$.

If $B_0 = \square$, then $H^0(B; G) \approx G$. Let $s = \phi^*(1) \in H^{r+1}(B; G)$. Then it can be proved that

$$\phi^*(x) = x \cup s = s \cup x$$

for every $x \in H^{m-r}(B; G)$ and that $2s = 0$ if r is even. Since this result will be used only in the present section, the proof is omitted. See [Serre 1; p. 470].

As an application of Gysin's sequences, let us consider the fiberings of spheres by spheres and study the structure of the integral cohomology ring of the base space. For this purpose, assume that X itself is a homology n-sphere for some $n > r$ and consider the cohomology ring $H^*(B)$. By the exactness of Gysin's cohomology sequence, we deduce that the homomorphism

$$\phi^* : H^{m-r-1}(B) \to H^m(B)$$

is a monomorphism if $m = 0$ or $m = n$, is an epimorphism if $m = 1$ or $m = n + 1$, and is an isomorphism for other values of m. It follows immediately that

$$H^m(B) \approx Z, \quad m \equiv 0 \ mod\ (r + 1), \quad 0 \leqslant m < n,$$
$$H^m(B) = 0, \quad m \not\equiv 0 \ mod\ (r + 1), \quad 0 \leqslant m < n.$$

The structure of $H^m(B)$ with $m \geqslant n$ depends on the relation between n and r. If

$$n = p(r + 1) + q, \quad 0 \leqslant q < r,$$

then it is easy to verify that the cohomology ring $H^*(B)$ is generated by three elements $1 \in H^0(B) \approx Z$, $s = \phi^*(1) \in H^{r+1}(B)$, and $t \in H^n(B)$ with $\omega^*(t)$ as a generator of $H^n(X) \approx Z$. More precisely, the cohomology group $H^*(B)$ is free abelian and has a free basis

$$\{ 1, s, s^2, \cdots, t, st, s^2t, \cdots \}, \tag{12.3}$$

where juxtaposition denotes cup product. If

$$n = p(r + 1) + r,$$

we have to consider the exact sequence

$$0 \xrightarrow{\phi^*} H^n(B) \xrightarrow{\omega^*} H^n(X) \xrightarrow{\psi^*} H^{p(r+1)}(B) \xrightarrow{\phi^*} H^{n+1}(B) \xrightarrow{\omega^*} 0,$$

where we have $H^n(X) \approx Z$ and $H^{p(r+1)}(B) \approx Z$. Since $H^n(B)$ is isomorphic with a subgroup of $H^n(X)$, it follows that either $H^n(B) \approx Z$ or $H^n(B) = 0$. If $H^n(B) \approx Z$, it follows that $\psi^* = 0$ and hence ω^* and ϕ^* are both isomorphisms. In this case, $H^*(B)$ is free abelian and has (12.3) as a free basis. If $H^n(B) = 0$, then it follows that $H^{n+1}(B)$ is a cyclic group of finite order $k \geqslant 1$. In this case, the groups $H^m(B)$, $m \geqslant n$, are given by

$$H^m(B) \approx Z_k, \quad m \equiv 0 \ mod\ (r + 1),$$
$$H^m(B) = 0, \quad m \not\equiv 0 \ mod\ (r + 1).$$

Hence $H^*(B)$ has a system of generators

$$\{ 1, s, s^2, \cdots \}, \tag{12.4}$$

where s^i is free if $i \leqslant p$ and s^i is of finite order k if $i > p$.

13. Wang's exact sequences

Let G denote either the additive group of integers or a field and assume that the base space B is a simply connected homology r-sphere over G for some $r \geqslant 2$. First, let us consider the case $B_0 = \square$.

Theorem 13.1. *There is an exact sequence*

$$\cdots \xrightarrow{\tau_*} H_{m-r+1}(F;G) \xrightarrow{\rho_*} H_m(F;G) \xrightarrow{\theta_*} H_m(X;G) \xrightarrow{\tau_*} H_{m-r}(F;G) \xrightarrow{\rho_*} \cdots$$

called Wang's homology sequence, where θ_* is induced by the inclusion $\theta : F \subset X$.

Proof. Since B is a homology r-sphere, it follows that E^2 contains only two rows which might be non-trivial, namely

$$E^2_{p,q} \approx H_p(B;G) \otimes_G H_q(F;G) \approx H_q(F;G), \quad (p = 0, r).$$

Applying (VII; 8.1) and one of the commutative triangles in § 7, we obtain the theorem. ∎

Similarly, one can establish the following

Theorem 13.2. *There is an exact sequence*

$$\cdots \xrightarrow{\tau^*} H^m(X;G) \xrightarrow{\theta^*} H^m(F;G) \xrightarrow{\rho^*} H^{m-r+1}(F;G) \xrightarrow{\tau^*} H^{m+1}(X,G) \xrightarrow{\theta^*} \cdots$$

called Wang's cohomology sequence, where θ^* is induced by the inclusion $\theta : F \subset X$.

The homomorphisms ρ^* define an endomorphism

$$\rho^* : H^*(F;G) \to H^*(F;G).$$

According to Leray, this endomorphism ρ^* is a derivation if r is odd and an anti-derivation if r is even, that is to say,

$$\rho^*(x \cup y) = \rho^*(x) \cup y + (-1)^{(r+1)p}\, x \cup \rho^*(y)$$

for each $x \in H^p(F;G)$ and $y \in H^*(F;G)$. The proof of this result is left to the reader. [Serre 1; p. 471].

Next, consider the case $B_0 = b_0$ and hence $X_0 = F$. Since B is a homology r-sphere, $H_r(B, b_0; G) \approx G$ and $H_m(B, b_0; G) = 0$ for $m \neq r$. Then E^2 and E^{*2} contain only one row which might be non-trivial, namely

$$E^2_{r,q} \approx H_q(F;G), \quad E^{*2}_{r,q} \approx H^q(F;G).$$

Hence, by (VIII; 8.1) we obtain the following

Theorem 13.3. *For every integer m, we have*

$$H_m(X,F;G) \approx H_{m-r}(F;G), \quad H^m(X,F;G) \approx H^{m-r}(F;G).$$

As an application of these theorems, let us consider the fiber space

$$X = [B; B, b_0]$$

over B with the initial projection $\omega : X \to B$ as defined in (III; § 10). Then we have

$$F = [B; b_0, b_0].$$

Since $\pi_1(B) = 0$, it follows from (IV; § 2) that F is pathwise connected. Therefore, we may apply (13.1) to this case. Since X is contractible, we obtain

$$H_m(F) \approx H_{m-r+1}(F)$$

for every $m \geqslant 1$. Together with $H_0(F) \approx Z$, this implies the following theorem of Morse.

Theorem 13.4. *If B is a simply connected homology r-sphere and $F = [B; b_0, b_0]$, then*

$$H_m(F) \approx Z \quad \textit{if } m \equiv 0 \textit{ mod } (r - 1),$$
$$H_m(F) = 0 \quad \textit{if } m \not\equiv 0 \textit{ mod } (r - 1).$$

For another application, let us consider the sphere bundles over spheres. Assume that the fiber F is a homology s-sphere for some $s \geqslant 1$ and consider the integral homology groups $H_m(X)$.

Assume that $s \geqslant 2$. Then, by the exactness of Wang's homology sequence, we deduce that the induced homomorphism

$$\theta_* : H_m(F) \to H_m(X)$$

is an epimorphism if $m = r - 1$ or $m = r + s - 1$, is a monomorphism if $m = r$ or $m = r + s$, and is an isomorphism for other values of m. Therefore, it remains to compute $H_m(X)$ for the four critical values $r - 1$, r, $r + s - 1$, $r + s$ of m.

First, let us compute $H_{r-1}(X)$. If $r - 1 \neq s$, then it follows that

$$H_{r-1}(X) = \theta_*[H_{r-1}(F)] = 0$$

since F is a homology s-sphere. If $r - 1 = s$, we define a numerical invariant k of this fibering as follows. Consider the homomorphism

$$\rho_* : H_0(F) \to H_s(F).$$

If $\rho_* = 0$, then k is defined to be zero; otherwise, the quotient group $H_s(F)/\rho_*[H_0(F)]$ is a finite cyclic group and k is defined to be the order of this finite cyclic group. Hence, if $r - 1 = s$, we have

$$H_{r-1}(X) \approx \begin{cases} Z, & \text{if } k = 0, \\ Z_k, & \text{if } k \neq 0. \end{cases}$$

Next, let us compute $H_r(X)$. From the exact sequence

$$0 \xrightarrow{\rho_*} H_r(F) \xrightarrow{\theta_*} H_r(X) \xrightarrow{\tau_*} H_0(F) \xrightarrow{\rho_*} 0,$$

it follows that

$$H_r(X) \approx \begin{cases} Z, & \text{if } r \neq s, \\ Z + Z, & \text{if } r = s. \end{cases}$$

Then, let us compute $H_{r+s-1}(X)$. Since $r > 1$, we have $H_{r+s-1}(F) = 0$. Hence, we obtain

$$H_{r+s-1}(X) = \theta^*[H_{r+s-1}(F)] = 0.$$

Finally, let us compute $H_{r+s}(X)$. In the following exact sequence

$$H_{r+s}(F) \xrightarrow{\theta_*} H_{r+s}(X) \xrightarrow{\tau_*} H_s(F) \xrightarrow{\rho_*} H_{r+s-1}(F),$$

we have $H_{r+s}(F) = 0 = H_{r+s-1}(F)$ and hence

$$H_{r+s}(X) \approx H_s(F) \approx Z.$$

For the remaining case $s = 1$, the induced homomorphism

$$\theta_* : H_m(F) \to H_m(X)$$

is an epimorphism if $m = r - 1$, is a monomorphism if $m = r + 1$, and is an isomorphism if $m < r - 1$ or $m > r + 1$. Therefore, it suffices to compute the group $H_m(X)$ for three critical values of m, namely, $m = r - 1$, r, and $r + 1$. The computation is similar to the case $s > 1$ and hence is left to the reader.

Note that the same results can be obtained by using Gysin's homology sequence instead of Wang's homology sequence.

14. Truncated exact sequences

Let G denote either the additive group of integers or a field. Assume that $X_0 = \square$ and that $\pi_1(B, b_0)$ operates simply on $H(F;G)$.

Theorem 14.1. *If $H_m(B, G) = 0$ for $0 < m < p$ and $H_m(F; G) = 0$ for $0 < m < q$, then we have an exact sequence*

$$H_{p+q-1}(F;G) \xrightarrow{\theta_*} \cdots \xrightarrow{T} H_m(F;G) \xrightarrow{\theta_*} H_m(X;G) \xrightarrow{\omega_*} H_m(B;G) \xrightarrow{T} H_{m-1}(F;G)$$

$$\to \cdots \xrightarrow{\omega_*} H_2(B;G) \xrightarrow{T} H_1(F;G) \xrightarrow{\theta_*} H_1(X;G) \xrightarrow{\omega_*} H_1(B;G) \xrightarrow{T} 0$$

where θ_* and ω_* are induced by the inclusion $\theta : F \subset X$ and the projection $\omega : X \to B$, and T is called the transgression.

Proof. According to (6.3), we have

$$E^2_{i,j} \approx H_i(B; G) \otimes_G H_j(F; G) + Tor_G[H_{i-1}(B; G), H_j(F; G)].$$

Therefore, $E^2_{i,j} = 0$ if $i \neq 0$, $j \neq 0$, and $i + j \leqslant p + q - 1$. It follows that, for each total degree m with $0 \leqslant m \leqslant p + q - 1$, E^2 contains only two terms which might not be zero, namely,

$$E^2_{m,0} \approx H_m(B; G), \quad E^2_{0,m} \approx H_m(F; G).$$

According to (VIII; 8.5), we have a fundamental exact sequence in which, by the results in (VIII; § 7), one can verify the assertions about the homomorphisms θ_*, ω_*, and T. ∎

Corollary 14.2. *If* $H_m(F; G) = 0$ *for each* $m > 0$, *then* $\omega_* : H_m(X; G) \approx H_m(B; G)$ *for each* $m \geqslant 0$.

Proof. Apply (14.1) with $p = 1$ and $q = \infty$. ∎

Corollary 14.3. *If* $H_m(B; G) = 0$ *for each* $m > 0$, *then* $\theta_* : H_m(F; G) \approx H_m(X; G)$ *for each* $m \geqslant 0$.

Proof. Apply (14.1) with $p = \infty$ and $q = 1$. ∎

Corollary 14.4. *If* $H_m(X) = 0$ *for each* $m > 0$ *and* $H_m(B; G) = 0$ *for* $0 < m < p$, *then we have* $T : H_m(B; G) \approx H_{m-1}(F; G)$ *for* $2 \leqslant m \leqslant 2p - 2$.

Proof. Applying (14.1) with $q = 1$, we find that $H_m(F; G) = 0$ for $0 < m < p - 1$. Then, apply (14.1) again with $q = p - 1$. ∎

Similar results hold for cohomology groups.

15. The spectral sequence of a regular covering space

Let X be a pathwise connected regular covering space over a locally pathwise connected base space B with a projection

$$\omega : X \to B$$

in the sense of (III; § 16). Then it follows readily that B is pathwise connected and X is locally pathwise connected. Pick a basic point $x_0 \in X$ and call $b_0 = \omega(x_0)$ the basic point in B. Then the basic fiber

$$F = \omega^{-1}(b_0)$$

is discrete and ω induces a monomorphism

$$\omega_* : \pi_1(X, x_0) \to \pi_1(B, b_0).$$

Since X is a regular covering space over B, the image

$$\chi(X, x_0) = \omega_*[\pi_1(X, x_0)]$$

is an invariant subgroup of $\pi_1(B, b_0)$ and the quotient group

$$\pi = \pi_1(B, b_0)/\chi(X, x_0)$$

acts on X as a group of right operators; see (III; 16.6). If $\xi \neq 1$ is any element in π, then ξ is a homeomorphism of X onto itself without fixed point.

Let W be a contractible (finite or infinite) simplicial complex (with weak topology) on which π acts freely as a group of simplicial homeomorphisms. For the existence of such a complex, see (VI; Ex. I). Let P denote the orbit space W/π with projection $\tau : W \to P$. Then P is a space of the homotopy type $(\pi, 1)$ and τ is the universal covering.

Following a procedure introduced by Ehresmann, we shall construct a space Y as follows. Let π operate on the left of the product space $W \times X$ by

$$\xi(w, x) = (\xi w, x\xi^{-1}), \quad \xi \in \pi, \ w \in W, \ x \in X.$$

Then Y is defined to be the orbit space $(W \times X)/\pi$. Define two maps

$$P \xleftarrow{l} W \times X \xrightarrow{r} B$$

by taking $l(w, x) = \tau(w)$ and $r(w, x) = \omega(x)$. Since obviously $l\xi = l$ and $r\xi = r$ for every $\xi \in \pi$, they induce two maps

$$P \xleftarrow{\lambda} Y \xrightarrow{\rho} B$$

One can verify that the maps $\lambda : Y \to P$ and $\rho : Y \to B$ are fiberings in the sense of (III; § 3). In fact, Y is a bundle space over P with X as fiber and λ as projection, and Y is also a bundle space over B with W as fiber and ρ as projection. These will be called the *bundle spaces associated with the regular covering space X over B.*

Let G be an abelian group on which π acts as left operators. Since P is a space of the homotopy type $(\pi, 1)$, it follows from (VI; Ex. G)

$$H_m(\pi; G) = H_m(P; G), \quad H^m(\pi; G) = H^m(P; G)$$

with local coefficients in G.

Throughout the remainder of the section, let G denote an abelian group on which the group π operates simply.

First, consider the associated fibering $\rho : Y \to B$. Since the fiber W in this fibering is contractible and hence acyclic, it follows from (14.2) that $\rho_* : H_m(Y) \approx H_m(B)$ for each $m \geqslant 0$ and therefore

$$\text{(15.1)} \qquad \rho_* : H_m(Y; G) \approx H_m(B; G), \quad (m \geqslant 0).$$

Next, consider the associated fibering $\lambda : Y \to P$ with fiber X. According to § 6 and § 7, this fibering determines a homology exact couple

$$\mathscr{C} = \langle D, E; i, j, k \rangle$$

and a spectral homology sequence $\{ E^n \}$ which will be referred to as those of the regular covering space X over B.

According to § 5, the fundamental group $\pi_1(P, p_0) \approx \pi$ acts on $H(X; G)$ as a group of left operators. Hence, by (6.2), we have

$$\text{(15.2)} \qquad E^2_{p,q} \approx H_p(P; H_q(X; G)) = H_p(\pi; H_q(X; G));$$

and, by § 7, the group $\mathscr{H}(\mathscr{C}) = H(Y; G) \approx H(B; G)$ is filtered with E^∞ as its associated graded group. More precisely, we have

$$H_m(B;G) = \mathscr{H}_{m,0}(\mathscr{C}) \supset \mathscr{H}_{m-1,1}(\mathscr{C}) \supset \cdots \supset \mathscr{H}_{0,m}(\mathscr{C}) \supset \mathscr{H}_{-1,m+1}(\mathscr{C}) = 0,$$
$$\mathscr{H}_{p,q}(\mathscr{C})/\mathscr{H}_{p-1,q+1}(\mathscr{C}) \approx E^\infty_{p,q}.$$

Furthermore, we have

$$R_q(\mathscr{C}) = H_q(X; G), \quad S_p(\mathscr{C}) = H_p(\pi; G)$$

together with two commutative triangles

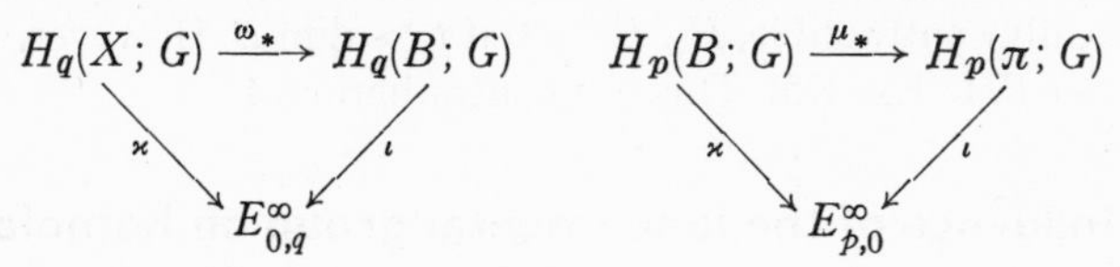

where ω_* is induced by the projection $\omega: X \to B$, $\mu_* = \lambda_* \rho_*^{-1}$, the $\varkappa$'s are epimorphisms, and the ι's are monomorphisms; μ_* will be called the *natural homomorphism*. As immediate consequences, $\mathscr{H}_{0,q}(\mathscr{C})$ is the image of ω_* and $\mathscr{H}_{p-1,1}(\mathscr{C})$ is the kernel of μ_*.

Similarly, we have the cohomology exact couple

$$\mathscr{C}^* = < D^*, E^*; i^*, j^*, k^*, >$$

and a spectral cohomology sequence $\{ E^{*n} \}$ of the regular covering space X over B. In particular,

$$E^{*2} \approx H^*(P; H^*(X; G)) = H^*(\pi; H^*(X; G)) \tag{15.3}$$

and $\mathscr{H}(\mathscr{C}^*) = H^*(B; G)$ is filtered with $E^{*\infty}$ as its associated graded group. Furthermore, $R_q(\mathscr{C}^*) = H^q(X; G)$, $S_p(\mathscr{C}^*) = H^p(\pi; G)$

and we have two commutative triangles

$$H^q(B; G) \xrightarrow{\omega^*} H^q(X; G) \qquad H^p(\pi; G) \xrightarrow{\mu^*} H^p(B; G)$$

$$\varkappa \searrow \quad \swarrow \iota \qquad\qquad \varkappa \searrow \quad \swarrow \iota$$

$$E^{*\infty}_{0,q} \qquad\qquad\qquad E^{*\infty}_{p,0}$$

with properties analogous to those in the homology case. If G is a commutative associative ring with unity element, then (15.3) is a ring isomorphism.

16. A theorem of P. A. Smith

In the present and the following two sections, we shall give a few of the applications of the spectral sequences obtained in § 15.

Theorem 16.1. *If a discrete group π acts freely on a locally contractible acyclic space X of finite dimension, then π has no element of finite order other than* 1.

Proof. Assume, on the contrary, that π has an element ξ of order $r \neq 1$. The subgroup of π generated by ξ is a cyclic group of order r and acts freely on X. Thus we may assume that π itself is a cyclic group of order r. Let $B = X/\pi$ denote the orbit space. Then X is a regular covering space of B with π as quotient group. This implies that B is locally contractible and finite dimensional.

Let $G = Z$ be the additive group of integers and let π operate simply on G. Consider the spectral homology sequence of § 15. Since X is acyclic, it follows that $E^2_{p,0} \approx H_p(\pi)$ and $E^2_{p,q} = 0$ if $q \neq 0$. Then, by (VIII; 8.3), we obtain

$$H_p(B) \approx H_p(\pi), \quad (p \geqslant 0).$$

Since B is locally contractible, $H_p(B) = 0$ if $p > \dim B$. However, $H_p(\pi) \approx \pi$ if p is odd. See (VI; Ex. K5). This is a contradiction. ▮

17. Influence of the fundamental group on homology and cohomology groups

Let B denote a pathwise connected space. Since the singular complex $S(B)$ has the same homotopy and homology structure as B, (V; Ex. I), we may assume without loss of generality that B is locally contractible. Then the universal covering space X of B is defined and the fundamental group $\pi = \pi_1(B)$ operates freely on X.

Theorem 17.1. *If $\pi_p(B) = 0$ for $1 < p < r$, then we have the natural isomorphisms*

$$\mu_*: H_p(B; G) \approx H_p(\pi; G), \quad \mu^*: H^p(\pi; G) \approx H^p(B; G)$$

for each $p < r$ and every coefficient group G on which π operates simply.

Proof. Consider the spectral homology sequence of § 15. Then the assumption implies that $H_q(X; G) = 0$ if $0 < q < r$. By (15.2), we have

$$E^2_{p,q} = 0, \quad \text{if } p + q < r \text{ and } q \neq 0.$$

Then it follows from (VIII; 8.5) and one of the commutative triangles in § 15 that μ_* is an isomorphism whenever $p < r$. By considering the spectral cohomology sequence, one can also show that μ^* is an isomorphism whenever $p < r$. ▮

Thus we see that the fundamental group of B determines the homology groups and the cohomology groups of B for all dimensions less than r if $\pi_p(B) = 0$ for every p satisfying $1 < p < r$.

To study the groups of the critical dimension r, let us consider the homomorphisms

$$\omega_*: H_r(X; G) \to H_r(B; G), \quad \omega^*: H^r(B; G) \to H^r(X; G)$$

induced by the projection $\omega : X \to B$. Denote by $\Sigma_r(B; G)$ the image of ω_* and by $\Lambda^r(B; G)$ the kernel of ω^*.

Theorem 17.2. *If $\pi_p(B) = 0$ for $1 < p < r$, then we have the exact sequences*

(*a*) $$H_r(X; G) \xrightarrow{\omega_*} H_r(B, G) \xrightarrow{\mu_*} H_r(\pi; G) \to 0,$$

(*b*) $$H^r(X; G) \xleftarrow{\omega^*} H^r(B; G) \xleftarrow{\mu^*} H^r(\pi; G) \leftarrow 0,$$

and hence $$H_r(B; G)/\Sigma_r(B; G) \approx H_r(\pi; G), \quad \Lambda^r(B; G) \approx H^r(\pi; G).$$

Proof. Consider the spectral homology sequence of § 15 and in particular the terms of E^2 with total degree $r - 1$ and r. Let

$$a_m = 0, \; b_m = m, \; c_m = m, \; d_m = 0, \quad (m = r - 1, r).$$

Then it is easily verified that the two-term condition $\{r-1, r; 2\}$ of (VIII; § 8) is satisfied. Since $E^2_{0,r-1} = 0$ and $E^2_{r,0} \approx H_r(B; G)$, we obtain an exact sequence

$$\text{(c)} \qquad H_r(B; G) \xrightarrow{\mu_*} H_r(\pi, G) \to 0.$$

According to § 15, the kernel of μ_* is $\mathscr{H}_{r-1,1}(\mathscr{C})$ and the image of $\omega_*: H_r(X; G) \to H_r(B; G)$ is $\mathscr{H}_{0,r}(\mathscr{C})$. Since E^∞ contains only two terms of total degree r which might be different from zero, namely, $E^\infty_{0,r}$ and $E^\infty_{r,0}$, it follows that $\mathscr{H}_{r-1,1}(\mathscr{C}) = \mathscr{H}_{0,r}(\mathscr{C})$. Hence we obtain the exact sequence (*a*). By considering the spectral cohomology sequence of § 15, one can also establish the exactness of the sequence (*b*). ∎

In the proof of (17.2), we did not make full use of the two-term condition $\{r-1, r; 2\}$. Indeed, we actually have longer exact sequences given in the following

Theorem 17.3. *If $\pi_p(B) = 0$ for $1 < p < r$, then we have the exact sequences:*

$$\text{(d)} \quad H_{r+1}(B;G) \xrightarrow{\mu_*} H_{r+1}(\pi;G) \xrightarrow{\varphi_*} J_\pi(H_r(X;G)) \xrightarrow{\psi_*} H_r(B;G) \xrightarrow{\mu_*} H_r(\pi;G) \to 0,$$

$$\text{(e)} \quad H^{r+1}(B;G) \xleftarrow{\mu^*} H^{r+1}(\pi;G) \xleftarrow{\varphi^*} I_\pi(H^r(X;G)) \xleftarrow{\psi^*} H^r(B;G) \xleftarrow{\mu^*} H^r(\pi;G) \leftarrow 0.$$

For the definition of the operators I_π and J_π, see (VI; Ex. K).

Proof. Applying the two-term condition $\{r-1, r; 2\}$ as in the proof of (17.2), we obtain instead of (c) a longer exact sequence

$$(i) \qquad E^2_{0,r} \xrightarrow{\psi_*} H_r(B; G) \xrightarrow{\mu_*} H_r(\pi; G) \to 0,$$

where $E^2_{0,r} \approx H_0(\pi; H_r(X; G))$ is isomorphic with $J_\pi(H_r(X; G))$ according to (VI; Ex. K1). For further extension of this exact sequence, we have to consider the terms of E^2 with total degree $r+1$. There are only three such terms which might be different from zero, namely

$$E^2_{0,r+1}, \; E^2_{1,r}, \; E^2_{r+1,0} \approx H_{r+1}(\pi; G).$$

Let $n \geqslant 2$. Since the differential operator d^n in E^n is of degree $(-n, n-1)$, the elements of $E^n_{0,r+1}$ and $E^n_{1,r}$ are cycles and d^n sends $E^n_{r+1,0}$ into $E^n_{0,r}$ iff $n = r+1$. Since every element of $E^n_{0,r}$ is a cycle, we get an exact sequence

$$(ii) \qquad E^{r+1}_{r+1,0} \xrightarrow{d^{r+1}} E^{r+1}_{0,r} \xrightarrow{\varkappa} E^{r+2}_{0,r} \to 0.$$

Since $E^2_{r+1,0} = E^{r+1}_{r+1,0}$, $E^2_{0,r} = E^{r+1}_{0,r}$, and $E^\infty_{0,r} = E^{r+2}_{0,r}$, we can combine (*i*) and (*ii*) so that we obtain an exact sequence

$$(iii) \qquad E^2_{r+1,2} \xrightarrow{\phi_*} E^2_{0,r} \xrightarrow{\psi_*} H_r(B; G) \xrightarrow{\mu_*} H_r(\pi; G) \to 0$$

with $\phi_* = d^{r+1}$. Since the kernel of d^{r+1} is obviously $E^{r+2}_{r+1,0} = E^\infty_{r+1,0}$ and since $E^2_{r+1,0} \approx H_{r+1}(\pi; G)$, one of the commutative triangles in § 15 gives an exact sequence

$$(iv) \qquad H_{r+1}(B; G) \xrightarrow{\mu_*} H_{r+1}(\pi; G) \xrightarrow{\phi_*} E^2_{0,r}.$$

Combining (*iii*) and (*iv*), we obtain the exact sequence (*d*). Similarly, one can deduce (*e*). ∎

Note. If G is the group of integers, then the groups $\Sigma_r(B; G)$ and $\Lambda^r(B; G)$ are essentially the groups $\Sigma_r(B)$ and $\Lambda^r(B)$ as defined in (VI; Ex. J). See also [Eilenberg–MacLane 1].

18. Finite groups operating freely on S^r

Let π be a finite group operating freely on the r-sphere $X = S^r$. Let $B = S^r/\pi$ denote the orbit space and $\omega: S^r \to B$ the projection. By the finiteness of π, one can easily show that S^r is the universal covering space of B with ω as projection and that π can be considered as the fundamental group of B. To apply the results of § 17, let us take the additive group Z of integers as the coefficient group and let π operate simply on Z.

By (17.1) and (17.3), we have the natural isomorphisms

$$(i) \qquad \mu_*: H_p(B) \approx H_p(\pi), \quad \mu^*: H^p(\pi) \approx H^p(B)$$

for each $p < r$ and the exact sequences

$$(ii) \qquad 0 \to H_{r+1}(\pi) \xrightarrow{\phi_*} J_\pi(H_r(S^r)) \xrightarrow{\psi_*} H_r(B) \xrightarrow{\mu_*} H_r(\pi) \to 0,$$

$$(iii) \qquad 0 \leftarrow H^{r+1}(\pi) \xleftarrow{\phi^*} I_\pi(H^r(S^r)) \xleftarrow{\psi^*} H^r(B) \xleftarrow{\mu^*} H^r(\pi) \leftarrow 0,$$

since dim $(B) = r$ and therefore $H_{r+1}(B) = 0 = H^{r+1}(B)$.

Proposition 18.1. *If there is an element $\xi \in \pi$ which changes the orientation of S^r, then r must be even.*

Proof. Since ξ changes the orientation of S^r, ξ is of order $h > 1$. The subgroup of π generated by ξ is a cyclic group of order h and operates freely on S^r. Therefore, we may assume that π itself is the cyclic group of order h generated by ξ. Since ξ changes the orientation of S^r, we have $\xi(x) = -x$ for every element x of the free cyclic group $H^r(S^r)$. It follows that $I_\pi(H^r(S^r)) = 0$ and the exact sequence (*iii*) reduces to

$$0 \leftarrow H^{r+1}(\pi) \leftarrow 0 \leftarrow H^r(B) \xleftarrow{\mu^*} H^r(\pi) \leftarrow 0.$$

This implies that

$$(iv) \qquad H^{r+1}(\pi) = 0, \quad \mu^*: H^r(\pi) \approx H^r(B).$$

By (VI; Ex. K5), the first equality implies that r must be even. ∎

Proposition 18.2. *If π contains an element $\xi \neq 1$ which preserves the orientation of S^r, then r must be odd.*

Proof. We may assume that π is the cyclic group generated by ξ. Since ξ preserves the orientation of S^r, π operates simply on $H_r(S^r)$. It follows that $J_\pi(H_r(S^r)) = H_r(S^r)$ and the exact sequence (*ii*) becomes

$$(v) \qquad 0 \to H_{r+1}(\pi) \to H_r(S^r) \to H_r(B) \to H_r(\pi) \to 0.$$

This implies that $H_{r+1}(\pi)$ is isomorphic to a subgroup of the free group $H_r(S^r)$ and hence it is also free. According to (VI; Ex. K5), we conclude that $H_{r+1}(\pi) = 0$ and so r must be odd. ∎

Proposition 18.3. *If r is even and π contains more than one element, then π is a cyclic group of order 2 and, for each p with $1 \leqslant p \leqslant r$, we have:*

$$H_p(B) = 0, \ H^p(B) \approx \pi, \quad \textit{if } p \textit{ is even};$$
$$H_p(B) \approx \pi, \ H^p(B) = 0, \quad \textit{if } p \textit{ is odd}.$$

Proof. Let $\xi \neq 1$ and $\eta \neq 1$ be any two elements in π, then, by (18.2), both ξ and η must change the orientation of S^r and hence $\xi\eta^{-1}$ preserves the orientation of S^r. Using (18.2) once more, we deduce that $\xi\eta^{-1} = 1$. This proves that π is cyclic of order 2. For the rest of the proposition, (*i*) and (*iv*) give all the homology and cohomology groups except $H_r(B)$. The latter can be computed either from the relations between homology and cohomology groups or as follows. In the exact sequence (*ii*), since both $H_{r+1}(\pi)$ and $J_\pi(H_r(S^r))$ are of order 2, ϕ_* must be an isomorphism. On the other hand, we have $H_r(\pi) = 0$. Therefore, it follows easily from the exactness that $H_r(B) = 0$. ∎

Proposition 18.4. *If r is odd and π is abelian, then π must be cyclic, $H_r(B) \approx Z \approx H^r(B)$, and for each p with $0 < p < r$ we have:*

$$H_p(B) = 0, \ H^p(B) \approx \pi, \quad \textit{if } p \textit{ is even};$$
$$H_p(B) \approx \pi, \ H^p(B) = 0, \quad \textit{if } p \textit{ is odd}.$$

Proof. According to (18.1), every element of π preserves the orientation of S^r. It follows that

$$J_\pi(H_r(S^r)) = H_r(S^r),$$

and hence the exact sequence (*ii*) becomes

$$(vi) \qquad 0 \to H_{r+1}(\pi) \xrightarrow{\phi_*} H_r(S^r) \xrightarrow{\omega_*} H_r(B) \xrightarrow{\mu_*} H_r(\pi) \to 0.$$

Since every element of π preserves the orientation of S^r, B is an orientable manifold of dimension r and hence we have $H_r(B) \approx Z \approx H^r(B)$. Assume that π is of order h. Then, by the definition of ω_*, one can easily see that we can choose generators $\alpha \in H_r(S^r)$ and $\beta \in H_r(B)$ such that $\omega_*(\alpha) = h\beta$. Hence the exact sequence (*vi*) implies that $H_r(\pi)$ is a cyclic group of order h.

According to (*i*) and (VI; Ex. K), it remains to prove that π is cyclic. Let us assume that π is not cyclic. Then there is a prime number p such that π contains a subgroup of the form $Z_p + Z_p$, where Z_p denotes the cyclic group of order p. We may assume that π itself is of the form $Z_p + Z_p$; then π is of order p^2. Therefore, $H_r(\pi)$ is a cyclic group of order p^2. Since $H_r(Z_p) \approx Z_p$, it follows from (VI; Ex. K) that Z_p is a direct summand of $H_r(\pi)$. This is impossible. ∎

Note. 1. If π is a discrete group operating freely on the r-sphere S^r, then the compactness of S^r implies that π is finite. 2. One can deduce all the homology and cohomology groups of the real projective spaces from (18.3) and (18.4).

EXERCISES

A. Cohomology with arbitrary coefficients

Let G be an abelian group and C denote the regular ∂-complex $C(X, X_0)$. Consider the group A^* of singular cochains of X modulo X_0 with coefficients in G. Then A^* is a graded group with $A^*_m = Hom(C_m, G)$ and a differential operator δ of degree 1. Since $Hom\,(C, G)$ is the direct product of the groups $Hom(C_m, G)$, [see E–S, p. 147], A^* is a subgroup of $Hom(C, G)$. Define a decreasing filtration $\{ A^{*p} \}$ in A^* by taking

$$A^{*p} = \{ f \in A^* \mid f(C^{p-1}) = 0 \}.$$

Then A^{*p} is a subgroup of the graded group A^* with $A^{*p}_m = A^{*p} \cap A^*_m$. Define a weight function w on A^* by taking $w(f)$, $f \in A^*$, to be the least upper bound of the integers p such that $f \in A^{*p}$. Prove that $0 \leqslant w(f) \leqslant \dim\,(f)$ for every non-zero homogeneous element $f \in A^*$. Therefore, with the filtration $\{ A^{*p} \}$, A^* becomes a regular δ-complex.

Study the associated exact couple

$$\mathscr{C}(A^*) = < D^*, E^*; i^*, j^*, k^* >$$

and its derived couples, as follows: prove that

$$E^*_{p,q} \approx Hom(C_p(B, B_0), H^q(F; G)),$$
$$E^{*2}_{p,q} \approx H^p(B, B_0; H^q(F; G)),$$

where $H^p(B, B_0; H^q(F; G))$ denotes the p-dimensional singular cohomology group of B modulo B_0 with local coefficients in $H^q(F; G)$; and that if G is a commutative associative ring with a unity element, then these are ring isomorphisms.

B. The spectral cohomology sequence

Consider the regular δ-complex A^* of the preceding exercise. Denote by

$$\mathscr{C}^n(A^*) = < D^{*n}, E^{*n}; i^{*n}, j^{*n}, k^{*n} >$$

the successive derived couples of the associated exact couple $\mathscr{C} = \mathscr{C}(A^*)$. Then the associated spectral sequence $\{ E^{*n} \mid n = 1, 2, \cdots \}$ of $\mathscr{C}$ will be called the *spectral cohomology sequence* of X modulo X_0 over G.

Prove that $\mathscr{H}(\mathscr{C}) = \mathscr{H}(A^*) = H^*(X, X_0; G)$ is filtered with $E^{*\infty}$ as its associated graded group; more explicitly,

$$H^m(X,X_0;G) = \mathscr{H}_{0,m}(\mathscr{C}) \supset \mathscr{H}_{1,m-1}(\mathscr{C}) \supset \cdots \supset \mathscr{H}_{m,0}(\mathscr{C}) \supset \mathscr{H}_{m+1,-1}(\mathscr{C}) = 0,$$
$$\mathscr{H}_{p,q}(\mathscr{C}) / \mathscr{H}_{p+1,q-1}(\mathscr{C}) \approx E^{*\infty}_{p,q}.$$

Hereafter, we shall use the notation $H^{p,q}(X, X_0; G) = \mathscr{H}_{p,q}(\mathscr{C})$.

Next, assume $X_0 = \square$ and prove that $R_q(\mathscr{C}) = H^q(F; G)$ and $S_p(\mathscr{C}) = H^p(B; G)$. Then we obtain the following commutative triangles

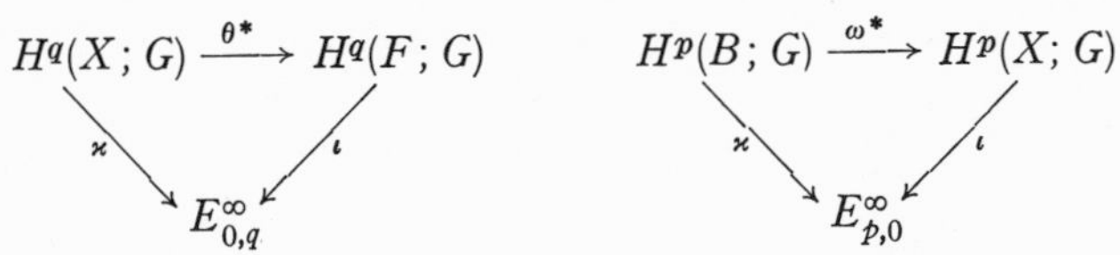

where θ^*, ω^* are induced by $\theta : F \subset X$ and $\omega : X \to B$, the $\varkappa$'s are epimorphisms, and the ι's are monomorphisms. Prove that $H^{1,q-1}(X; G)$ is the kernel of θ^* and $H^{p,0}(X; G)$ is the image of ω^*.

Now, consider the transgression. According to (VIII; Ex. H), there are equivalent definitions:

(1) The transgression is the differential operator $d^{*m} : E^{*m}_{0,m} \to E^{*m}_{m,0}$.

(2) In the following homomorphisms

$$H^{m-1}(F; G) \xrightarrow{\delta} H^m(X, F; G) \xleftarrow{\omega^*} H^m(B, G), \quad (m \geqslant 2),$$

let M denote the image of ω^* and K the kernel of ω^*. Then the transgression is the homomorphism

$$T : \delta^{-1}(M) \to H^m(B; G)/K$$

defined as in (VIII; Ex. H).

A comparison of these definitions shows that $E^{*m}_{m,0}$ is isomorphic to the image of $H^m(B; G)$ in $H^m(X, F; G)$ under the induced homomorphism ω^*.

Let G be the group of integers mod 2. Prove that the transgression T commutes with the square operations and the reduced powers of Steenrod.

Investigate analogously the transgression in the spectral homology sequence of § 7.

C. The maximal cycle theorem

In addition to the assumptions of § 3, assume that G is a field. Then prove the following

Theorem. If $H_m(B, B_0; G) = 0$ for each $m > p$ and $H_m(F; G) = 0$ for each $m > q$, then we have:

$$H_m(X, X_0; G) = 0, \quad (m > p + q);$$
$$H_{p+q}(X, X_0; G) \approx H_p(B, B_0; G) \otimes_G H_q(F; G).$$

Deduce the following assertions:

1. If $G \neq 0$ and $H_m(X; G) = 0$ for each $m > 0$, then at least one of the following three statements must be true:

(*a*) $H_m(B; G) = 0 = H_m(F; G)$ for each $m > 0$.

(*b*) $H_m(B; G) \neq 0$ for infinitely many values of m.

(*c*) $H_m(F; G) \neq 0$ for infinitely many values of m.

2. If a euclidean space $X = R^n$ has a bundle structure over a base space B with a connected fiber F, then both B and F are acyclic.

D. An isomorphism theorem

In addition to the assumptions of § 3, assume that $X_0 \neq \square$ and hence $B_0 \neq \square$. Prove the following

Theorem. If $H_m(B, B_0; G) = 0$ for each $m < p$ and $H_m(F; G) = 0$ for $0 < m < q$, then we have

$$\omega_* : H_m(X, X_0; G) \approx H_m(B, B_0; G), \quad (m < p + q),$$

and an exact sequence

$$H_{p+q+1}(X, X_0; G) \xrightarrow{\omega_*} H_{p+q+1}(B, B_0; G) \xrightarrow{\phi} H_p(B, B_0; G) \otimes_G H_q(F; G)$$
$$\xrightarrow{\psi} H_{p+q}(X, X_0; G) \xrightarrow{\omega_*} H_{p+q}(B, B_0; G) \to 0.$$

Establish a similar theorem for the spectral cohomology sequence.

E. Some special cases of § 15

Prove the following assertions:

1. If π is a free cyclic group and acts simply on the homology groups $H_m(X; G), m \geqslant 0$, then $H_m(B; G)$ is an extension of $H_m(X; G)$ by $H_{m-1}(X; G)$. Hence, in particular, if G is a field, then $H_m(B; G)$ is the direct sum of $H_m(X; G)$ and $H_{m-1}(X; G)$.

2. If π is a finite group of order h and G is a field of characteristic zero or prime to h, then

$$H_m(B; G) \approx J_\pi(H_m(X; G)), \quad H^m(B : G) \approx I_\pi(H^m(X; G)).$$

3. If π is the direct product of a free cyclic group and a finite group of order h, G is a field of characteristic zero or prime to h, and π acts simply on the homology groups $H_m(X; G)$, $m \geqslant 0$, then $H_m(B; G) \approx H_m(X; G) + H_{m-1}(X; G)$.

F. Relations between the cohomology algebras of a space and its space of loops

Let B be a pathwise connected and simply connected space and b_0 a given point in B. Let $X = [B; B, b_0]$. Then, X is a contractible fiber space over B with fiber $F = \Lambda(B)$, the space of all loops in B with b_0 as basic point. Prove the following assertions:

1. If K is a field and the cohomology algebra $H^*(F; K)$ is isomorphic to an exterior algebra over K generated by an element of odd degree n, then the cohomology algebra $H^*(B; K)$ is isomorphic to a polynomial algebra generated by an element of even degree $n + 1$. [Serre 1; p. 501].

2. If K is a field of characteristic zero and the cohomology algebra $H^*(F; K)$ is isomorphic to a polynomial algebra generated by an element of even degree n, then the cohomology algebra $H^*(B; K)$ is isomorphic to an exterior algebra generated by an element of odd degree $n + 1$. [Serre 1; p. 501].

3. If K is a field and the cohomology algebra $H^*(B; K)$ is isomorphic to a polynomial algebra generated by an element of even degree $n \geqslant 2$, then the cohomology algebra $H^*(F; K)$ is isomorphic to an exterior algebra generated by an element of odd degree $n - 1$. [Serre 1; p. 492].

4. If K is a field of characteristic zero and the cohomology algebra

$H^*(B; K)$ is isomorphic to an exterior algebra generated by an element of odd degree n, then the cohomology algebra $H^*(F; K)$ is isomorphic to a polynomial algebra generated by an element of odd degree $n - 1$. [Serre 1; p. 489].

5. If K is a field of characteristic zero and the cohomology algebra $H^*(B; K)$ is isomorphic to an exterior algebra generated by an element of even degree $n \geqslant 2$, then the cohomology algebra $H^*(F; K)$ is isomorphic to the tensor product of an exterior algebra generated by an element of degree $n - 1$ and a polynomial algebra generated by an element of degree $2(n - 1)$. [Serre 1; p. 489].

6. If K is a field of characteristic $p \neq 0$ and $H^m(B; K) \approx H^m(S^n; K)$ for every $m \leqslant p(n - 1) + 1$, where $n \geqslant 3$ is an odd integer, then the subspace of the cohomology algebra $H^*(F; K)$ formed by the elements of degree not exceeding $p(n - 1)$ admits a homogeneous basis which consists of the elements $\{ 1, y, y^2, \cdots, y^{p-1}, z \}$, where [Serre 1; p. 494]

$$deg\,(y) = n - 1, \; deg\,(z) = p(n - 1), \; y^p = 0.$$

7. If K is a field of characteristic $p \neq 0$ and the subspace of the cohomology algebra $H^*(B; K)$ formed by the elements of degree not exceeding pq, where $q \geqslant 2$ is an even integer, admits a homogeneous basis which consists of the elements $\{ 1, y, y^2, \cdots, y^{p-1}, z \}$, where

$$deg\,(y) = q, \; deg\,(z) = pq, \quad y^p = 0,$$

then the subspace of $H^*(F; K)$ formed by the elements of degree not exceeding $pq - 2$ admits a homogeneous basis which consists of the elements $\{ 1, u, v \}$, where [Serre 1; p. 495]

$$deg\,(u) = q - 1, \quad deg\,(v) = pq - 2.$$

G. The cohomology algebra of (Z, n)

Let Z be the free cyclic group. Prove that the cohomology algebra $H^*(Z)$ over the ring of integers is isomorphic to the exterior algebra generated by an element of degree 1 and the cohomology algebra $H^*(Z, 2)$ over the ring of integers is isomorphic to the polynomial algebra generated by an element of degree 2.

Next, let K be a field of characteristic zero. By means of the relations 1 and 2 in Ex. F, prove that the cohomology algebra $H^*(Z, n; K)$ over K is isomorphic to an exterior algebra generated by an element of degree n if n is odd and is isomorphic to a polynomial algebra generated by an element of degree n if n is even.

H. The cohomology algebra of $\Lambda(S^n)$

Consider the sphere S^n of dimension $n \geqslant 2$ and study the cohomology algebra of the space $\Omega^n = \Lambda(S^n)$ of loops in S^n with s_0 as basic point. Let $T = [S^n; S^n, s_0]$. Then T is a contractible fiber space over S^n with projection $\omega : T \to S^n$ and fiber $\Omega^n = \omega^{-1}(s_0)$. From (13.2), deduce an isomorphism

$$\rho^* : H^m(\Omega^n) \approx H^{m-n+1}(\Omega^n)$$

for each m; ρ^* is a derivation or an anti-derivation according as n is odd or even. Thus, a homogeneous basis $\{ e_i \}$, $i = 0, 1, 2, \cdots$, of $H^*(\Omega^n)$ is given by

$$e_0 = 1, \; e_i = \rho^{*-1}(e_{i-1}), \; i = 1, 2, \cdots,$$

with dim $(e_i) = i(n - 1)$. Prove the following

Theorem. The multiplicative structure of the integral cohomology algebra $H^*(\Omega^n)$ *is given by the rule* $e_p e_q = c_{p,q} e_{p+q}$, *where* $c_{p,q}$ *is an integer given as follows:*

(*i*) *If n is odd, then*

$$c_{p,q} = \frac{(p + q)!}{p!\, q!}.$$

(*ii*) *If n is even, then* $c_{p,q} = 0$ *if both p and q are odd; otherwise*

$$c_{p,q} = \frac{[(p + q)/2]!}{[p/2]!\, [q/2]!},$$

where $[x]$ denotes the largest integer not exceeding x.

Then prove

1. If n is odd, then $(e_1)^p = (p!)e_p$. If n *is* even, then $(e_1)^2 = 0$, $(e_2)^p = (p!)e_{2p}$, and $e_1 e_{2p} = e_{2p} e_1 = e_{2p+1}$.

2. If n is even, then $H^*(\Omega^n)$ is isomorphic to the tensor product of the algebras $H^*(S^{n-1})$ and $H^*(\Omega^{2n-1})$.

3. Let K be a field of characteristic zero. If n is odd, then $H^*(\Omega^n; K)$ is isomorphic to a polynomial algebra generated by an element of degree $n - 1$. If n is even, then $H^*(\Omega^n; K)$ is isomorphic to the tensor product of an exterior algebra generated by an element of degree $n - 1$ and a polynominal algebra generated by an element of degree $2(n - 1)$.

4. Let K be a field of characteristic $p \neq 0$. If n is odd, then $H^*(\Omega^n; K)$ is isomorphic to a polynomial algebra with infinitely many generators f_i, $(i = 0, 1, \cdots)$, modulo the ideal generated by $(f_i)^p$, $(i = 0, 1, \cdots)$, f_i being of degree $p^i(n - 1)$.

5. The cohomology algebra $H^*(S^{n-1} \times \Omega^{2n-1})$ is isomorphic to the tensor product $H^*(S^{n-1}) \otimes H^*(\Omega^{2n-1})$.

I. The connective fiber spaces of S^3

For each $n \geqslant 3$, let X_n denote an n-connective fiber space over S^3. Verify the following results on the cohomology algebra $H^*(X_n; Z_2)$ due to [Serre 3]:

1. For the dimensions $\leqslant 11$, $H^*(X_3; Z_2)$ has a homogeneous basis $\{ 1, a, b, c, d \}$ where *dim* $(a) = 4$, *dim* $(b) = 5$, *dim* $(c) = 8$, *dim* $(d) = 9$, and where $b = Sq^1 a$, $c = a^2$, $d = ab$.

2. For the dimensions $\leqslant 8$, $H^*(X_4; Z_2)$ has a homogeneous basis $\{ 1, e, f, g, h, i \}$ where *dim* $(e) = 5$, *dim* $(f) = 6 =$ *dim* (g), *dim* $(h) = 7$, *dim* $(i) = 8$, and where $f = Sq^1 e$, $h = Sq^1 g = Sq^2 e$, $i = Sq^2 f$, $Sq^2 g = 0$.

3. For the dimensions $\leqslant 7$, $H^*(X_5; Z_2)$ has a homogeneous basis $\{ 1, j, k \}$ where *dim* $(j) = 6$, *dim* $(k) = 7$, and where $Sq^1 j = 0$, $Sq^2 j = 0$.

4. $H^7(X_6; Z_2)$ has a basis formed by a single element m with $Sq^1 m \neq 0$.

CHAPTER X

CLASSES OF ABELIAN GROUPS

1. Introduction

Our principal remaining object is the computation of certain of the groups $\pi_m(S^n)$. To facilitate these computations, a digression on Serre's class theory of abelian groups is necessary. It is to this digression that the present chapter is devoted; the study of the groups $\pi_m(S^n)$ is deferred to the next (and final) chapter.

In §§ 2–6, we introduce the formal definitions. In §§ 7–10, certain immediate topological consequences are obtained; the most important of these is the generalized Hurewicz theorem, from which one deduces that the homotopy groups of a simply connected finite polyhedron are finitely generated.

2. The Definition of Classes

A collection $\mathscr{C}$ of abelian groups is called a *class* if the following four conditions are satisfied:

(CG1) $\mathscr{C}$ contains a group which consists of a single element.

(CG2) If a group A is isomorphic to some group in $\mathscr{C}$, then A is in $\mathscr{C}$.

(CG3) If a group A is a subgroup or a quotient group of some group in $\mathscr{C}$, then A is in $\mathscr{C}$.

(CG4) If an abelian group A is an extension of a group in $\mathscr{C}$ by a group in $\mathscr{C}$, then A is in $\mathscr{C}$.

Examples of classes are listed as follows. Verifications are left to the reader.

(1) The class $\mathscr{A}$ of all abelian groups.

(2) The class $\mathscr{O}$ of all groups consisting of a single element.

(3) The class $\mathscr{A}_f$ of all finitely generated abelian groups.

(4) The class $\mathscr{F}$ of all finite abelian groups.

(5) The class of all finite abelian groups with order not divisible by any prime number of a given family of prime numbers. In particular, if the given family consists of all prime numbers except p, this reduces to the class $\mathscr{A}_p$ of all finite abelian groups of order p^r, $r = 0, 1, 2, \cdots$.

(6) The class $\mathscr{T}$ of all torsion groups. An abelian group A is called a *torsion group* if every element of A is of finite order.

Other examples of classes will be given in the next section and in Ex. A at the end of the chapter.

It is an easy exercise to prove that a non-empty collection $\mathscr{C}$ of abelian groups is a class iff, for every exact sequence $L \to M \to N$ of abelian groups, the following condition is satisfied:

(CG5) $L \in \mathscr{C}$ and $N \in \mathscr{C}$ imply $M \in \mathscr{C}$.

Furthermore, it can also be verified that every class $\mathscr{C}$ has the following properties:

(CG6) If $A \in \mathscr{C}$ and $B \in \mathscr{C}$, then $A + B \in \mathscr{C}$.

(CG7) If $A \in \mathscr{C}$ and B is finitely generated, then $A \otimes B$ and $Tor(A, B)$ are in $\mathscr{C}$.

(CG8) If A is finitely generated and $A \in \mathscr{C}$, then $H_m(A) \in \mathscr{C}$ for every m. See (VI; Ex. G).

It is, of course, necessary to observe that a class $\mathscr{C}$ cannot be a set; thus the usual precautions must be taken to avoid contradictions. See [E–S; p. 120].

3. The Primary Components of Abelian Groups

For any given prime number p, the *p-primary component* of an abelian group A is defined to be the subgroup of A consisting of the elements of order p^r, $r = 0, 1, 2, \cdots$. If A is a torsion group, then it is the direct sum of its p-primary components for all p. Hence, in order to compute a certain torsion group A, it suffices to compute all of its primary components. See [Ka].

If, for some prime number p, an abelian group A reduces to its p-primary component, that is, if all elements of A have orders which are powers of p, then A is said to be a *p-primary group*. A finitely generated abelian group is p-primary iff it is of order p^r for some r. It is evident that, for a given prime number p, the p-primary groups form a class $\mathscr{P}_p$.

Let F be a given family of prime numbers. Then the torsion groups with null p-primary components for each prime number p in the family F constitute a class. If F contains all prime numbers, then this class reduces to the class $\mathscr{O}$; if F consists of all primes except a given prime number p, then it becomes the class $\mathscr{P}_p$; if F is empty, then it is the class $\mathscr{T}$.

4. The $\mathscr{C}$-Notions on Abelian Groups

For a given homomorphism $f: A \to B$, let $Im(f)$, $Ker(f)$, and $Coker(f)$ denote respectively the image, kernel, and cokernel of f; this latter is defined by

$$Coker(f) = B/Im(f).$$

The following sequence is obviously exact:

$$0 \to Ker(f) \to A \xrightarrow{f} B \to Coker(f) \to 0.$$

Furthermore, any pair of homomorphisms

$$A \xrightarrow{f} B \xrightarrow{g} C$$

gives rise to a *natural exact sequence* $S(f, g)$:

$$0 \to Ker(f) \to Ker(gf) \to Ker(g) \to Coker(f) \to Coker(gf) \to Coker(g) \to 0.$$

In the applications of class theory, the groups in a class $\mathscr{C}$ are usually to be neglected in a certain sense. Thus we are led to the following terminology: Let $\mathscr{C}$ be a given class. A group A is said to be *$\mathscr{C}$-null* if $A \in \mathscr{C}$. Let $f : A \to B$ be a homomorphism. Then f is said to be a *$\mathscr{C}$-monomorphism* if $Ker(f) \in \mathscr{C}$, a *$\mathscr{C}$-epimorphism* if $Coker(f) \in \mathscr{C}$, and a *$\mathscr{C}$-isomorphism* if it is both a $\mathscr{C}$-monomorphism and a $\mathscr{C}$-epimorphism. If a $\mathscr{C}$-isomorphism $f : A \to B$ exists, then A is said to be *$\mathscr{C}$-isomorphic* to B. If $\mathscr{C}$ is the class $\mathscr{O}$, these notions coincide with the corresponding classical notions; and it is more or less obvious that, for an arbitrary class $\mathscr{C}$, these notions have the same formal properties as the classical notions. The detailed statement of these facts is deferred to Ex. B at the end of the chapter.

Two abelian groups A and B are said to be *$\mathscr{C}$-equivalent* if there exists an abelian group L with two $\mathscr{C}$-isomorphisms $f : L \to A$ and $g : L \to B$.

Proposition 4.1. *Two abelian groups* A *and* B *are $\mathscr{C}$-equivalent if there exists an abelian group M and two $\mathscr{C}$-isomorphisms $h : A \to M$ and $k : B \to M$.*

Proof. *Sufficiency.* Let L denote the subgroup of the direct sum $A + B$ consisting of the elements (a, b) such that $h(a) = k(b)$. Define homomorphisms $f : L \to A$ and $g : L \to B$ by $f(a, b) = a$ and $g(a, b) = b$. Then one can verify that f and g are $\mathscr{C}$-isomorphisms.

Necessity. Let A and B be $\mathscr{C}$-equivalent. Then there exists an abelian group L together with two $\mathscr{C}$-isomorphisms $f : L \to A$ and $g : L \to B$. Let M denote the quotient group of the direct sum $A + B$ over the subgroup consisting of the elements $(f(l), g(l))$ for all $l \in L$. Let $p : A + B \to M$ denote the natural projection. Define homomorphisms $h . A \to M$ and $k : B \to M$ by $h(a) = p(a, 0)$ and $k(b) = p(0, b)$. Then one can verify that h and k are $\mathscr{C}$-isomorphisms. ∎

The relation of being $\mathscr{C}$-equivalent is obviously reflexive and symmetric. It is also transitive.

To verify this, assume that A, B and B, C are both $\mathscr{C}$-equivalent. By definition and (4.1), there are two abelian groups L, M and four $\mathscr{C}$-isomorphisms

$$f : L \to A, \; g : L \to B, \; h : B \to M, \; k : C \to M.$$

Since $hg : L \to M$ is a $\mathscr{C}$-isomorphism, it follows that A and C are $\mathscr{C}$-equivalent. This proves the transitivity and the relation of being $\mathscr{C}$-equivalent is an equivalence relation.

5. Perfectness and Completeness

The classes of abelian groups with which we will deal in the sequel usually satisfy some further conditions described as follows.

A class $\mathscr{C}$ is said to be *perfect* if $A \in \mathscr{C}$ implies that $H_m(A) \in \mathscr{C}$ for every $m > 0$. $\mathscr{C}$ is said to be *complete* if $A \in \mathscr{C}$ implies that $A \otimes B \in \mathscr{C}$ and $Tor(A, B) \in \mathscr{C}$ for every B. $\mathscr{C}$ is said to be *weakly complete* if $A \in \mathscr{C}$ and $B \in \mathscr{C}$ imply that $A \otimes B \in \mathscr{C}$ and $Tor(A, B) \in \mathscr{C}$. $\mathscr{C}$ is said to *strongly complete* if every finite or infinite direct sum of groups in $\mathscr{C}$ is also in $\mathscr{C}$. Every complete class is obviously weakly complete, and it can be verified that every strongly complete class is complete and perfect. See Ex. C at the end of the chapter.

The usefulness of these completeness conditions can be illustrated by the following

Proposition 5.1. *If $\mathscr{C}$ is a complete class, X a pathwise connected space, X_0 a subspace of X, and $G = \{ G_x \mid x \in X \}$ a local system of groups in X with each $G_x \in \mathscr{C}$, then $H_m(X, X_0; G)$ is in $\mathscr{C}$ for each $m \geqslant 0$.*

Proof. Pick $x_0 \in X$ as in (IX; § 2) and denote G_{x_0} also by G. Then $H_m(X,X_0;G)$ is isomorphic with a quotient group of some subgroup of $C_m(X,X_0) \otimes G$ which is in $\mathscr{C}$. Hence $H_m(X,X_0; G)$ is in $\mathscr{C}$. ∎

As to the examples of classes (1)–(6) in § 2, one can verify that (1), (2), (6) are strongly complete, and that (3), (4), (5) are prefect and weakly complete but not complete. The classes in § 3 are all strongly complete; in particular, the class $\mathscr{P}_p$ of the p-primary groups is strongly complete. For an example of a class which is perfect and complete but not strongly complete, see Ex. A2, at the end of the chapter.

In the sequel, we will deal with two different kinds of applications. For the first kind, it suffices to assume that the class $\mathscr{C}$ involved is weakly complete (and, sometimes, perfect). For the second kind, we have to assume that $\mathscr{C}$ is complete. However, this difference is not of much practical importance. In fact, the homotopy and homology groups considered in the applications are usually finitely generated; and if $\mathscr{C}$ is a given class and $\mathscr{C}_f$ is the class consisting of those abelian groups all of whose finitely generated subgroups are in $\mathscr{C}$, then $\mathscr{C}_f$ is strongly complete and

$$\mathscr{C}_f \cap \mathscr{A}_f = \mathscr{C} \cap \mathscr{A}_f.$$

6. Applications of Classes to Fiber Spaces

Let us go back to the notation of (IX; § 3) and assume that $\pi_1(B, b_0)$ operates simply on the homology and cohomology groups of the fiber F. Unless otherwise stated, the coefficient group G is the group of integers and hence is omitted from the notations.

Let $\mathscr{C}$ be any given class of abelian groups. Let us consider the spectral homology sequence of (IX; § 7).

Lemma 6.1. *If, for some* $n \geqslant 1$, $E^n_{p,q} \in \mathscr{C}$, *then* $E^\infty_{p,q} \in \mathscr{C}$.

Proof. Since $E^{n+1}_{p,q}$ is the quotient group of a subgroup of $E^n_{p,q}$, it is in $\mathscr{C}$. Then it follows by finite induction that $E^\infty_{p,q}$ is in $\mathscr{C}$. ∎

Lemma 6.2. *If, for a given pair* (p, q), $E^\infty_{i,j} \in \mathscr{C}$ *whenever* $i + j = p + q$ *and* $i \leqslant p$, *then* $H_{p,q}(X, X_0) \in \mathscr{C}$. *In particular, if* $E^\infty_{i,j} \in \mathscr{C}$ *whenever* $i + j = p + q$, *then* $H_{p+q}(X, X_0) \in \mathscr{C}$.

Proof. The lemma follows from the fact that $H_{i,j}(X, X_0)$ is an extension of $H_{i-1,j+1}(X, X_0)$ by $E^\infty_{i,j}$ and that $H_{0,m}(X, X_0) = E^\infty_{0,m}$. ∎

Proposition 6.3. *If* $\mathscr{C}$ *is a weakly complete class and if, for some integer* $r > 0$, $H_m(B, B_0) \in \mathscr{C}$ *and* $H_m(F) \in \mathscr{C}$ *whenever* $0 < m \leqslant r$, *then we have* $H_m(X, X_0) \in \mathscr{C}$ *whenever* $0 < m \leqslant r$.

Proof. According to the above lemmas, it suffices to show that $E^2_{p,q} \in \mathscr{C}$ whenever $0 < p + q \leqslant r$. By (IX; 6.3),

$$E^2_{p,q} \approx H_p(B, B_0) \otimes H_q(F) + Tor(H_{p-1}(B, B_0), H_q(F)).$$

If $p > 1$, $q > 0$, and $0 < p + q \leqslant r$, the weak completeness of $\mathscr{C}$ implies $E^2_{p,q} \in \mathscr{C}$. If $q = 0$ and $0 < p \leqslant r$, we have $E^2_{p,0} \approx H_p(B, B_0) \in \mathscr{C}$ since $H_0(F) \approx Z$.

To verify the proposition for the cases $p = 0$ and $p = 1$, let us first assume $B_0 \neq \square$. Then $H_0(B, B_0) = 0$ and hence

$$E^2_{0,m} = 0, \ E^2_{1,m-1} \approx H_1(B, B_0) \otimes H_{m-1}(F) \in \mathscr{C}$$

whenever $0 < m \leqslant r$. Next, assume that $B_0 = \square$. Then $H_0(B, B_0) \approx Z$ and hence

$$E^2_{0,m} \approx H_m(F) \in \mathscr{C}, \quad E^2_{1,m-1} \approx H_1(B, B_0) \otimes H_{m-1}(F) \in \mathscr{C}$$

whenever $0 < m \leqslant r$. ∎

Proposition 6.4. *If* $\mathscr{C}$ *is a complete class and if, for some integer* $p > 0$ *and* $q > 0$, $H_m(B, B_0) \in \mathscr{C}$ *whenever* $m \geqslant p$ *and* $H_m(F) \in \mathscr{C}$ *whenever* $m \geqslant q$, *then* $H_m(X, X_0) \in \mathscr{C}$ *whenever* $m \geqslant p + q$.

Proof. By the completeness of $\mathscr{C}$, it follows that $E^2_{i,j} \in \mathscr{C}$ whenever, $i + j \geqslant p + q$. Hence the proposition is an immediate consequence of the above lemmas. ∎

Throughout the remainder of the section, we assume that $B_0 \neq \square$ and hence $H_0(B, B_0) = 0$.

Theorem 6.5. *If* $\mathscr{C}$ *is a weakly complete class and if, for some integers* $p > 0$ *and* $q > 0$, *we have* $H_1(B, B_0) = 0$, $H_m(B, B_0) \in \mathscr{C}$ *whenever* $1 < m < p$, *and* $H_m(F) \in \mathscr{C}$ *whenever* $0 < m < q$, *then the induced homomorphism*

$$\omega_* : H_m(X, X_0) \to H_m(B, B_0)$$

is a $\mathscr{C}$-isomorphism whenever $m \leqslant r$ and is a $\mathscr{C}$-epimorphism whenever $m = r + 1$, where $r = Inf(p, q + 1)$.

Proof. The kernel of ω_* is $H_{m-1,1}(X, X_0)$ according to (IX; § 7). Therefore, in order to prove that ω_* is a $\mathscr{C}$-monomorphism for $m \leqslant r$, it suffices to show that $E^2_{i,j} \in \mathscr{C}$ whenever $i + j \leqslant r$ and $j \geqslant 1$. Since

$$E^2_{i,j} \approx H_i(B, B_0) \otimes H_j(F) + Tor(H_{i-1}(B, B_0), H_j(F)),$$

it follows that $E^2_{0,j} = 0, E^2_{1,j} = 0$, and $E^2_{i,j} \in \mathscr{C}$ whenever $i > 1, j \geqslant 1$, and $i + j \leqslant r$. This proves that ω_* is a $\mathscr{C}$-monomorphism for $m \leqslant r$.

It remains to prove that ω_* is a $\mathscr{C}$-epimorphism whenever $m \leqslant r + 1$. By (IX; § 7), the image of ω_* is $E^\infty_{m,0}$. Furthermore, we have

$$E^\infty_{m,0} = E^{m+1}_{m,0} \subset \cdots \subset E^{n+1}_{m,0} \subset E^n_{m,0} \subset \cdots \subset E^2_{m,0} = H_m(B, B_0),$$

$$E^n_{m,0}/E^{n+1}_{m,0} \approx d^n(E^n_{m,0}) \subset E^n_{m-n,n-1}, \ (2 \leqslant n \leqslant m).$$

Since $E^n_{m-n,n-1}$ is of total degree $m - 1 \leqslant r$, it is in $\mathscr{C}$ for $2 \leqslant n \leqslant m$. This implies that the cokernel $E^2_{m,0}/E^{m+1}_{m,0}$ of ω_* is in $\mathscr{C}$. ∎

Note. If $p < q + 1$, the condition $H_1(B, B_0) = 0$ may be replaced by $H_1(B, B_0) \in \mathscr{C}$. In fact, only this is used in the proof of $E^2_{1,r-1} \in \mathscr{C}$.

Theorem 6.6. *If $\mathscr{C}$ is a complete class and if, for some integers $p > 0$ and $q > 0$, we have $H_m(B, B_0) \in \mathscr{C}$ whenever $0 < m < p$ and $H_m(F) \in \mathscr{C}$ whenever $0 < m < q$, then the induced homomorphism*

$$\omega_* : H_m(X, X_0) \to H_m(B, B_0)$$

is a $\mathscr{C}$-isomorphism whenever $m \leqslant r$ and is a $\mathscr{C}$-epimorphism whenever $m = r + 1$, where $r = p + q - 1$.

The proof of this theorem is analogous to that of (6.5).

In the remainder of the section, we are concerned with the important special case where the subspace B_0 consists of a single point b_0. Then we have

$$H_0(B, B_0) = 0, \ H_m(B, B_0) \approx H_m(B), \ (m \geqslant 1).$$

Proposition 6.7. *If $\mathscr{C}$ is a weakly complete class, if $H_m(X) \in \mathscr{C}$ for each $m > 0$, and if $H_1(B) = 0$ and $H_m(B) \in \mathscr{C}$ whenever $1 < m < p$ for some given integer $p > 0$, then $H_m(F) \in \mathscr{C}$ whenever $0 < m < p - 1$ and the homomorphisms*

$$H_{p-1}(F) \xleftarrow{\partial} H_p(X, F) \xrightarrow{\omega_*} H_p(B, b_0) \approx H_p(B)$$

define a $\mathscr{C}$-equivalence of $H_{p-1}(F)$ and $H_p(B)$.

Proof. We shall prove this proposition by means of induction. The case $p = 1$ is trivial. Assume that $p > 1$ and the proposition is true for $p - 1$. Then, by the hypothesis of induction, we have $H_m(F) \in \mathscr{C}$ whenever $0 < m < p - 2$ and $H_{p-2}(F)$ is $\mathscr{C}$-equivalent to $H_{p-1}(B)$ and hence is in $\mathscr{C}$.

Applying (6.5) with $q = p - 1$ and $B_0 = b_0$, we deduce that $H_p(X, F)$ is $\mathscr{C}$-isomorphic to $H_p(B, b_0)$ under ω_*. In the exact sequence

$$\cdots \to H_p(X) \to H_p(X, F) \xrightarrow{\partial} H_{p-1}(F) \to H_{p-1}(X) \to \cdots,$$

we have $H_p(X) \in \mathscr{C}$ and $H_{p-1}(X) \in \mathscr{C}$. This implies that ∂ is a $\mathscr{C}$-isomorphism. ∎

Proposition 6.8. *If $\mathscr{C}$ is a complete class, $H_m(X) \in \mathscr{C}$ for each $m > 0$, and $H_m(B) \in \mathscr{C}$ whenever $0 < m < p$ for some given integer $p > 0$, then $H_m(F) \in \mathscr{C}$ whenever $0 < m < p - 1$ and the homomorphisms*

$$H_m(F) \xleftarrow{\partial} H_{m+1}(X, F) \xrightarrow{\omega_*} H_{m+1}(B, b_0) \approx H_{m+1}(B)$$

define a $\mathscr{C}$-equivalence of $H_m(F)$ and $H_{m+1}(B)$ whenever $p - 1 \leqslant m < 2p - 2$.

The proof of this theorem is analogous to that of (6.7).

A space X is said to be *$\mathscr{C}$-acyclic* if $H_m(X) \in \mathscr{C}$ for each $m > 0$.

Theorem 6.9. *If $\mathscr{C}$ is a weakly complete class, $H_1(B) = 0$, and two of the spaces X, B, F are $\mathscr{C}$-acyclic, then so is the third.*

Proof. If the two spaces are B and F, then it follows from (6.3) with $r = \infty$ and $B_0 = \square$ that X is $\mathscr{C}$-acyclic.

If the two spaces are X and B, then it follows from (6.7) with $p = \infty$ that F is $\mathscr{C}$-acyclic.

If the two spaces are X and F, then we shall prove $H_p(B) \in \mathscr{C}$ by induction on p. The case $p = 1$ is trivial. Assume $p > 1$ and $H_m(B) \in \mathscr{C}$ if $1 < m < p$. By (6.7), $H_p(B)$ is $\mathscr{C}$-equivalent to $H_{p-1}(F)$ and hence $H_p(B) \in \mathscr{C}$. ∎

As an application of these results, let us consider the special case where $X = [B; B, b_0]$ and $\omega : X \to B$ is the initial projection. In this case, the fiber F becomes the space

$$\Lambda(B) = [B; b_0, b_0]$$

of all loops in B with b_0 as basic point. Assume that B is simply connected and hence $\Lambda(B)$ is pathwise connected. Therefore, we may apply the results of this section. Since $[B; B, b_0]$ is contractible, the following theorem is an immediate consequence of (6.7) and (6.8).

Theorem 6.10. *If $\mathscr{C}$ is a class and $H_m(B) \in \mathscr{C}$ whenever $0 < m < p$, then $H_m(\Lambda(B))$ is $\mathscr{C}$-equivalent to $H_{m+1}(B)$ for the following values of m:*

(i) $0 < m < p$ if $\mathscr{C}$ is weakly complete.

(ii) $0 < m < 2p - 2$ if $\mathscr{C}$ is complete.

Therefore, for a weakly complete class $\mathscr{C}$, $\Lambda(B)$ is $\mathscr{C}$-acyclic iff B is $\mathscr{C}$-acyclic.

In particular, if B is a space of the homotopy type (π, n) with $n > 1$, then $\Lambda(B)$ is a space of the homotopy type $(\pi, n - 1)$. Thus, we may apply (6.10) to this special case and obtain

$$H_m(\pi, n - 1) \approx H_{m+1}(\pi, n), \quad 0 < m < 2n - 2.$$

Furthermore, we have the following obvious

Proposition 6.11. *For an abelian group π, an integer $n > 1$, and a weakly complete class $\mathscr{C}$, the following two statements are equivalent*:

(*i*) $H_m(\pi) \in \mathscr{C}$ *for each* $m > 0$.

(*ii*) $H_m(\pi, n) \in \mathscr{C}$ *for each* $m > 0$.

7. Applications to n-connective fiber spaces

Let B be a pathwise connected space and b_0 a given point in B. According to (V; § 8), we may construct inductively a sequence of spaces (B, n), $n = 0, 1, 2, \cdots$, and a sequence of maps

$$\beta_n : (B, n) \to (B, n-1), \quad n = 1, 2, 3, \cdots,$$

as follows. Let $(B, 0) = B$. For each $n > 0$, let (B, n) be an n-connective fiber space over $(B, n-1)$ with projection β_n. This system $\{ (B, n), \beta_n \}$ will be referred to as a *connective system* of the space B. If B is locally pathwise connected and semi-locally simply connected, then of course we may take $(B, 1)$ to be the universal covering space over B with β_1 as the projection.

Now let $\{ (B, n), \beta_n \}$ be any connective system of B. Since (B, n) is an n-connective fiber space over an $(n-1)$-connected space $(B, n-1)$ with β_n as projection, it follows that the fiber of this fibering is a space F_n of the homotopy type $(\pi_n(B), n-1)$. Next, let

$$\omega_n = \beta_1\beta_2\cdots\beta_n : (B, n) \to B, \quad n = 1, 2, \cdots,$$

then it is clear that (B, n) is an n-connective fiber space over B with ω_n as projection.

Applying (6.9) and (6.11) to the fibering β_n, we obtain the following

Proposition 7.1. *If $\mathscr{C}$ is a perfect and weakly complete class, $n > 1$, and $\pi_n(B) \in \mathscr{C}$, then the following two statements are equivalent*:

(*i*) (B, n) *is $\mathscr{C}$-acyclic.*

(*ii*) $(B, n-1)$ *is $\mathscr{C}$-acyclic.*

Applying (6.6) with $q = \infty$ to the fibering β_n, we obtain the following

Proposition 7.2. *If $\mathscr{C}$ is a perfect and complete class, $n > 1$, and $\pi_n(B) \in \mathscr{C}$, then*

$$(\beta_n)_* : H_m(B, n) \to H_m(B, n-1)$$

is a $\mathscr{C}$-isomorphism for each $m \geqslant 0$.

Finally, if B is $(n-1)$-connected, then we may take $(B, m) = B$ for every $m \leqslant n-1$. The following proposition will be used in the sequel.

Proposition 7.3. *Let $\mathscr{C}$ be a perfect and weakly complete class. If B is $(n-1)$-connected, $\pi_n(B) \in \mathscr{C}$, $n > 1$, and $p > n$ is an integer such that $H_m(B) \in \mathscr{C}$ whenever $n \leqslant m < p$, then the induced homomorphism*

$$(\beta_n)_* : H_m(B, n) \to H_m(B)$$

is a $\mathscr{C}$-isomorphism for $m \leqslant p$ and is $\mathscr{C}$-epimorphism for $m = p + 1$.

Proof. Since the fiber $F = F_n$ of the fibering β_n is of the homotopy type $(\pi_n(B), n-1)$, it follows from (6.11) that $H_m(F) \in \mathscr{C}$ for every $m > 0$. Let $B_0 \subset B$ consist of a single point, then we have $H_1(B, B_0) = 0$ and $H_m(B, B_0) \in \mathscr{C}$ for each $m < p$. Thus we may apply (6.5) with $q = \infty$. Hence the induced homomorphism

$$(\beta_n)_{\#} : H_m(X, F) \to H_m(B, B_0), \quad X = (B, n),$$

is a $\mathscr{C}$-isomorphism whenever $m \leqslant p$ and is a $\mathscr{C}$-epimorphism whenever $m = p + 1$. Since $H_m(F) \in \mathscr{C}$ for each $m > 0$, it follows from the homology sequence that

$$j : H_m(X) \to H_m(X, F)$$

is a $\mathscr{C}$-isomorphism for every $m > 0$. Finally, using the isomorphisms

$$k : H_m(B) \approx H_m(B, B_0), \quad m > 0,$$

we obtain $(\beta_n)_* = k^{-1}(\beta_n)_{\#}j$. ▮

8. The generalized Hurewicz theorem

Theorem 8.1. *Let $\mathscr{C}$ be a perfect and weakly complete class. If X is a simply connected space and $n \geqslant 2$ is an integer such that $\pi_m(X) \in \mathscr{C}$ whenever $1 < m < n$, then the natural homomorphism*

$$h_m : \pi_m(X) \to H_m(X)$$

is a $\mathscr{C}$-isomorphism whenever $0 < m \leqslant n$ and is a $\mathscr{C}$-epimorphism whenever $m = n + 1$.

Proof. We are going to prove this theorem by induction on n. If $n = 2$, then this is implied by the usual Hurewicz theorem since $\pi_0(X) = 0$ and $\pi_1(X) = 0$. See (V; 4.4) and (V; Ex. C).

Let $p > 2$ be an integer and assume that the theorem is true for $n < p$. Let us prove the theorem for $n = p$. By the inductive hypothesis, it follows that h_m is a $\mathscr{C}$-isomorphism whenever $0 < m < p$ and is a $\mathscr{C}$-epimorphism for $m = p$. It remains to prove that h_p is a $\mathscr{C}$-monomorphism and h_{p+1} is a $\mathscr{C}$-epimorphism.

Consider a connective system $\{ (X, r), \beta_r \}$ of the space X as defined in § 7. Since X is simply connected, we may assume $(X, 1) = X$. Then $(X, p-1)$ is a $(p-1)$-connective fiber space over X with

$$\omega = \beta_2\beta_3\cdots\beta_{p-1} : (X, p-1) \to X$$

as projection. Thus we obtain a commutative rectangle

$$\begin{array}{ccc} \pi_m(X, p-1) & \xrightarrow{g_m} & H_m(X, p-1) \\ \downarrow \omega_* & & \downarrow \omega_{\#} \\ \pi_m(X) & \xrightarrow{h_m} & H_m(X) \end{array}$$

where ω_*, $\omega_\#$ are induced by ω, and g_m, h_m are the natural homomorphisms.

By the definition of $(p-1)$-connective fiber space, ω_* is an isomorphism for $m \geqslant p$. Since $(X, p-1)$ is (p—1)-connected, it follows from the usual Hurewicz theorem that g_p is an isomorphism and g_{p+1} is an epimorphism. Hence, it suffices to prove that $\omega_\#$ is a $\mathscr{C}$-isomorphism for $m = p$ and a $\mathscr{C}$-epimorphism for $m = p + 1$.

Since $\omega = \beta_2\beta_3 \cdots \beta_{p-1}$, it suffices to prove that, for each $r = 2, 3, \cdots$, $p - 1$, the induced homomorphism,

$$(\beta_r)_\# : H_m(X, r) \to H_m(X, r-1)$$

is a $\mathscr{C}$-isomorphism for $m = p$ and is a $\mathscr{C}$-epimorphism for $m = p + 1$. For this purpose, let $B = (X, r-1)$. Then, B *is* $(r-1)$-connected and we may take a connective system of B with $(B, r-1) = B$ and $(B, r) = (X, r)$. Since B is simply connected and $\pi_m(B) \in \mathscr{C}$ whenever $1 < m < p$, it follows from the inductive hypothesis that $H_m(B) \in \mathscr{C}$ whenever $0 < m < p$. Therefore, by (7.3), $(\beta_r)_\#$ is a $\mathscr{C}$-isomorphism for $m = p$ and is a $\mathscr{C}$-epimorphism for $m = p + 1$. ∎

Corollary 8.2. *Let $\mathscr{C}$ be a perfect and weakly complete class. If X is a simply connected space and $n \geqslant 2$ is an integer such that $H_m(X) \in \mathscr{C}$ whenever $1 < m < n$, then $\pi_m(X) \in \mathscr{C}$ whenever $0 < m < n$.*

This corollary follows from (8.1) by finite induction on m. Therefore, if X is simply connected and $\mathscr{C}$-acyclic, then X is *$\mathscr{C}$-aspherical*, i.e. $\pi_m(X) \in \mathscr{C}$ for all $m > 1$. In particular, we have the following

Corollary 8.3. *The homotopy groups of any simply connected finitely triangulable space are finitely generated.*

9. The relative Hurewicz theorem

Theorem 9.1. *Let $\mathscr{C}$ be a perfect and complete class, X a simply connected space, and X_0 a simply connected subspace of X. If $\pi_2(X, X_0) = 0$ and $n \geqslant 2$ is an integer such that $\pi_m(X, X_0) \in \mathscr{C}$ whenever $2 < m < n$, then the natural homomorphism*

$$h_m : \pi_m(X, X_0) \to H_m(X, X_0)$$

is a $\mathscr{C}$-isomorphism whenever $2 \leqslant m \leqslant n$ and is a $\mathscr{C}$-epimorphism whenever $m = n + 1$.

Proof. We prove this theorem by induction on n. If $n = 2$, the theorem is true since $H_2(X, X_0) = 0$ and h_3 is an isomorphism by (V; Ex. C). It follows from the hypothesis of induction that h_m is a $\mathscr{C}$-isomorphism whenever $2 \leqslant m < n$. Hence $H_m(X, X_0) \in \mathscr{C}$ whenever $0 < m < n$. It remains to prove that h_n is a $\mathscr{C}$-isomorphism and h_{n+1} is a $\mathscr{C}$-epimorphism.

Pick a point $x_0 \in X_0$ and consider the space of paths $Y = [X; X, x_0]$. Then, Y is a fiber space over X with projection $\omega : Y \to X$ defined by

$\omega(\xi) = \xi(0)$ for each $\xi \in Y$. Let $Y_0 = [X; X_0, x_0]$. Then Y_0 is pathwise connected and
$$\pi_m(Y_0) \approx \pi_{m+1}(X, X_0), \quad (m \geqslant 1).$$
Hence, $\pi_1(Y_0) = 0$ and $\pi_m(Y_0) \in \mathscr{C}$ for $2 \leqslant m < n-1$. By (8.1), the natural homomorphism $g_m : \pi_m(Y_0) \to H_m(Y_0)$ is a $\mathscr{C}$-isomorphism for $m = n-1$ and is a $\mathscr{C}$-epimorphism for $m = n$.

Since $F = \omega^{-1}(x_0) = \Lambda(X)$ is pathwise connected and $H_m(X, X_0) \in \mathscr{C}$ whenever $0 < m < n$, we may apply (6.6) with $p = n$ and $q = 1$. Hence the induced homomorphism $\omega_{\#} : H_m(Y, Y_0) \to H_m(X, X_0)$ is a $\mathscr{C}$-isomorphism for $m = n$ and is a $\mathscr{C}$-epimorphism for $m = n+1$. In the diagram

$$\begin{array}{ccccc}
\pi_m(X, X_0) & \xleftarrow{\omega_*} & \pi_m(Y, Y_0) & \xrightarrow{\partial_*} & \pi_{m-1}(Y_0) \\
\downarrow h_m & & \downarrow k_m & & \downarrow g_{m-1} \\
H_m(X, X_0) & \xleftarrow{\omega_{\#}} & H_m(Y, Y_0) & \xrightarrow{\partial_{\#}} & H_{m-1}(Y_0)
\end{array}$$

we obtain $h_m = \omega_{\#}\partial_{\#}^{-1} g_{m-1}\partial_*\omega_*^{-1}$. Therefore, h_m is a $\mathscr{C}$-isomorphism for $m = n$ and is a $\mathscr{C}$-epimorphism for $m = n+1$. ∎

Corollary 9.2. *If X and X_0 are simply connected, $\pi_2(X, X_0) = 0$, and $H_m(X, X_0)$ is in a perfect and complete class $\mathscr{C}$ whenever $2 < m < n$, then $\pi_m(X, X_0)$ is also in $\mathscr{C}$ whenever $2 < m < n$.*

Remark. The theorem (9.1) does not hold if we merely assume that $\mathscr{C}$ is perfect and weakly complete. For example, let $X = A \times B$ and $X_0 = A \times b$, where A and B are simply connected, B is $\mathscr{C}$-acyclic, and $b \in B$. By making various choices of A, one can see that (9.1) is true for a given class $\mathscr{C}$ iff $\mathscr{C}$ is perfect and complete. Similarly, (8.1) is true iff $\mathscr{C}$ is perfect and weakly complete.

10. The Whitehead theorem

Theorem 10.1. *Let $\mathscr{C}$ be a perfect and complete class. If X and Y are simply connected spaces, $f : X \to Y$ is a map such that $f_* : \pi_2(X) \to \pi_2(Y)$ is an epimorphism, and $n \geqslant 2$ is a given integer, then the following two statements are equivalent:*

(1) *$f_* : \pi_m(X) \to \pi_m(Y)$ is a $\mathscr{C}$-isomorphism for $m < n$ and is a $\mathscr{C}$-epimorphism for $m = n$.*

(2) *$f_{\#} : H_m(X) \to H_m(Y)$ is a $\mathscr{C}$-isomorphism for $m < n$ and is a $\mathscr{C}$-epimorphism for $m = n$.*

Proof. Consider the mapping cylinder Z_f of the map f. By (I; § 12), both X and Y can be naturally imbedded in Z_f and Y is then a strong deformation retract of Z_f. Thus the map f is decomposed into the composition ri of the inclusion map $i : X \subset Z_f$ and a strong deformation retraction $r : Z_f \to Y$. Since r induces isomorphisms on homotopy and homology groups, (1) and (2) are equivalent respectively to the following two statements:

(1′) $i_* : \pi_m(X) \to \pi_m(Z_f)$ *is a* $\mathscr{C}$*-isomorphism for* $m < n$ *and is a* $\mathscr{C}$*-epimorphism for* $m = n$.

(2′) $i_\# : H_m(X) \to H_m(Z_f)$ *is a* $\mathscr{C}$*-isomorphism for* $m < n$ *and is a* $\mathscr{C}$*-epimorphism for* $m = n$.

Then it follows from the homotopy sequence and the homology sequence of (Z_f, X) that (1′) and (2′) are equivalent respectively to the following two statements:

(1″) $\pi_m(Z_f, X) \in \mathscr{C}$ *whenever* $2 \leqslant m \leqslant n$.

(2″) $H_m(Z_f, X) \in \mathscr{C}$ *whenever* $2 \leqslant m \leqslant n$.

By (9.1), we conclude that (1″) and (2″) are equivalent. This entails the equivalence of (1) and (2). ∎

EXERCISES

A. Examples of Classes of Abelian Groups.

In addition to the classes given in § 2 and § 3. we give the following examples:

1. The class of all abelian groups with power not exceeding a given infinite cardinal number $\aleph_\alpha$. In particular, if $\aleph_\alpha$ is the power of the set of natural numbers, this reduces to the class $\mathscr{A}_\omega$ of all countable abelian groups. Verify that this class is perfect and weakly complete but not complete.

2. The class of all abelian groups A such that there is an integer N depending on A with $Na = 0$ for every $a \in A$. Verify that this class is perfect and complete but not strongly complete.

3. The class of all abelian groups satisfying the descending chain condition. Verify that this class is perfect and weakly complete but not complete.

B. Composed Homomorphisms

By using the natural exact sequence $S(f, g)$ of two homomorphisms $f : A \to B$ and $g : B \to C$, prove the following six assertions for a given class $\mathscr{C}$:

1. If f and g are $\mathscr{C}$-monomorphisms, then so is gf.
2. If f and g are $\mathscr{C}$-epimorphisms, then so is gf.
3. If gf is a $\mathscr{C}$-monomorphism, then so is f.
4. If gf is a $\mathscr{C}$-epimorphism, then so is g.
5. If gf is a $\mathscr{C}$-monomorphism and f is a $\mathscr{C}$-epimorphism, then g is a $\mathscr{C}$-monomorphism.
6. If gf is a $\mathscr{C}$-epimorphism and g is a $\mathscr{C}$-monomorphism, then f is a $\mathscr{C}$-epimorphism.

C. On Perfectness and Completeness

Prove the following assertions:

1. For any class $\mathscr{C}$ of abelian groups, the following three statements are equivalent:

(a) $\mathscr{C}$ is complete.

(b) $A \in \mathscr{C}$ implies $A \otimes B \in \mathscr{C}$ for every abelian group B.

(c) For any $A \in \mathscr{C}$, every finite or infinite direct sum of groups isomorphic to A is in $\mathscr{C}$.

2. Every strongly complete class is perfect and complete.

It is unknown whether there is a class which is not perfect or not weakly complete.

D. The C-Generalization of the "Five" Lemma

Prove that the "five" lemma, [E–S; p. 16], remains true modulo a class $\mathscr{C}$. Precisely, if we have two exact sequences each with five terms and five homomorphisms of the groups of the first sequence into the corresponding groups of the second sequence with the commutativity relations being satisfied, and if the four extreme homomorphisms are $\mathscr{C}$-isomorphisms, then the middle homomorphism is also a $\mathscr{C}$-isomorphism.

E. The C-Inverse Homomorphism Theorem

Consider a class $\mathscr{C}$ and an exact sequence

$$A_1 \xrightarrow{f_1} A_2 \xrightarrow{f_2} A_3 \xrightarrow{f_3} A_4 \xrightarrow{f_4} A_5$$

Assume that there exist two homomorphisms $g_1 : A_2 \to A_1$, and $g_4 : A_5 \to A_4$ such that the endomorphisms $f_1 g_1$ and $f_4 g_4$ are $\mathscr{C}$-isomorphisms. Define a homomorphism $h: A_3 + A_5 \to A_4$ by taking $h(x, y) = f_3(x) + g_5(y)$. Prove that h is a $\mathscr{C}$-isomorphism.

F. The Products of C-Equivalent Groups

Let $\mathscr{C}$ be a complete class. Prove that, if A and B are $\mathscr{C}$-equivalent respectively to A' and B', then $A \otimes B$ and $Tor(A, B)$ are $\mathscr{C}$-equivalent respectively to $A' \otimes B'$ and $Tor(A', B')$.

G. C-Exact Sequences

Let $\mathscr{C}$ be a class and A, B two subgroups of an abelian group G. We say that A and B are *$\mathscr{C}$-equal* if the inclusion homomorphisms $A \cap B \to A$ and $A \cap B \to B$ are $\mathscr{C}$-isomorphisms. Replacing equality by $\mathscr{C}$-equality, one can define the notion of a *$\mathscr{C}$-exact sequence.*

1. Establish the elementary properties of $\mathscr{C}$-exact sequences as in [E–S; p. 50].

2. Generalize the results in (VIII; § 8) to obtain various fundamental $\mathscr{C}$-exact sequences.

H. On Induced Homomorphisms

Let X, Y be simply connected spaces, $f : X \to Y$ a map such that $f_* : \pi_2(X) \to \pi_2(Y)$ is an epimorphism. Assume that the homology groups are finitely generated.

1. Let $\mathscr{F}$ denote the class of all finite abelian groups, $\mathscr{T}$ the class of all torsion groups, and G a field of characteristic zero. Prove that the following four statements are equivalent:

(a) $f_{\#} : H_m(X) \to H_m(Y)$ is an $\mathscr{T}$-isomorphism for $m < n$ and is an $\mathscr{F}$-epimorphism for $m = n$.

(b) $f_{\#} : H_m(X) \to H_m(Y)$ is a $\mathscr{F}$-isomorphism for $m < n$ and is a $\mathscr{T}$-epimorphism for $m = n$.

(c) $f_{\#} : H_m(X; G) \to H_m(Y; G)$ is an isomorphism for $m < n$ and is an epimorphism for $m = n$.

(d) $f^{\#} : H^m(Y, G) \to H^m(X; G)$ is an isomorphism for $m < n$ and is a monomorphism for $m = n$.

2. Let $\mathscr{F}_p$ denote the class of all finite abelian groups of order not divisible by a given prime number p, $\mathscr{T}_p$ the class of all torsion groups with null p-primary component, and G a field of characteristic p. Prove that the following four statements are equivalent:

(a) $f_{\#} : H_m(X) \to H_m(Y)$ is an $\mathscr{F}_p$-isomorphism for $m < n$ and is an $\mathscr{F}_p$-epimorphism for $m = n$.

(b) $f_{\#} : H_m(X) \to H_m(Y)$ is a $\mathscr{T}_p$-isomorphism for $m < n$ and is a $\mathscr{T}_p$-epimorphism for $m = n$.

(c) $f_{\#} : H_m(X; G) \to H_m(Y; G)$ is an isomorphism for $m < n$ and is an epimorphism for $m = n$.

(d) $f^{\#} : H^m(Y; G) \to H^m(X; G)$ is an isomorphism for $m < n$ and is a monomorphism for $m = n$.

The usefulness of these two propositions is that, on many occasions, we may replace the calculus mod $\mathscr{C}$ by the calculus with coefficients in a field. Secondly, since $\mathscr{T}$ and $\mathscr{T}_p$ are perfect and complete, we may apply (10.1). Finally, the statements (b), (c), (d) are equivalent even if the homology groups are not finitely generated.

CHAPTER XI

HOMOTOPY GROUPS OF SPHERES

1. Introduction

Finally we come to the determination of certain of the homotopy groups of spheres. The calculations are particularly based on the results of the previous two chapters, and, since they are quite technical, we will not attempt to summarize them here. However, in the course of the development, several topics of independent interest appear: Freudenthal's suspension theorem (stated in § 2 and proved in §§ 2–5), pseudo-projective spaces and Stiefel manifolds (§§ 10–11), and the Hopf invariant of a map $f: S^{2n-1} \to S^n$ (§ 14).

We have already seen that if $r < 0$ then $\pi_{n+r}(S^n)$ is the zero group, and, that if $r = 0$ then this group is free cyclic; and now in §§ 15–17 we shall settle the cases $r = 1, 2, 3$, and 4. A brief report of the cases $5 \leqslant r \leqslant 15$ is given in the final section.

2. The suspension theorem

Let $n > 1$ and consider the n-sphere S^n as the equator of the $(n+1)$-sphere S^{n+1} with u and v denoting respectively the north and south poles of S^{n+1}. Pick a point s_0 in S^n and consider the space

$$W = \Lambda(S^{n+1})$$

of loops in S^{n+1} with s_0 as basic point.

There is a natural imbedding $i: S^n \to W$ described as follows. For each $x \in S^n$, $i(x)$ is the loop in S^{n+1} joining s_0 to u, u to x, x to v, and v back to s_0, all by shortest geodesic arcs. That i is a homeomorphism of S^n into W is obvious. Furthermore, the loop $i(s_0)$ is homotopic to the degenerate loop $w_0 \in W$ which maps I into s_0 by means of a natural homotopy; in other words, the points w_0 and $i(s_0)$ of W are connected by a natural path σ in W.

Hereafter, we shall identify x and $i(x)$ for every $x \in S^n$. Thus, S^n becomes the subspace $i(S^n)$ of W and $i: S^n \to W$ reduces to the inclusion map. For each $m > 0$, i induces a homomorphism

$$i_*: \pi_m(S^n, s_0) \to \pi_m(W, s_0),$$

the path σ induces an isomorphism

$$\sigma_*: \pi_m(W, s_0) \approx \pi_m(W, w_0),$$

and, according to (IV; 2.2), we have an isomorphism

$$h_* : \pi_m(W, w_0) \approx \pi_{m+1}(S^{n+1}, s_0).$$

Composing i_*, σ_*, and h_*, we obtain a homomorphism

$$\Sigma = h_* \sigma_* i_* : \pi_m(S^n, s_0) \to \pi_{m+1}(S^{n+1}, s_0)$$

for each $m > 0$, called the *suspension*. One can verify that this definition is equivalent to the more general one given in (V; § 11) for this special case.

Theorem 2.1. (The Suspension Theorem). *The suspension Σ is an isomorphism if $m < 2n - 1$ and is an epimorphism if $m = 2n - 1$.*

Proof. Since σ_* and h_* are isomorphisms, it suffices to prove that i_* is an isomorphism if $m < 2n - 1$ and is an epimorphism if $m = 2n - 1$. Thus, according to Whitehead theorem, (X; 10.1), it suffices to prove that the induced homomorphism

$$i_\# : H_m(S^n) \to H_m(W)$$

is an isomorphism if $m < 2n - 1$ and is an epimorphism if $m = 2n - 1$. Since

$$H_m(W) \approx Z, \quad \text{if } m \equiv 0 \ mod(n),$$

$$H_m(W) = 0, \quad \text{if } m \not\equiv 0 \ mod(n),$$

by (IX; 13.4), it remains to prove the following

Lemma 2.2. $i_\# : H_n(S^n) \approx H_n(W)$.

This lemma will be proved in the next three sections; we conclude this section with one immediate consequence of the suspension theorem. Let U and V denote respectively the north and south hemispheres of S^{n+1}; then, using (V; 11.1), we have the following

Corollary 2.3. *The excision homomorphism*

$$e_* : \pi_m(U, S^n) \to \pi_m(S^{n+1}, V)$$

is an isomorphism whenever $2 \leqslant m < 2n$ and is an epimorphism whenever $m = 2n$.

The identity map on S^n extends to a map $f : (E^{n+1}, S^n) \to (U, S^n)$. If we compose this map with the excision $e : (U, S^n) \subset (S^{n+1}, V)$, we obtain a map

$$g = ef : (E^{n+1}, S^n) \to (S^{n+1}, V).$$

Since V is contractible to the point s_0, g is homotopic in (S^{n+1}, V) to a map

$$h : (E^{n+1}, S^n) \to (S^{n+1}, s_0)$$

which represents a generator of $\pi_{n+1}(S^{n+1}, s_0)$. For any element α of $\pi_m(S^n, s_0)$, choose a map $\phi : S^m \to S^n$ which represents α. The map ϕ has an extension

$$\psi : (E^{m+1}, S^m) \to (E^{n+1}, S^n).$$

Then it can be seen that the composed map $h\psi$ represents the element $\Sigma(\alpha)$ in $\pi_{m+1}(S^{n+1}, s_0)$.

3. The canonical map

Consider the space of paths $X = [S^{n+1}; s_0, S^{n+1}]$. According to (III; § 13), X is a fiber space over S^{n+1} with a projection $\omega : X \to S^{n+1}$ defined by $\omega(x) = x(1)$ for every path $x \in X$ and with fiber $W = \omega^{-1}(s_0)$.

Let U and V denote the north and the south hemispheres of S^{n+1} respectively and let

$$X_0 = \omega^{-1}(S^n), \quad X_u = \omega^{-1}(U), \quad X_v = \omega^{-1}(V).$$

We are going to define a map

$$\varkappa : (U \times W, S^n \times W) \to (X_u, X_0)$$

which will be called the *canonical map*.

For each point $b \in U$, let $\gamma(b) \in X_u$ denote the path joining s_0 to u and then u to b by geodesic arcs. The assignment $b \to \gamma(b)$ defines a cross-section $\gamma : U \to X_u$. Then $\varkappa$ is defined by $\varkappa(b, f) = f\cdot\gamma(b)$ for each $b \in U$ and $f \in W$, where $f\cdot\gamma(b)$ denotes the product of the paths f and $\gamma(b)$. As a consequence of the construction, we have

$$\omega\varkappa(b, f) = b, \quad (b \in U, f \in W).$$

Lemma 3.1. *The canonical map $\varkappa$ is a homotopy equivalence.*

Proof. Let $\lambda : (X_u, X_0) \to (U \times W, S^n \times W)$ be the map defined by

$$\lambda(x) = (\omega(x), x\cdot[\gamma\omega(x)]^{-1}), \quad (x \in X_u).$$

where $[\gamma\omega(x)]^{-1}$ denotes the reverse of the path $\gamma\omega(x)$. Then we have

$$\varkappa\lambda(x) = \varkappa(\omega(x), x\cdot[\gamma\omega(x)]^{-1}) = (x\cdot[\gamma\omega(x)]^{-1})\cdot\gamma\omega(x),$$
$$\lambda\varkappa(b,f) = \lambda(f\cdot\gamma(b)) = (b, [f\cdot\gamma(b)]\cdot[\gamma(v)]^{-1}).$$

Hence, $\varkappa\lambda$ and $\lambda\varkappa$ are both homotopic to the identity maps. ∎

Next, let us define a map

$$\mu : S^n \times W \to W$$

by taking $\mu(b, f) = f\cdot i(b)$ for each $b \in S^n$ and $f \in W$, where $i : S^n \to W$ denotes the imbedding in § 2.

Lemma 3.2. *The map μ is homotopic in X_v to the map $\nu : S^n \times W \to X_0$ defined by $\nu = \varkappa \mid S^n \times W$.*

Proof. Intuitively, a homotopy of μ to ν is accomplished by "unwinding" the path $i(b)$ to half its original length. More precisely, define a homotopy $h_t : S^n \to X_v$, $(0 \leqslant t \leqslant 1)$, by taking

$$[h_t(b)](s) = [i(x)]\left(\frac{s}{1+t}\right), \quad (b \in S^n, s \in I, t \in I).$$

Then $h_0 = i$ and $h_1 = \gamma$. Define a homotopy $k_t : S^n \times W \to X_v$, $(0 \leqslant t \leqslant 1)$, by taking

$$k_t(b, f) = f \cdot h_t(b), \quad (b \in S^n, f \in W, t \in I).$$

Then $k_0 = \mu$ and $k_1 = \nu$. ∎

4. Wang's isomorphism ρ_*

In the present section, we shall construct for each integer $q \geqslant 0$ an isomorphism

$$\rho_* : H_n(S^n) \otimes H_q(W) \approx H_{n+q}(W).$$

Let $m = n + q + 1$. The construction of ρ_* will be made in six steps as follows.

Step 1. Since the south hemisphere V of S^{n+1} is contractible to the point s_0, an application of the covering homotopy theorem proves that W is a strong deformation retract of X_v. Hence the inclusion map induces an isomorphism

$$\xi : H_m(X, W) \approx H_m(X, X_v).$$

Step 2. The inclusion map induces a homomorphism

$$\eta : H_m(X_u, X_0) \to H_m(X, X_v).$$

Let

$$D_u = S^{n+1} \setminus v, \; D_v = S^{n+1} \setminus u, \; D = D_u \cap D_v;$$

$$Y_u = \omega^{-1}(D_u), \; Y_v = \omega^{-1}(D_v), \; Y = Y_u \cap Y_v.$$

Since Y_u and Y_v are open sets whose union is X, the excision theorem holds and hence the inclusion map induces an isomorphism

$$H_m(Y_u, Y) \approx H_m(X, Y_v).$$

Since U, V, S^n are strong deformation retracts of D_u, D_v, D respectively, an application of the covering homotopy theorem proves that X_u, X_v, X_0 are strong deformation retracts of Y_u, Y_v, Y respectively. Hence η is an isomorphism.

Step 3. Since the canonical map $\varkappa$ is a homotopy equivalence, it induces an isomorphism

$$\varkappa_* : H_m(U \times W, S^n \times W) \approx H_m(X_u, X_0).$$

Step 4. By the Künneth theorem, we get an isomorphism

$$\zeta : H_{n+1}(U, S^n) \otimes H_q(W) \approx H_m(U \times W, S^n \times W).$$

Step 5. Since X is contractible, we have an isomorphism

$$\partial : H_m(X, W) \approx H_{m-1}(W).$$

Step 6. Since U is contractible, we have an isomorphism

$$H_{n+1}(U, S^n) \approx H_n(S^n).$$

Taking tensor products, we obtain an isomorphism

$$\theta : H_{n+1}(U, S^n) \otimes H_q(W) \approx H_n(S^n) \otimes H_q(W).$$

Composing these steps, we get an isomorphism

$$\rho_* = \partial\xi^{-1}\eta\varkappa_*\zeta\theta^{-1} : H_n(S^n) \otimes H_q(W) \approx H_{n+q}(W).$$

This isomorphism ρ_* is the same as the homomorphism ρ_* in Wang's exact sequence (IX; 13.1) for the fibering $\omega : X \to S^{n+1}$. See also [Wang 2].

5. Relation between ρ_* and $i_\#$

The space W of loops has a continuous multiplication

$$M : W \times W \to W$$

defined in (III; § 11). The total homology group

$$H(W) = \sum_{m=0}^{\infty} H_m(W)$$

becomes a ring under the *Pontrjagin multiplication* defined as follows.

Let $\alpha \in H_p(W)$ and $\beta \in H_q(W)$. By the Künneth theorem, α and β determine a unique element $\alpha \times \beta$ of $H_{p+q}(W \times W)$. The map M induces a homomorphism

$$M_* : H_m(W \times W) \to H_m(W)$$

for every m. Then the *Pontrjagin product* of α and β is defined to be the element

$$\alpha \cdot \beta = M_*(\alpha \times \beta) \in H_{p+q}(W).$$

Proposition 5.1. *For every* $\alpha \in H_n(S^n)$ *and* $\beta \in H_q(W)$, *we always have*

$$\rho_*(\alpha \otimes \beta) = \beta \cdot i_\#(\alpha),$$

where $i_\# : H_n(S^n) \to H_n(W)$ *denotes the homomorphism induced by the imbedding* $i : S^n \to W$ of § 2.

Proof. Consider the diagram

$$\begin{array}{ccccccc}
H_m(U \times W, S^n \times W) & \xrightarrow{\varkappa_*} & H_m(X_u, X_0) & \xrightarrow{\eta} & H_m(X, X_v) & \xleftarrow{\xi} & H_m(X, W) \\
\downarrow \partial & & \downarrow \partial & & \downarrow \partial & & \downarrow \partial \\
H_{m-1}(S^n \times W) & \xrightarrow{\nu_*} & H_{m-1}(X_0) & \xrightarrow{\tau} & H_{m-1}(X_v) & \xleftarrow{\sigma} & H_{m-1}(W)
\end{array}$$

where σ and τ are induced by inclusion maps and the homomorphisms ∂ are boundary operators. The rectangules are all commutative and hence

$$\partial\xi^{-1}\eta\varkappa_* = \sigma^{-1}\tau\nu_*\partial. \tag{1}$$

By (3.2), we have

$$\tau\nu_* = \sigma\mu_*. \tag{2}$$

By the Künneth theorem, we have an isomorphism

$$\chi : H_n(S^n) \otimes H_q(W) \approx H_{n+q}(S^n \times W)$$

and a commutative rectangle

$$\begin{array}{ccc} H_{n+1}(U, S^n) \otimes H_q(W) & \xrightarrow{\zeta} & H_{n+q+1}(U \times W, S^n \times W) \\ \downarrow{\scriptstyle\theta} & & \downarrow{\scriptstyle\partial} \\ H_n(S^n) \otimes H_q(W) & \xrightarrow{\chi} & H_{n+q}(S^n \times W). \end{array}$$

Hence we obtain

$$\chi = \partial\zeta\theta^{-1}. \tag{3}$$

Using (1), (2) and (3), we deduce

$$\rho_* = \partial\xi^{-1}\eta\varkappa_*\zeta\theta^{-1} = \sigma^{-1}\tau\nu_*\partial\zeta\theta^{-1} = \mu_*\partial\zeta\theta^{-1} = \mu_*\chi.$$

Then it follows from the definition of μ that

$$\rho_*(\alpha \otimes \beta) = \mu_*\chi(\alpha \otimes \beta) = \beta \cdot i_{\#}(\alpha). \quad \blacksquare$$

In particular, if $q = 0$, then $H_0(W)$ is a free cyclic group generated by the element e represented by w_0 as a 0-cycle of W. For each $\alpha \in H_n(S^n)$, we have

$$i_{\#}(\alpha) = e \cdot i_{\#}(\alpha) = \rho_*(\alpha \otimes e).$$

This proves Lemma 2.2.

6. The triad homotopy groups

Consider the space of paths

$$T = [W; S^n, s_0].$$

By (IV; 3.1), we have

$$\pi_m(T) = \pi_{m+1}(W, S^n)$$

for every m. Hence the homotopy sequence of the pair (W, S^n) gives rise to an exact sequence

$$\cdots \to \pi_m(T) \to \pi_m(S^n) \xrightarrow{\Sigma} \pi_{m+1}(S^{n+1}) \to \pi_{m-1}(T) \to \cdots.$$

This is essentially the suspension sequence of the triad $(S^{n+1}; U, V)$, $\pi_m(T)$ being essentially the triad homotopy group $\pi_{m+2}(S^{n+1}; U, V)$. See (V; §§ 10–11).

Because of this exact sequence, it is desirable to determine the triad homotopy groups $\pi_m(T)$. The following lemma is an immediate consequence of the suspension theorem (2.1).

Lemma 6.1. $\pi_m(T) = 0$ *for every* $m \leqslant 2n - 2$.

To determine the higher homotopy groups of T, let us study the space of paths

$$Q = [W; S^n, W]$$

which is of the same homotopy type as S^n. Consider the projection $\omega : Q \to W$ defined by $\omega(\sigma) = \sigma(1)$ for every $\sigma \in Q$; then Q becomes a fiber space over W with fiber $\omega^{-1}(s_0) = T$.

Since $H_m(W) = 0$ whenever $0 < m < n$ and $H_m(T) = 0$ whenever $0 < m < 2n - 1$, we have by (IX; 14.1) an exact sequence

$$H_{3n-2}(T) \to \cdots \to H_{m+1}(Q) \to H_{m+1}(W) \to H_m(T) \to H_m(Q) \to \cdots \to H_1(W) \to 0.$$

Since Q is of the same homotopy type as S^n, we have $H_{m+1}(Q) = 0 = H_m(Q)$ for every $m > n$. This implies that $H_m(T) \approx H_{m+1}(W)$ whenever $n < m < 3n - 2$. Hence, we deduce the following

Lemma 6.2. *$H_{2n-1}(T) \approx Z$ and $H_m(T) = 0$ whenever $2n - 1 < m < 3n - 2$.*

Choose a map $f : S^{2n-1} \to T$ which represents a generator of the free cyclic group $\pi_{2n-1}(T) \approx H_{2n-1}(T)$. Then f induces an isomorphism

$$f_{\#} : H_{2n-1}(S^{2n-1}) \approx H_{2n-1}(T).$$

Then, by (6.1) and (6.2), it follows that $f_{\#} : H_m(S^{2n-1}) \approx H_m(T)$ for every $m < 3n - 2$. An application of Whitehead's theorem proves that the induced homomorphism

$$f_* : \pi_m(S^{2n-1}) \to \pi_m(T)$$

is an isomorphism if $m < 3n - 3$ and is an epimorphism if $m = 3n - 3$. Hence we have proved the following

Proposition 6.3. *$\pi_m(T)$ is isomorphic to $\pi_m(S^{2n-1})$ for every $m < 3n - 3$ and $\pi_{3n-3}(T)$ is isomorphic to a quotient group of $\pi_{3n-3}(S^{2n-1})$.*

7. Finiteness of higher homotopy groups of odd-dimensional spheres

In the present section, we are concerned with an odd-dimensional sphere S^n. Since the homotopy groups of the 1-sphere S^1 are completely computed in (IV; § 2), we may assume that $n \geqslant 3$.

Consider an n-connective fiber space X over S^n with a projection $\omega : X \to S^n$. By definition,

$$\pi_m(X) = 0, \quad (m \leqslant n),$$

$$\omega_* : \pi_m(X) \approx \pi_m(S^n), \quad (m > n).$$

Then it follows that the fiber F is a space of the homotopy type $(Z, n - 1)$.

By (X; 8.3), $\pi_m(S^n)$ is a finitely generated abelian group for every m. An application of the generalized Hurewicz theorem proves that $H_m(X)$ is finitely generated for every m.

Let K be a field of characteristic zero. Since $n - 1$ is even, it follows from (IX; Ex. G) that the cohomology algebra $H^*(F; K)$ is isomorphic to a polynomial algebra over K generated by an element of degree $n - 1$. Let $\alpha \in H^{n-1}(F; K)$ be a generator of $H^*(F; K)$.

Consider Wang's cohomology sequence (IX; 13.2):

$$\cdots \to H^m(X; K) \to H^m(F; K) \xrightarrow{\rho^*} H^{m-n+1}(F; K) \to H^{m+1}(X; K) \to \cdots,$$

where ρ^* is a derivation since n is odd. Since $H^n(X; K) = 0 = H^{n-1}(X; K)$, ρ^* sends $H^{n-1}(F; K)$ isomorphically onto $H^0(F; K)$. Therefore, $\rho^*(\alpha)$ is a non-zero element of $H^0(F; K) \approx K$. Now let us prove that

$$\rho^* : H^{p(n-1)}(F; K) \approx H^{(p-1)(n-1)}(F; K)$$

for every positive integer p. In fact, the vector space $H^{p(n-1)}(F; K)$ over K admits α^p as a basis. Since ρ^* is a derivation, we have

$$\rho^*(k\alpha^p) = pk\rho^*(\alpha)\alpha^{p-1}, \quad (k \in K).$$

Hence ρ^* is an isomorphism for every $m > 0$. Then an exactness argument proves that $H^m(X; K) = 0$ for every $m > 0$. Since $H_m(X)$ is finitely generated and K is of characteristic zero, this implies that $H_m(X)$ is finite for every $m > 0$. An application of the generalized Hurewicz theorem (X; 8.1) proves that $\pi_m(X)$ is finite for every m. Thus, we have proved the following

Theorem 7.1. *If S^n is an odd-dimensional sphere and $m > n$, then $\pi_m(S^n)$ is finite.*

8. The iterated suspension

The natural imbedding $S^{n+1} \subset \Lambda(S^{n+2})$ of § 2 induces an imbedding

$$j : \Lambda(S^{n+1}) \to \Lambda^2(S^{n+2}).$$

Composing with the natural imbedding $i: S^n \to \Lambda(S^{n+1})$, we obtain an imbedding

$$k = ji : S^n \to \Lambda^2(S^{n+2}).$$

For each m, k induces a homomorphism

$$k_* : \pi_m(S^n, s_0) \to \pi_m(\Lambda^2(S^{n+2}), s_0).$$

As in § 2, we have a natural isomorphism

$$l_* : \pi_m(\Lambda^2(S^{n+2}), s_0) \approx \pi_{m+2}(S^{n+2}, s_0).$$

Proposition 8.1. *$l_* k_*$ is equal to the iterated suspension Σ^2.*

Proof. By § 2, there is an isomorphism

$$\alpha = h_*\sigma_* : \pi_m(\Lambda(S^{n+1}), s_0) \approx \pi_{m+1}(S^{n+1}, s_0).$$

Similarly, there are isomorphisms

$$\beta : \pi_m(\Lambda^2(S^{n+2}), s_0) \approx \pi_{m+1}(\Lambda(S^{n+2}), s_0),$$
$$\gamma : \pi_{m+1}(\Lambda(S^{n+2}), s_0) \approx \pi_{m+2}(S^{n+2}, s_0).$$

Then $l_* = \gamma\beta$ and $k_* = j_* i_*$. The proposition is a consequence of the commutativity of the diagram:

$$\begin{array}{ccccc}
\pi_m(S^n, s_0) & \xrightarrow{i_*} & \pi_m(\Lambda(S^{n+1}), s_0) & \xrightarrow{j_*} & \pi_m(\Lambda^2(S^{n+2}), s_0) \\
 & \searrow^{\Sigma} & \downarrow{\alpha} & & \downarrow{\beta} \\
 & & \pi_{m+1}(S^{n+1}, s_0) & \xrightarrow{i_*} & \pi_{m+1}(\Lambda(S^{n+2}), s_0) \\
 & & & \searrow^{\Sigma} & \downarrow{\gamma} \\
 & & & & \pi_{m+2}(S^{n+2}, s_0).
\end{array}$$ ∎

The following proposition is an immediate consequence of (8.1) and (2.1).

Proposition 8.2. *The homomorphism k_* is an isomorphism if $m < 2n - 1$ and is an epimorphism if $m = 2n - 1$.*

If we study the p-primary components instead of the whole homotopy groups, then we can deduce more detailed information from the iterated suspension Σ^2.

Theorem 8.3. *Let $n \geqslant 3$ be an odd integer, p a prime number, and $\mathscr{C}$ the class of all finite abelian groups of order prime to p. Then the iterated suspension*

$$\Sigma^2 : \pi_m(S^n) \to \pi_{m+2}(S^{n+2})$$

is a $\mathscr{C}$-isomorphism if $m < p(n + 1) - 3$ and is a $\mathscr{C}$-epimorphism if $m = p(n + 1) - 3$.

Proof. According to (8.1), it suffices to prove the theorem for the homomorphisms

$$k_* : \pi_m(S^n) \to \pi_m(\Lambda^2(S^{n+2}))$$

induced by the natural imbedding $k : S^n \to \Lambda^2(S^{n+2})$.

Let K be a field of characteristic p. Then k induces the homomorphisms

$$k^\# : H^m(\Lambda^2(S^{n+2}); K) \to H^m(S^n; K).$$

By Whitehead theorem (X; 10.1) and (X; Ex. H2), it suffices to show that $k^\#$ is an isomorphism for every $m \leqslant p(n + 1) - 3$.

By (8.2) k_* is an isomorphism if $m < 2n - 1$ and is an epimorphism if $m = 2n - 1$. An application of (X; 10.1) and (X; Ex. H2) proves that $k_\#$ is an isomorphism for $m < 2n - 1$. By (IX; Ex. F6 and F7), we have

$$H^m(\Lambda^2(S^{n+2}); K) = 0, \quad (n < m \leqslant p(n + 1) - 3).$$

Since $n \geqslant 3$, it follows that $k^\#$ is an isomorphism for every $m \leqslant p(n+1) - 3$. ∎

Corollary 8.4. *If $n \geqslant 3$ is an odd integer, p a prime number, and $m < n + 4p - 6$, then the p-primary components of $\pi_m(S^n)$ and $\pi_{m-n+3}(S^3)$ are isomorphic.*

Proof. We shall prove the corollary by induction on n. When $n = 3$, there is nothing to prove. Assume that $q \geqslant 5$ is an odd integer and the corollary is true for every odd integer n with $3 \leqslant n < q$.

By (8.3), the p-primary components of $\pi_m(S^q)$ and $\pi_{m-2}(S^{q-2})$ are isomorphic if $m - 2 < p(q - 1) - 3$. Since $q \geqslant 5$, we have $(p - 1)(q - 5) \geqslant 0$ and hence

$$q + 4p - 8 \leqslant p(q - 1) - 3.$$

This implies $m - 2 < p(q - 1) - 3$ whenever $m < q + 4p - 6$.

By the hypothesis of induction, the p-primary components of $\pi_{m-2}(S^{q-2})$ and $\pi_{m-q+3}(S^3)$ are isomorphic if $m < q + 4p - 6$. Hence, the corollary is also true for $n = q$. ∎

9. The p-primary components of $\pi_m(S^3)$

The corollary (8.4) reveals the importance of finding the p-primary components of the homotopy groups of the 3-sphere S^3.

Consider a 3-connective fiber space X over S^3 with a projection $\omega : X \to S^3$ and fiber F which is a space of homotopy type $(Z, 2)$.

Lemma 9.1. *The integral homology groups of X are as follows: $H_m(X) = 0$ if m is odd; $H_{2n}(X)$ is cyclic of order n for every $n > 0$. Thus, the first few homology groups are:*

$$Z, 0, 0, 0, Z_2, 0, Z_3, 0, Z_4, 0, Z_5, \cdots$$

Proof. Consider Wang's cohomology sequence (IX; 13.2):

$$\cdots \to H^m(X) \to H^m(F) \xrightarrow{\rho^*} H^{m-2}(F) \to H^{m+1}(X) \to \cdots$$

where ρ^* defines a derivation of $H^*(F)$. By (IX; Ex. G), $H^*(F)$ is isomorphic to a polynomial algebra over the ring of integers generated by an element of degree 2.

Since $\pi_m(X) = 0$ for $m \leqslant 3$, we have $H^m(X) = 0$ for $m \leqslant 3$. Hence we obtain

$$\rho^* : H^2(F) \approx H^0(F) = Z.$$

Let α denote element of $H^2(F)$ with $\rho^*(\alpha) = 1$. Then α generates the algebra $H^*(F)$ and $\rho^*(\alpha^n) = n\alpha^{n-1}$.

Let $m = 2n$ with $n \geqslant 2$. The sequence becomes

$$0 \to H^{2n}(X) \to H^{2n}(F) \xrightarrow{\rho^*} H^{2n-2}(F) \to H^{2n+1}(X) \to 0.$$

Since $H^{2n}(F)$ is free cyclic with α^n as generator and $\rho^*(\alpha^n) = n\alpha^{n-1}$, it follows that ρ^* is a monomorphism and its cokernel is cyclic of order n. Then the exactness implies that

$$H^{2n}(X) = 0, \quad H^{2n+1}(X) \approx Z_n.$$

The lemma follows from this and the duality between homology and cohomology. ∎

Theorem 9.2. *If p is a prime number, then the p-primary component of $\pi_m(S^3)$ is 0 if $m < 2p$ and is Z_p if $m = 2p$.*

Proof. Let $\mathscr{C}$ denote the class of all finite abelian groups of order prime to p. By (9.1), we have $H_m(X) \in \mathscr{C}$ whenever $0 < m < 2p$. An application of the generalized Hurewicz theorem proves that $\pi_m(X) \in \mathscr{C}$ whenever $0 < m < 2p$ and $\pi_{2p}(X)$ is $\mathscr{C}$-isomorphic to Z_p. Since X is a 3-connective fiber space over S^3, we have $\pi_m(S^3) \approx \pi_m(X)$ for each $m > 3$. This implies the theorem. ∎

Corollary 9.3. *If $n \geqslant 3$ is an odd integer and p a prime number, then the p-primary component of $\pi_m(S^n)$ is 0 if $m < n + 2p - 3$ and is Z_p if $m = n + 2p - 3$.*

Proof. Since $p \geqslant 2$, we have $n + 2p - 3 < n + 4p - 6$. Then it follows (8.4) that the p-primary components of $\pi_m(S^n)$ and $\pi_{m-n+3}(S^3)$ are isomorphic. ∎

10. Pseudo-projective spaces

If we adjoin to the n-sphere S^n an $(n+1)$-cell E^{n+1} by means of a map $\phi : \partial E^{n+1} = S^n \to S^n$ of degree h as in (I; § 7), we obtain a space

$$P = P_h^{n+1}$$

which is called a *pseudo-projective space*, [A–H; p. 266]. We shall assume that $h > 0$; in this case, the homology groups of P are

$$H_0(P) \approx Z, \quad H_n(P) \approx Z_h;$$

$$H_m(P) = 0, \quad m \neq 0, m \neq n.$$

Lemma 10.1. *For every $m < 2n - 1$, we have an exact sequence*

$$0 \to \pi_m(S^n) \otimes Z_h \to \pi_m(P) \to Tor(\pi_{m-1}(S^n), Z_h) \to 0.$$

Proof. The map ϕ extends to a map $\psi : E^{n+1} \to P$ in an obvious way. We obtain a commutative diagram:

$$\begin{array}{ccccccccc}
\cdots \to \pi_{m+1}(P,S^n) & \xrightarrow{\partial} & \pi_m(S^n) & \to & \pi_m(P) & \to & \pi_m(P,S^n) & \xrightarrow{\partial} & \pi_{m-1}(S^n) \to \cdots \\
\uparrow \psi_* & & \uparrow \phi_* & & & & \uparrow \psi_* & & \uparrow \phi_* \\
\pi_{m+1}(E^{n+1},S^n) & \xrightarrow{\partial} & \pi_m(S^n) & & & & \pi_m(E^{n+1},S^n) & \xrightarrow{\partial} & \pi_{m-1}(S^n)
\end{array}$$

where the top row is the homotopy sequence of the pair (P, S^n) and

$$\phi_* : \pi_m(S^n) \to \pi_m(S^n), \quad \psi_* : \pi_m(E^{n+1}, S^n) \to \pi_m(P, S^n)$$

are induced by the maps ϕ and ψ. Since ϕ is of degree h, we have $\phi_*(\alpha) = h\alpha$ for each $\alpha \in \pi_m(S^n)$ whenever $m < 2n - 1$; on the other hand, ψ_* is an isomorphism for every $m \leqslant 2n - 1$. See Ex. A and Ex. B at the end of the chapter. Hence we may replace the homotopy sequence of (P, S^n) by the exact sequence

$$\pi_m(S^n) \xrightarrow{\phi_*} \pi_m(S^n) \to \pi_m(P) \to \pi_{m-1}(S^n) \xrightarrow{\phi_*} \pi_{m-1}(S^n)$$

for every $m < 2n - 1$. Since the kernel and the cokernel of $\phi_* : \pi_m(S^n) \to \pi_m(S^n)$ are isomorphic to $Tor(\pi_m(S^n), Z_h)$ and $\pi_m(S^n) \otimes Z_h$ respectively, the exactness of this sequence implies the lemma. ∎

Let X denote a 3-connective fiber space over S^3 with projection $\omega : X \to S^3$. Since the p-primary component of $\pi_{2p}(X)$ is cyclic of order p, there exists a map $f : S^{2p} \to X$ which represents a generator $[f]$ of this p-primary component of $\pi_{2p}(X)$.

Consider the pseudo-projective space $P = P_p^{2p+1}$. Then $S^{2p} \subset P$. Since $p\,[f] = 0$, f can be extended to a map $g : P \to X$. Composing with $\omega : X \to S^3$, we get a map $\chi = \omega g : P \to S^3$ which induces the homomorphisms

$$\chi_* : \pi_m(P) \to \pi_m(S^3).$$

Lemma 10.2. *χ_* is a monomorphism for $m < 4p - 1$ and sends $\pi_m(P)$ onto the p-primary component of $\pi_m(S^3)$ for $m \leqslant 4p - 1$.*

Proof. It suffices to prove the lemma for the induced homomorphisms

$$g_* : \pi_m(P) \to \pi_m(X).$$

It follows from the generalized Hurewicz theorem (X; 8.1) that $\pi_m(P)$ is a p-primary group for every m. Hence g_* sends $\pi_m(P)$ into the p-primary component of $\pi_m(X)$.

Let $\mathscr{C}$ denote the class of all finite abelian groups of order prime to p. It remains to prove that g_* is a $\mathscr{C}$-isomorphism for $m < 4p - 1$ and is a $\mathscr{C}$-epimorphism for $m = 4p - 1$.

From the construction of g, one can see that $g_{\#} : H_{2p}(P) \approx H_{2p}(X)$. Then, by (9.1), $g_{\#} : H_m(P) \to H_m(X)$ is a $\mathscr{C}$-isomorphism for $m < 4p$. An application of the Whitehead theorem (X; 10.1) completes the proof. ∎

This lemma reveals the importance of finding the homotopy groups of $P = P_p^{2p+1}$.

Theorem 10.3. *If p is a prime number and $m \leqslant 4p - 2$, then $\pi_m(P) = 0$ if m is different from $2p$, $4p - 3$, and $4p - 2$, while:*

$$\pi_{2p}(P) \approx Z_p;$$

$$\pi_{4p-3}(P) \approx Z_p;$$

$$\pi_{4p-2}(P) \approx Z_p, \quad \text{if } p > 2.$$

Proof. According to Hurewicz's theorem, we have $\pi_m(P) = 0$ for each $m < 2p$ and $\pi_{2p}(P) \approx Z_p$. Applying (10.1) with $n = 2p$, $h = p$, and $m \leqslant 4p - 3$, we obtain an exact sequence

$$0 \to \pi_m(S^{2p}) \otimes Z_p \to \pi_m(P) \to Tor(\pi_{m-1}(S^{2p}), Z_p) \to 0.$$

By the suspension theorem (2.1), we have $\pi_m(S^{2p}) \approx \pi_{m-1}(S^{2p-1})$ for every $m \leqslant 4p - 3$. Since $2p - 1 \geqslant 3$, it follows from (9.3) that the p-primary component of $\pi_{m-1}(S^{2p-1})$ is 0 if $m < 4p - 3$ and is Z_p if $m = 4p - 3$. Hence we obtain:

$$\pi_m(P) = 0, \quad (2p < m < 4p - 3);$$

$$\pi_{4p-3}(P) \approx Z_p.$$

By (10.2), the p-primary component of $\pi_m(S^3)$ is 0 whenever $2p < m < 4p - 3$ and is Z_p if $m = 4p - 3$. Then, by (8.4), we deduce that, if $n \geqslant 3$ is odd, the p-primary component of $\pi_m(S^n)$ is 0 whenever $n + 2p - 3 < m < n + 4p - 6$. By (2.1), $\pi_m(S^{2p}) \approx \pi_{m+1}(S^{2p+1})$ for every $m \leqslant 4p - 2$. Hence, the p-primary component of $\pi_{m+1}(S^{2p+1})$ is 0 whenever $4p - 3 < m < 6p - 6$.

If $p > 2$, then $4p - 2 < 6p - 6$ and hence the p-primary component of $\pi_{4p-2}(S^{2p})$ is 0. By (10.1) with $n = 2p$, $h = p$, and $m = 4p - 2$, we get $\pi_{4p-2}(P) \approx Z_p$. ∎

Corollary 10.4. *If p is a prime number, then the p-primary component of $\pi_m(S^3)$ is 0 if $2p < m < 4p - 3$ and is Z_p is $m = 4p - 3$. If $p > 2$, then the p-primary component of $\pi_{4p-2}(S^3)$ is Z_p.*

Corollary 10.5. *If $n \geqslant 3$ is an odd integer and p a prime number, then the p-primary component of $\pi_m(S^n)$ is 0 if $n + 2p - 3 < m < n + 4p - 6$ and that of $\pi_{n+4p-6}(S^n)$ is 0 or Z_p.*

11. Stiefel manifolds

Let $n \geqslant 4$ be an even integer and consider the Stiefel manifold $V = V_{n+1,2}$ of all unit tangent vectors on S^n, (III; Ex. G). Then, V is simply connected and its homology groups are as follows:

$$H_0(V) \approx Z, \quad H_{n-1}(V) \approx Z_2, \quad H_{2n-1}(V) \approx Z$$

and all other homology groups are zero, [Stiefel 1, 2] and [S; p. 132].

Since V is the tangent bundle of S^n, it is a fiber space over S^n with a projection $\omega: V \to S^n$ and fibers homeomorphic to S^{n-1}. According to (V; § 6), this fibering gives an exact sequence

$$\cdots \to \pi_m(V) \xrightarrow{\omega_*} \pi_m(S^n) \xrightarrow{d_*} \pi_{m-1}(S^{n-1}) \xrightarrow{\tau_*} \pi_{m-1}(V) \to \cdots.$$

This exact sequence is usually used to deduce properties of the homotopy groups of V. Here, on the contrary, it will be used to study the groups $\pi_m(S^n)$.

Let u_m denote the generator of $\pi_m(S^m)$ represented by the identity map. Consider the following part of the sequence:

$$\cdots \to \pi_n(V) \xrightarrow{\omega_*} \pi_n(S^n) \xrightarrow{d_*} \pi_{n-1}(S^{n-1}) \xrightarrow{\tau_*} \pi_{n-1}(V) \to 0.$$

By Hurewicz's theorem, $\pi_{n-1}(V) \approx Z_2$. Since τ_* is an epimorphism, the exactness of the sequence implies that the image of d_* is a subgroup of $\pi_{n-1}(S^{n-1})$ of index 2. Hence we deduce that $d_*(u_n) = \pm 2u_{n-1}$. In fact, it is known that $d_*(u_n) = 2u_{n-1}$, but we shall not need this refinement.

The structure of the homomorphism d_* with $m > n$ is described by the following

Lemma 11.1. *If X is a fiber space over S^n with a pathwise connected fiber F, then the homomorphism d_* in the exact homotopy sequence*

$$\cdots \to \pi_m(X) \to \pi_m(S^n) \xrightarrow{d_*} \pi_{m-1}(F) \to \pi_{m-1}(X) \to \cdots$$

sends the suspension $\Sigma(\alpha)$ of any element $\alpha \in \pi_{m-1}(S^{n-1})$ into the composition $d_*(u_n) \bigcirc \alpha$.

If a map $k: S^{n-1} \to F$ represents $d_*(u_n)$, then k induces a homomorphism $k_*: \pi_{m-1}(S^{n-1}) \to \pi_{m-1}(F)$ which gives $k_*(\alpha) = d_*(u_n) \bigcirc \alpha$ for each α in

$\pi_{m-1}(S^{n-1})$. Hence the relation stated in the lemma means that the following triangle is commutative:

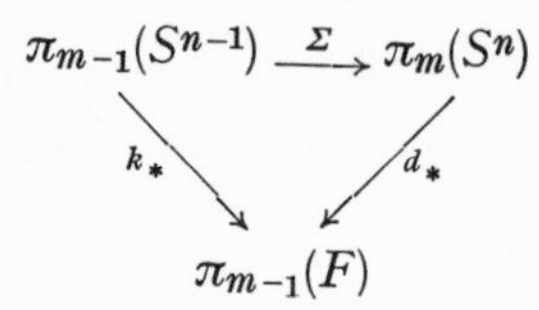

Proof. The projection $\omega : X \to S^n$ induces an isomorphism $\omega_* : \pi_m(X, F) \approx \pi_m(S^n)$. By definition, d_* is ω_*^{-1} followed by the boundary homomorphism $\partial : \pi_m(X, F) \to \pi_{m-1}(F)$.

A representative map $k : S^{n-1} \to F$ of $d_*(u_n)$ can be constructed as follows. Let $h : (E^n, S^{n-1}) \to (S^n, s_0)$ be a map which represents u_n. It follows from the covering homotopy theorem that there exists a map $H : (E^n, S^{n-1}) \to (X, F)$ such that $\omega H = h$. Then the restriction $k = H \mid S^{n-1}$ represents $d_*(u_n)$.

Let $\alpha \in \pi_{m-1}(S^{n-1})$. It remains to prove $k_*(\alpha) = d_*\Sigma(\alpha)$. Let $\phi : S^{m-1} \to S^{n-1}$ represent α; then $k_*(\alpha)$ is represented by $k\phi$. Extend ϕ to a map $\psi : (E^m, S^{m-1}) \to (E^n, S^{n-1})$. By § 2, $h\psi$ represents $\Sigma(\alpha)$. Since $\omega H\psi = h\psi$, $H\psi$ represents $\omega_*^{-1}\Sigma(\alpha)$ and hence $k\phi = H\psi \mid S^{m-1}$ represents $d_*\Sigma(\alpha)$. ∎

Next, let us study the homotopy groups of the Stiefel manifold $V = V_{n+1,2}$. Using the generalized Hurewicz theorem, we can deduce that:

(1) $\pi_m(V)$ is finitely generated.

(2) $\pi_m(V) = 0$ if $m < n - 1$.

(3) $\pi_{n-1}(V) \approx Z_2$.

(4) $\pi_m(V)$ is a finite 2-primary group whenever $n - 1 < m < 2n - 1$.

(5) $\pi_{2n-1}(V)$ is isomorphic to the direct sum of Z and a finite 2-primary group.

Lemma 11.2. *If $\mathscr{C}$ denotes the class of all finite 2-primary groups, then there exists a map $q : S^{2n-1} \to V$ such that the induced homomorphism*

$$q_* : \pi_m(S^{2n-1}) \to \pi_m(V)$$

is a $\mathscr{C}$-isomorphism for every m.

Proof. By (5), there is a map $q : S^{2n-1} \to V$ which represents a free element α of $\pi_{2n-1}(V)$ such that the free cyclic subgroup generated by α is of index some power of 2. Since the natural homomorphism of $\pi_{2n-1}(V)$ into $H_{2n-1}(V)$ is a $\mathscr{C}$-isomorphism, it follows that the induced homomorphism

$$q_{\#} : H_m(S^{2n-1}) \to H_m(V)$$

is a $\mathscr{C}$-isomorphism for $m = 2n - 1$ and hence for every m. An application of Whitehead's theorem (X; 10.1) proves the lemma. ∎

We list the following additional results on the homotopy groups of V; these are immediate consequences of (10.2).

(6) $\pi_m(V)$ is finite if $m > 2n - 1$.

(7) If p is an odd prime number, then the p-primary component of $\pi_m(V)$ is isomorphic to that of $\pi_m(S^{2n-1})$.

12. Finiteness of higher homotopy groups of even-dimensional spheres

Theorem 12.1. *If S^n is an even-dimensional sphere and m is an integer such that $m > n$ and $m \neq 2n - 1$, then $\pi_m(S^n)$ is finite and $\pi_{2n-1}(S^n)$ is isomorphic to the direct sum of Z and a finite group.*

Proof. Since $\pi_m(S^2) \approx \pi_m(S^3)$ for every $m > 2$, the theorem is true for $n = 2$. Hence we may assume $n \geqslant 4$ and apply the results of § 11. If $m > n$ and $m \neq 2n - 1$, then both $\pi_m(V)$ and $\pi_{m-1}(S^{n-1})$ are finite. Therefore, the exactness of the sequence $\pi_m(V) \to \pi_m(S^n) \to \pi_{m-1}(S^{n-1})$ implies that $\pi_m(S^n)$ is finite.

To study the critical case $m = 2n - 1$, let $\mathscr{C}$ denote the class of all finite abelian groups. It follows from the exact sequence that

$$\omega_* : \pi_{2n-1}(V) \to \pi_{2n-1}(S^n)$$

is a $\mathscr{C}$-isomorphism. This implies that $\pi_{2n-1}(S^n)$ is isomorphic to the direct sum of Z and a finite group. ▮

13. The p-primary components of homotopy groups of even-dimensional spheres

In the present section, we are concerned with the p-primary components of the homotopy groups $\pi_m(S^n)$ of an even-dimensional sphere S^n. Since $\pi_m(S^2) \approx \pi_m(S^3)$ for every $m \geqslant 3$, we may restrict ourselves to the case $n \geqslant 4$ and apply the results of § 11.

Consider the following part of the exact sequence appearing in § 11:

$$\pi_{m+1}(S^n) \xrightarrow{d_*} \pi_m(S^{n-1}) \xrightarrow{\tau_*} \pi_m(V) \xrightarrow{\omega_*} \pi_m(S^n) \xrightarrow{d_*} \pi_{m-1}(S^{n-1}).$$

Using the homomorphism ω_* and the suspension Σ, we define a homomorphism

$$\Gamma : \pi_m(V) + \pi_{m-1}(S^{n-1}) \to \pi_m(S^n)$$

by setting $\Gamma(\alpha, \beta) = \omega_*(\alpha) + \Sigma(\beta)$ for each $\alpha \in \pi_m(V)$ and $\beta \in \pi_{m-1}(S^{n-1})$.

Lemma 13.1. *If $\mathscr{C}$ denotes the class of all finite 2-primary groups, then Γ is a $\mathscr{C}$-isomorphism.*

Proof. According to (11.1), the suspension Σ followed by d_* is the induced endomorphism k_* on $\pi_{m-1}(S^{n-1})$ of a map $k : S^{n-1} \to S^{n-1}$ of degree $d = \pm 2$. By Ex. A6 at the end of the chapter, k_* is a $\mathscr{C}$-automorphism. Hence the lemma follows as a consequence of (X; Ex. E). ▮

Theorem 13.2. *If $\mathscr{C}$ denotes the class of all finite 2-primary groups, then the homotopy group $\pi_m(S^n)$ of an even-dimensional sphere S^n is $\mathscr{C}$-isomorphic to the direct sum of $\pi_m(S^{2n-1})$ and $\pi_{m-1}(S^{n-1})$.*

Proof. Since $\pi_m(S^2) \approx \pi_m(S^3)$ for every $m \geqslant 3$, the theorem holds for $n = 2$. If $n \geqslant 4$, then (13.2) is a direct consequence of (13.1) and (11.2). ∎

The importance of (12.1) and (13.2) is that the calculation of the homotopy groups of an even-dimensional sphere, except for their 2-primary components, reduces to that of the homotopy groups of odd-dimensional spheres. Precisely, we have the following

Corollary 13.3. *If n is even and p an odd prime, then the p-primary component of $\pi_m(S^n)$ is isomorphic to the direct sum of those of $\pi_m(S^{2n-1})$ and $\pi_{m-1}(S^{n-1})$.*

14. The Hopf invariant

In order to strengthen (13.2), we propose to present Serre's version of the notion of Hopf invariant.

Consider the n-sphere S^n and a given point $s_0 \in S^n$. Let $\Omega^n = \Lambda(S^n)$ denote the space of loops in S^n with s_0 as basic point. Consider the natural isomorphism

$$j : \pi_{2n-2}(\Omega^n) \approx \pi_{2n-1}(S^n)$$

and the natural homomorphism

$$h : \pi_{2n-2}(\Omega^n) \to H_{2n-2}(\Omega^n) \approx Z.$$

Let $f : S^{2n-1} \to S^n$ be a given map representing an element $[f] \in \pi_{2n-1}(S^n)$. The *Hopf invariant* of f is the integer $H(f)$ uniquely determined by

$$hj^{-1}([f]) = H(f)u_2,$$

where u_2 denotes the generator $\rho_*{}^2(1)$ of $H_{2n-2}(\Omega^n)$ determined by the homomorphisms ρ_* in (IX; 13.1). For other definitions of Hopf invariant, see Ex. C at the end of the chapter.

When n is odd, $H(f)$ is always zero; when n is even, there exists a map f with $H(f) = 2$; if $n = 2$, 4 or 8, then there exists a map f with $H(f) = 1$, namely the Hopf maps of (III; § 5). See [Hopf 2] and [S; p. 113].

Theorem 14.1. *Let n be an even integer and $f : S^{2n-1} \to S^n$ a map with Hopf invariant $H(f) = k \neq 0$. Let $\mathscr{C}$ denote the class of all finite abelian groups of order dividing some power of k. If*

$$\chi_f : \pi_{m-1}(S^{n-1}) + \pi_m(S^{2n-1}) \to \pi_m(S^n)$$

is the homomorphism defined by

$$\chi_f(\alpha, \beta) = \Sigma(\alpha) + f_*(\beta), \quad \alpha \in \pi_{m-1}(S^{n-1}),\ \beta \in \pi_m(S^{2n-1}),$$

then χ_f is a $\mathscr{C}$-isomorphism for every $m > 1$.

Proof. The theorem is obvious if $n = 2$. Hence we assume $n \geqslant 4$.

The map f defines a map $g : \Omega^{2n-1} \to \Omega^n$ in the obvious way and g induces a homomorphism

$$g^* : H^*(\Omega^n) \to H^*(\Omega^{2n-1})$$

of the cohomology algebras with integral coefficients. According to (IX; Ex. H), $H^*(\Omega^n)$ admits a homogeneous basis $\{ a_i \}$ with $\dim (a_i) = i(n-1)$ for each $i = 0, 1, \cdots$ such that

$$a_0 = 1, (a_1)^2 = 0, (a_2)^p = (p!)a_{2p}, a_1 a_{2p} = a_{2p} a_1 = a_{2p+1}.$$

Similarly, $H^*(\Omega^{2n-1})$ admits a homogeneous basis $\{ b_i \}$ with $\dim (b_i) = i(2n-2)$ for each $i = 0, 1, \cdots$ such that

$$b_0 = 1, \quad (b_1)^p = (p!)b_p.$$

Since $H(f) = k$, it follows that $g^*(a_2) = kb_1$. Then we get

$$(p!)g^*(a_{2p}) = g^*(a_2{}^p) = k^p(b_1)^p = k^p(p!)b_p;$$

and therefore $g^*(a_{2p}) = k^p b_p$. Since obviously $g^*(a_{2p+1}) = 0$, it follows that g^* is completely determined.

Now let $i : S^{n-1} \to \Omega^n$ denote the natural imbedding in § 2 and consider the induced homomorphism

$$i^* : H^*(\Omega^n) \to H^*(S^{n-1}).$$

By (2.2), $e = i^*(a_1)$ is a generator of $H^{n-1}(S^{n-1})$.

By means of the multiplication in Ω^n, we may define a map $\phi : S^{n-1} \times \Omega^{2n-1} \to \Omega^n$ by setting $\phi(x, w) = i(x) \cdot g(w)$ for each $x \in S^{n-1}$ and $w \in \Omega^{2n-1}$. By (IX; Ex. H), $H^*(S^{n-1} \times \Omega^{2n-1})$ is naturally isomorphic to the tensor product $H^*(S^{n-1}) \otimes H^*(\Omega^{2n-1})$. This enables us to determine the induced homomorphism

$$\phi^{\#} : H^*(\Omega^n) \to H^*(S^{n-1} \times \Omega^{2n-1})$$

as follows:

$$\phi^{\#}(a_{2p}) = k^p \cdot 1 \otimes b_p,$$

$$\phi^{\#}(a_{2p+1}) = \phi^{\#}(a_1)\phi^{\#}(a_{2p}) = k^p e \otimes b_p.$$

Hence, $\phi^{\#} : H^m(\Omega^n) \to H^m(S^{n-1} \times \Omega^{2n-1})$ is a monomorphism and its cokernel is finite of order equal to a power of k for every dimension m. By duality, this implies that $\phi_{\#} : H_m(S^{n-1} \times \Omega^{2n-1}) \to H_m(\Omega^n)$ is a $\mathscr{C}$-isomorphism for every m. An application of the Whitehead theorem shows that the induced homomorphism

$$\phi_* : \pi_m(S^{n-1} \times \Omega^{2n-1}) \to \pi_m(\Omega^n)$$

is a $\mathscr{C}$-isomorphism for every m.

The group $\pi_m(S^{n-1} \times \Omega^{2n-1})$ is isomorphic to the direct sum $\pi_m(S^{n-1}) + \pi_m(\Omega^{2n-1})$; on the other hand, $\pi_m(\Omega^n) \approx \pi_{m+1}(S^n)$. Then, the construction of ϕ shows that ϕ_* reduces to the homomorphism χ_f. ∎

Corollary 14.2. *If $H(f) = \pm 1$, then χ_f is an isomorphism and hence the suspension $\Sigma : \pi_{m-1}(S^{n-1}) \to \pi_m(S^n)$ is a monomorphism for every $m > 1$.*

Because of the existence of a map f with $H(f) = 2$, (14.1) implies (13.2).

15. The groups $\pi_{n+1}(S^n)$ and $\pi_{n+2}(S^n)$

Since $\pi_m(S^1) = 0$ for each $m > 1$ and $\pi_m(S^2) \approx \pi_m(S^3)$ for every $m > 2$, we may assume $n \geqslant 3$.

Theorem 15.1. *$\pi_{n+1}(S^n)$ is cyclic of order 2 for every $n \geqslant 3$.*

Proof. Let X denote a 3-connective fiber space over S^3. Then, by (9.1), we have

$$\pi_4(S^3) \approx \pi_4(X) \approx H_4(X) \approx Z_2.$$

By the suspension theorem (2.1), we deduce

$$\pi_4(S^3) \approx \pi_5(S^4) \approx \cdots \approx \pi_{n+1}(S^n) \approx \cdots.$$

Hence $\pi_{n+1}(S^n) \approx Z_2$ for every $n \geqslant 3$. ∎

One constructs the generator of $\pi_{n+1}(S^n)$ as follows. Let us consider the Hopf map $p : S^3 \to S^2$ defined in (III; § 5). According to (V; § 6), p represents the generator of $\pi_3(S^2)$. Since the suspension $\Sigma : \pi_3(S^2) \to \pi_4(S^3)$ is an epimorphism, the suspended map $\Sigma p : S^4 \to S^3$ represents the generator of $\pi_4(S^3)$. Then the generator of $\pi_{n+1}(S^n)$ is represented by the $(n-2)$-times iterated suspension $\Sigma^{n-2}p$ of the Hopf map p.

Theorem 15.2. *$\pi_{n+2}(S^n)$ is cyclic of order 2 for every $n \geqslant 3$.*

Proof. Applying (10.1) with $n = 4$, $h = 2$, $m = 5$, we obtain an exact sequence

$$0 \to Z_2 \otimes Z_2 \to \pi_5(P_2^5) \to Tor(Z, Z_2) \to 0.$$

Since $Z_2 \otimes Z_2 \approx Z_2$ and $Tor(Z, Z_2) = 0$, this implies $\pi_5(P_2^5) \approx Z_2$. Then, by (10.2), the 2-primary component of $\pi_5(S^3)$ is isomorphic to Z_2. Since the p-primary component of $\pi_5(S^3)$ is 0 if $p > 2$ by (9.2), it follows that $\pi_5(S^3) \approx Z_2$.

Since the Hopf map $S^7 \to S^4$ is of Hopf invariant 1, we may apply (14.2) with $n = 4$ and $m = 6$. Hence, (2.1) and (14.2) imply that $\Sigma : \pi_5(S^3) \approx \pi_6(S^4)$. Thus, $\pi_6(S^4) \approx Z_2$.

Finally, by (2.1). we deduce

$$\pi_6(S^4) \approx \pi_7(S^5) \approx \cdots \approx \pi_{n+2}(S^n) \approx \cdots.$$

Hence $\pi_{n+2}(S^n) \approx Z_2$ for every $n \geqslant 3$. ∎

To obtain the generator for $\pi_{n+2}(S^n)$, let $i : S^4 \to P_2^5$ denote the imbedding given by the definition of P_2^5. Then, by the proof of (10.1), the generator of $\pi_5(P_2^5)$ is represented by the composition of i and $\Sigma^2 p : S^5 \to S^4$. Composing with the map $\chi : P_2^5 \to S^3$ in § 10, we obtain a representative map $\chi i\, \Sigma^2 p$ for the generator of $\pi_5(S^3)$. This implies that χi represents the generator of $\pi_4(S^3)$ and hence is homotopic to $\Sigma p : S^4 \to S^3$. Therefore, the generator of $\pi_5(S^3)$ is represented by

$$q = \Sigma p \bigcirc \Sigma^2 p : S^5 \to S^3,$$

where $p : S^3 \to S^2$ denotes the Hopf map. Then it follows that the generator of $\pi_{n+2}(S^n)$ is represented by the $(n-3)$-times iterated suspension $\Sigma^{n-3}q$ of q for every $n \geqslant 3$.

Corollary 15.3. $\pi_3(S^2) \approx Z, \pi_4(S^2) \approx Z_2, \pi_5(S^2) \approx Z_2$.

By (V; § 6), the generators of these cyclic groups are represented by respectively the maps

$$p : S^3 \to S^2,\ p \bigcirc \Sigma p : S^4 \to S^2,\ p \bigcirc \Sigma p \bigcirc \Sigma^2 p : S^5 \to S^2.$$

16. The groups $\pi_{n+3}(S^n)$

Theorem 16.1. $\pi_6(S^3) \approx Z_{12}$.

Proof. Applying (10.1) with $n = 4$, $h = 2$, $m = 6$, we obtain an exact sequence

$$0 \to Z_2 \otimes Z_2 \to \pi_6(P_2^5) \to Tor(Z_2, Z_2) \to 0.$$

Since $Z_2 \otimes Z_2 \approx Z_2$ and $Tor(Z_2, Z_2) \approx Z_2$, $\pi_6(P_2^5)$ is isomorphic to an extension of Z_2 by Z_2 and hence has 4 elements. Hence, by (10.2), the 2-primary component of $\pi_6(S^3)$ has 4 elements.

By (9.2), the 3-primary component of $\pi_6(S^3)$ is isomorphic to Z_3 and the p-primary component of $\pi_6(S^3)$ is 0 for every prime $p > 3$. It follows that $\pi_6(S^3)$ has 12 elements and hence is isomorphic to either Z_{12} or $Z_2 + Z_6$.

Suppose that $\pi_6(S^3) \approx Z_2 + Z_6$. Let X denote a 5-connective fiber space over S^3, then

$$H_6(X) \approx \pi_6(X) \approx \pi_6(S^3) \approx Z_2 + Z_6,$$

and it follows from the universal coefficient theorem [E–S; p. 161] that

$$H^6(X; Z_2) \approx Hom\,(H_6(X); Z_2) \approx Z_2 + Z_2.$$

This contradicts to (IX; Ex. I); hence, we conclude that $\pi_6(S^3) \approx Z_{12}$. ∎

Examination of the first paragraph of the proof reveals that the composition of the maps

$$S^6 \xrightarrow{\Sigma^3 p} S^5 \xrightarrow{\Sigma^2 p} S^4 \xrightarrow{\Sigma p} S^3$$

represents an element of $\pi_6(S^3)$ of order 2. A generator of $\pi_6(S^3)$ is represented by the characteristic map $\xi : S^6 \to S^3$ of the fiber bundle $Sp(2)$ over S^6 with $Sp(1)$ as fiber, [Borel and Serre 1; p. 442]. For the definition of the characteristic map, see [S; p. 97].

Corollary 16.2. $\pi_6(S^2) \approx Z_{12}$.
A generator of $\pi_6(S^2)$ is represented by the composed map $p\xi : S^6 \to S^2$.

Theorem 16.3. $\pi_7(S^4) \approx Z + Z_{12}$.

Proof. Let us denote by $q : S^7 \to S^4$ the Hopf map in (III; § 5). Since $H(q) = 1$, we may apply (14.2) with $n = 4$, $m = 7$, and $f = q$. Thus, we obtain an isomorphism

$$\chi_q : \pi_6(S^3) + \pi_7(S^7) \approx \pi_7(S^4).$$

Since $\pi_7(S^7) \approx Z$ and $\pi_6(S^3) \approx Z_{12}$, this proves the theorem. ∎

From the preceding proof, it follows that q represents the generator of the free component Z of $\pi_7(S^4)$ and the suspended map $\Sigma\xi : S^7 \to S^4$ represents an element of order 12 which generates the torsion component Z_{12} of $\pi_7(S^4)$.

Theorem 16.4. $\pi_{n+3}(S^n) \approx Z_{24}$ *if* $n \geqslant 5$.

Proof. By the suspension theorem (2.1), Σ maps $\pi_7(S^4)$ onto $\pi_8(S^5)$. According to Ex. D at the end of the chapter, the kernel of Σ is the free cyclic subgroup of $\pi_7(S^4)$ generated by the Whitehead product $[e, e]$, where e denotes the generator of $\pi_4(S^4)$ represented by the identity map on S^4.

On the other hand, it follows from a theorem on characteristic maps, [S; p. 121], that

$$[e, e] = 2[q] - \varepsilon \Sigma [\xi]$$

where $\varepsilon = \pm 1$ depends on the conventions of orientation. Hence, in $\pi_8(S^5)$, we have

$$\Sigma^2 [\xi] = \varepsilon 2 \Sigma [q].$$

This implies that $\pi_8(S^5)$ is isomorphic to Z_{24} with $\Sigma [q]$ as a generator.

Finally, by the suspension theorem (2.1), we deduce

$$\pi_8(S^5) \approx \pi_9(S^6) \approx \cdots \approx \pi_{n+3}(S^n) \approx \cdots.$$

Hence $\pi_{n+3}(S^n) \approx Z_{24}$ for every $n \geqslant 5$. ∎

Obviously, a generator of $\pi_{n+3}(S^n)$, $n \geqslant 5$, is represented by the $(n-4)$-times iterated suspension $\Sigma^{n-4}q : S^{n+3} \to S^n$ of the Hopf map $q : S^7 \to S^4$.

17. The groups $\pi_{n+4}(S^n)$

Theorem 17.1. $\pi_7(S^3) \approx Z_2$.

Proof. By (9.2), the p-primary component of $\pi_7(S^3)$ is 0 for every prime $p > 3$. By (10.4), the 3-primary component of $\pi_7(S^3)$ is also 0. Hence $\pi_7(S^3)$ is a 2-primary group.

Next, consider a 6-connective fiber space X over S^3. Then $\pi_7(S^3) \approx \pi_7(X) \approx H_7(X)$ and hence we have

$$Hom\, (\pi_7(S^3), Z_2) \approx H_7(X; Z_2).$$

By (IX; Ex. I4), $H^7(X; Z_2) \approx Z_2$. This implies that $\pi_7(S^3)$ is isomorphic to a cyclic group Z_q with $q = 2^h$, $h \geqslant 1$. If $h > 1$, every homomorphism of $\pi_7(S^3)$ into Z_2 can be factored into

$$\pi_7(S^3) \to Z_4 \to Z_2.$$

Then it follows from the exact sequence in (VIII; Ex. I7) that $Sq^1\alpha = 0$ for every element $\alpha \in H^7(X; Z_2)$. This contradicts (IX; Ex. I4). Hence $h = 1$ and $\pi_7(S^3) \approx Z_2$. ∎

According to [Hilton 2; p. 549], the two maps

$$\xi \bigcirc \Sigma^4 p : S^7 \to S^3, \quad \Sigma p \bigcirc q : S^7 \to S^3$$

are both essential. Hence they are homotopic and represent the non-zero element of $\pi_7(S^3)$.

Corollary 17.2. $\pi_7(S^2) \approx Z_2$.

The non-zero element of $\pi_7(S^2)$ is represented by the homotopic maps

$$p \bigcirc \xi \bigcirc \Sigma^4 p : S^7 \to S^2, \quad p \bigcirc \Sigma p \bigcirc q : S^7 \to S^2.$$

Theorem 17.3. $\pi_8(S^4) \approx Z_2 + Z_2$.

Proof. As in the proof of (16.3), we obtain an isomorphism

$$\chi_q : \pi_7(S^3) + \pi_8(S^7) \approx \pi_8(S^4).$$

Since $\pi_7(S^3) \approx Z_2$ and $\pi_8(S^7) \approx Z_2$, the theorem is proved. ∎

The group $\pi_8(S^4)$ is generated by two elements α and β of order 2. α is represented by the homotopic maps

$$\Sigma\,(\xi \bigcirc \Sigma^4 p) : S^8 \to S^4, \quad \Sigma\,(\Sigma\, p \bigcirc q) : S^8 \to S^4,$$

and β is represented by $q \bigcirc \Sigma^5 p : S^8 \to S^4$.

Theorem 17.4. $\pi_9(S^5) \approx Z_2$.

Proof. Consider the following part of the suspension sequence in § 6:

$$\pi_8(T) \xrightarrow{\phi} \pi_8(S^4) \xrightarrow{\Sigma} \pi_9(S^5) \xrightarrow{\psi} \pi_7(T) \xrightarrow{\phi} \pi_7(S^4) \xrightarrow{\Sigma} \pi_8(S^5) \to 0.$$

As mentioned in the proof of (16.4), the kernel of $\Sigma : \pi_7(S^4) \to \pi_8(S^5)$ is a free cyclic group. By (6.3), $\pi_7(T) \approx \pi_7(S^7) \approx Z$. It follows from the exactness of the sequence that $\phi : \pi_7(T) \to \pi_7(S^4)$ is a monomorphism and hence $\Sigma : \pi_8(S^4) \to \pi_9(S^5)$ is an epimorphism.

Since $\pi_8(T) \approx \pi_8(S^7) \approx Z_2$, the kernel of $\Sigma : \pi_8(S^4) \to \pi_9(S^5)$ contains at most two elements. On the other hand, consider the element α of $\pi_8(S^4)$ represented by $\Sigma\,(\xi \bigcirc \Sigma^4 p) : S^8 \to S^4$. In $\pi_9(S^5)$, we have

$$\Sigma(\alpha) = \Sigma^2[\xi] \bigcirc \Sigma^6[p] = (\varepsilon 2 \Sigma[q]) \bigcirc (\Sigma^6[p]) = (\varepsilon\, \Sigma[q]) \bigcirc (2\Sigma^6[p]) = 0.$$

Hence the kernel of $\Sigma : \pi_8(S^4) \to \pi_9(S^5)$ consists of exactly two elements, namely, 0 and α. This implies that $\pi_9(S^5) \approx Z_2$. ∎

The non-zero element of $\pi_9(S^5)$ is $\Sigma(\beta)$ represented by the map $\Sigma(q \bigcirc \Sigma^5 p) : S^9 \to S^5$. On the other hand, the Whitehead product $[e, e]$ of the generator e of $\pi_5(S^5)$ is also non-zero and hence $[e, e] = \Sigma(\beta)$, [Serre 3; p. 230].

Theorem 17.5. $\pi_{n+4}(S^n) = 0$ *if* $n \geqslant 6$.

Proof. By the suspension theorem (2.1), Σ maps $\pi_9(S^5)$ onto $\pi_{10}(S^6)$. According to the delicate suspension theorem in Ex. D at the end of the chapter, $\Sigma\,[e, e] = 0$. Hence we obtain $\pi_{10}(S^6) = 0$. Finally, by (2.1), we deduce

$$\pi_{10}(S^6) \approx \pi_{11}(S^7) \approx \cdots \approx \pi_{n+4}(S^n) \approx \cdots.$$

Hence $\pi_{n+4}(S^n) = 0$ for every $n \geqslant 6$. ∎

18. The groups $\pi_{n+r}(S^n)$, $5 \leqslant r \leqslant 15$

In this final section of the book, we will list the groups $\pi_{n+r}(S^n)$ for the cases $r = 5, 6, 7, 8$. For more detailed information, see [Serre 5, 6]. H. Toda has computed the groups $\pi_{n+r}(S^n)$ for $9 \leqslant r \leqslant 15$. We will not list his results here; the interested reader should refer to [Toda 1, 2] with recent corrections given in [Toda 3].

$r = 5$.

$$\begin{aligned}
&\pi_7(S^2) \approx Z_2 \\
&\pi_8(S^3) \approx Z_2 \\
&\pi_9(S^4) \approx Z_2 + Z_2 \\
&\pi_{10}(S^5) \approx Z_2 \\
&\pi_{11}(S^6) \approx Z \\
&\pi_{n+r}(S^n) = 0, \qquad (n \geqslant 7).
\end{aligned}$$

$r = 6$.

$$\begin{aligned}
&\pi_8(S^2) \approx Z_2 \\
&\pi_9(S^3) \approx Z_3 \\
&\pi_{10}(S^4) \approx Z_{24} + Z_2 \\
&\pi_{n+6}(S^n) \approx Z_2, \qquad (n \geqslant 5).
\end{aligned}$$

$r = 7$.

$$\begin{aligned}
&\pi_9(S^2) \approx Z_3 \\
&\pi_{10}(S^3) \approx Z_{15} \\
&\pi_{11}(S^4) \approx Z_{15} \\
&\pi_{12}(S^5) \approx Z_{30} \\
&\pi_{13}(S^6) \approx Z_{60} \\
&\pi_{14}(S^7) \approx Z_{120} \\
&\pi_{15}(S^8) \approx Z + Z_{120} \\
&\pi_{n+7}(S^n) \approx Z_{240}, \qquad (n \geqslant 9).
\end{aligned}$$

$r = 8$.

$$\begin{aligned}
&\pi_{10}(S^2) \approx Z_{15} \\
&\pi_{11}(S^3) \approx Z_2 \\
&\pi_{12}(S^4) \approx Z_2 \\
&\pi_{13}(S^5) \approx Z_2 \\
&\pi_{14}(S^6) \approx Z_{24} + Z_2 \\
&\pi_{15}(S^7) \approx Z_2 + Z_2 + Z_2 \\
&\pi_{16}(S^8) \approx Z_2 + Z_2 + Z_2 + Z_2 \\
&\pi_{17}(S^9) \approx Z_2 + Z_2 + Z_2 \\
&\pi_{n+8}(S^n) \approx Z_2 + Z_2, \qquad (n \geqslant 10).
\end{aligned}$$

EXERCISES

A. The distributive laws

Let $\alpha \in \pi_m(S^n, s_0)$ and $\beta \in \pi_n(X, x_0)$. If α and β are represented by the maps

$$f : (E^m, S^{m-1}) \to (S^n, s_0), \quad g : (S^n, s_0) \to (X, x_0)$$

respectively, that the composed map gf represents an element γ of $\pi_m(X, x_0)$. Prove:

1. The element γ depends only on the elements α, β and will be called the *composition* $\beta \bigcirc \alpha$ of α and β.

2. *The right distributive law.* For a given $\beta \in \pi_n(X, x_0)$, the assignment $\alpha \to \beta \bigcirc \alpha$ defines a homomorphism. In fact, this is the induced homomorphism g_*.

3. *The left distributive law.* If X is an H-space with x_0 as homotopy unit or if α is the suspension $\Sigma(\delta)$ of some element $\delta \in \pi_{m-1}(S^{n-1}, s_0)$, then the assignment $\beta \to \beta \bigcirc \alpha$ defines a homomorphism. [S; p. 122].

In particular, let $(X, x_0) = (S^n, s_0)$. Consider a map $g : (S^n, s_0) \to (S^n, s_0)$ of degree d and study its induced homomorphism

$$g_* : \pi_m(S^n, s_0) \to \pi_m(S^n, s_0).$$

Prove:

4. $g_*(\alpha) = d\alpha$ for every $\alpha \in \pi_m(S^n, s_0)$ if $n = 1, 3, 7$ or if $m < 2n - 1$.

5. If $m = 3$ and $n = 2$, then $g_*(\alpha) = d^2\alpha$ for every $\alpha \in \pi_3(S^2, s_0)$. [Hopf 1].

6. If $\mathscr{C}$ denotes the class of all finite abelian groups of order dividing a power of d, then g_* is a $\mathscr{C}$-automorphism for every $m > 0$ and $n \geqslant 3$.

B. On relative (n + 1)-cells

Let (X, A) be a relative $(n + 1)$-cell obtained by adjoining E^{n+1} to A by means of a map $g : S^n \to A$. Then g has an extension $f : (E^{n+1}, S^n) \to (X, A)$ defined by $f(x) = x$ for every $x \in E^{n+1} \setminus S^n = X \setminus A$. This map f is called the *characteristic map* of (X, A). If we identify A to a single point, we obtain an $(n + 1)$-sphere S^{n+1} as quotient space with projection $h : X \to S^{n+1}$. Choose $s_0 \in S^n$ and let $x_0 = f(s_0) \in A$. Use s_0 and x_0 as basic points of the homotopy groups. Verify that the rectangle

$$\begin{array}{ccc} \pi_m(E^{n+1}, S^n) & \xrightarrow{f_*} & \pi_m(X, A) \\ \big\downarrow{\scriptstyle \partial} & & \big\downarrow{\scriptstyle h_*} \\ \pi_{m-1}(S^n) & \xrightarrow{\Sigma} & \pi_m(S^{n+1}) \end{array}$$

is commutative, i.e. $h_* f_* = \Sigma \partial$. Prove

1. If Σ is a monomorphism, so is f_*. If Σ is an epimorphism, so is h_*.

2. If $m < 2n$, then f_* is a monomorphism, h_* is an epimorphism, and $\pi_m(X, A)$ decomposes into the direct sum of $Im(f)_*$ and $Ker(h_*)$.

3. If A is r-connected for some $r \leqslant n$, then f_* is an epimorphism whenever $m \leqslant n + r$. See [J. H. C. Whitehead 6; p. 14] and [Hilton 1; p. 464].

As an application of these results, consider the relative $(p + q)$-cell (X, A) with $X = S^p \times S^q$ and $A = S^p \vee S^q$. By 2 and (V; 3.1), we obtain natural isomorphisms

$$\pi_m(S^p \times S^q, S^p \vee S^q) \approx \pi_{m+1}(S^{p+q-1}) + Ker(h_*),$$

$$\pi_{m-1}(S^p \vee S^q) \approx \pi_{m-1}(S^p) + \pi_{m-1}(S^q) + \pi_{m-1}(S^{p+q-1}) + Ker(h_*)$$

for every $m < 2p + 2q - 2$. Finally, by 3, $Ker(h_*) = 0$ if $m \leqslant p + q + min\,(p, q) - 2$.

C. Other definitions of the Hopf invariant

In § 14, we gave Serre's definition of the Hopf invariant $H(f)$ of a given map

$$f : S^{2n-1} \to S^n, \quad n \geqslant 2.$$

There are a few different but equivalent definitions given by various authors listed below. Each of these definitions has a natural generalization but the various generalized Hopf invariants are not necessarily equivalent.

1. *Hopf's definition.* By using a homotopy if necessary, we may assume that f is a simplicial map relative to some triangulations. Choose a pair of distinct interior points, u, v of n-simplexes of S^n. Then $f^{-1}(u), f^{-1}(v)$ are disjoint $(n - 1)$-manifolds in S^{2n-1}; and H(f) is the linking number of the $(n - 1)$-cycles $f^{-1}(u)$ and $f^{-1}(v)$. See [Hopf 2] and [S; p. 113]. This definition can be generalized to manifolds.

2. *Steenrod's definition.* Let M denote the mapping cylinder of f with $S^n \subset M$ and $S^{2n-1} \subset M$. The integral cohomology algebra $H^*(M, S^{2n-1})$ has a homogeneous basis $\{ a, b \}$ with $dim\,(a) = n$ and $dim\,(b) = 2n$. Then, $H(f)$ is the integer defined by $a^2 = H(f)b$. See [Steenrod 3; p. 983] and [Serre 2; p. 286]. This definition can be generalized to the maps $f : S^{n+i-1} \to S^n$ as follows. In this case, $H^*(M, S^{n+i-1}; Z_2)$ has a homogeneous basis $\{ a, b \}$ with $dim\,(a) = n$ and $dim\,(b) = n + i$. Then, $H(f) \in Z_2$ is defined by $Sq^i a = H(f)b$.

3. *Whitehead's definition.* Identifying the equator S^{n-1} of S^n to a single point, we obtain a quotient space $S^n \vee S^n$ with projection $g : S^n \to S^n \vee S^n$. The composed map gf represents an element $[gf]$ of $\pi_{2n-1}(S^n \vee S^n)$. Then $H(f)$ is obtained by projecting $[gf]$ into the direct summand $\pi_{2n-1}(S^{2n-1}) \approx Z$ of $\pi_{2n-1}(S^n \vee S^n)$. Since

$$\pi_m(S^n \vee S^n) \approx \pi_m(S^n) + \pi_m(S^n) + \pi_m(S^{2n-1}) + Ker(h_*)$$

for every $m \leqslant 4n - 4$, this definition can be generalized to the maps

$f : S^m \to S^n$ for $m \leqslant 4n - 4$ in an obvious way. See [G. W. Whitehead 2] and [Hilton 1]. Thus, for each $m \leqslant 4n - 4$, we obtain a *Hopf homomorphism*

$$H : \pi_m(S^n) \to \pi_m(S^{2n-1}).$$

Among the properties of this generalization of Hopf invariants, verify the following few:

(*i*) If $\alpha \in \pi_m(S^n)$, $m \leqslant 3n - 3$, and $\beta_1, \beta_2 \in \pi_n(X)$, then

$$(\beta_1 + \beta_2) \bigcirc \alpha = \beta_1 \bigcirc \alpha + \beta_2 \bigcirc \alpha + [\beta_1, \beta_2] \bigcirc H(\alpha).$$

(*ii*) If $\alpha \in \pi_{m-1}(S^{n-1})$ with $m \leqslant 4n - 4$, then $H\,\Sigma(\alpha) = 0$.

(*iii*) If n is odd and $m \leqslant 4n - 4$, then $2H(\alpha) = 0$ for every $\alpha \in \pi_m(S^n)$.

(*iv*) If $\alpha \in \pi_m(S^n)$ with $m \leqslant 4n - 4$ and $\Sigma(\alpha) = 0$, then $H(\alpha) = 0$ if n is odd $H(\alpha) \in 2\pi_m(S^{2n-1})$ if n is even.

4. *Hilton's definition.* If we identify the subset $S^n \vee S^n$ of the product space $S^n \times S^n$ to a single point s_0, we obtain a $2n$-sphere S^{2n} as quotient space with projection

$$h : (S^n \times S^n, S^n \vee S^n) \to (S^{2n}, s_0).$$

For each $m > 0$, define a homomorphism

$$H^* : \pi_m(S^n) \to \pi_{m+1}(S^{2n})$$

by taking H^* to be composition of the sequence

$$\pi_m(S^n) \xrightarrow{g_*} \pi_m(S^n \vee S^n) \xrightarrow{p} \pi_{m+1}(S^n \times S^n, S^n \vee S^n) \xrightarrow{h_*} \pi_{m+1}(S^{2n}),$$

where g_*, h_* are induced homomorphisms and p denotes the projection of $\pi_m(S^n \vee S^n)$ onto its direct summand $\pi_{m+1}(S^n \times S^n, S^n \vee S^n)$. Prove the following relations:

(*i*) $H^* = \Sigma H$ whenever $m \leqslant 4n - 4$. [Hilton 1; p. 473] and [Hilton 3; p. 166].

(*ii*) If $\alpha \in \pi_n(S^q)$, $\beta \in \pi_m(S^n)$, then

$$H^*(\alpha \bigcirc \beta) = H^*(\alpha) \bigcirc \Sigma(\beta) + \Sigma^q(\alpha) \bigcirc \Sigma^n(\alpha) \bigcirc H^*(\beta).$$

(*iii*) If n is odd and $\alpha \in \pi_m(S^n)$, then $2H^*(\alpha) = 0$.

(*iv*) If n is even, $\alpha \in \pi_m(S^n)$ is of odd order, and $H^*(\alpha) = 0$, then $\alpha = \Sigma(\beta)$ for some $\beta \in \pi_{m-1}(S^{n-1})$.

D. The delicate suspension theorem

The suspension theorem (2.1) is the easy part of Freudenthal's result and is usually called the crude suspension theorem. The delicate part of Freudenthal's result has been slightly strengthened to the following form, [G. W. Whitehead 2]:

1. The image of $\Sigma : \pi_{2n}(S^n) \to \pi_{2n+1}(S^{n+1})$ is the subgroup of $\pi_{2n+1}(S^{n+1})$ consisting of the elements of Hopf invariant zero.

2. The kernel of $\Sigma : \pi_{2n-1}(S^n) \to \pi_{2n}(S^{n+1})$ is the cyclic subgroup of $\pi_{2n-1}(S^n)$ generated by the Whitehead product $[e, e]$, where e denotes the generator of $\pi_n(S^n)$ represented by the identity map. If n is even, $[e, e]$ has Hopf invariant 2 and therefore has infinite order. If n is odd, $2[e, e] = 0$, and $[e, e] = 0$ iff there is an element of $\pi_{2n+1}(S^{n+1})$ with Hopf invariant 1.

BIBLIOGRAPHY

BOOKS *and Mimeographed Notes*

[A–H] ALEXANDROFF, P. und HOPF, H.: Topologie I, Springer, Berlin, 1935.

[B] BOURBAKI, N.: Éléments de mathématiques. Hermann, Paris, 1939–1948.

[C–E] CARTAN, H., and EILENBERG, S.: Homological Algebra. Princeton Univ. Press, 1956.

[Che] CHEVALLEY, C.: Theory of Lie Groups I. Princeton Univ. Press, 1946.

[E–S] EILENBERG, S., and STEENROD, N. E.: Foundations of Algebraic Topology. Princeton Univ. Press, 1952.

[Hi] HILTON, P. J.: An Introduction to Homotopy Theory. Cambridge Univ. Press, 1953.

[H] HU, S. T.: Homotopy Theory I. Dittoed Technical Reports, Tulane University, 1950.

[H–W] HUREWICZ, W., and WALLMAN, H.: Dimension Theory. Princeton Univ. Press, 1941.

[Ka] KAPLANSKY, I.: Infinite Abelian Groups. Univ. of Michigan Press, Ann Arbor, 1954.

[K] KELLEY, J. L., General Topology. D. van Nostrand Co. Inc., New York, 1955.

[L_1] LEFSCHETZ, S.: Topology. (Amer. Math. Soc. Coll. Publ., Vol. 12), 1930.

[L_2] LEFSCHETZ, S.: Algebraic Topology. (Amer. Math. Soc. Coll. Publ., Vol. 27). 1942.

[L_3] LEFSCHETZ, S.: Topics in Topology. (Annals of Math. Studies, No. 10), 1942.

[M_1] MORSE, M.: The Calculus of Variations in the Large. (Amer. Math. Soc. Coll. Publ., Vol. 18), 1934.

[M_2] MORSE, M.: Introduction to Analysis in the Large, 2nd ed. Mimeographed, Institute for Advanced Study, Princeton, 1951.

[S–T] SEIFERT, H., und THRELFALL, W.: Lehrbuch der Topologie. Teubner, Leipzig, 1934.

[S] STEENROD, N. E.: The Topology of Fibre Bundles. Princeton Univ. Press, 1951.

[V] VEBLEN, O.: Analysis Situs. (Amer. Math. Soc. Coll. Publ., Vol. 5, Part 2), 2nd ed., 1931.

PAPERS

ARENS, R.

1. *Topologies for homeomorphism groups*. Amer. J. Math., 68 (1946), 593–610.
2. *A topology for spaces of transformations*. Ann. of Math., 47 (1946), 480–495.

BLAKERS, A. L. and MASSEY, W. S.

1. *The homotopy groups of a triad*, I, II, III. Ann. of Math., 53 (1951), 161–205, 55 (1952), 192–201, 58 (1953), 409–417.
2. *Products in homotopy theory*. Ann. of Math., 58 (1953), 295–324.

BOREL, A., and SERRE, J.-P.

1. *Groupes de Lie et puissances réduites de Steenrod*. Amer. J. Math., 75 (1953), 409–448.

BORSUK, K.

1. *Sur les retracts*. Fund. Math., 17 (1931), 152–170.
2. *Über eine Klasse von lokal zusammenhängenden Räumen*. Fund. Math., 19 (1932), 220–240.

3. *Sur les groupes des classes de transformations continues.* C. R. Acad. Sci., Paris, 202 (1936), 1400–1403.

BRUSCHLINSKY, N.

1. *Stetige Abbildungen und Bettische Gruppen der Dimensionszahlen* 1 *und* 3. Math. Ann. 103 (1934), 525–537.

CARTAN, H.

1. *Une théorie axiomatique des carrés de Steenrod.* C. R. Acad. Sci., Paris, 230 (1950), 425–427.

CHEN, C.

1. *A note on the classification of mappings of a* $(2n - 2)$*-dimensional complex into an* n*-sphere.* Ann. of Math., 51 (1950), 238–240.

CURTIS, M. L.

1. *The covering homotopy theorem.* Proc. Amer. Math. Soc., 7 (1956), 682–684.

DOWKER, C. H.

1. *Mapping theorems for non-compact spaces.* Amer. J. Math., 69 (1947), 200–240.

DUGUNDJI, J.

1. *An extension of Tietze's theorem.* Pacific J. Math., 1 (1951), 353–367.

EILENBERG, S.

1. *Cohomology and continuous mappings.* Ann. of Math., 41 (1940), 231–251.
2. *Singular homology theory.* Ann of Math., 45 (1944), 407–447.
3. *Homology of spaces with operators,* I. Trans. Amer. Math. Soc., 61 (1947), 378–417; errata, 62 (1947), 548.
4. *On the problems of topology.* Ann. of Math., 50 (1949), 247–260.

EILENBERG, S., and MACLANE, S.

1. *Relations between homology and homotopy groups of spaces.* Ann. of Math., 46 (1945), 480–509. II, Ann. of Math., 51 (1950), 514–533.
2. *Acyclic models.* Amer. J. Math., 75 (1953), 189–199.

EILENBERG, S., and ZILBER, J. A.

1. *Semi-simplicial complexes and singular homology.* Ann. of Math., 51 (1950), 499–513.

FOX, R. H.

1. *On homotopy type and deformation retracts.* Ann. of Math., 44 (1943), 40–50.
2. *On fibre spaces* I, II. Bull. Amer. Math. Soc., 49 (1943) 555–557, 733–735.
3. *On topologies for function spaces.* Bull. Amer. Math. Soc., 51 (1945), 429–432.

GIEVER, J. B.

1. *On the equivalence of two singular homology theories.* Ann. of Math., 51 (1950), 178–191.

GRIFFIN, J. S., JR.

1. *Theorems on fibre spaces.* Duke Math. J., 20 (1953), 621–628.

HANNER, O.

1. *Some theorems on absolute neighborhood retracts.* Arkiv Math., Svenska Vetens. Akad., 1 (1951), 389 408.

HILTON, P. J.

1. *Suspension theorems and the generalized Hopf invariant.* Proc. London Math. Soc. (3), 1 (1951), 462–492.
2. *The Hopf invariant and homotopy groups of spheres.* Proc. Camb. Phil. Soc., 48 (1952), 547–554.
3. *On the Hopf invariant of a composite element.* J. London Math. Soc., 29 (1954), 165–171.

HOPF, H.

1. *Über die Abbildungen der dreidimensionalen Sphäre auf die Kugelfläche.* Math. Ann., 104 (1931), 637–665.

2. *Über die Abbildungen von Sphären auf Sphären niedrigerer Dimension.* Fund. Math., 25 (1935), 427–440.

HU, S. T.

1. *Inverse homomorphisms of the homotopy sequence.* Indagationes Math., 9 (1947), 169–177.
2. *On spherical mappings in a metric space.* Ann. of Math., 48 (1947), 717–734.
3. *A theorem on homotopy extension.* Doklady Akad. Nauk. S.S.S.R., 57 (1947), 231–234.
4. *An exposition of the relative homotopy theory.* Duke Math. J., 14 (1947), 991–1033.
5. *Mappings of a normal space into an absolute neighborhood retract.* Trans. Amer. Math. Soc., 64 (1948), 336–358.
6. *Extension and classification of the mappings of a finite complex into a topological group or an n-sphere.* Ann of Math., 50 (1949), 158–173.
7. *Extensions and classification of maps.* Osaka Math. J., 2 (1950), 165–209.
8. *On generalising the notion of fibre spaces to include the fibre bundles.* Proc. Amer. Math. Soc., 1 (1950), 756–762.
9. *Cohomology and deformation retracts.* Proc. London Math. Soc., (2), 53 (1951), 191–219.
10. *On the realizability of homotopy groups and their operations.* Pacific J. Math., 1 (1951), 583–602.
11. *On products in homotopy groups.* Univ. Nac. del Tucuman, Revista, Ser. A, 8 (1951), 107–119.
12. *The homotopy addition theorem.* Ann. of Math., 58 (1953), 108–122.

HUEBSCH, W.

1. *On the covering homotopy theorem.* Ann. of Math., 61 (1955), 555–563.

HUREWICZ, W.

1. *Beiträge zur Topologie der Deformationen* I–IV. Proc. Akad. Wetensch., Amsterdam, 38 (1935), 112–119, 521–528, 39 (1936), 117–126, 215–224.
2. *On the concept of fiber space.* Proc. Nat. Acad. Sci. U.S.A., 41 (1955), 956–961.

HUREWICZ, W., and STEENROD, N. E.

1. *Homotopy relations in fibre spaces.* Proc. Nat. Acad. Sci. U.S.A., 27 (1941) 60–64.

JACKSON, J. R.

1. *Comparison of topologies on function spaces.* Proc. Amer. Math. Soc. 3 (1952), 156–158.
2. *Spaces of mappings on topological products with applications to homotopy theory.* Proc. Amer. Math. Soc., 3 (1952), 327–333.

KAN, D. M.

1. *Abstract Homotopy* I–IV. Proc. Nat. Acad. Sci. U.S.A., 41 (1955), 1092–1096, 42 (1956), 255–258, 419–421, 542–544.

KURATOWSKI, C.

1. *Quelques problèmes concernant les espaces métriques non-séparables.* Fund. Math. 25 (1935), 534–545.

LERAY, J.

1. *L'anneau spectral et l'anneau filtré d'homologie d'un espace localement compact et d'une application continue.* J. Math. Pures Appl., (9), 29 (1950), 1–39.
2. *L'homologie d'un espace fibré dont la fibre est connexe.* J. Math. Pures Appl. (9), 29 (1950), 169–213.

LIAO, S. D.

1. *On non-compact absolute neighborhood retracts.* Acad. Sinica, Science Record, 2 (1949), 249–262.

MASSEY, W. S.

1. *Exact couples in algebraic topology.* I–V. Ann. of Math., 56 (1952), 363–396, 57 (1953), 248–286.

2. *Products in exact couples.* Ann. of Math., 59 (1954), 558–569.
3. *Some problems in algebraic topology and the theory of fibre bundles.* Ann. of Math., 62 (1955), 327–359.

MILNOR, J.
1. *Construction of universal bundles* I–II. Ann. of Math., 63 (1956), 272–284, 430–436.

MOORE, J. C.
1. *Some applications of homology theory to homotopy problems.* Ann. of Math., 58 (1953), 325–350.
2. *On homotopy groups of spaces with a single non-vanishing homology group.* Ann. of Math., 59 (1954), 549–557.

OLUM, P.
1. *Obstructions to extensions and homotopies.* Ann. of Math., 52 (1950), 1–50.

PETERSON, F. P.
1. *Some results on cohomotopy groups.* Amer. J. of Math., 78 (1956), 243–257.

SERRE, J.-P.
1. *Homologie singulière des espaces fibrés.* Ann. of Math., 54 (1951), 425–505.
2. *Groupes d'homotopie et classes de groupes abélians.* Ann. of Math., 58 (1953), 258–294.
3. *Cohomologie modulo 2 des complexes d'Eilenberg–MacLane.* Comm. Math. Helv. 27 (1953), 198–232.
4. *Sur les groupes d'Eilenberg–MacLane.* C. R. Acad. Sci., Paris, 234 (1952), 1243–1245.
5. *Sur la suspension de Freudenthal.* C. R. Acad. Sci., Paris, 234 (1952), 1340–1342.
6. *Quelques calculs de groupes d'homotopie.* C. R. Acad. Sci., Paris, 236 (1953), 2475–2477.

SPANIER, E. H.
1. *Borsuk's cohomotopy groups.* Ann. of Math., 50 (1949), 203–245.

STEENROD, N. E.
1. *Homology with local coefficients.* Ann. of Math., 44 (1943), 610–627.
2. *Products of cocycles and extensions of mappings.* Ann. of Math., 48 (1947), 290–320.
3. *Cohomology invariants of mappings.* Ann. of Math., 50 (1949), 954–988.

STIEFEL, E.
1. *Richtungsfelder und Fernparallelismus in Mannigfaltigkeiten.* Comm. Math. Helv., 8 (1936), 3–51.

TODA, H.
1. *Calcul de groupes d'homotopie des sphères.* C. R. Acad. Sci., Paris, 240 (1955) 147–149.
2. *Le produit de Whitehead et l'invariant de Hopf.* C. R. Acad. Sci., Paris, 241 (1955), 849–850.
3. *p-primary components of homotopy groups. I. Exact sequences in the Steenrod Algebra, II. Mod p Hopf invariant.* Mem. Coll. Sci. Univ. Kyoto, Ser. A. (1958).

VERMA, S.
1. *Relation between abstract homotopy and geometric homotopy.* Dissertation, Wayne State University, 1958.

WALLACE, A. D.
1. *The structure of topological semigroups.* Bull. Amer. Math. Soc., 61 (1955), 95–112.

WANG, H. C.
1. *Some examples concerning the relations between homology and homotopy groups.* Indagationes Math., 9 (1947), 384–386.
2. *The homology groups of the fibre bundles over a sphere.* Duke Math. J., 16 (1949), 33–38.

WHITEHEAD, G. W.
1. *On spaces with vanishing low-dimensional homotopy groups.* Proc. Nat. Acad. Sci., U.S.A., 34 (1948), 207–211.
2. *A generalization of the Hopf invariant.* Ann. of Math., 51 (1950), 192–237.
3. *On the Freudenthal theorems.* Ann. of Math., 57 (1953), 209–228.

WHITEHEAD, J. H. C.
1. *Simplicial spaces, nuclei and m-groups*. Proc. London Math. Soc. (2), 45 (1939), 243–327.
2. *On adding relations to homotopy groups*. Ann. of Math., 42 (1941), 409–428.
3. *On the groups $\pi_r(V_{n,m})$ and sphere bundles*. Proc. London Math. Soc. (2), 48 (1944), 243–291. Corrigendum, 49 (1947), 478–481.
4. *Combinatorial homotopy* I, II. Bull. Amer. Math. Soc., 55 (1949), 213–245, 453–496.
5. *On the realizability of homotopy groups*. Ann. of Math., 50 (1949), 261–263.
6. *Note on suspension*. Quart. J. Math., Oxford (2), 1 (1950), 9–22.
7. *A certain exact sequence*. Ann. of Math., 52 (1950), 51–110.
8. *On the theory of obstructions*. Ann. of Math., 54 (1951), 66–84.

WOJDYSLAWSKI, M.
1. *Rétractes absolus et hyperespaces des continus*. Fund. Math., 32 (1939), 184–192.

Index

H

I

K

L

M

T

U

W